Printed in the United States of America.
Version: 1.0

ISBN-13: 978-1-64331-142-5 (print)
ISBN-13: 978-1-64331-143-2 (ePub)
ISBN-13: 978-1-64331-144-9 (Kindle)

ISA
67 T. W. Alexander Drive
P.O. Box 12277
Research Triangle Park, NC 27709

Library of Congress Cataloging-in-Publication Data
Lipták, Béla.
Controlling the future - controlling nonindustrial processes: preventing climate and other disasters / by Béla Lipták.
p. cm.
Durham: International Society of Automation, 2023.
Includes bibliographical references and index.
ISBN: 9781643311425 (hardback)
LLCN: 2023938381
1. Automation and control. 2. Climate change. 3. Global warming. 4. Artificial intelligence.
https://lccn.loc.gov/2023938381

Controlling the Futur

Controlling Nonindustria
Processes: Preventin
Climate and Other Disaster

Béla Liptá

This work is dedicated to my grandchildren, Ivan, Katie, Ava, Emmi, Adam, Salome, Cyrus, Barbara, and Stanley, and to their generation, which I hope will make a reality out of its contents.

Contents

About the Author

Béla Lipták was born in 1936 in Hungary. As a Technical University student, he participated in the revolution against the Soviet occupation. He escaped and entered the United States as a refugee in 1956. In 1959, he received an engineering degree from Stevens Institute of Technology; he received a master's degree from the City College of New York in 1962; and later, he did graduate work at Pratt Institute.

In 1960, he became the chief instrument engineer at Crawford and Russell, where he led the automation of dozens of industrial plants for more than a decade. In 1969, he published the multivolume *Instrument and Automation Engineers' Handbook,* which is currently in its fifth edition. In 1975, he received his professional engineering license and founded a consulting firm named Béla Lipták Associates PC, which provides design and consulting services in the fields of energy-related automation and industrial safety. Over the years, he lectured at many universities around the world, including Yale University, where he was an adjunct professor in 1987.

Béla Lipták published more than 300 technical articles on climate change, global warming, and the automation of the new infrastructure they require (https://www.controlglobal.com/home/contact/11392257/bla-liptk), as well as more than 20 books on various aspects of automation, safety, and energy technology.

In 1973, he was elected a Fellow of the International Society of Automation (ISA); in 1995, he received ISA's Technical Achievement Award; and in 2001, he was inducted into *Control* magazine's Process Automation Hall of Fame. He was also the keynote speaker at many domestic and overseas conventions.

Introduction

In the evolution of humankind, energy has played a critical role. From the discovery of fire in the early Stone Age to the eighteenth century, humans used wood as a source of heat energy. The switch from wood to coal and the development of a steam engine that could drive machinery by James Watt—working at the University of Glasgow—triggered the beginning of the Industrial Age. One of the key elements of Watt's design was the automatic control of speed through the invention of the pressure gauge, which thereby triggered the beginning of process control. The invention of the electric generator in 1832 by Michael Faraday enabled the transportation and distribution of energy as well as automatic control of the use of that energy by modern devices. This system worked for the next two centuries, but then came global warming (GW) and the desperate search for more fossil fuels, obtained by fracking them from the land and scraping them from the bottom of the oceans. Climate change was proven when process control, through careful measurement of carbon in the atmosphere and temperature in the seas, determined that Nature is telling us, "Enough—the age of fossil fuels is over!"

Today, the greenhouse gas concentration in the air is the highest ever, and 80% of our primary energy still comes from burning fossil fuels. The present emission rate equals the weight of five Empire State Buildings every hour, and the extra energy warming Earth equals that of five Hiroshima bombs per second.

The Russian war on Ukraine is only the first energy war, others are coming, and they will only stop when the use of fossil fuels is stopped. The conversion from the fossil energy economy to a hydrogen-based one is essential, not only because we want to prevent future energy wars but because as long as we are *only reducing* our carbon emissions, we are only reducing the rate at which the amount of accumulated carbon in the atmosphere will keep rising and, therefore, GW will keep rising.

When applying the rules of process control to analyze the nature of the global heat-balance process, one must consider the unique characteristics of this process. These characteristics—essential but often neglected by the traditional climate models—are the consequences of the multiple-mass effect, the positive feedback effect, and the already-accumulated integral error in the process.

- **Multiple-mass effect** – The Earth consists of four very different heat-absorbing bodies. The oceans absorb 94% of the total extra heat received through GW, whereas the atmosphere, the land, and the melting of the ice absorb only 2% each. From the perspective of process control analysis, we know that the speed of response (gain) of heat-transfer systems increases as the mass of the object decreases. We also know that the mass of the oceans is about a hundred times greater than that of the atmosphere. Therefore, it is no surprise that the ocean surfaces are warming much more slowly (half as fast) than the atmosphere. The fact that the deeper we go below the ocean's surface, the faster the temperature rises has nothing to do with the relative masses; it is caused by "upwelling." Therefore, it is because of the drastically different masses of the oceans and the atmosphere that we are already experiencing the atmospheric effects of GW (drought, fires, etc.). In contrast, the oceanic responses (changes in ocean currents, degassing, coral deaths, etc.) are just beginning to be felt.

- **Positive feedback effects (self-acceleration)** – Another fundamental difference between simple heat transfer and the climate change process is that GW is the sum of many subprocesses that are triggered at different times and that, after being triggered, are self-accelerating. One of these is the reflectivity (albedo) drop of the planet, which is caused by the melting of highly heat-reflecting ice, which is then replaced by highly heat-absorbing land and water surfaces. Today, 30% of the solar energy reaching Earth is being reflected back into outer space, and if this amount decreased by only 1% (to 29%), that would add more heat to the planet than all of the GW caused by humans. This feedback process is accelerating because melting increases temperature, which further increases melting. The permafrost effect has the same characteristics of self-accelerating feedback because when permafrost melts, it releases immense amounts of methane, which further increases GW and accelerates melting. Degassing of the oceans (caused by the decrease in CO_2 solubility with increasing temperature) and wildfires are similarly self-accelerating, positive feedback subprocesses, and others could also be listed.

- **Continued heating caused by CO_2** accumulated in the past century – In "regular" heat-balance processes, if the heat input decreases, the process temperature is also reduced. This is not the case with the GW process because 410 ppm CO_2 equivalent (CO_2e) is already in the air, whereas the preindustrial stable concentration was 280 ppm CO_2e. This means that, even if we stop emitting CO_2, GW will continue (although at a lower rate) because the present 410 ppm concentration exceeds the preindustrial concentration by 130 ppm. It should be understood that when people talk about reaching "net-zero" conditions, they do not mean the elimination of all carbon emissions, only about half of that. That is, they mean to reduce emissions to the state at which the CO_2 concentration is no longer rising, but GW will still increase, only at a lower rate. Net-zero steps are not enough to stop GW and start moving toward preindustrial conditions; cooling is also needed. Cooling can be achieved by planting forests, converting our energy economy to a completely "green" one, or "whitening the human footprint" by increasing the amount of heat reflected back into space.

I have spent the past 60 years designing industrial process control systems.[1] For decades, I took it for granted that what we learned from controlling industrial processes could be used only in industry. By the time of this writing, I had realized that I was wrong! I realized that all processes, be they biological, cultural, or any other, have features and characteristics in common with industrial ones. I realized that if we analyzed the dynamics of nonindustrial processes, we could determine where they would lead if left uncontrolled and what it would take to bring them under control. The other basic rule I have learned is that to control a process, any process, one must fully understand its "personality," that is, its dynamics.

When I was born, the field of automation and process control was limited to such simple tasks as thermostats automatically maintaining comfort in our homes and toilet tanks automatically refilling. Today, our automation applications include tasks such as guiding rockets and smart cars, optimizing nuclear power plants, and guiding the behavior of robots with artificial intelligence (AI) software packages. Yet, when I was teaching automation and control (A&C) at Yale University, my course was offered by the chemical engineering department. When I published the five editions of my A&C handbook, my book was handled by the electrical engineering department of the publisher. This was not because these institutions had something against our profession. It was because they did not even know that the process control engineering profession existed.

---

1. Some of the text and illustrations in this book were originally published in *Control* magazine, www.controlglobal.com.

It is time to change that. It is time to understand that professionals knowledgeable about the fields of, say, chemical, electrical, mechanical, and other aspects of a process do not need to understand the totality of the process; only the process control engineer must understand its totality. It is time for technical universities, for example, to establish A&C departments that issue PhDs to A&C engineers. In this book, I demonstrate the contributions we can make to understanding such nonindustrial processes as the safe development of AI and protection against the consequences of GW.

The goal of this book is to evaluate the nature, that is, the dynamic personality, of the AI and GW processes; to examine their positive and negative feedback; and to analyze their lag-dominant nature, rates of change, tipping points, and interactions with other processes. Only when this analysis is completed will I predict the behavior of these nonindustrial processes using the industrial model-based control (MBC) approach.

In addition, I discuss the past and likely future of both fission and fusion nuclear energy because they are considered by some as green energy sources. I emphasize that natural gas (NG) can only be considered to be a "transitional or bridge" fuel because when burned, it still emits 53 kg of CO_2 per million Btu produced. In contrast, oil and coal emit 73 and 91 kg of CO_2 Btu, respectively.

However, CO_2 is not the only greenhouse gas emitted when NG is used. In fact, methane, the primary component of NG, also increases GW by leaking. Specifically, for every cubic foot of NG drawn from underground reservoirs, about 1.4% escapes into the atmosphere. The climate impact of this immense amount of NG is substantial because methane is more than 25 times as potent as CO_2 in trapping heat in the atmosphere. The impact of the sum of the burning and leaking effects of NG can reach that of coal; therefore, NG is not a green fuel.

I.1 Artificial Intelligence

In 2022, we witnessed the effects of cyberwarfare in Ukraine. In 2021, we learned how AI in the wrong hands can make us vulnerable to cybercriminals attacking our critical infrastructure, including electric grids, energy pipelines, water supplies, mass transit, and nuclear power plants. We also learned that 1000 hackers at the Russian Foreign Intelligence Service (SVR) broke into government and private servers in the 2021 SolarWinds attack. AI can be used for manipulative advertising and to develop racially biased operations in public assistance services, not to mention its potential for eventually deciding whether an individual will or will not be hired. Probably even worse, the Internet can confuse people to the point that they accept living in a "truthless" age when it is OK to lie, and it can brainwash people to the point that their confidence in their own institutions or even in their constitutional democracy is in doubt. In the hands of lobbyists for the fossil-fuel industry, AI can become a tool to question the very existence of GW, to convert this existential threat to all humankind to a political issue, and to delay effective action until it is too late.

The potential of AI is practically unlimited. Hacking software packages can target journalists or dissenters. Hacking is becoming the spying tool of the twenty-first century because it is less expensive and harder to defend against than human spies. AI can also be used by "hackers for hire," as was exemplified by the DarkSide attack on the Colonial Pipeline in 2020, which resulted in the payment of millions of dollars in ransom. Still another dangerous application of AI is the use of autonomous drones that can attack military targets with or without human supervision or control.

Therefore, nationwide or even global regulations are required to protect against the misuse of AI. The time has arrived for a "Geneva Convention" to set the legal limits for the allowable uses for and applications of AI.

I.2 Climate Change

Climate change "deniers" argue that today's positive 1.1°C (2.0°F) GW (anomaly) is not that important, that it is just a variation in the weather. NASA[2] (Figure I-1) has shown that the global temperature anomaly during the past half-million years never exceeded +2°C (+3.6°F) and when it dropped to –3.0°C (–5.4°F), North America was covered with a mile-thick layer of ice.

Climate change has been accelerating, and humankind's "energy footprint" on Earth has increased 30-fold during the past couple of centuries. This caused a 7% increase in the global temperature, from a mean of 14.0°C (57.2°F) to over 15.1°C (59.2°F) today. (Some data suggest that the global average temperature has actually risen to 15.6°C [60.1 °F].[3]) In this book, I show that, although GW is a well-understood heat-balance process and the rules of process control apply to its behavior, we must also realize that the "manipulated variable" is human behavior, which is neither logical nor predictable.

This was a difficult book to write because I tried to write it simultaneously for two audiences. One is the community of my own colleagues, who are familiar with the science of process control but want to learn more about the nature of the nonindustrial GW and AI development processes. The other larger group consists of readers who are not familiar with control theory or control jargon but are interested in the conclusions of an analysis made from a process control perspective. Therefore, I skip using A&C

---

2. "Sea Level Change Over 400,000 Years," Center for Environmental Studies, Florida Atlantic University, http://www.ces.fau.edu/nasa/impacts/i4-sea-change/explanation1a.php. *Graph source:* J. Englander, author of *Moving to higher ground: Rising sea level and the path forward*, Science Bookshelf, 2021.
3. "Global Climate Report – May 2021," NOAA National Centers for Environmental Information, June 2021, https://www.ncdc.noaa.gov/sotc/global/202105.

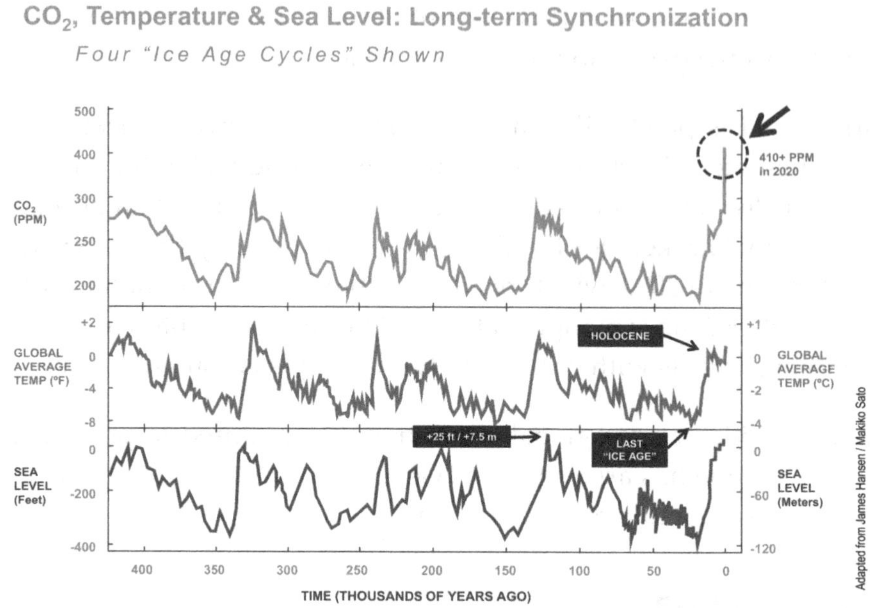

Figure I-1. Temperature anomaly during the past half-million years.

Graph source: J. Englander, author of *Moving to higher ground: Rising sea level and the path forward.* Creative Commons 4.0 with Attribution license, www.johnenglander.net.

jargon and differential equations. I define unfamiliar terms, use everyday language, and minimize showing detailed controls throughout this book.

I.3 Nuclear Energy

The topic of nuclear energy is discussed in detail in Chapter 3, Section 3.5, so I include only a few introductory comments about it here.

The currently operating nuclear power plants utilize a reaction in which the nucleus of a Uranium-235 atom splits into two or more smaller nuclei. Some consider them to be transitional or "half-green" sources of energy because they do not emit carbon dioxide. I disagree and disapprove of their continued use, mostly for economic and safety reasons as follows:

1. I consider nuclear power plants to be potential nuclear bombs because they cannot be fully protected from cyberattacks. I also believe this risk will worsen as their sizes are reduced and their numbers multiply.

2. I consider all currently operating designs unsafe because they will all melt down if the energy sources that supply cooling water to the reactors fail. In Chapter 3, Section 3.5.2.4, I describe a design that theoretically would be safe because it would not require any external energy source to provide cooling.

Rather, it would depend only on gravity and thermal expansion for its operation, and neither of these can fail. No such design exists today.

3. No permanent disposal site exists anywhere in the world to store used fuel rods.
4. The capital cost[4] per KW of the power generation capacity of nuclear plants is 2 to 3 times that of fossil generation and 4 to 5 times that of solar or wind power plant generation.

The other method of generating nuclear energy is by fusing hydrogen atoms into helium. Fusion requires temperatures of about 100 million kelvin (approximately six times hotter than the Sun's core). This process has been investigated for a century, and in August 2021, the scientists at Lawrence Livermore National Laboratory (LLNL) triggered fusion for 100 trillionths of a second (1×10^{-10} s) by a process using giant lasers. The energy generated would have been enough to supply a 60 W light bulb for 5 minutes. Moreover, the energy obtained was only 70% of the energy needed to produce it.

In addition to laser fusion units, doughnut-shaped tokamaks, which use magnetic fields to compress hydrogen fuel, have also been investigated. To date, this design produced even less energy than the laser design of the LLNL.

I oppose the development and use of fusion-type nuclear plants as well. In this case, my reasons are not only about safety, but safety is certainly included among them:

- Fusion reactors operate at a minimum of 100 million kelvin, a temperature at which all materials on Earth evaporate. Therefore, this process would become extremely dangerous if its containment (often magnetic) failed.
- This process is not operational, and we cannot wait to find out if it ever will be.
- Its costs are prohibitive.

I.4 What Questions Are Answered?

In this book, I attempt to provide answers to questions such as the following:

- Is heat-balance evaluation a realistic approach for analyzing the GW process?
- Are established goals such as those in the Paris Agreement and the United Nations recommendations realistic? Can we stay under 1.5°C (2.7°F) or 2.0°C (3.6°F) of warming just by cutting our carbon emissions?

---

4. "Cost of Renewable Energy vs. Fossil Fuels & Nuclear," Better Meets Reality website, updated February 21, 2023, https://bettermeetsreality.com/the-cost-of-renewable-energy-solar-wind-hydro-etc-vs-fossil-fuels-nuclear/.

- Will GW continue to increase even if we eliminate all human-generated carbon emissions?
- Can we cut net carbon emissions by 50% by 2030 and reduce emissions to zero by 2050?
- How much warming will occur if we burn all fossil-fuel reserves?
- Can and should direct cooling be applied in addition to reducing the heating effect of the emitted greenhouse gases?
- What are the tipping points in the GW process? When will they occur, and what will be their consequences? Can they trigger a runaway process?
- Are the GW and AI processes self-regulating, integrating, lag-dominant, or dead-time-dominant and self-intensifying processes? Do these processes have positive feedback? What subprocesses produce this positive feedback?
- Can we quantify the feedback of surface reflectivity (albedo) effects such as permafrost melting, ocean degassing, and excessive cloud formation?
- What are the effects of the different time scales of ice melting on land and water?
- What are the characteristics of a control loop in which human behavior is the manipulated variable?
- Is nuclear energy a safe option for reducing carbon emissions?

The basic structure of control loops can be described by the simple example of regulating the comfort in our homes. That loop consists of five components: (1) the process being controlled, which is the home itself; (2) the sensor that detects the temperature in the home; (3) the set point, which sets the desired temperature; (4) the controller that measures the difference between the actual and desired temperatures; and (5) the manipulated variable, namely, the on-off status of the furnace or air conditioner.

I.5 Controlling the Future

Figure I-2 shows the five components of an imaginary control loop of the future. The structure of this loop is much more complex than that operated by the home thermostat because, for example, the error (the difference between the actual and desired conditions) is a function of judgment. In addition, there is a significant difference between this future "control loop" and the control of home comfort because, in the latter case, the controller and the final control element (thermostat and furnace or air conditioner) are different devices, whereas in the future control loop, they are *the same*. This is the

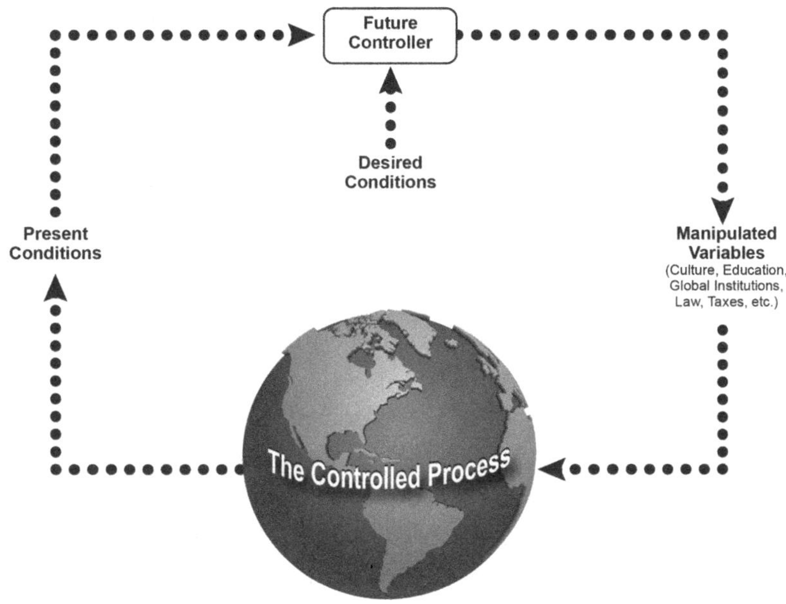

Figure I-2. Controlling the future.

case because the controller is humankind, and the manipulated variable is human behavior. Our leaders function as the controllers by setting the goals for a desired future and deciding which variables should be manipulated to change the behavior of humankind to achieve these goals.

In the past, Nature was the controller of our evolution. During the past few centuries, the configuration of this control loop has changed. Today, to some extent, humankind is controlling its own future. This does not mean that Nature is no longer in overall control. No, it still is! Humankind did not change the rules of Nature. However, it did start to modify its own behavior and thereby change its own environment, thus becoming the controller of the survival of human civilization.

During the decades when I was designing control systems for industrial processes, the most important thing I learned was that to control a process, one must fully understand it. This does not mean that, in the case of GW, one must understand the behavior of all the subprocesses to the same extent as, say, climate experts or ocean current specialists do, but it does mean that the personality and the dynamics of both the overall process and its contributing subprocesses must be fully understood.

When designing controls in the process industry, we first determine the nature of the process to be controlled: its lag and dead times (the time it takes for a manipulation to start having an effect); its gain (the rate at which it will change after the lag or dead time has passed); its inertia, whether it is a continuous or a batch process (in other words, whether it will reach a new steady state or an end point); whether it is linear or nonlinear (is the rate at which it is changing fixed or not); whether it is self-controlling or out of control; and so on. Once we fully understand the nature of the process, we perform the same analysis for the manipulated variable to determine its characteristics and the size and speed of the impact it must have on the controlled variable. Only after both analyses are completed would we start configuring the MBC algorithms for such a lag-dominant control loop.

The purpose of this book is to try to perform a similar analysis for our future control system by gaining an understanding of the speed, size, and feedback of the many subprocesses that influence the behavior of the GW and AI loops. I start by discussing the nature and potential consequences of the evolution of AI, and only after that do I discuss the road on which GW is taking us. I demonstrate that, today, both loops are out of control and that this is dangerous because AI can destroy our moral fiber, and GW can destroy our physical environment. To quote President Vladimir Putin of Russia:

Whoever becomes the leader in this sphere will become the ruler of the world.

In this book, I analyze the control of both of these seemingly unrelated processes: the development of AI and the control of GW. It is important to understand what these processes have to do with each other and how they interact.

AI is a tool that can be both useful and harmful, and the stakes are too high to allow it to become harmful. I consider some of the ways in which AI can be helpful in our efforts to stop climate change. It can model and quantify the consequences of inaction in terms of both their scale and timing, and it can do the same to evaluate the nature, timing, speed, and scale of the corrective actions we must take to guarantee a safe and clean future for our grandchildren. In addition, AI can free up the immense workforce needed to accomplish this Marshall Plan-scale effort required.[5]

The ways in which AI can be useful are somewhat obvious and will become even more evident as the reader turns the pages of this book. What is less obvious is the harm AI can do if it gets into the wrong hands. Cyber propagandists can use social media to spread extremism, erode social trust, and influence elections. By changing people's perceptions of the world, cyber propagandists can manipulate the choices people make on many topics, including climate change. They can confuse people to

5. The Marshall Plan was a program created in 1948 by the United States to provide aid to Western Europe following the devastation of World War II. For more information, see: "Marshall Plan," History website, updated November 1, 2022, https://www.history.com/topics/world-war-ii/marshall-plan-1.

the point that they cannot tell facts from lies, make society insensitive to lies, and brainwash people to the point where their confidence in their own institutions and even in their constitutional democracy is destroyed. In the hands of the wrong people, AI can become a tool to question the very existence of GW, convert this existential threat to all of humankind into a political issue, and delay effective action until it is too late. In this book, I analyze how the evolution of AI influences our behavior and how our behavior changes the climate.

I.6 Historical Background

In 1900, only 3% of homes in the United States had electricity.[6] Today, optimists say that, by the turn of the next century, most of our great-grandchildren will live in "zero-energy" homes, their self-driving cars will run on hydrogen, and perils such as nuclear weapons and GW will be remembered only in the history books. Pessimists say otherwise, and no one knows who is right. I know that if humankind is to transition to a sustainable lifestyle, it will take an immense Marshall Plan–type effort. It will be a challenge comparable to going to the Moon, and the process control profession can help smooth this transition.

We understand very little about the road we will be traveling. All of this is new. Science as such is very young! We learned of the theory of relativity a little more than 100 years ago and about the big bang theory, or the fact that gaseous oxygen is created only by life (photosynthesis) and is present only on Earth, even more recently. We know that scientists have sequenced the genome of the chimpanzee and found that humans are 96% similar to the great ape species. So what? All we know is that evolution is in progress, and although 96% of today's human genes are similar to those of chimpanzees, the remaining 4% makes a significant difference. During the period when only Nature controlled our evolution, we made progress. Now that we are beginning to share in that control, we do not really know if we are safe drivers. We do not know if humans are capable of protecting their own physical or cultural environment or caring about the needs of future generations. What we do know is that we must avoid the slippery slopes that lead to devolution (Figure I-3).

This book describes the first steps that should be taken to get off this slippery slope. It shows how AI can help protect our cultural and moral environments and how a carbon-free energy economy can protect our physical environment. The goal of this book is to demonstrate that applying the knowledge obtained in developing our industrial

---

6. Arthur G. Woolf, "The Residential Adoption of Electricity in Early Twentieth-Century America," *The Energy Journal* 8, no. 2 (April 1987): 19–30 (extrapolating from Table 1, page 21).

Figure I-3. Can our evolutionary period be followed by a period of devolution into "clicksters"? Will AI be only a smart tool, an obedient servant of man, or will it be allowed to become a tool of cultural devolution?

control technology can also be useful in gaining a better understanding of the dynamics of these nonindustrial processes and in determining how much time we have available to make the required corrections.

I.7 Structure of This Book

In Chapter 1, "Controlling AI," I analyze the evolution of AI as a process. In the next two chapters, I analyze the nature and dynamics of GW, which requires MBC because GW is interacting with many subprocesses; this analysis includes the interaction between the AI and GW loops. In some ways, the main purpose of this book is to explain how to keep this interaction positive, that is, how to utilize the evolution of AI to assist in controlling GW.

In Chapter 2, "Global Warming Processes," I describe the many interacting processes that cause or are triggered by the warming of the planet, including their sizes, response speeds, inertias, and feedback interactions.

In Chapter 3, "Green Energy Sources," I describe the types and dynamics of carbon-free, green energy processes.

In Chapter 4, "Storing Green Energy," I focus on making intermittently available energy sources, such as solar and wind energy, continuously available by enabling the energy they generate to be stored and transported. I end this chapter with a description of my reversible fuel cell (RFC), which I hope will become one of the building blocks of our sustainable energy future.

In Chapter 5, "Summary of the Process Control Analysis," I present some of the conclusions of the process control analysis in this book, comment on some of the projections that have been made, and explain why we have less time for corrective actions than currently believed.

In Chapter 6, "Data Summary," I provide a means for quickly looking up various data values relating to the AI and GW processes. Some of this data could be inaccurate because the impacts of the Ukraine war (still in progress as of this writing) are not considered. Also included is an alphabetical list of all the abbreviations used in this book and their meanings.

In Chapter 7, "Conclusions," I provide a fairly detailed summary of the many results of this analysis and the many steps that can be taken to correct the present trends and prevent reaching the various "tipping points." I also describe what the next century will look like if we protect our physical and cultural environments by using AI for the right purposes and if we stop GW by using inexhaustible and clean energy sources to convert our lifestyle into a sustainable and stable one.

In Chapter 6, "Data [illegible] of the data values relating to the [illegible] be used [illegible] [illegible] and their meanings.

In Chapter 7, [illegible] of this analysis [illegible]

1

Controlling AI

Table of Contents

> *The rise of powerful AI will be either the best, or the worst, thing ever to happen to humanity. We do not yet know which.*
>
> – Stephen Hawking

Artificial intelligence (AI) is mostly based on the use of neural networks, which have been used by the process control profession for a long time. It was founded as an academic discipline in 1956 and refers to the intelligence of machines, as opposed to the natural intelligence of animals and humans. Machines can behave intelligently if programmed to do so and are usually programmed to perform goal-directed behavior. Depending on how machines are programmed, they can serve ethical or unethical goals and can impact the cultural environment. In the 2020s, we observed the arrival of the truthless age when, in 2022 for example, Russia spread the claim that attacking an independent nation is only a "special military operation." These events, along with those of 2021, clearly indicate that both nationwide and global sets of rules and regulations are needed to forbid the harmful or criminal use of AI. The time has arrived for a digital agreement "akin to the Geneva Convention" to set the legal limits for the allowable use and applications of this new and powerful tool. Yet, as only a tool, AI can never develop its own ethics. It can only be the repository of our moral values and seldom adds to what we already know.

In this chapter, I apply the rules we have learned in industrial process control to controlling the nonindustrial process of developing a new human tool called artificial intelligence (AI). As in configuring all control loops, here, too, the first step is to identify the controller and its set point. Here, the controller is humanity, and the set points are the limits that can keep AI useful. We must keep in mind that behind every robot and algorithm, there is always a human being and that this human must be held responsible for the uses allowed for this new tool.

Today, AI is all around us—filtering our news, recommending whom we should date, and making decisions about our finances and health. In this chapter, my goal is to show that AI development is out of control and that we can take action to bring it under control. I not only focus on the practical aspects of this development, such as how AI must be prevented from assisting cyberterrorism and how robotics must be prevented from creating global unemployment, but also emphasize that the age of Big Brother is just around the corner. This tool can watch our every step and change our cultural environment to one in which truth does not matter and moral and ethical values do not even exist.

1.1 Adapting to an Artificial Environment

1.1.1 Controlling the Evolution of AI

We have evolved to control the planet, not because humans are the strongest, fastest, or largest creatures, but because we are the smartest of all God's creations. We have created a tool that, in some ways, can become more capable than its creator, which can be helpful or get us into deep trouble.

Human or "normal" intelligence is the totality of all mental processes we use to acquire and apply knowledge. When we delegate some of that intelligence to machines, they acquire an "artificial intelligence." In this sense, the AI in our toolbox is different from all previously acquired tools. It is a tool that can influence human culture, a tool that can influence or even guide our own future evolution.

AI grew out of automation and the science of process control, which, over the past century, has evolved to the point where we can measure just about anything, and we can develop algorithms to evaluate just about any number of measurement values and initiate any variety of actions in response to what we have detected.[1] It was only near the end of World War II that John von Neumann proved that the computer could perform complex calculations. Yet, today, only 75 years later, computers not only perform calculations but have also acquired intelligence. The speed at which we moved from the analog into the digital age is remarkable. When I started as a process control engineer, I did not expect that a gadget that speaks only two words ("yes" and "no") could one day become not only intelligent but, in some ways, smarter than humans. Today, this tool provides us with practically unlimited memory capacity and immense speed to implement even the most complex process control algorithms.

I hesitated for a long time before deciding to write this book because I was not sure—and I still am not—if the direction of the development of AI is controllable. *Should AI be allowed to do all it can, or should it be limited to doing only what it should?*

Process control experience tells me that leaving a poorly understood process uncontrolled can lead to disaster. It also tells me that controlling a process requires an understanding of its characteristics and a knowledge of both its desirable destination

---

1. Stephen Hawking demonstrated this. He was almost completely paralyzed, except for the working of his brain and the motion of one muscle on his cheek, but he was still able to make great contributions to theoretical physics and our understanding of black holes, quantum mechanics, and general relativity. He communicated his remarkable findings by stopping a moving cursor on a computer screen using an infrared sensor to detect the movement of the muscle on his cheek.

(set point) and the road to get there. My attempt in this book is to provide some answers to such questions, but the list of questions is incomplete and vaguely defined, and it represents only a hesitant step toward the goal of bringing the evolution of AI under control.

If left uncontrolled, AI can evolve in both desirable and undesirable directions. AI-supported cyber weapons used by terrorists can already turn our nuclear power plants into atomic bombs, sink uncrewed oil drilling platforms, attack the water supply or electric grid, and even manipulate social attitudes by spreading hate or falsehoods. In this respect, social media can also be harmful when it stirs up controversy and creates a polarized cultural climate.

Out-of-control machine intelligence can take us into uncharted and uncontrolled territory. On the one hand, it can change our culture and lifestyle, and on the other hand—through genetic engineering—it can affect human evolution itself. We headed in that direction when we started to believe that knowledge does not require thinking, only a click of our fingers, or when we began to think that it is OK to spend 10 hours a day staring at a screen and only 15 minutes exercising.

It is up to us to force the development of AI in the right direction so that it serves us by making life better, easier, and happier on this Earth. In other words, we must stay in control of AI and make sure that it does only what it *should* and not whatever it *can* (Figure 1-1). This is important because, as Hawking said, "there is no deep difference between what can be achieved by a biological brain and what can be achieved by a computer. It, therefore, follows that computers can, in theory, emulate human intelligence—and exceed it."

Figure 1-1 Evolution or devolution, are we bringing up a generation of "clicksters"?

However, I do not wholly agree with Hawking if one considers that "human intelligence" includes the emotional and spiritual aspects. In my view, computers will never have a conscience, a sense of self, or moral and ethical values of their own; rather, their actions will always depend on the algorithms loaded into them by humans. In other words, they will remain dual-purpose tools that can do both good and harm, depending on who programmed them. AI development must therefore be directed and controlled by global institutions to keep AI from becoming a harmful tool.

The slippery slope on which we are traveling becomes steeper when we look at genetic engineering. That process might also start with harmless and desirable goals such as aging reversal through in vitro fertilization. However, if it is left uncontrolled, the natural next step could be inserting new genetic traits into human eggs or sperm, cloning human beings, or even trying to create a "perfect race." In other words, uncontrolled AI could spell an end to natural evolution. Therefore, this type of AI development must not be allowed until and unless its goals are clearly defined and controlled.

AI is useful and convenient as a substitute for some of our muscle and routine mental functions, but we must stay in control of it. Otherwise, disasters such as the malfunctions of Boeing 737 Max aircraft will continue. An Ethiopian Airline crash on March 10, 2019, killed 157, and only 5 months later, a Lion Air crash in Indonesia claimed 189 lives. These failures were caused by Boeing violating two fundamental laws of process control: (1) The control system did not use redundant sensors, in this case, three airflow sensors to measure the angle of the wing relative to the airflow, known as the angle of attack. (2) The control system did not allow pilots to overrule the automatic controls, an ability that I consider to be an essential feature of any autonomous vehicle.

Humans must remain in control of all AI-assisted systems. We must stay in control because all control loops fail, and there is nothing more dangerous than an out-of-control automatic control loop that the operator cannot overrule.

Asking whether the system should follow the instructions of the computer or those of the human operator is the wrong question. In industry, we have learned that both can fail. Safety will be served only if, when the two instructions disagree, we look at both and choose to execute the safer one (while automatically starting to evaluate the cause of the difference). In other words, no matter how sophisticated AI algorithms become, human operators must always stay in control, but they must also be prevented from giving unsafe instructions.

This is easier said than done. It requires fully understanding the process being controlled. This is why I started this chapter by saying that the very first step in developing a control system is to study the process until we fully understand it.

A comparison of nonindustrial and industrial processes reveals that the response times of the two can be dramatically different. In industry, when the heat input into a distillation column is increased or the control rods in a nuclear reactor are lifted, for example, these processes respond in minutes because their time constants are short, and their capacities are relatively small. On the other hand, if we try to use AI to modify human culture or if we change the rate at which carbon is emitted into the atmosphere, nothing will happen for years or even decades. When the consequences start to become noticeable, they will initially evolve slowly, but later, they can accelerate and become uncontrollable because of positive feedback or other causes.

Nature's rules controlled human evolution for millions of years. The Creator placed Nature in control of all evolutionary processes on this planet. Today, humankind must come up with solutions to the problem we have created; otherwise, Nature will take charge. It would be much better if our combined wisdom prevented that from happening because, if we wait for Nature to "close" this control loop, the "throttling" will be neither smooth nor gradual, and it will certainly not be empathetic. Today, the future of our civilization is in our own hands, and until now, we have allowed it to drift out of control. Today, a number of industrial (nuclear, plastics, etc.) and nonindustrial (such as the development of AI or the limiting of GW) processes are out of control.

1.1.2 Nature and Effects of AI

As I have already emphasized, before "letting AI loose" we must concentrate on getting this dual-use technology under control. We must understand that, although AI can do a great deal of good, it can also cause a U-turn in human evolution, taking us onto a slippery road that leads in an unknown direction.

Therefore, AI must be nothing more than a fast and smart tool under full human control. We must always keep in mind that AI and all robots are not necessarily beneficial. One of the purposes of this chapter is to discuss the steps we must take to "herd" AI applications in a safe and useful direction to prevent AI's evolution from harming our civilization.

Today, robots and instant communication are making our lives easier, but the ability to make thousands of complex decisions in a millisecond both helps and harms us. The computer started a new age by introducing a new, artificial factor into our real environment, and adapting to its presence is influencing our own evolution. In the past, the totality of our mental processes enabled us to adapt to our environment on this planet. This has been the case since the end of the last ice age, but this evolutionary period is ending because of our own actions. Our activity does not endanger the planet, but it could endanger human civilization. The development of 5G, the fifth generation of cellular networks, followed by "deep AI," opens a new chapter in our evolution. As of today, this development process is out of control.

When we designed the first gadgets that made industry safer and more efficient, we did not realize where this would lead. Next, we designed smart instruments that could self-diagnose and request their own recalibration or maintenance when needed. This road eventually led to robots, and today, we are beginning to realize that these human creations can be more than mechanical workers. In fact, they can outsmart their human owners in some respects (e.g., hackers through the Internet can access advanced controls in industry, in administrative or financial institutions, or even smart home devices).

According to our industrial control experience, nonlinear, high-capacity, and accelerating processes can be unpredictable, and the out-of-control spread of AI is such a process. AI-operated robots could become uncontrollable if they became superintelligent and acquired the skills of self-learning. In addition, we also know that AI is already serving violence, as some 50 nations are preparing for space wars or developing robot soldiers who no longer need human direction. Although these soldiers will never possess human intelligence and will never have a conscience, emotions, or ethical standards of their own, they could break free from human control.

1.1.3 Risks of Uncontrolled AI Development

The Internet was born to provide universal connectivity and ready access to an unlimited supply of data and information. It achieved those goals, but today, it is being used to serve both good and bad goals. The Internet is simply a tool, just as a knife is a tool, one that can help both a cook and a bank robber. Just as we do not blame knives if they are used in a crime, we should not blame the Internet for the harm it can do. However, we should severely punish those who are using it for theft, spreading hatred and lies, or subversion. Today, nation-states and cybercriminals both commit cyber sabotage and cyber espionage, and there are no international institutions or legal systems to stop them.

The Internet is evolving today as an out-of-control process that can damage our industrial, economic, and political systems or even alter our culture and civilization. If we were using process control language, we would say that the process of integrating AI into our culture is an uncontrolled "herding process." As we observe this process, we see that its impact is accelerating, and its effects are increasingly felt in our daily life as the applications of AI are multiplying.

AI's impact will depend on how we control it. We are already seeing AI's dual-use capability. For example, AI can be beneficial in finding accident victims by detecting the locations of their cell phones, but it can also be used by the secret police in a dictatorship to track down opposition leaders.

This worries some intelligent people, including Bill Gates and Elon Musk, because, in their view, hyperintelligent machines in the wrong hands could do anything through tactics such as brainwashing, fermenting division and hate, and spreading misinformation. Therefore, safeguards are needed to ensure that AI does not get out of control. AI does not need explosives to destroy or disrupt industry, and because robots have no sense of self, no ego, and no conscience, they can do immense harm. Therefore, the user of this tool needs to be controlled.

1.1.4 Processes Affected by AI

One cannot control a process by manipulating only a part of it. In other words, global rules and global enforcement are needed to control global processes. In the case of AI development, effective global institutions do not exist, and nation-states are not ready to form them.

It took thousands of years for our social structures to evolve from tribal groupings to nation-states, and globalization has just started. Effective global institutions already exist in some areas of communications, culture, and economics, but it will take decades, if not centuries, for them to emerge to control AI applications. While global regulations do not exist, the European Commission of the European Union has published an overarching regulatory framework titled the Artificial Intelligence Act;[2] in January 2021, the US Congress enacted the National AI Initiative Act;[3] and in March 2022, China passed a regulation[4] governing companies requiring the use of moral, ethical, accountable,

---

2. "Regulatory framework proposal on artificial intelligence," European Commission, accessed November 15, 2022, https://digital-strategy.ec.europa.eu/en/policies/regulatory-framework-ai.
3. For more information on the National Artificial Intelligence Initiative Act of 2020, see: https://www.congress.gov/116/crpt/hrpt617/CRPT-116hrpt617.pdf#page=1210.
4. Rogier Creemers, Graham Webster, and Helen Toner, "Translation: Internet Information Service Algorithmic Recommendation Management Provisions – Effective March 1, 2022," Stanford University website, January 10, 2022, accessed November 15, 2022, https://digichina.stanford.edu/work/translation-internet-information-service-algorithmic-recommendation-management-provisions-effective-march-1-2022/.

and transparent use of AI. These are the right first steps, but global standards still do not exist.

1.1.4.1 Economy

When analyzing the economy as a process, one would conclude that it is a highly interacting, multivariable process. One might argue that, in addition to using traditional process control tools, one should consider chaos theory. One would probably conclude that envelope control (i.e., keeping the operations of a system within the safe range of limits of several variables) is best suited for anticipating and controlling the economy because the overall economic well-being of a society is a function of several variables of similar importance (unemployment, inflation, trade and tariffs, corporate profits, interest rates, budget deficits, etc.) AI is an ideal tool for implementing such an envelope control system.

Whereas industrial processes can be described by heat and material balances, this is not the case with the economy because one of its key variables is human emotion. Computers can measure, manipulate, calculate, and memorize, but they cannot have and do not understand emotion. In the case of the economy, human emotion—our optimism or pessimism about the future—is a key factor.

It seems that, in the past, the global economy had an overall oscillation period of about 50 years (Figure 1-2). These cycles start with a phase of rising prosperity and end with a phase of recession or depression. It seems to me that the first phase is triggered by new inventions that create optimism about the future and cause a willingness to spend money by investing in new technologies and giving loans with the expectation that the economy will expand. Naturally, the lengths of these overall cycles are influenced by wars, trade agreements, and other factors, but it still seems that they are triggered by new inventions.

Loans for the economy are like oxygen for a scuba diver. As long as the oxygen supply lasts, the dive proceeds. Once the oxygen of loans starts to run out, recession starts and depression arrives when the oxygen (optimism) is gone and the "dive" is over. Then, the diver must come up and refill the oxygen tank. In the case of the economy, invention is what refills the tank of optimism. Such a spread of optimism can be triggered by imaginative undertakings such as the Marshall Plan, inventions such as the computer, or the AI-supported conversion to renewable energy technologies.

I believe that the next cycle of prosperity will start when the next generation begins to convert our lifestyle into a sustainable one or, in process control terms, when we agree on configuring a global control loop for economic advancement. This beginning will

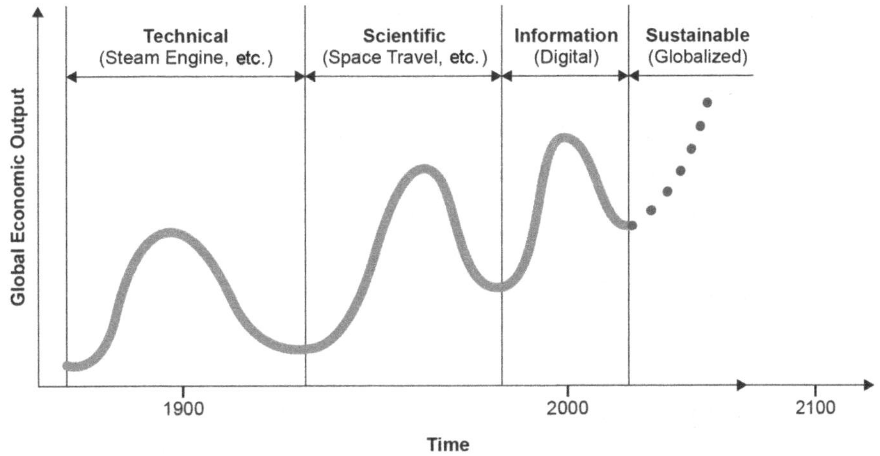

Figure 1-2. Overall, economic cycles seem to be triggered by new inventions. It seems to me that, today, we are at the beginning of a cycle consisting of economic globalization and a transition from a fossil-fuel- and nuclear-based energy economy into a "green" one that will result in a sustainable lifestyle on the planet.

be triggered by such new technological approaches as converting to renewable energy and establishing global institutions to guide our evolution.

One of the manipulated variables in controlling the economy is the upbringing of the next generation. In other words, to trigger a new cycle of innovation-based optimism that fuels sustained material and emotional growth, we must concentrate on education and research and development (R&D) so that the new inventions will aim at solving long-term global problems (cultural and material) and not just short-term financial goals such as profiting from extending the life of the fossil-fuel and nuclear-energy industries.

1.1.4.2 Culture

From a process control perspective, culture is a lag-dominant process. It takes 10 to 20 years for a child to form values. In the beginning, every child's soul is a blank page on which moral and ethical standards can be inscribed by our churches, our schools, the media, and examples from their role models, most importantly, their parents. To understand the evolution of culture as a controllable process, we must also realize that culture is the sum total of all of the beliefs and convictions shared by the members of society.

Considering social behavior as a process, we know that, for a society to function smoothly, at least three generations should share the same outlook on life and the same moral and ethical values. When the prevailing culture changes more quickly than this, cultural tensions arise as conflicts evolve between the generations (in process control terms: one control loop generates positive or negative feedback on another). Positive feedback is desirable (e.g., elimination of prejudices, understanding of environmental concerns and interrelationships), whereas negative feedback can cause social tension (e.g., intolerance of those with a different skin color, religion, sexual orientation, etc., and acceptance of amoral behavior, such as lying). Whatever the cause of the tension between generations, decoupling the majority and minority views is necessary if society is to function smoothly and effectively. Such "decoupling" in human processes requires increasing the influence of the manipulated variable, which is usually education.

This decoupling is easily accomplished in industrial control loops, but it is difficult to achieve between cultures. One of the goals of decoupling is eliminating the fears of change and of "otherness." This is resisted by isolationism and best served by maximizing contacts among groups. This might sound like a political program, but it is only an explanation of how industrial process control achieves decoupling.

In the case of nonindustrial processes, the interests of various segments of society often collide. As long as group interests can be subordinated to the shared ethical and moral standards of society ("one nation under God," "all men are created equal," etc.), decoupling can be achieved. On the other hand, problems will arise when national and global attitudes or interests collide.

The conflicts provoked by globalization are often caused by the dead time in the changing of cultural attitudes, which can take decades or centuries. Our loyalty to "our culture," meaning our language, customs, music, and the traditions of our extended family (our nation), harms no one. It is a form of healthy loyalty. Yet, this loyalty should not stand in the way of developing other loyalties. A person with multiple loyalties is a richer and happier person.[5]

Understanding and accepting the importance of developing additional loyalties will be the great challenge to a successful globalization process. Understanding that globalization does not endanger national cultures will be the key to accepting and developing a multicultural world. Like AI, globalization is a dual-purpose process because it can lead to a multicultural world or to a uniformly commercialized, materialistic, and valueless global culture.

---

5. Edward Teller, who wrote the preface to my first book, said that a healthy individual does not switch loyalties, but collects them. For example, falling in love with a woman should not cause you to stop loving your mother, he said. He used to say, "I am a 300% guy: 100% American, 100% Hungarian, and 100% Jewish, and it is good to be home in many cultures. It is good to be a 300% person!"

Compared to cultural processes, the time constants of other global processes are even longer. The preservation of Nature's billions of species and the time constants of oceanic processes can exceed thousands of years; therefore, the consequences of our pollution might survive for many centuries.[6] Naturally, the same holds true for emitting pollution into ground waters, the atmosphere, and, metaphorically, human culture.

1.1.5 An Intelligent Tool in Our Hands

Intelligence is the ability to solve problems; consciousness is the ability to feel things (hate, love, pleasure). We want to develop computers and robots that are smart, but we must realize that consciousness cannot be programmed. Attempting to do so is dangerous. We cannot and should not try to play God; we must accept that tools cannot have a conscience, emotions, or moral values. These mechanical devices should only be operated by humans loading instructions into them. They cannot apply any moral judgment of their own; they can only carry out our instructions. Humankind must stay in control, and AI must remain what it is: a tool.

Why do I say all of this? I believe that superintelligent tools in the hands of humankind can eventually lead to a deeper understanding of the universe. On the other hand, it is also possible that AI can be used to destroy our trust in science and our ability to distinguish facts from lies or even to develop a cultural attitude in which we no longer care about what is true or right. We do not yet know the future effects of AI on our culture. We know that people like Elon Musk, Bill Gates, and others are worried about letting AI loose. They believe that the spread of uncontrolled AI endangers human civilization, and I agree.

I know that robots could replace most blue-collar jobs; however, I also know that "robotization" will create many new jobs that will be needed to design and maintain the components of the new technologies being created. On the other hand, I am also beginning to believe that AI could cause humans to become intellectually lazy and detached from Nature or even from culture itself. Obviously, AI in the wrong hands can also do immense harm, but that does not worry me because that danger is not new, and we have figured out how to handle such dangers in the past. After all, humankind has faced and survived many evils, including tyrants, fanatics, and the dangers posed by nuclear weapons.

---

6. For example, some estimates predict that, by 2030, the weight of plastics in the oceans will exceed the weight of fish.

One hopes that we will figure out a way to live with AI and keep it benevolent or, at least, neutral. But is hope good enough? Can we keep developing a machine that can have superior intelligence and no conscience? I do not think so! The laws of process control tell us that we must control this development, that—in the same way that we must protect our physical environment from pollution and warming—we must also protect our cultural environment from the potential effects of deculturation and psychological manipulation. This effort requires joint and conscious action, which is lacking today.

1.2 Neural Networks

Neural networks (NNs) and deep learning have been the most influential technologies of the past decade. Applications of NNs include Amazon's Alexa AI-powered assistant, Google's translation service, and Apple's Face ID iPhone lock. NNs have also contributed to AI breakthroughs in a wide variety of other fields, from the self-driving cars being able to "see" to the ability to diagnose skin and breast cancers.

In industrial controls, NNs have long been used to control multivariable processes. For example, the physical properties of distillates can be predicted by using about a dozen measurements as inputs into a back-propagation NN having three or more layers (some hidden) of nodes and links, each having different gains[7] (also referred to as weights or biases[8]) (Figure 1-3). The experience gained by using NNs in industrial control applications is very useful in developing AI networks.

Biological NNs are the basis of the operation of both animal and human brains. Animal brains learn to perform tasks by considering examples, generally without being programmed with task-specific rules.

The differences between human and animal brains include the following:[9]

- A human brain can be divided into three main areas: the forebrain, midbrain, and hindbrain (see Figure 1-4). This feature is absent in most primitive animals.
- The portion of the brain devoted to vision is much larger in humans than in other animals.

---

7. Gain or weight is the ratio of the change in the output of a node to the change in the input to that node. The connections between the nodes have weights (positive or negative) so that they can add to or subtract from (excite or inhibit, respectively) the output of the receiving node.

8. Biases are systematic measurement errors.

9. "Difference Between Human Brain and Animal Brain," DifferenceBetween.com website, June 28, 2012, accessed November 1, 2022, https://www.differencebetween.com/difference-between-human-brain-and-vs-animal-brain/.

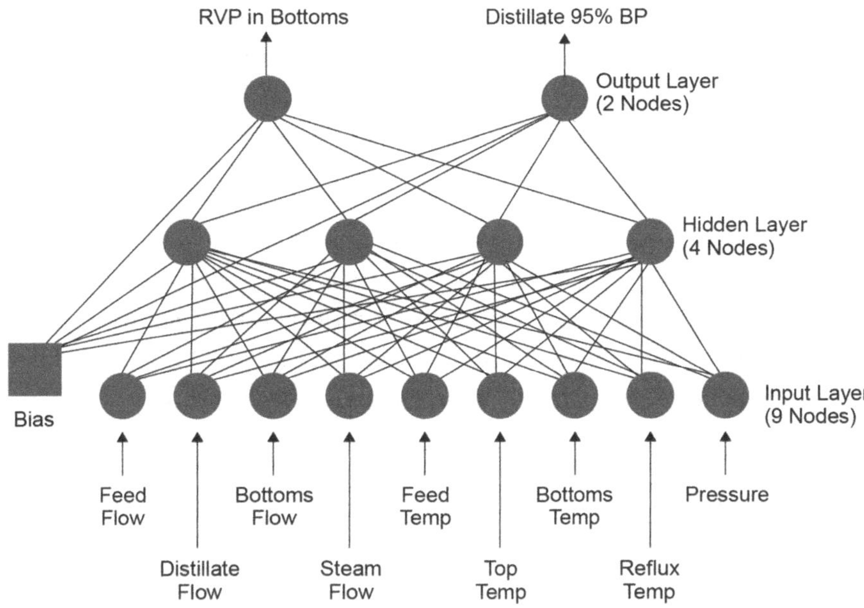

Figure 1-3. In an industrial application, the physical properties of the products of a distillation process can be predicted by a three-layer network. In this network, the inputs of the nodes increase or decrease the output of the receiving node according to the weight of the connection. Abbreviations: BP, boiling point; RVP, Reid vapor pressure.

Source: Courtesy of *Control* magazine. Béla Lipták, "AI's Impact Will Depend on How We Control It," Figure 2, July 17, 2018. https://www.controlglobal.com/articles/2018/ais-impact-will-depend-on-how-we-control-it/.

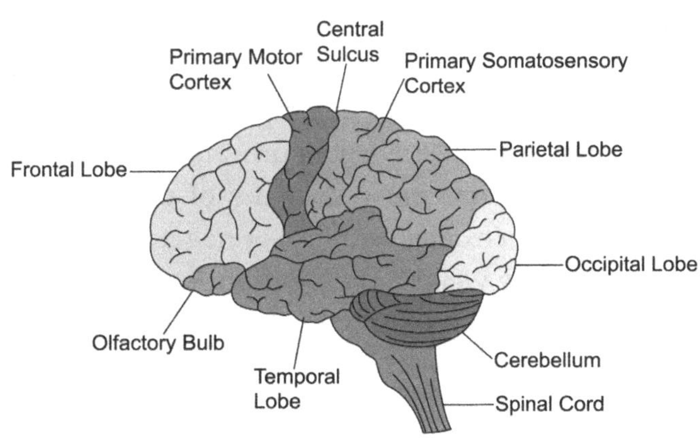

Figure 1-4. The cerebral cortex consists of a mass of deeply folded, rippled, convoluted tissue.[10]

- Several areas of the human brain control language, a skill that is unique to humans.

---

10. For more information, see: "Chapter 3: Physiological Bases of Behavior," *BSN 1A1-2: General Psychology Blog*, July 24, 2013, accessed November 1, 2022, http://ginevinbsn.blogspot.com/2013/07/chapter-3-physiological-bases-of.html.

- Corticalization, or the transfer of cognitive functions from the primitive areas of the brain to the cerebral cortex, is the most dominant feature of the human brain when compared with animals' brains, especially with other vertebrates. It creates new cortical structures that enable new and more complex functions to be performed.
- Humans have gained many characteristic abilities that are unique to humans only because of the development of the brain. Examples include humor, advanced planning, decision-making, an appreciation of beauty, questioning the meaning of life, an appreciation of mortality, adaptation to unsuitable environments (e.g., deserts and frozen lands), morality, religion, and worship.

Compared to the animal brain, the human brain is Nature's smartest and most complex creation, many functions of which remain a mystery. Every function of the brain involves electrical currents and chemical reactions running across a vast number of neurons.

The human brain contains about 100 billion neurons, arranged in depths of six, seven, or more layers and connected by many billions of nerves, each having its own gain (also called weight). Figure 1-5 illustrates the signal flow through a neuron, from inputs at the dendrites to outputs at the axon terminals.

The nerves establish electrical connections between the neurons (brain cells), but we do not yet know exactly how all of these connections add up to higher reasoning because, as of yet, this complex circuitry is mostly incomprehensible. We know that NNs are usually fully connected, meaning that every neuron unit is connected to every other unit on both sides. The connections between pairs of units have different weights, which can be positive (if one unit excites another) or negative (if one unit suppresses or inhibits another). The higher the weight, the more influence one unit has on another. (This corresponds to the way actual brain cells trigger one another across tiny gaps called synapses.)

The average brain contains about 100 billion neurons (brain cells). The neurons are about 10 μm (0.01 mm) in size and make up only about 10% of the cells in the brain. The rest, called neuroglia or glial cells, support the neurons and provide them with nutrients for growth and operation. In a computer, the equivalent of a neuron is a switching device called a *transistor.* A postage-stamp-sized integrated circuit in a microprocessor might contain from 50 million to 2 billion transistors.

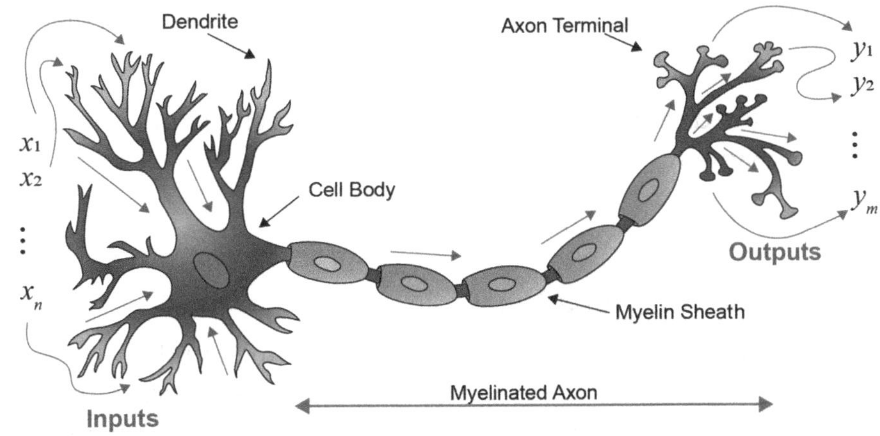

Figure 1-5. The structure of the neuron (brain cell) includes the cell body; the dendrites, which carry the inputs (x_1 to x_n) into the cell body; and the myelinated axon leading from the cell body to the output terminals, called axons (y_1 to y_m).

Source: Derived from "Neuron.svg" by Quasar Jarosz, September 16, 2018, which is licensed under CC BY-SA 3.0, https://en.wikipedia.org/wiki/Biological_neuron_model#/media/File:Neuron3.png.

1.2.1 The Human Brain versus the Computer

In the development of AI, it is important to clarify the differences between the mechanisms of neural information processing and those of AI. According to Chris Chatham, these differences flow from the differences between the nature and operation of the brain and those of the computer,[11] including the following:

1. Brains are analog; computers are digital.
2. Computers access information in memory by polling a memory address; brains search memories using cues.
3. The brain is a massively parallel machine; computers are modular and serial.
4. Processing speed is not fixed in the brain; there is no system clock.
5. Short-term memory is not like RAM.
6. Computers are the hardware that runs software; there is no "mind software" running on brains.
7. Synapses are far more complex (electrochemical) than computer logic gates (electrical).

---

11. George Dvorsky, "Chris Chatham: 10 Important Differences Between Brains and Computers," *Sentient Developments Blog*, May 22, 2011, http://www.sentientdevelopments.com/2011/05/chris-chatham-10-important-differences.html.

8. Computers use processors and memory for different functions; there is no such distinction in the brain.

9. Computers are designed, built, and are of fixed architecture; the brain is a self-organizing system.

10. Computers have no body; brains do.

1.2.2 Scientific and Religious Views

Applications of AI are multiplying and are already affecting our daily life at a startling pace. This is a new experience for humanity. Throughout the ages, humans have not only struggled for survival but they have also struggled to understand the universe and its purpose. This search progressed on two roads, the spiritual and the scientific. Those on the spiritual road assumed that understanding the universe was beyond the abilities of humans, whereas those on the road of science decided to try it anyway. Scientists focused on learning the laws guiding the universe.

When scientists such as Aristotle, Copernicus, Galileo, Sir Isaac Newton, Albert Einstein, and Stephen Hawking discovered such things as relativity, the existence of black holes, and the continuous expansion of the universe, they demonstrated that we have the ability to understand some of the laws that give order to the universe. This suggests that the spiritual and scientific perspectives are not in conflict. In fact, these views seem to be merging, as they both seem to agree that human beings contain a nonphysical component that science cannot explain and that religions call the soul.

It seems to me that human personhood also has an immaterial component that only God could create. It seems that human evolution is not a strictly material process but one in which the immaterial "soul" also plays a major role. Deep down, we all know that we are more than biological robots. In the end, the greatest threat humans face is to forget this! I believe it is up to us to prevent those who are already misusing AI from brainwashing the coming generations into believing that there are "alternative facts," that the truth does not matter, that there is no right and wrong, and that there is no soul inside us, only an unsatisfiable ego.

General Douglas MacArthur put it wisely:[12]

---

12. "Japan's Surrender," in *The Reports of General MacArthur: The Campaigns of MacArthur in the Pacific, Volume I* (Washington, DC: Department of the Army, 1994 reprint), chapter 14, 467, https://history.army.mil/books/wwii/MacArthur%20Reports/MacArthur%20V1/ch14.htm#b13.

If we do not devise some greater and more equitable system, Armageddon will be at our door. The problem is basically theological and involves a spiritual recrudescence and improvement of human character that will synchronize with our almost matchless advances in science, art, literature, and all material and cultural developments of the past two thousand years. It must be of the spirit if we are to save the flesh.

1.2.3 Artificial Neural Networks

An artificial neural network (ANN) is an interconnected network of nodes, similar to the vast network of neurons in the human brain. The core components of ANNs are called artificial neurons, which receive inputs that change their internal state (activation) and their outputs, which are functions of the inputs received. Each neuron receives inputs from several other neurons, multiplies them by assigned weights, adds them, and passes the sum to one or more other neurons. Some artificial neurons might apply an activation function to the output before passing it to the next neuron (Figure 1-6).

The working of the ANN mimics that of a biological NN, but they are not the same. In an ANN, the outputs of certain neurons become the inputs of other neurons, forming a directed, weighted graph. The weights of the connections and the functions that compute the activation can be modified by a process called learning, which is governed by learning rules.

A fully connected NN consists of input units (the first layer on the left of Figure 1-7), followed by one or more hidden layers, and finally, an output layer on the right. All units are connected to all other units on either side of the layers. Inputs are fed in from the left, activate the hidden units in the middle, and generate outputs on the right. The ANN can change the strength (weight) of the individual connection between any two

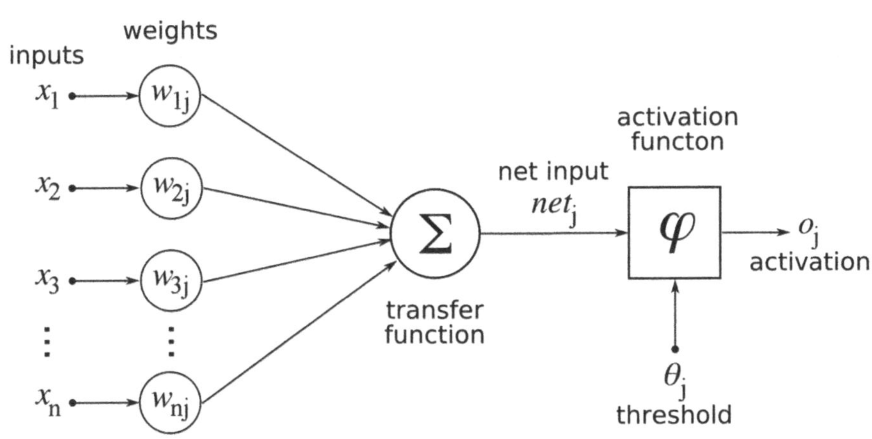

Figure 1-6. The structure of an artificial neuron is the basic component of ANNs.

Source: "ArtificialNeuronModel english.png" by Chrislb, July 14, 2005, is licensed under CC BY-SA 3.0, https://commons.wikimedia.org/w/index.php?curid=224555/.

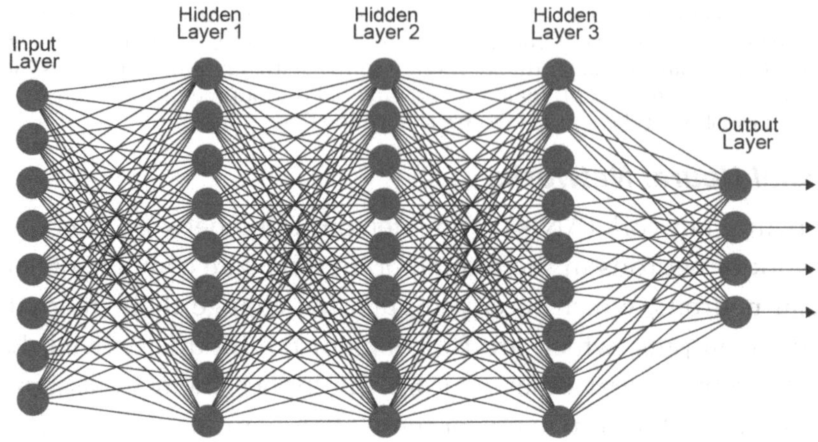

Figure 1-7. Illustration of a self-learning ANN with three hidden layers.

units. An ANN learns by gradually adjusting the strengths of the various connections until the outputs match the observed object or process.

Although ANNs mimic biological NNs, the two types of networks are completely different. The brain is capable of thoughts, memories, and feelings because its neurons are densely interconnected in complex, *parallel* ways (each neuron is connected to perhaps 10,000 of its neighbors). In contrast, the transistors of an ANN are wired in simple *serial* chains (each artificial neuron is connected to maybe two or three others in basic arrangements known as logic gates). It is this structural difference that causes computers and brains to "think" differently. Not only is the number of brain cells greater, but the brain cells are connected in richer and more complex ways.

Thus, whereas computers can store vast amounts of information and manipulate this information according to instructions from human programmers (Figure 1-8), our brains function differently. They can be fast or slow (we might need to "sleep on" some complex or emotional issues), and even when everyday influences and information

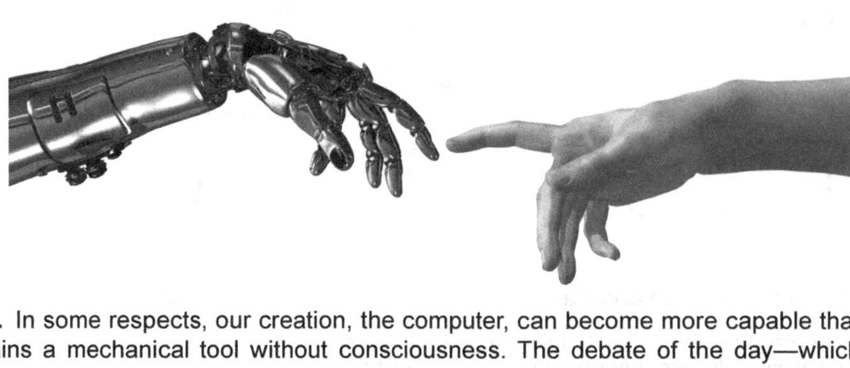

Figure 1-8. In some respects, our creation, the computer, can become more capable than its creator, but it remains a mechanical tool without consciousness. The debate of the day—which will not be resolved soon—concerns whether an ANN can ever overcome that limitation and gain consciousness.

guide us, we act on such inputs according to our consciousness and emotions, which ANNs cannot do.

Some believe if the number of transistors is high enough and they are densely interconnected, then such a deep ANN will be able to learn on its own, recognize patterns, and make decisions in a human-like way. In other words, they believe that we will not need to program such a network, as it will learn all by itself, just like a human brain! (My view is that robots will never be able to do that and will never have consciousness.)

1.2.4 Training an ANN

The operation of an ANN is different from that of a human brain in that the ANN uses silicon and wires to duplicate the dendrites and neurons working in the brain. As stimuli from the external environment are accepted by dendrites, in the same way, the inputs create electric impulses that travel through the NN. An ANN consists of several nodes that behave as neurons. The nodes are connected by links (wires) for communication. Nodes take input data to perform small operations on trained data, and the results of these operations are passed to other nodes (neurons). The output at a node is called its node value.

The learning process of an ANN is similar to the feedforward control used in industry, as illustrated in Figure 1-3, except that the number of hidden layers is larger. Let's look at how it works.

As explained earlier, the measurements during the learning process (when the ANN is being trained) enter through the input layer nodes and travel through the node(s) in the hidden layer(s). The signals are multiplied by the "weight" of each connection, which can be positive or negative. (For example, in the case of distillation, the amount of a component in the distillation product can be increased or decreased if a particular measurement on the input side increases or decreases.) The nodes in each hidden layer perform a weighted sum of all of the signals received from the previous layer to the left, and they "trigger" when that sum reaches a certain limit. The nodes in one hidden layer trigger the corresponding units in the next layer until the signals arrive at the output layer on the right (Figure 1-9). This is the feedforward phase of the training, which is the first part of the learning process.

This feedforward phase is followed by a feedback phase, which improves and modifies the feedforward model. This is similar to how children learn. Take, for example, the process of a young boy learning to play ping-pong. In the beginning, all the child knows is to hit the ball, so he applies this arbitrary feedforward model and hits the ball. If the ball misses the table, he receives feedback instruction (for example, his

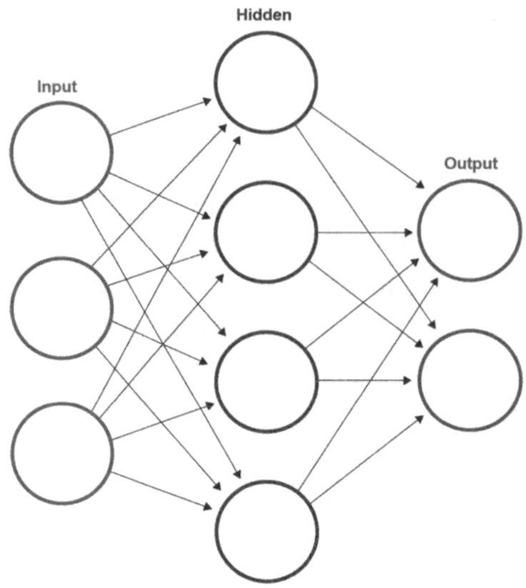

Figure 1-9. In an ANN, each node output is an input signal (weighted connection) to all nodes in the next layer to the right.

father saying, "aim for the table"). The child remembers these instructions and adjusts his approach (adjusts the feedforward model). The next time, the ball ends up in the net, and he again receives a feedback instruction: "Get it over the net." In the child's mind, each feedback instruction refines the original feedforward model (until he is ready for the Olympics).

The ANN learns in the same way as a child, by comparing the actual outcome with the desired outcome and using the difference to make the changes needed to reduce that difference the next time. This process is similar to how a control loop works. The main difference between how an ANN is trained and how a child learns is speed. The determination of who will win at the Olympics—the robot or the child?—is a function of the number of measurements serving their models (the measured portion of the factors affecting the movement of the ball) and the accuracy of each of those measurements.

As we will see later, smart cars are trained in the same manner, by employing self-learning capabilities. In such super- and hyper-AI systems, there might be six hidden layers in the feedforward network. As these systems learn, they evaluate the causal chain, and on this basis, they can build (learn) a seventh layer of links in their NNs.

Self-learning AI development is a rapidly changing and improving process with the aim of achieving human-competitive performance on certain tasks. This development is beyond the scope of this work; but readers who are interested in it can refer to the "Deep Learning" article by IBM Cloud Education.[13]

If the ultimate goal is to achieve autonomous control, which requires no human supervision, multiple control-loop layers are utilized like the hierarchical control systems in industry. These systems are remarkably fast. In the case of flight control, they "tick" (decompose the aircraft behavior into manageable chunks) as fast as 32,000 times per second, whereas higher-level loops may cycle once per second. The most common control mechanism used in these layers is the proportional-integral-derivative (PID) controller, which calculates precise inputs for the electronic speed controllers and motors.

1.3 AI in Cybersecurity

Although cyberterrorism is very dangerous, effective safety steps are being developed to protect against it, and eventually, we will overcome most of it. AI-based cyber sabotage is more dangerous than the earlier forms because it is easier for the saboteur to implement, and its effects can be more widespread. Still, we must protect our civilization against it.

Some say that World War III will be fought utilizing drones and cyberattacks and that AI, because it is a dual-purpose tool, will play a significant role both as a weapon of attack and as a means of defense. Such cyberwar is different from previous wars because its targets are not limited to military or physical objects but include economic, informational, political, and cultural targets as well. In fact, computer malware attacks can be used against all systems that use computers.

The tools of malware attacks and the technology for protecting against them are both progressing very rapidly, and because the "bad" guys are just as smart as the "good" ones, neither side will ever "win." This weapon will continue to evolve and will probably be with us for the foreseeable future, as has happened with most other weapons in the past. In addition, as AI becomes more powerful, malicious computer worms will start to think, learn, and adapt on their own.

In the past, bombs could not think for themselves; they just dropped. Today's malicious computer worms can think. Malware attacks can already turn nuclear reactors

---

13. "Deep learning is a subset of machine learning, which is essentially a neural network with three or more layers. These neural networks attempt to simulate the behavior of the human brain—albeit far from matching its ability—allowing it to 'learn' from large amounts of data. While a neural network with a single layer can still make approximate predictions, additional hidden layers can help to optimize and refine for accuracy." *Source:* IBM Cloud Education, May 1, 2020, accessed November 21, 2022, https://www.ibm.com/cloud/learn/deep-learning.

into bombs, sink uncrewed oil drilling platforms, attack electric grids, flood the sky with drones, and flood the Internet with disinformation, including fake photos and videos, in an effort to psychologically manipulate people and thereby attack most institutions, including democracy itself.

Such attacks have already included shutting down power grids in Ukraine and petrochemical plants in Saudi Arabia. Israel and the United States successfully attacked a nuclear enrichment plant in Iran, and unknown attackers shut down an airport in England and crippled some newspaper printing presses in California. Attacks by Iran on the US financial system (operation Ababil) and Saudi Arabia's ARAMCO (cyberattack Shamoon) have also occurred. The list goes on.

In 2018, the White House Council of Economic Advisers reported that malicious cyberactivity in 2016 cost about $100 billion to the US economy. Although the number of cyberattacks per year is rising, it is still "only" in the hundreds, which—considering the number of industrial control systems (ICSs) exposed to such attacks—is surprisingly small. Reports say that cybersecurity threats are increasing in frequency, scale, sophistication, and severity. For example, in 2021, hackers targeted a major gas provider in April and the world's largest meat producer in May. In general, hackers are after ransom, but in other cases, they intend to steal information.

However, despite several recent cyberattacks by nation-states against private-sector targets, the US Department of Homeland Security's Industrial Control Systems Cyber Emergency Response Team (ICS-CERT) estimates the likelihood of a catastrophic attack to be remote.

Although malicious computer worms can reach the programmable logic controller (PLC) and supervisory control and data acquisition (SCADA) interfaces of industrial plants, ICS-CERT reported that there were only 39 cyber incidents involving ICSs in 2010; this number grew to 290 by 2016. In addition, there was little evidence that sophisticated threat actors were conducting focused R&D programs to develop ICS exploitation capabilities. However, a small number of ICS-focused campaigns used BlackEnergy and Havex malware, so vast commercial research efforts have been directed toward identifying ICS vulnerabilities.

These attacks are increasing rapidly. For example, a 2021 cybersecurity report[14] revealed that the number of ICS-CERT advisories increased dramatically in 2021, as

14. "ICS-CERT advisories affecting ICS environments show significant increase in 2021," Industrial Cyber website, February 1, 2022, https://industrialcyber.co/threats-attacks/ics-cert-advisories-affecting-ics-environments-show-significant-increase-in-2021/.

attackers had a larger arsenal to use and launch potential threats in an ICS environment. The report stated that 389 ICS-CERT advisories occurred in 2021, rising from the 2020 figure of 249 advisories, reflecting the largest year-to-year growth in the history of the ICS-CERT program.

In general, ICS-CERT advises industrial computer users to

- never connect to the Internet without a firewall,
- prohibit direct access to control systems by business and information technology (IT) staff,
- require different logins and passwords for business and control departments,
- require multifactor authentication codes,
- allow data from control systems only to go out through network demilitarized zones (DMZs) and not back in, and
- perform a thorough security assessment.

ICS-CERT offers a variety of resources, risk assessment tools, training, and other services that organizations in the process industries can use to improve their cybersecurity. All of these tools are available and described on the ICS-CERT website.[15]

Although ICS-CERT reports in previous decades were fairly optimistic, the situation is actually getting worse. Examples include Russian-sponsored attacks by the Cozy Bear and Fancy Bear groups[16] (linked to Russian intelligence agencies) who were responsible for the attack on the US presidential campaign in 2016 and the 2020 data breach at the US departments of Homeland Security, Agriculture, and Commerce. Other nations have also used cyberattacks. For example, a drone attack by the United States killed Iran's top general in January 2020, and similar attacks took place in 2021 in Afghanistan.

Hostile but nonmilitary attacks have included many attacks on electric grids, water supplies, and industrial sites. The largest attack occurred on October 30, 2019, at India's

---

15. "CISA'S role in Industrial Control Systems," Cybersecurity & Infrastructure Security Agency website, accessed February 28, 2022, https://www.cisa.gov/uscert/ics.

16. For more information, see: "Cozy Bear," Wikipedia, accessed February 28, 2022, https://en.wikipedia.org/wiki/Cozy_Bear. "Fancy Bear," Wikipedia, accessed February 28, 2022, https://en.wikipedia.org/wiki/Fancy_Bear.

largest nuclear power plant, the Kudankulam.[17] To mention a few other examples, North Korean cyber activity has targeted the Indian Space Research Organisation's National Remote Sensing Centre and the Indian National Metallurgical Laboratory, and North Korea has conducted network reconnaissance on other laboratories and research centers.

Since the recent Kashmir crisis, there has been an increase in cyberattacks from Pakistan on India. India has also been responding with its own cyber operations against Pakistan.

Cyberattacks for the purpose of obtaining personal records and data have also multiplied. The following list describes just a few that occurred in 2019:

- In early January 2019, Singapore's Minister of Health stated that hackers had managed to steal data on more than 14,000 HIV-positive patients.
- In April 2019, Facebook discovered that more than 540 million records were stored on two unsecured Amazon Web Services (AWS) servers.
- In April 2019, Georgia Institute of Technology IT technicians discovered that data on 1.3 million students and former students had vanished.
- In early May 2019, a spokesperson for the Office of the Australian Commissioner revealed that the personal records of 10 million citizens had been exposed.
- In June 2019, Quest Diagnostics revealed that the medical data of more than 11 million patients had been leaked.
- A cyberattack on a database storing personal information about the Ecuadorian populace is said to have leaked sensitive information on more than 20 million people.
- More than 750 million mortgage deal documents in digital form were accessed by hackers when they launched a cyberattack on the database of First American Corporation. The breached data included tax records, social security numbers, digital transaction receipts, driver's license images, and bank account numbers.
- The Oklahoma Department of Securities witnessed a breach in 2019 as hackers accessed over a million files related to several FBI investigations.

---

17. Alexander Campbell and Vickram Singh, "Lessons from the cyberattack on India's largest nuclear power plant," *Bulletin of the Atomic Scientists*, November 14, 2019, https://thebulletin.org/2019/11/lessons-from-the-cyberattack-on-indias-largest-nuclear-power-plant/.

1.3.1 Methods of Protection

1.3.1.1 Analog Instruments

Old industrial plants that are still controlled by analog pneumatic systems are relatively safe from cyberattacks. I say "relatively" because, even in these plants, control systems have been gradually updated over the years and some have become hybrids as the newer control loops or components have usually been electronic (analog and later digital). In other cases, a computerized layer of monitoring devices has been added to make supervising the performance of old controls more convenient.

1.3.1.2 Isolation from the Internet

Malicious computer worms usually travel through the Internet. Theoretically, protection against this type of hacking should be relatively simple because the entry of such a worm can be prevented by eliminating all direct and indirect connections to the Internet or other networks. However, this is easier said than done because it is hard to isolate every single component in the control network.

1.3.1.3 Firewalls

Firewalls are software filters designed to recognize certain types of viruses or threatening communications and prevent them from entering individual computers or networks. Firewalls are controlled by access control lists that allow or prevent access to certain types of communications. Network firewalls usually safeguard the sensitive internal data of entire companies or organizations, whereas a host-based firewall works similarly but is stored locally on a single computer. Host-based firewalls protect individual computers from malware, cookies, email viruses, pop-up windows, and so on. Similarly, mobile devices can also be protected by firewalls, but no firewall can protect ICSs from sophisticated online attackers.

The US infrastructure is vulnerable to cyberattacks in a number of industrial sectors. Although switching the controls to manual might be possible in some cases, that option is often not available, or the systems might not be able to switch to manual control easily. In many industrial plants, the risks of attack are also increased by the control systems having remote access without adequate isolation and boundary protection.

Companies are working to develop a process to detect when malware has breached a plant's firewalls and proprietary protocols before it can damage a plant's control systems. Their goal is to restore control operations by immediately reversing any change the malware might have made. Unfortunately, neither of these methods (blocking malware from entering or neutralizing it when already present) provides foolproof protection against sophisticated attacks.

1.3.1.4 Disruption of the Global Positioning System

Advanced monitors often depend on global positioning system (GPS) satellites to track their phasor measurement units (PMUs), which provide signals from the atomic clocks to the grid with millisecond accuracy. This might make such grids as the power and aircraft industry, financial markets, and other civilian sectors vulnerable to hacking attacks. Attacks could include jamming to interrupt transmissions between satellites and ground receivers and "spoofing," which delivers bogus data to ground receivers. According to the Department of Homeland Security (DHS), as the smart grid is being developed in the electricity subsector, it becomes increasingly reliant on PMUs, which increases its vulnerability to GPS disruption.

Synchrophasor technology collects and reports grid data 100 times faster than older technologies. This helps protect against blackouts because problems evolving on the grid can be detected much sooner. For example, with older technologies, power frequency readings or power flow calculations could be in error, causing the grid to overheat in the affected area as a result of overloaded lines or overloaded transformers.

Applications that rely on GPS include uncrewed aerial and ground systems; self-driving cars, ships, and car tracking units (which are also used for mileage monitoring and insurance purposes); cell phone towers; financial systems; and the electrical grid itself. For these reasons, grid operators should have redundant sources of time signals. In addition, GPS attacks can flow up to affect dependent software on other systems that rely on GPS signals.

1.3.2 Air-Gap Worms

Even if a computer is not connected to the Internet, it can transmit data using light, sound, vibrations, or even heat. These transmissions can also be exploited in carrying out cyberattacks. Therefore, one should no longer assume that a computer that is isolated from the Internet, Ethernet, and all other networks and does not have a Wi-Fi connection is safe. It is not! The more sophisticated and malicious higher-level computer worms (Stuxnet, Flame, Duqu, Agent.BTZ, etc.) need no road to travel on. They are called air-gap worms because they travel through the air.

One such malicious computer worm (Stuxnet) was used by Israel and the George W. Bush administration to stop Iran's nuclear program by destroying the controls of a large number of its ultracentrifuges at Natanz. Stuxnet entered the targets through the computers' universal serial bus (USB) flash drives and infected the PLCs controlling the centrifuges. At that time, Iran did not have the capability for air-gap worms, so it could not retaliate. According to security expert Dr. John Michener:[18]

> However, even with a more robust air-gap in place, there are multiple ways for attackers to piggyback off physical access methods.
>
> The most common method is to infect a USB drive (or other removable media) that a contractor or employee is likely to connect to the secure computer. This is how the Stuxnet malware was able to disrupt Iran's nuclear program. Other real-world examples of malware jumping the air-gap by first infecting USBs include Agent.BTZ, SymonLoader, Rain Maker, Brutal Kangaroo, and Cottonmouth.

Air-gap-type worms can enter not only through USB flash drives but also through any of the digital control components in a plant's network. Such components can include wireless sensors (through which the worm enters by faking the sensor signals), speakers, and microphones. It is also possible to inject malware into any of the operator's mobile devices (including iOS devices such as iPhones, iPads, and iPod touches, as well as Android devices), which can then attack control computers.

For too many years, industry has falsely assumed that current measures for preventing air-gap attacks provide sufficient defense against malicious intrusions inside critical systems. Yet, in the first air-gap attack (the Israeli–US joint attack on the Iranian ultracentrifuges), the controls of the plant were isolated from the Internet but were still infected by the Stuxnet worm. Since the Stuxnet attack, other methods of attack have been developed. For example, the ANGRYNEIGHBOR family of covert listening devices (i.e., bugs) work even if the target device is not online.

According to the *Times of Israel* news service, malware can also be inserted into cell phones and used to attack computers in the vicinity of the phone. In this way, a hacker can insert malware into a particular computer through that cell phone from any place on the planet to remotely access data stored on a targeted system, even if it is isolated from the Internet.

In the case of the cell phone method, the attacker sends a text message that appears legitimate and harmless to the mobile device of an unsuspecting employee inside a sensitive installation. Naturally, this innocent-looking message hides a link to malware specifically developed to implement the air-gap network technique. In this way, the malicious code is installed on the phone. Once installed, it scans for electromagnetic waves, which can be manipulated to build a network connection using FM frequencies

---

18. John Michener, "Beating the Air-Gap: How Attackers Can Gain Access to Supposedly Isolated Systems," Energy Central website, August 24, 2018, accessed November 1, 2022, https://energycentral.com/c/iu/beating-air-gap-how-attackers-can-gain-access-supposedly-isolated-systems.

to install a virus onto a computer or server. The phone then connects to the system at the FM frequency, and the virus "sucks the information out of the server and uses the phone's cellular network connection to transmit the data back to [the] hackers. All that is needed is physical proximity to the system. The team said that one to six meters is enough."[19]

1.3.3 Cyberattacks by Nation-States

Adversaries are becoming more advanced in reconnaissance, network penetration, detection evasion, and similar capabilities. AI-based cyberattacks are also used as effective propaganda. In addition to attacking the physical infrastructure, several nations use these means to misinform the people of other nations. I have already mentioned some such cases, so here, I only briefly mention the activities of a few nations that are involved in such attacks through cyber propaganda or assaults on industrial systems such as electric grids.

1.3.3.1 Ukraine

According to the US ICS-CERT, the 2015 cyberattacks on Ukraine caused power outages in its electrical grid. The attackers used a spear-phishing attack—an email or electronic communications scam in which victims are tricked into opening spurious emails and downloading malware—to steal credentials and connect to the local electric utility's virtual private network (VPN) and remote desktop software to manipulate human-machine interface (HMI) controls. Power was restored in 4 to 6 hours by switching to manual control, and the system remained under manual control for months.

Before and during the 2022 war, Ukrainian officials reported attacks on about 70 government websites, including those of the Ministry of Foreign Affairs, the Cabinet of Ministers, and the Security and Defense Council. Most of the sites were restored within hours of the attacks. On February 15, 2022, another cyberattack on Ukraine took down multiple government and bank services, and later, these attacks expanded and intensified.

1.3.3.2 Russia

The efforts of Russia to influence the outcome of the 2016 and 2020 elections in the United States are well known. These efforts aimed to make a highly polarized electorate even more polarized and to undermine faith in the election systems. The hackers

---

19. David Shamah, "With new hack, cellphone can get data out of computers," *Times of Israel*, June 9, 2014, accessed November 1, 2022, https://www.timesofisrael.com/with-new-hack-cellphone-can-get-data-out-of-computers/.

did this by flooding the Internet with disinformation, including fake photographs and videos, to manipulate people psychologically.

According to the US DHS, Russian hackers have also been able to exploit insufficiently secure vendors in the power industry to access privileged, air-gapped systems that they support in electric utilities in the United States. In 2018, the DHS reported that, during the past year, Russia's military intelligence agency had infiltrated the control rooms of power plants across the United States. In theory, this could enable the hackers to take over the plants or shut down the grid, as they did in Ukraine in 2015 and 2016 and during the 2022 war. In these cases, *implants* (hard-to-find malware) were installed in utility operating systems to bring down the power systems.

According to the DHS, Russian hackers have also breached power plants through the networks of contractors that provide software to the utility company's systems. They then proceeded to try to trick utility operators into changing their passwords. According to the DHS, hundreds of US utilities were compromised between 2016 and 2018.

These Russian attacks have continued for years. The method of attack is to find vendors with access to a utility's networks because they were hired to update software or run diagnostics and target them with hacking tools to get into the computer networks. Once inside, they steal the users' credentials, thereby obtaining direct access to the utility's networks. In this way, the hackers learn how the networks, the hardware and software, and the administrative accounts controlling the network are configured.

In a more recent case, hackers attached their malware to a software update package coming from SolarWinds, based in Austin, Texas. This company produces a network- and applications-monitoring platform called Orion, which is used by some 18,000 customers in the government and private sectors. The tainted software update packages were distributed between March and June 2020, which also allowed the hackers to access the network of the cybersecurity firm FireEye. It took 6 months to discover that this breach had occurred. It will probably take another 6 months to learn what has been taken and how much of it was top-secret material.

1.3.3.3 Korea

North Korean hackers have used GPS-based navigation aids to jam GPS signals. By 2019, South Korea completed an alternative land-based coastal radio transmitter system called eLoran (enhanced long-range navigation), which (it hopes) will provide reliable alternative position and timing signals for navigation. In the past, some attacks lasted for nearly a week and affected the signal reception of more than 1000 aircraft and 700 ships. Aircraft traffic was not affected because, in that case, GPS signals are normally used only as a backup and not as the primary navigation tool.

There are many scenarios in which spoofing (altering) GPS signals can lead to disastrous consequences, such as taking a ship or a drone off course, as occurred in 2013 when the US Navy reported the intentional jamming of the GPS signals of one of its vessels sailing near Iranian territorial waters. More recently, North Korea attacked the GPS signal reception of more than 1000 South Korean aircraft and 700 ships. GPS vulnerability poses security and commercial risks, especially for ships whose crews are not familiar with traditional navigation techniques and using paper charts.

1.3.4 Electric Grid Protection

The US power industry has been active in advising utility operators about recommended methods of protection against cyberattacks based on DHS reports. According to Dr. Michener,[20] air-gap networks are particularly vulnerable. Michener addresses three issues that create this vulnerability: (1) It is common to assume that air-gap networks are not connected to the Internet when in fact, they may contain dual-homed computers and servers, rogue Wi-Fi routers, and legacy and backup connections to the Internet. (2) The network may be vulnerable because of the levels of access provided to contractors and system administrators. (3) Contractors and vendors use remote desktop VPNs to monitor and adjust equipment at electric utilities remotely. Remote access is a back channel for hackers to exploit.

In the following sections, I quote some of the advice[21] published by Energy Central on their Power Industry Network on protection against cyberattacks.

1.3.4.1 Compromised Users and Malicious Insiders

Dr. John Michener states:

> Attackers could also utilize insiders within the utility, from engineers to executives, in order to conduct malicious operations. These may be insiders who are actually compromised, but more commonly, this involves insiders whose personal devices and/or corporate accounts have been compromised, allowing the attacker to insert malware targeted at devices and systems within the protected environment.
>
> Blackmail and bribery are both risks which ought to be considered, particularly given the growing importance of the cyber domain to nation-states—who want to have a "plan in place" to strike at the US if they should ever be attacked themselves.

---

20. John Michener, "Beating the Air-Gap."
21. John Michener, "Beating the Air-Gap."

> However, it is not unrealistic for cybercriminals to use this method as well, particularly when one considers the increased focused criminals are now giving to cyber extortion campaigns. Note that devices outside of the protected environment typically have far fewer network, operational, or physical security protections and may be surreptitiously physically tampered to make the device a remote trojan.

1.3.4.2 Supply Chain

According to Dr. Michener:

> Computer equipment and software that is vulnerable or infected prior to being delivered to the utility is another threat the industry needs to prepare for.
>
> This type of effort would almost always be organized by a nation-state, but it could also be exploited later on by a number of downstream actors, once the vulnerability becomes more generally known.
>
> So-called supply chain attacks are real. A few recent examples include Juniper's firewall backdoors and vulnerabilities in Cisco ASA devices and Fortinet FortiGate firewall. Researchers at the University of Michigan also demonstrated how backdoors could be hidden in computer chips during the manufacturing process.
>
> The recent attack against the Ukrainian power grid involved compromising a third-party software package used by the utility. The developer of the software was compromised by the attackers. When the updated software package was distributed to the utility, the backdoor was installed for later use.

1.4 Robots

Robots are an outgrowth of process control, and therefore, all control knowledge that has been accumulated in automating regular industries also applies to robots. The only difference is that the controller is the AI "brain" that simultaneously receives a large number of measurements and, in response, manipulates a large number of variables (motors, magnets, springs, etc.) in a manner identical to that of industrial controllers throttling control valves or positioning moving parts. Therefore, robots are multivariable controllers, a natural outgrowth of industrial automation.

When the first instruments were designed to make industry safer and more efficient, no one realized where this effort would lead. Next, we designed "smart" instruments so that they would "self-diagnose;" in other words, they would determine if something was wrong with themselves. This road eventually led to robots, and today, we are beginning to realize that there is a possibility for these human creations to become even more than just tools.

First, we believed that robots are good because they can do boring things, or they can do things more accurately and more quickly than humans can. Later, we realized that robots could also go to places that are inhospitable for humans, such as Mars, war zones, or the bottom of the oceans. Now, we are beginning to realize that AI can also change our lifestyle! Today, not only can AI-brained robots build cars but they can

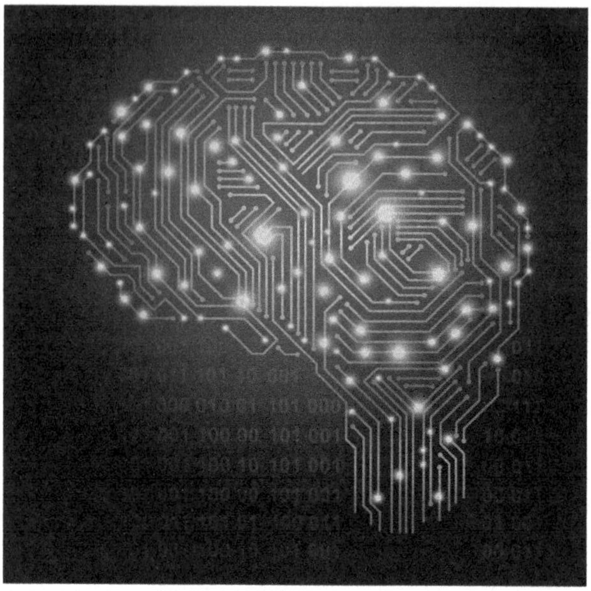

Figure 1-10. Can this brain do more than just regurgitate what we put into it?

also drive the cars they build. We realize that these machines of practically unlimited memory and speed can analyze data and execute logic and could eventually self-improve their own intelligence. Some believe that AI can not only build and drive cars but also invent better ones (see Figure 1-10). If so, several questions arise: Why could they not design smarter robots? Why could they not improve their own software? Why could they not gain superhuman intelligence?

We have just started on this slippery slope where a tool can become a servant and eventually do what it wants. We must realize that this development needs to be controlled because not only can AI analyze our genetic code but it could possibly make inheritable changes to the human genome.

1.4.1 Robot Types

The computational capacity of computers is doubling almost yearly, and the capabilities of AI and the applications of robots are also exploding. In the following pages, I discuss some of these applications—such as self-driving cars—focusing on aspects in which process control experience can be useful (sensors, safety, etc.), but I will make only a few general remarks on the direction of this technology.

For decades, people have predicted that machines would make workers obsolete and cause unemployment. To date, this has not turned out to be the case. Machines have replaced human muscles and a few of the routine functions of our minds, but they have not made workers obsolete. Human tasks have just shifted to other industries,

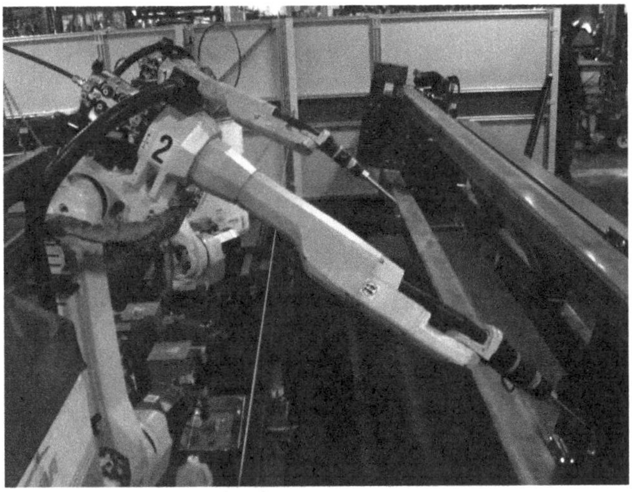

Figure 1-11. A typical industrial robot used for welding in factories.

Source: "FANUC 6-axis welding robots.png" by Phasmatisnox, May 27, 2009, is licensed under CC BY 3.0, https://commons.wikimedia.org/w/index.php?curid=79553342.

such as service industries or industries that design and maintain robots, and so, on balance, automation has not created unemployment. AI could also do to white-collar jobs what steam power did to blue-collar ones during the Industrial Revolution, and, eventually, robots could almost completely replace some of the blue-collar workforce (Figure 1-11). However, we live in a time when these displaced workers will have other, better jobs because they will be needed to carry out the immense project of converting our energy economy into a renewable one.

1.4.2 Industrial Robots

Industrial robots are primarily used in manufacturing. They are usually programmable and capable of moving along three axes. They can be grouped according to the types of motion and positioning they are able to perform into categories such as articulated, Cartesian, cylindrical, and polar. In the context of general robotics, most of these types of robots fall into the category of robotic arms. Their applications include welding, painting, assembly, and pick and place for tasks such as manufacturing printed circuit boards, packaging and labeling, palletizing, product inspection, material handling, and testing, all accomplished with precision and at high speed. The most commonly used robot configurations are articulated robots, selective compliance articulated robot arm (SCARA) robots, delta robots, and Cartesian coordinate robots (gantry robots or x–y–z robots).

Industrial robots can also be distinguished by their flexibility. A majority are rather inflexible; they are programmed to carry out specific actions repeatedly without variation and with a high degree of accuracy. These actions are determined by the program,

Figure 1-12. Mini robots.[22]

which specifies the direction, acceleration, velocity, deceleration, and distance of a series of coordinated motions. On the other hand, more flexible robots can be provided with precise guidance through sensors, such as machine vision subsystems acting as visual sensors, that are linked to powerful computers or controllers that are often provided with AI.

Robots can be autonomous or semiautonomous and can serve many applications in addition to industrial ones. These include, but are not limited to, medical, military, miniature (Figure 1-12), microscopic, and humanoid applications. Semiautonomous designs can perform such tasks as finding power sources or, in military applications, independently choosing targets. Autonomous robots are expected to proliferate in the coming decade, with home robotics and autonomous car applications likely to be the most common.

1.4.3 Humanoid Robots

By mimicking a lifelike appearance or automating movements, a robot may convey a sense of intelligence. Humanoid robots have a body shape built to resemble the human body. The design may be for functional purposes, such as interacting with human tools (Figure 1-13) and environments, or for experimental purposes, such as studying bipedal locomotion. Humanoid robots have a torso, a head, two arms, and two legs, although some humanoid robots may model only part of the body, for example, from the waist up. Some humanoid robots also have heads designed to replicate human facial features such as eyes and mouths and to aesthetically resemble humans. Others

---

22. For more information, see: Mindy Weisberger, "Bug-Sized Robot Competitors to Swarm DARPA's 'Robot Olympics,'" Live Science website, July 21, 2018, accessed November 1, 2022, https://www.livescience.com/63129-darpa-tiny-robot-olympics.html.

Figure 1-13. Humanoid robot.
Source: Editorial credit: vapadi / Shutterstock.com.

have five-fingered hands and feet, usually for maintaining balance and achieving complex motions.

Humanoid robots are not necessarily designed to accomplish a task but to mimic the abilities of human beings. They are designed to imitate human movements (Figure 1-14); they have rib cages, backbones, and rotating joints. The result is surprisingly life-like movement in similarly lifelike forms. For example, some humanoid robots can do push-ups, pull-ups, and backbends and even hit a badminton birdie. Humanoid robots

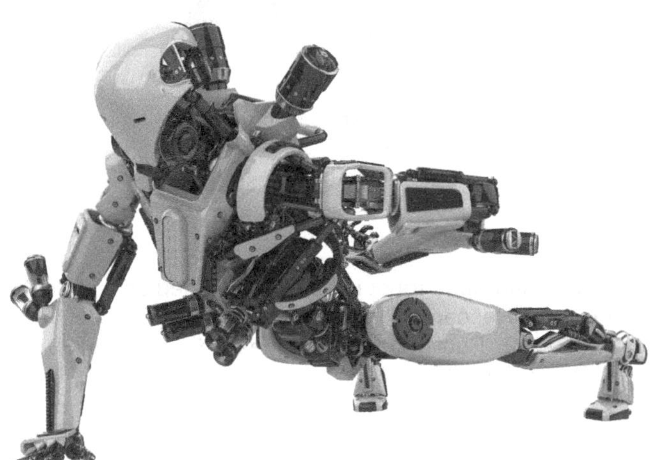

Figure 1-14. Humanoid robots can perform surprisingly lifelike movements.

can also play a role in automobile crash tests, where they can provide a realistic depiction of what happens to humans when they undergo such trauma.

Currently, some 40 nations are developing robot soldiers. Some of these machines do not need human direction because they can think for themselves. Such killer robots can choose targets on their own and, depending on their software, might also be able to distinguish civilian from military targets or enemy soldiers from friendly forces.

The global AI arms race has already started. In the past 10 years, approximately $120 billion, if not more, has been spent on developing and improving robot warriors (Figure 1-15) that have no moral judgment and no consciousness or feelings such as mercy. Today, nations are mainly developing these machines in secret and have not placed limits on their natures or capabilities. There are plans to introduce these machines into space and even to provide them with miniaturized nuclear weapons.

Figure 1-15. Military robot.

Source: "Bio-inspired Big Dog quadruped robot is being developed as a mule that can traverse difficult terrain.tiff" by Defense Advanced Research Projects Agency (DARPA), derived from DARPA Strategic Plan (2007).pdf, August 23, 2012, Public Domain, accessed February 28, 2022, https://commons.wikimedia.org/w/index.php?curid=20798332.[23]

23. For more information, see: "Robot," Wikipedia, accessed August 29, 2022, https://en.wikipedia.org/wiki/Robot.

One does not need to be a process control engineer to realize that these developments are out of control and that international regulations must be established and enforced in the fields of both nuclear and AI-based weapons development.

1.5 Smart Electric Cars

The data provided in this section is valid only at the time of this writing because the capabilities of electric vehicles and the infrastructure they require are rapidly changing. Also in this section, I briefly mention their energy storage methods (battery and hydrogen). In Chapter 4, Section 4.4, "Battery Storage," and Section 4.5, "The Hydrogen Age," I return to the subject of electric cars in more detail.

The public perception is that switching to battery electric vehicles (BEVs), also called plug-in electric vehicles (PEVs), will decrease carbon emissions. However, this belief is mistaken because batteries are most often recharged with "gray electricity," which is electricity generated in carbon-emitting power plants or by other carbon-emitting processes. Today, only about 4% of BEVs use "green" electricity, and a similarly low percentage of fuel-cell-operated electric vehicles (FCEVs) use green hydrogen. However, these percentages are slowly rising as families install solar panels on their homes and use solar-roof-generated electricity to recharge their electric cars. (In California, for example, there are 1.3 million homes with solar panels, and the use of energy storage in new homes is compulsory.)

Ten years ago, only about 25,000 BEVs were on the roads. Today, they number 5.6 million, and by 2030, their number is projected to rise to about 200 million (10% of the total number of vehicles). Currently, the global number of FCEVs (including buses and trucks) is roughly 100,000, which is small but still larger than the number of BEVs a decade ago. The increase in the number of BEVs is limited by infrastructure availability. For example, today, in just the United States, there are almost 170,000 regular gas stations, whereas the global number of BEV charging stations in 2019 was only 25,000, and the number of hydrogen fueling stations was about 500. This will drastically change because the Biden administration plans to build a national network of 500,000 electric vehicle charging stations by 2030. The 2022 data[24] is given in the footnote.

BEVs are superior to FCEVs in terms of costs and infrastructure availability. In Europe, before the war in Ukraine, drivers paid about €9 to €12 to drive 100 km in an FCEV, whereas driving a BEV cost only €2 to €7 per 100 km (depending on electricity prices

---

24. Steven Loveday, "A Comprehensive Guide to U.S. Public EV Charging Networks," *US News and World Report*, April 22, 2022, accessed August 29, 2022, https://cars.usnews.com/cars-trucks/features/ev-charging-stations.

and subsidies in individual countries). The advantages of FCEVs include that their hydrogen fuel source is free and unlimited, they emit only water, and their driving range is higher. In addition, their weights, refueling times, and noise levels are all lower than those of BEVs.

A growing number of electric vehicles (EVs) have been provided with some degree of self-driving capability. These electric cars are mostly powered by batteries, although a much smaller portion of them employ hydrogen fuel cells.

Electric cars are good for local driving when they can be charged during the night. They are convenient if the owners do not drive more than 200 miles a day. The typical range of battery electric cars is between 250 and 350 miles. Most electric cars can be recharged from a 240 V outlet overnight. If the owner's daily commute is only 20 miles, a 120 V outlet will suffice. Today, charging at a 240 V outlet for 1 hour can add the electricity needed to drive 25 miles. The shortest time required to reach 80% of full capacity requires a supercharger, which can complete charging in about 20 minutes, but is offered by only a few designs, such as the Porsche Taycan and the Audi e–tron (Table 1-1).

In the United States, the number of regular gas stations is dropping (200,000 in 1990, 115,000 in 2020) while the number of EV charging stations is rising (56,000 in 2022, of which over 5,000 were privately owned). As to driving distances, regular EVs require refilling every 200 to 300 miles. Regarding the types of charging stations in operation, about 90% are level 2, and only 10% are DC fast-charging types. At a level 2 charging station, it takes about 30 minutes to charge enough electricity to drive 20 miles or about 8 hours for a full charge. At DC supercharging stations, refilling takes about 1 hour. To save time, it is best to charge just enough to make it to the next charging station. Charging time is also reduced if the car is refilled when the fuel level drops to about 10% and if charging stops when the level reaches 80%. Hydrogen-fueled EVs can travel 400 to 500 miles on a full tank, and their refueling takes only 3 to 4 minutes. However, as of 2023, there were only about 1,000 hydrogen refueling stations globally. Today, the majority of fuel-cell-operated EVs are trucks, buses, or trains that require larger fuel tanks so that they can travel longer distances without refueling.

Table 1-1. Times required for BEV batteries to recharge.

Charging Level	Supply Voltage (V)	Distance That Can Be Driven After 1 Hour of Charging Miles) [km]
Level 1	120	3 to 5 [5 to 8]
Level 2	206 to 240	12 to 80 [20 to 125]
Level 3	400	180 [300]
Supercharge	900	300 [500] (after 15 minutes)

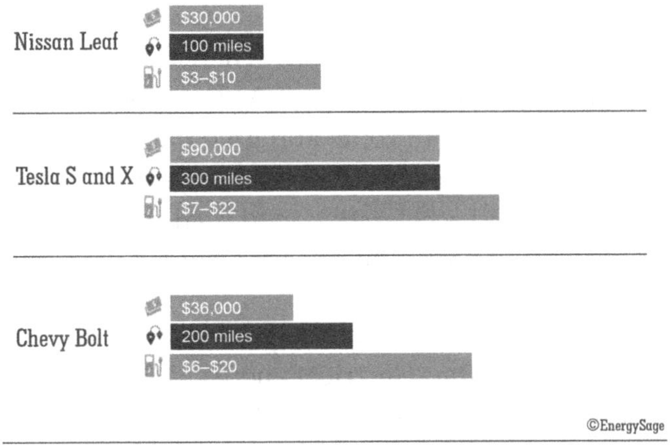

Figure 1-16. Costs and driving ranges of some of the battery-operated electric cars.

Source: "Charging your EV: how do EV charging stations work?," EnergySage website, last updated November 21, 2019. https://www.energysage.com/electric-vehicles/charging-your-ev/.

Figure 1-16 shows the approximate driving ranges and purchase and refueling costs of a few popular PEV designs. Regarding refueling times, the full charging time of battery-operated electric cars, when charged at home, is approximately 8 hours (all night), but this can be reduced by going to supercharging stations, where the charging time is less than 1 hour.[25] (In contrast, the fill-up time of a hydrogen-fuel-cell-operated electric car is only 3 minutes.)

Some manufacturers, such as Tesla, have the goal of increasing the driving distance per charge, whereas others (e.g., GM, Nissan, BMW, and Volkswagen) place less emphasis on driving distance per charge because they consider their primary market to be commuters traveling shorter distances.

The electricity consumption of these cars is about 0.4 kWh per mile, which corresponds to about $0.08 per mile based on $0.20 per kWh electricity cost. This fuel cost is about the same as that of a 50 miles per gallon consumption of an internal combustion vehicle at 2022 gas prices. The batteries are usually warrantied to last for about 100,000 miles; beyond that, the replacement cost is about $5,000.

The most popular electric car models include the Nissan Leaf, BMW i3, Mercedes-Benz B-Class, Chevrolet Bolt EV, Ford Focus Electric, Tesla Model S, Chevrolet Spark EV, Tesla Model X, Tesla Model 3, and Mitsubishi i-MiEV. The models with the highest horsepower include the Tesla Model X, Tesla Model 3, Ford Mustang Mach-E GT

25. As of this writing, the full charging time for a Nissan Leaf is 4 to 8 hours at home and 0.5 hour at a fast charging station, that for a Tesla is 6 to 10 hours at home and 1 hour with a supercharger; and that for a Chevy Bolt is about 10 hours at home and 1.5 hours at a fast charging station.

(460 hp), and Porsche Taycan (560 hp). In terms of acceleration, the Porsche Taycan can accelerate from zero to 60 mph in 3.8 seconds, whereas the Faraday Future requires only 2.2 seconds. In terms of providing special AI-based features, the Mercedes EQC can identify the most energy-efficient route and give the location of the closest charging station, whereas the Byton can provide facial recognition of the driver and precool the car interior. These power and acceleration values and features are progressing rapidly, although features like facial recognition still need much improvement.

Pure-hydrogen-powered electric cars have the potential to convert 70% to 80% of the energy content in the hydrogen into electrical energy. After that—as for battery-powered cars—it is still necessary to convert the electrical energy into mechanical work at an efficiency of about another 80%, so the overall efficiency of FCEVs is about 60%. A 60% efficiency rate has been reported for Honda's FCX concept vehicle.

In terms of price, the National Renewable Energy Laboratory (NREL) estimates that if the cost of gasoline is $3.50 per gallon, the cost of operation of the average gasoline-fueled car is about $0.13 per mile, which is nearly twice that of PEVs, but only slightly more than half that of FCEVs. Moreover, a team at the Massachusetts Institute of Technology (MIT) evaluated the cost of PEV electric cars and found that, although their initial cost is higher over a lifetime, they are less expensive than gas-burning cars. Naturally, these values rise when the cost of fossil fuels or electricity rises and drops if the cost of hydrogen drops.

Some advantages of FCEVs include their short refueling times (about 3 minutes); longer driving distances (about 500 miles); and absence of space-consuming and heavy batteries (in Tesla cars, the battery weight ranges from 1000 to 1200 pounds). Also, if the hydrogen is made with green energy, they have zero carbon emissions. This is not the case with PEV electric cars because their batteries are recharged with electricity that is often generated by carbon-emitting power plants. The main disadvantages of FCEVs are that they are still in the testing stage of development, their initial cost is higher than that of BEVs, they are not yet mass-produced, and the available network of filling stations (Figure 1-17) is very limited. It is projected that PEVs will still dominate the electric vehicle market in the next decade or so, but in applications where larger energy supplies need to be stored on the vehicle (such as trucks, buses, and trains), FCEVs will gradually take over.

In the following sections, I cover the development of smart cars, their safety history, and the debate over their operating philosophies. I discuss the state of the art of controlling self-driving cars beginning with their control loops as they exist today. I then suggest improvements to their sensors and operating algorithms based on the experience and knowledge accumulated in the processing industries.

Figure 1-17. Hydrogen filling station. As of 2019, hydrogen filling stations were available in only a few cities in about half of the states in the United States. This illustrates the difference in the scale of preparation for the future use of FCEV and BEV electric cars because the Biden administration is proposing building 500,000 BEV electric vehicle charging stations by 2030.

Source: Editorial credit: Rob Crandall / Shutterstock.com.

1.5.1 History and Development

In the transportation industry, the use of automation has evolved slowly. As of today, trains are still mostly under manual control, airplanes are usually controlled in the semiautomatic mode because autopilots are available but can be overruled by the pilot, and only space travel and military applications use fully automatic control.

The use of automatic process control in automobiles and trucks started later than in other industries and has picked up speed only in the past two decades. The goals of this technology are both saving lives and reducing transportation costs. Vehicle automation is also expected to bring mobility to millions who cannot afford to own cars.

Until the beginning of the twenty-first century, most cars were controlled entirely by the drivers, who were assisted by a few sensors (mirrors, speedometer, and GPS after 1990). In 2019, there were about 275 million cars in the United States and about 1.2 billion worldwide, expected to grow to 2 billion by 2035. Regarding safety, the current fatality rate in the United States is about 1.2 per 100 million miles traveled; in 2018, 36,500 people died in car accidents. In 2020, vehicles with Tesla Autopilot engaged in the United States reported 0.2 accidents per million miles. The US average in the same year was nine times higher at 2.0 accidents per million miles driven.

Tesla has been releasing Autopilot safety data to show that its Autopilot is safer than human drivers under average driving conditions. The company updated these numbers to show that driving on Autopilot control is nearly nine times safer than driving without it.[26]

Semiautonomous (autopilot) control systems were developed during the previous decade. Their sensors include ultrasonic sensors, LiDAR (light detection and ranging or laser imaging, detection, and ranging), radar, cameras, and GPS. These systems control two variables: speed and direction. They can park and maintain lane position, legal speed, and distance from other cars, but they operate in a "hands-on-the-wheel" mode (Figure 1-18), meaning that the driver can take control at any time. Semiautonomous vehicles are assumed to be safer than driver-controlled vehicles, but how much safer is still debated.

Driverless (autonomous) automobile control systems are still under development, although those designed by Google have logged some 1.5 million miles in California; the Chevrolet Bolt is "running around" in San Francisco; and Ford, Apple, and others are planning to launch autonomous vehicle models soon. Currently, testing is primarily done by the manufacturers. It would be preferable, however, if they were government-tested in standardized testing centers that included simulated streets, cars, traffic lights, road signs, tunnels, and moving and stationary objects, including robots behaving like people and animals.

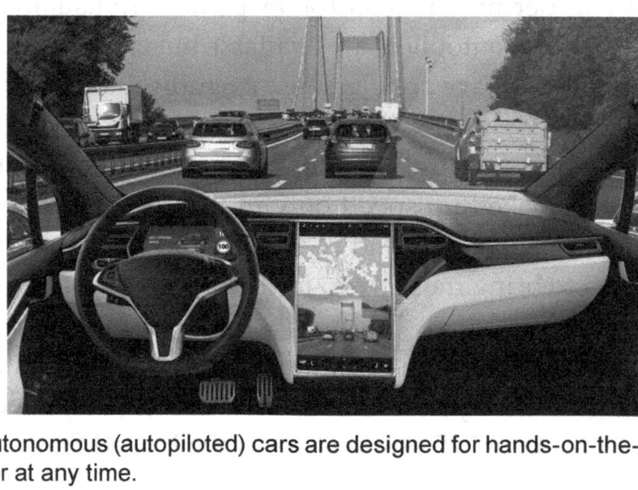

Figure 1-18. Semiautonomous (autopiloted) cars are designed for hands-on-the-wheel driving, allowing the driver to take over at any time.

---

26. Jameson Dow, "Tesla updates Autopilot safety numbers; almost 9x safer than average driving," Electrek website, accessed March 1, 2022, https://electrek.co/2019/10/23/tesla-autopilot-safety-9x-safer-than-average-driving/.

The first markets for autonomous robotic cars are expected to be ride-hailing, ride-sharing (Uber, Lyft), and delivery-type applications using commercial fleets. In addition to increased safety, it is assumed that such vehicles would have reduced operating costs because they can be used full-time and drivers' salaries can be eliminated. Fiat and Chrysler are working with Google, General Motors is in a partnership with Lyft, and Volvo has teamed up with Uber to develop autonomous cars. The first applications will probably be trucks on highways traveling from exit to exit.

1.5.2 Tesla Accident in Florida

In this section, I use the example of a fatal Tesla accident in Florida to illustrate the roles played by the three loop components: sensors, control algorithms, and final control elements. I discuss the contributions these components made to the accident and how these early design errors can be corrected. I should also mention that an unresolved debate is in process concerning the legal limits of where the responsibilities of the car manufacturers (hardware, such as sensors, and final control elements, such as brakes) end and those of the "brain" of the car (the control software) begin.

The 2016 crash in Florida occurred on a divided highway when a white tractor-trailer in the westbound lane made a left turn, crossing in front of oncoming traffic in the eastbound lane where the Tesla was traveling. The question of whether neither the driver nor the autopilot attempted to brake or the brake actuators were defective is unresolved. All that is known is that the Tesla did not brake (Figure 1-19).

The failure to detect the white tractor-trailer was probably because neither the camera of the autopilot nor the driver's eyes could see it against the brightly lit sky. The

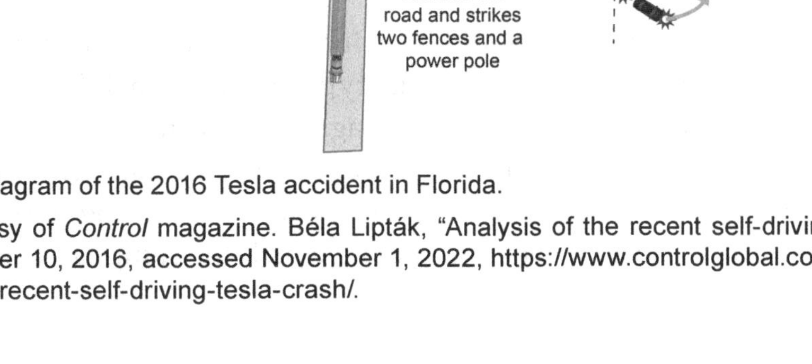

Figure 1-19. Diagram of the 2016 Tesla accident in Florida.

Source: Courtesy of *Control* magazine. Béla Lipták, "Analysis of the recent self-driving Tesla crash," Figure 3, October 10, 2016, accessed November 1, 2022, https://www.controlglobal.com/articles/2016/analysis-of-the-recent-self-driving-tesla-crash/.

Figure 1-20. The Tesla software received a radar signal from the back of the truck and, for some (unknown to me) reason, assumed that it was a stationary object above the road, such as an overpass or a road sign.

radar would likely have seen it if it had been properly focused, but because the truck appeared to be a stationary object above the road, the control algorithm could have assumed that it was not a vehicle but rather an overpass or a road sign. As a result, the Tesla hit the center of the trailer at windshield height and came to rest at the side of the road after hitting a fence and a power pole (Figure 1-20).

From a process control point of view, all three components of the control loop could have been defective. The camera sensor should have seen the white truck against the bright sky, but it did not; the control software should have distinguished the truck from a road sign, but it did not; and finally, the brake actuator was possibly defective.

It is unknown whether the autopilot was on or off at the time of the accident, but it does not matter. If it was on, it failed. In this case, the driver had the means to stop the car, but either he was distracted or the override did not work. My view is that a human override should always be provided, but it should be applied only when its operation is safe. For example, if the driver is drunk and turns the wheel into oncoming traffic, the AI software should block that instruction from being carried out.

If I were designing the controls, I would have included two basic features. The first is that all sensors would have been the "voting" type, meaning that three sensors would detect each measurement, and when one disagreed with the "majority," the others would take over and the defective one would be reported for maintenance. The second feature would have been "safe override." This means that, during normal operation, the car would be under normal override, during which the computer drives it, but the driver can override the automatic controls. On the other hand, if a probable accident is evolving, safe override overrides both the computer and the driver and brings the car to the safest condition possible.

1.5.3 Autopilot or Driverless?

Smart cars have great lifesaving potential because their safety does not depend on the condition of the driver, who can be tired, sleepy, panicked, drunk, under the influence of drugs, angry, or inexperienced, for example. On the other hand, when the decision-making is complicated, such as distinguishing between a pedestrian trying to hitch a ride and a police officer flagging you down, the human operator can probably do a better job because such decisions are too "fuzzy" to be effectively enshrined in computer code.

It has been argued that a great advantage of autopilots is that, after an accident, when it has been carefully analyzed and its cause determined, corrective action can be applied not only to that vehicle but also to the whole fleet through a software update. This is important. For example, in the case of the Florida accident, one might suggest that the driver be reminded to use sunglasses or to activate the radar, but that would have made driving safer for only that one car. The debate concerning self-driving cars is not over the question of whether they are safer; it is about what level of computer involvement is best. Should the driver be kept "in the control loop" or not? Should the car have steering wheels and brake pedals or not? In other words, the debate is between the advocates of autopilot and autonomous designs.

It seems to me that, over time, the same evolution that occurred in other industries will take place here too, where we gradually moved from the age of semiautomatic (PID) control to higher levels of automation. The transportation industry will follow its own learning curve, but that curve could rise more quickly if this industry took advantage of what was learned in the processing industries.

The smart cars on the road today use autopilots and require that drivers keep their hands on the steering wheel so they can take over control when needed. Driverless cars are being developed and tested, but they will not be used until more testing is performed. Manufacturers such as GM and Audi are also taking an incremental approach by developing cars that are capable of being gradually updated from an autopilot mode of operation to a driverless mode. Others are starting with limited application goals. An example is Google (working with Otto), which plans to market trucks that will be self-driving only on highways. The development of entirely driverless cars is still in the testing or field trial stage (in places such as Pittsburgh, Pennsylvania; San

Francisco and Palo Alto, California; Miami, Florida; and Detroit, Michigan); they are not expected to be available until 2022 or later.

Google, Apple, and Uber are developing and testing autonomous driverless cars. They argue that using autopilot is less safe than the fully autonomous mode because, if the driver is passive (hands on the steering wheel but inactive) when the autopilot is on, the driver will be slow to take over. In other words, the driver cannot be expected to concentrate while inactive and cannot snap back and make split-second decisions when needed. Studies have found that the average time needed for "awake" drivers to respond is 17 seconds. A car moving at 65 mph travels a distance longer than five football fields during that time.

Naturally, if a car or truck is driverless, its operating costs are reduced, as no salary must be paid to the driver. Still, as of the foreseeable future, the use of driverless cars will be limited to fixed-route applications, such as grocery or mail deliveries, if used in that way at all.

My overall view is that we can and should make our digital tools as good as we can so they can effectively substitute for some of our routine mental functions. Nevertheless, we must stay in control of these tools. We must not use automobiles without steering wheels and brakes. We must understand that there is nothing more unpredictable, and therefore more dangerous, than an out-of-control automatic control loop. Some airplane accidents demonstrate how dangerous it is to block pilots from overruling defective autopilot actions. This must not be repeated in the case of automobiles. No matter how sophisticated AI algorithms and the corresponding sensors are, they can all fail, and when (not if, but when!) that happens, the driver must be able to take control.

It is also interesting to note that the major process control firms (ABB, Emerson, Invensis, Siemens, etc.) are not focusing on this market. They have developed very few sensors or software packages for self-driving car applications. Instead, new companies are starting to fill this void, such as Nirenberg Neuroscience (machine vision), SAIPS (computer vision), and Velodyne LiDAR (radar-based laser sensors).

In the case of automatic versus human control, an ethical question arises: During an evolving accident, whose safety is more important? A human driver would usually respond by trying to keep the passengers in the car safe, but a marketed software package cannot be allowed to do that. Such software must consider the life of pedestrians or passengers in other cars to be equally important. Therefore, when the appropriate government agencies set safety rules for self-driving cars, they will have more than just technical questions to consider.

1.5.4 Sensors and Controls

In addition to computers, self-driving cars are equipped with LiDAR, radar, ultrasonic sensors, and video cameras (as many as 12 on some cars). Tesla vehicles use cameras instead of LiDAR; Uber, Waymo, and Toyota use both. The advantage of LiDAR is extreme precision, whereas the advantage of cameras is their ability to identify an object, such as distinguishing a plastic bag from a bump in the road. The main reason for the two approaches is money. Originally LiDAR cost $7,500, whereas cameras cost about $100. Today the cost of LiDAR is about $1,000. In the case of the more limited LiDARs introduced by Luminar and Velodyne, such as the Velabit, the cost can drop further, but it remains to be seen how many of these will be needed per vehicle. My view is that both cameras and LiDAR should be used, and their information should be fed into a neural network to obtain an accurate and three-dimensional color view of the surroundings.

The primary function of the sensors on a self-driving car is to view the totality of the car's surroundings, which they do by breaking it into observation zones. In Figure 1-21, the numbered zones are as follows: (1) protection zone detected by 360° rotating LiDAR; (2) rear collision protection and lane change assistance zones; (3) self-parking function; (4) traffic light, lane departure, collision avoidance, and emergency brake-assist functions, and (5) cross-traffic and pedestrian detection function.

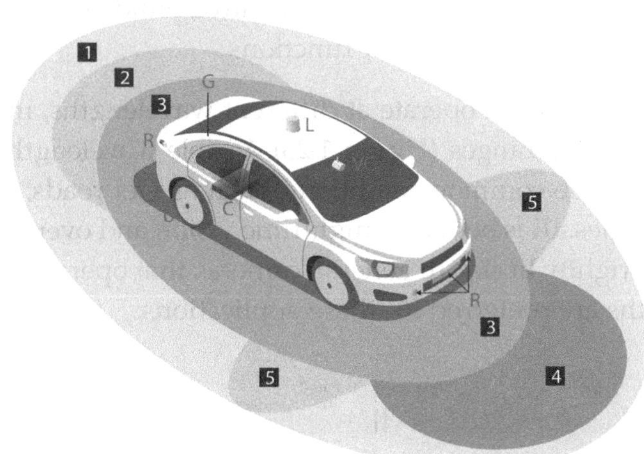

Figure 1-21. Types of sensors used by first-generation smart cars and their observation zones. The instruments on this car include a computer (C), a backup camera (G), a LiDAR system (L), a radar system (R), an ultrasonic sensor (U), and a video camera (VC) (some cars use as many as 12 cameras). The numbered observation zones are as follows: (1) a protection zone detected by 360° rotating LiDAR; (2) a rear collision protection and lane change assistance zone; (3) a self-parking function; (4) traffic light, lane departure, collision avoidance, and emergency brake-assist functions; and (5) a cross-traffic and pedestrian detection function.

Source: Courtesy of *Control* magazine. Béla Lipták, "Controlling smart cars, part 2," Figure 1, November 14, 2016, https://www.controlglobal.com/articles/2016/controlling-smart-cars-part-2/.

These systems and software packages can safely handle simple tasks such as changing lanes, stopping at red lights, parking, and keeping safe distances between vehicles. As I already mentioned, they cannot yet distinguish between situations such as a pedestrian trying to hitch a ride or a police officer flagging down a car. Such fuzzy conditions have not yet been effectively enshrined in computer code, whereas human drivers can usually recognize them.

Next, I describe each type of sensor used in smart cars in detail.

- **Radiation sensors** – Radiation sensors are the most commonly used sensors. They operate using a variety of wavelengths (Figure 1-22), can use focusing lasers and change their viewing angles, are less sensitive to interference, have improved accuracies and ranges, and have single-path or multipath designs. Among these features, comparative insensitivity to interference is probably the most important. The interference insensitivity varies with the radiation wavelength used, but all radiation sensors have some sensitivity to interference. Potential sources of interference include dust, rain, snow, solar radiation, and nighttime conditions, such as oncoming reflector lights in the opposite lane.

  At the time of the accident described in Section 1.5.2, "Tesla Accident in Florida," Tesla used cameras and orientation and wheel-hub sensors, but not LiDAR. These sensors assist in automatic cruise control, blind-spot and lane-assist monitoring, and parking, among other functions.

- **Cameras** – Cameras can operate at different wavelengths, including the infrared (IR) and visible ranges (Figure 1-23). Visible-wavelength cameras do very poorly under some common conditions, such as wet roads, reflective surfaces, or low Sun angles. IR is better for nighttime vision and overcoming interference caused by a brightly lit sky. In general, cameras that operate using wider ranges of wavelengths are preferred for these applications.

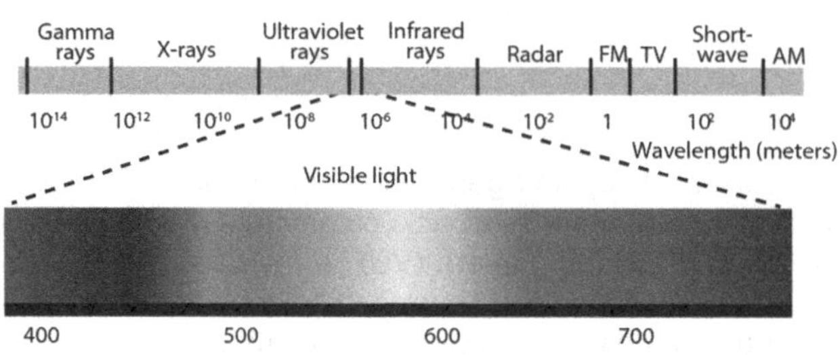

Figure 1-22. Potential wavelength ranges of sensors. (See color images after page 306.)

Source: Courtesy of *Control* magazine. Béla Lipták, "Controlling smart cars, part 3," Figure 2, January 19, 2017, https://www.controlglobal.com/articles/2017/controlling-smart-cars-part-3/.

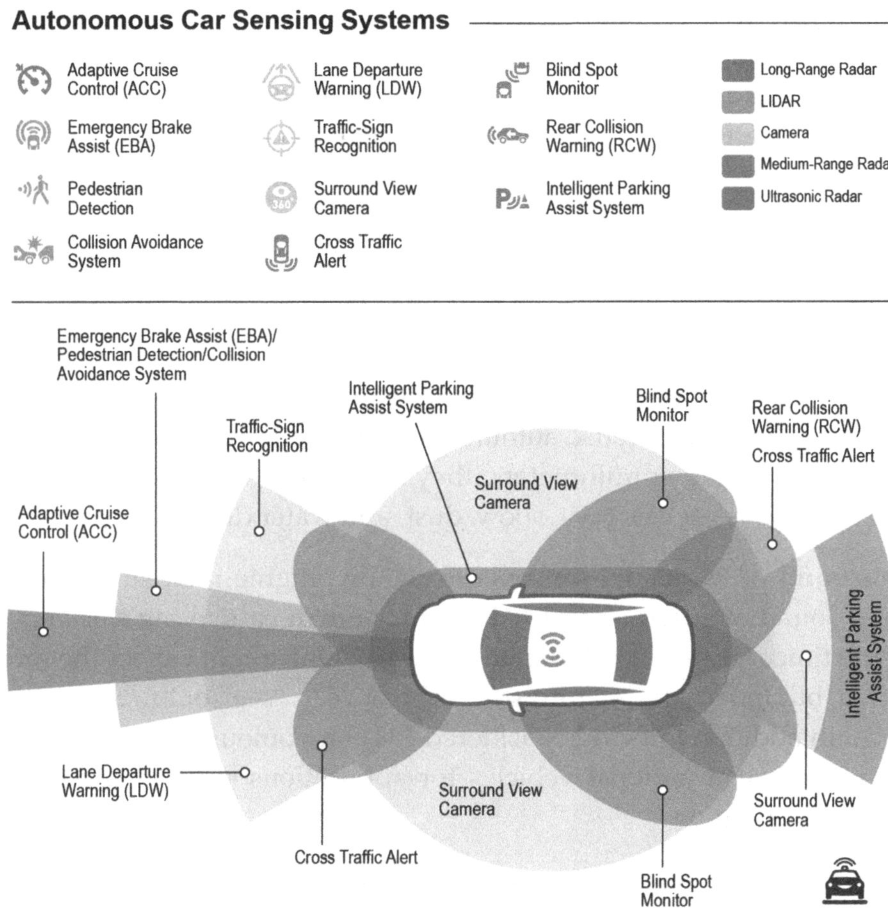

Figure 1-23. In autonomous driverless cars, different wavelength ranges are used for different measurement functions, including camera, radar, and LiDAR. (See color images after page 306.)

- **LiDAR** – Almost every self-driving car company uses LiDAR, including Uber, Waymo, and Toyota, but not Tesla. LiDAR measures distance by detecting the time it takes for an emitted signal to return to its origin. This method is the same as that of radar, but instead of radio-wave frequencies, it usually operates in the IR frequency range. LiDAR technology is extremely accurate. It can detect the position of objects at a 100 m distance to a precision of a few millimeters. These pulsed laser systems are usually insensitive to natural light. Their life expectancy is high, up to 100,000 hours (11 to 12 years). Both photocell and phototransistor detectors can be used. Phototransistors and photodiodes are most sensitive in the IR region and are faster than photocells. These sensors can distinguish different objects and thus draw conclusions about the type of object being viewed. Older generations of these sensors performed poorly in bad weather.

- **Radar** – The value of radar sensors is limited because their resolution is relatively low, but radar is good for looking through dust or mist. It can measure speed and distance and operate in either frequency-modulated or pulse-reflection-time detection mode. Both of these modes detect the time of flight of the microwave (radar) signal. The first measures the difference in frequencies between the emitted and returned signals; the second operates by measuring the transit time. The return signal's strength depends on the detected surface's reflective properties and the dielectric constant of the air through which the signals travel.

  Radar signals are good at detecting objects that reflect electromagnetic radiation (e.g., metal objects). Because automotive radar systems operate at wavelengths on the order of a few millimeters, they are also good at ignoring interference from conditions such as mist, snow, dust, and water droplets in fog.

- **Ultrasound** – Ultrasound sensors operate by pinging their surroundings with a loud sound at a frequency too high for human ears to register and measuring the time it takes for this signal to return. One might expect the range to be limited, but in fact, Tesla's ultrasonic sensors can "see" out to a range of nearly 500 m. In addition to being considered for autonomous cars, these sensors are also used in conventional vehicles for applications such as self-parking and blind-spot detection.

- **GPS** – GPS has good potential, particularly in vehicle-to-vehicle (V2V) communication. The federal safety standard mandating this V2V communications system has already been developed. If all vehicles (or at least trucks) had V2V capabilities, they could communicate their position and mobility information, which could not only prevent accidents but also ease traffic congestion, among other possibilities. I hope all vehicles will soon use GPS to inform all other vehicles in the area of their location, direction, and speed and to inform drivers of road construction or maintenance, accidents, and traffic jams. Today's GPS uses radio signals that measure the locations of vehicles to an accuracy of about 5 to 10 m, but, over time, the precision of GPS signals is likely to improve to about 0.5 m.

- **Software** – The 2016 Tesla accident showed that the control software of self-driving vehicles must be improved in a number of areas. It seems that excessive emphasis has been placed on avoiding false braking. It also seems that, during the accident, the radar sensor falsely assumed that the object in front of the automobile was a road sign when, in fact, it could not have been because of its low elevation. In addition, if the cameras had operated in the IR frequency range, Sun interference (in the ultraviolet range) would have been less of a problem and the system would have correctly identified the object as a white truck trailer.

The system's overall performance could also have been improved through the use of self-diagnosing and redundant sensors. In addition, it is advisable to use availability detectors on the brakes and other controlled components to verify that they are working properly and are accessible. In terms of redundancy, it is best to use two sensors with different operation principles (such as LiDAR and radar) to measure the same variable and use a control algorithm that responds to the less safe measurement (i.e., the most pessimistic option). Voting systems using three sensors are also valuable.

Moreover, AI software can be used to train the sensor computers to understand the visual world.

1.5.5 Other Vehicles

1.5.5.1 Trucks

Crashes involving large trucks kill nearly 4000 people each year in the United States alone, according to Department of Transportation statistics. Copilot, a package of advanced driver assistance systems (ADASs), offers lane maintenance, cruise control, and collision avoidance for commercial trucks. Like Tesla's Autopilot, these systems require an alert human driver to pay attention at all times. Similarly, companies like Waymo are already testing self-driving trucks with highway-only control systems. These trucks have not been trained to drive on city streets, where pedestrians, cyclists, narrow roads, and oncoming traffic make driving exponentially more difficult.

1.5.5.2 Motorcycles

Motorcyclists will soon benefit from the same driving aids already available for cars. In fact, radar-based driving aids are already being added to motorcycles. For example, Bosch is cooperating with Ducati and KTM to bring features such as adaptive cruise control to its bikes. Harley-Davidson might soon follow suit by adding driver-assist features to its electric motorcycles, which have a range of 110 mi and accelerate to 60 mph in just 3.5 seconds.

1.5.6 Safety

I believe that GPS-based V2V communication will be integrated into the control software as a safety backup system in the coming generations of self-driving cars. The safety of self-driving cars is likely to improve as time passes because the vehicle control systems will learn from each accident. Through this continuous improvement process, they can eventually safely navigate under challenging conditions such as construction, bad weather, and darkness. This is probably true, but we do not know how long it will take. We know that, according to some estimates, the development of the "go-anywhere self-driving car" promised by Google will take 20 years; others believe even that prediction is optimistic.

I am not sure that the performance of a software package will ever be able to compete with the skills displayed by a Manhattan taxi driver on a rainy night. I also know that improvement will be faster if the transportation industry takes advantage of the existing hardware and know-how available in other industries. Some believe that ride-sharing smart car designs will enter the market, improve driving safety, and provide less expensive transportation even for those who do not know how to drive and do not have a license. Naturally, two downsides of this change will be the loss of jobs for taxi and limousine drivers and a decrease in car sales.

The sky is the limit for coming innovations, such as replacing car keys with voice, face, or mobile phone recognition sensors.

1.6 Drones

Uncrewed aerial vehicles (UAVs) have been around for a long time. The first designs, unpiloted balloons, were launched by Austria in an attack against Venice in 1849. Their travel direction was controlled only by the wind, and the wind blew some of the balloons backward so they ended up bombing their own lines.[27] Much has changed since the wind blew back these exploding balloons.

UAVs were originally designed for tasks that were too dirty or dangerous for humans. Their first applications were mainly military, but their use is rapidly expanding. Today, all aircraft without human pilots are operated by ground-based controllers. They have various degrees of autonomy when onboard computers are provided.

1.6.1 Permits and Licenses

Basically, drones are flying robots. The US Federal Aviation Administration (FAA) is overseeing their development to make sure that they will safely fly in US airspace. Drones have transformed the battlefield in Iraq, Iran, Pakistan, Yemen, Somalia, and Afghanistan. US Customs and Border Protection agents have long been using drones to prevent illegal border crossings. In addition, an increasing number of hobbyists have mini drones taking off around the country. In most states, hobbyists must keep their drones within their line of sight and below a certain altitude. In the United States, the use of larger drones must receive an OK from the government on a case-by-case basis. One of the problems remaining is that there are no clear rules for new types of drones as they come on the market.

---

27. We should be careful not to repeat that outcome with our military drones. They are better controlled than the Austrian balloons were 200 years ago, but still have no consciousness and are unable to tell friend from foe or civilian from soldier.

Today, some companies are considering using drones to deliver goods, and others are talking about self-driving flying taxis and other applications. The first year the FAA issued a commercial drone permit was 2006. Between 2015 and 2016, the number of FAA drone permits increased from 1000 to 3100, and according to the FAA, more than 860,000 drones were registered by 2021. Most of these are in recreational use (530,000), but approximately 330,000 commercial drones are also registered.[28] Drones over a certain weight limit (usually 250 g) must be registered with the FAA. Drones are usually provided with integral GPS controls to prevent them from entering restricted areas, such as airports, where they could interfere with air traffic.

Various countries and US states have developed different rules for regulating drone use. For example, to fly a drone in California as a commercial pilot (i.e., for work or business purposes), one must obtain a Remote Pilot Certificate by passing the FAA's Aeronautical Knowledge Test to prove that one knows the requirements of the FAA's Small Unmanned Aircraft Systems Rule (Part 107). The FAA plans to establish locations around the country for certification testing.

The FAA concluded that commercial drones are safe and has developed the proper regulatory measures for using recreational drones. To fly a drone as a hobbyist (i.e., for fun or pleasure) in the state of California, one is required to follow the FAA's Special Rule for Model Aircraft.

Until now, it has not been clear whether drones would have to fly in their own section of the sky, but now, it seems that they will eventually be joining other aircraft in the same airspace (which I believe is a mistake). To safely integrate drones into the US airspace, new policies, guidelines, and procedures will need to be developed not only for drone pilots but also for airplane pilots, air traffic control crew members, mechanics, and anyone else involved with sending drones into the sky. Privacy is also a major concern, considering that drones can be outfitted with cameras and sensors that can collect all types of information about people who are just minding their own business at home.

It seems that Congress will eventually have to develop rules that will consider the protection of civil liberties in this area. This development will be a slow, meticulous process that will eventually ensure that all safety measures are considered before the drones take over.

---

28. "UAS [Unmanned Aircraft Systems] by the Numbers," FAA, last modified December 20, 2021, https://www.faa.gov/uas/resources/by_the_numbers/.

1.6.2 Operation and Applications

The software guiding drones includes layers of algorithms for planning, trajectory generation or regulation, and other features. The goal of path planning is to find an optimal path for the vehicle to follow while meeting the objectives of avoiding obstacles and considering fuel requirements. Trajectory generation involves the generation of the required control maneuvers that need to be taken to follow the given path from one location to another. Trajectory regulation involves the control (often PID) of the vehicle's position within some tolerance of the desired path.

In consumer drones, GPS communication provides accurate position information and autopilot features. Wi-Fi provides the means for the drone to transmit data within various distances. Bluetooth can also be used for data transmission within a limited radius. For larger-range communication at a lower rate, a frequency of 900 MHz (433 MHz in Europe) can be used. These days, remote controls do not require handheld transmitters as they can be controlled by smartphones.

For longer distances, radio control (RC) is used. The first RC designs (in the 1950s) were single-channel units requiring high-voltage batteries, but as transistors became common, the batteries were eliminated. Following this advance, multichannel units became available, allowing the control of a minimum of three control dimensions (yaw, pitch, and motor speed). More recently, pulse-code modulation (PCM) (working with digital signals) technology became available for applications including high-end hobby drone systems. The older variety of these digital designs operated 72 MHz receivers and radios, whereas today's designs use 2.4 GHz RC systems.

Available drone applications include programmable flight routes and first-person video controls. In addition to the RC transmitter controller, some more advanced drones can also view the drone position (using GPS) and monitor flight time, battery life, and flight speed. The more advanced flight modes allow one to find a lost drone or program a precise flight route. Drones have GPS, so they can return home after losing connection with the controller or flying out of the controller's range. In addition, unique applications are also available, such as having a drone follow the owner at a specified distance and altitude to transmit a view of the surroundings.

1.6.3 Civilian Applications

In addition to hobby systems, drones are used in many civilian applications. The number of these applications is too many to list, but they include such diverse fields as disaster relief, archaeology, conservation (monitoring pollution and combating poaching), law enforcement, crime and terrorism prevention, aerial surveillance (Figure 1-24), filmmaking, scientific research, surveying, and cargo transport.

Figure 1-24. Survey drone with an automatic camera.

There are many other potential applications, such as using drones to deliver goods to customers.

Drones can potentially reduce the time, costs, and energy needed for many everyday deliveries. All civilian drone applications are similar and differ only in the purposes they serve. Therefore, I mention only a few here and give a detailed description of only one civilian drone application: searching for and evaluating the size and other characteristics of oil spills (Section 1.6.4, "Oil Slick Evaluation").

1.6.3.1 Crowded Skies

In December 2018 at Gatwick Airport in England, 800 flights were canceled because some drones entered the airspace. In California, a hobby drone flying above a wildfire forced officials to ground firefighting aircraft because of similar safety concerns. According to the FAA, more than 1000 pilots have reported drone sightings in the past year. Drone strikes are more dangerous than bird strikes because of their size, and I am sure that accidents will keep occurring until firm and uniform laws are established.

Drone accidents, crashes, and attacks are becoming everyday events. Some of these events show that drones can be used for more than recreational and commercial purposes. For example, on Monday, January 26, 2015, a drone crash-landed on the White House lawn. A similar event occurred during a Christian Democratic Party campaign in September 2014, when a Parrot AR drone crashed in front of German Chancellor Angela Merkel. These and other similar incidents raise concerns about the potential for the use of weaponized drones.[29]

29. Conner Forrest, "17 drone disasters that show why the FAA hates drones," TechRepublic, June 13, 2018, accessed August 29, 2022, https://www.techrepublic.com/article/12-drone-disasters-that-show-why-the-faa-hates-drones/.

Figure 1-25. When drones are used, safety should be the primary goal.

Closer to the ground, Amazon, Google, and DHL are all planning quick delivery systems for medications and other much less urgently needed goods (Figure 1-25), but the development of official policies concerning drone usage is not keeping up with the advances in drone technology. In my view, drone policy should give primary consideration to safety, and the upcoming rules should prohibit drone flights in cities and crowded areas.

1.6.3.2 Need for Regulations

Drone-related privacy invasion issues are beginning to be addressed in most states, and about half have already passed laws restricting drone usage. When an 18-year-old teenager (Austin Haughwout) in Connecticut built drones that could fire guns and shoot flames, the incident triggered the realization that there are no regulations to ban mounting weapons on drones.

1.6.3.3 Insurance

Similarly, there are unresolved issues related to drone aviation liability and worker's compensation issues. Some insurance firms have prepared specific coverage packages tailored for uncrewed aircraft accidents. Other firms believe that the standard homeowner's insurance policy already covers the falling of a drone onto a house or a car, just as it does any other falling object.

1.6.4 Oil Slick Evaluation

If equipped with the proper sensors, drones can quickly assess and map the surface of petrochemical leaks from point sources. Doing this requires flight operations expertise, mapping-related hardware, and software for data analysis. Fiber-optic laser photodetectors can be dropped into the water to detect oil/water interfaces. Such drones are already in limited use, as their size and weight problems have largely been solved.

The same drone designs with optical and other sensors can be used to locate people, floating objects, and underwater debris in cases of ocean accidents or rescue efforts. Drones can also be used for leak detection in the case of oil pipeline inspections (Figure 1-26), as well as flare stack and gas emission monitoring.

When detecting ocean spills (Figure 1-27), drone measurements can provide information on the rate and direction of oil movement and assist in guiding cleanup and control efforts by drift prediction modeling. Underwater remote sensing instruments and technologies based on acoustics, fluorescence polarization, in situ fluorometry, nondispersive IR spectroscopy, and in situ mass spectrometry are also being developed for use on remotely operated vehicles (ROVs) and autonomous underwater vehicles (AUVs).

Hydrocarbon detection can be achieved using a variety of optical sensors, thermal IR imagers, airborne laser fluorosensors, laser radar devices, and so on. These families of sensors are also suited for AUV applications. They can be combined into all-in-one cameras and integrated area sensor (IAS) units with 360° field views and can be controlled using Google's Android platform.

Laser fluorosensors operate by emitting radiation at a particular wavelength that is absorbed by oil. In response, the oil emits visible fluorescence, which is then detected by a laser sensor. This fluorescence detection technique measures the spectrum and

Figure 1-26. Oil pipeline leak-inspection drone.

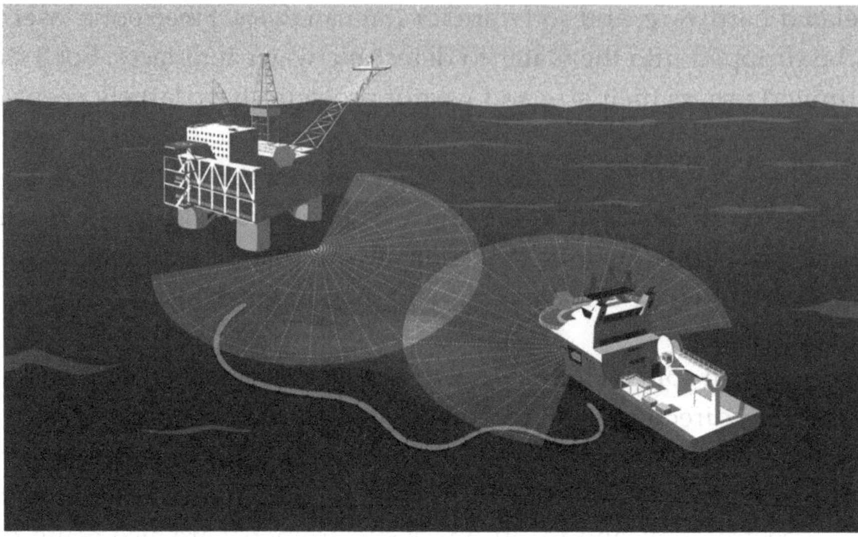

Figure 1-27. Oil spill detection.

Source: Miros Group, "A Record Delivery of 8 Oil Spill Detection Systems to Petrobras," Miros Group website, November 28, 2018, accessed November 15, 2022, https://www.miros-group.com/blog/a-record-delivery-of-8-oil-spill-detection-systems-to-petrobras/.

the decay rate of the emitted fluorescence. Thus, light, heavy, and other oils can be identified separately.

Oil reflects sunlight in the UV range at low thicknesses (<0.1 μm). UV and IR images have been superimposed to yield relative oil thickness maps, but the resulting thicknesses are not useful for spill cleanup techniques. For oil cleanup, thickness measurements of 0.5 to 10 mm are required, and these thicknesses are about three orders of magnitude greater than those that can be mapped. The use of UV reflection in oil spill remote sensing is decreasing because of the low relevance of thin slicks to oil spill cleanup.

Thermal imaging by long-wavelength IR (LWIR) sensors can detect the thickest portion of an oil slick. As such, this technique can be used to measure the size of the oil slick and to focus on the thickest portion of the slick (Figure 1-28), which is necessary when planning the cleanup effort.

Radar sensors operate by detecting the reduction in the backscattering of radar signals caused by oil. Radar-based detector systems can identify the position of a spill and track its drift to direct the recovery vessels to the location of most of the oil. The main advantage is that they can operate all day under almost all visibility conditions.

The longer-range goal for oil slick evaluation is to integrate several optical and microwave sensors while continuing to reduce sensor size, complexity, and cost. Sensors

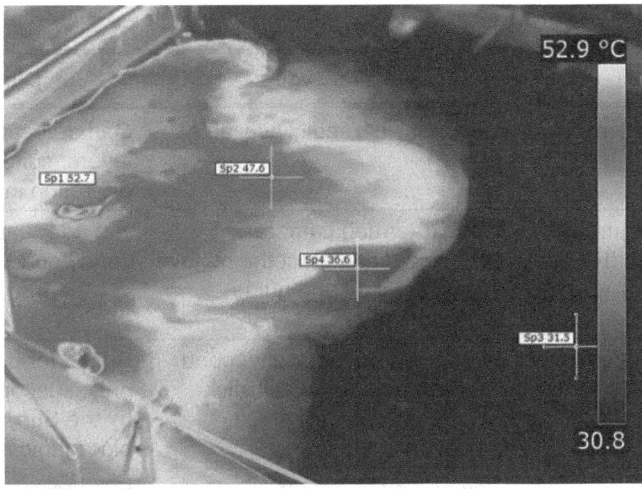

Figure 1-28. Simultaneous detection of visible and LWIR images of oil on water.

Source: Courtesy of *Control* magazine. Béla Lipták, "Ask the Experts: Oil Slick Sensors for Drones?," Figure 2, December 28, 2016. https://www.controlglobal.com/articles/2016/ask-the-experts-oil-slick-sensors-for-drones/.

that are less expensive, are more energy efficient, utilize solid-state laser technology, and operate with true real-time processing are also needed to meet that goal.

In Table 1-2, I summarize some of the features, capabilities, and limitations of drone-based oil slick detection applications.

1.6.5 Military Applications

As mentioned earlier, the first variations of military drones were uncrewed balloons, built by Austrian forces, filled with explosives, and allowed to be blown by the wind to attack Venice in 1849. More than 100 years later, in 1971, the modern military drone was invented by John Stuart Foster Jr., a former head of the Lawrence Livermore National Laboratory (then called the Lawrence Radiation Laboratory). Based on his design, two prototypes were built by the Defense Advanced Research Projects Agency (DARPA) in 1973. They could stay in the air for only 2 hours and carry only 28 pounds of load.

In the 1973 Yom Kippur War, Israel used US-made target drones against Egypt. This effort was successful because no pilots were lost on the Israeli side, whereas Egypt used up all of its antiaircraft missiles by firing at the drones. Later, Israel developed lighter, glider-type drones, the model most often used today. The US Hunter and Pioneer UAVs are direct descendants of the Israeli models. In general, drones have demonstrated their value for the military, including in search and rescue, reconnaissance, demining, and target practice (Figure 1-29).

Table 1-2. Oil slick sensors.[30]

Sensor Type	Advantages	Disadvantages
Visible-light (0.4 to 0.7 μm) camera	Inexpensive, readily available	Can detect thick but not thin oil layers; suited for daytime operation in good weather only
LWIR (thermal, 8 to 12 μm) or MWIR (mid-wavelength IR, 3 to 5 μm) camera	LWIR (thermal) cameras preferred; can detect the relative thickness of oil and guide cleanup efforts	Not adequate in rainy, humid, or sandy conditions
UV (10 to 400 nm) laser camera	High oil UV reflectability; good for mapping the overall slick	Suited for daytime operation only; susceptible to interference by the Sun, wind, and bioorganic materials
Laser fluorosensor	Day and night operation; distinguishes light, medium, and heavy oils	Expensive
Radar (1 mm to 1 m) sensor	Suited for day or night use for fast detection of large slicks from high altitudes	Expensive, suited only for surface speeds of 2 to 7 m/s

Israel pioneered using UAVs for real-time surveillance, electronic warfare, and decoys. The images and radar decoying capabilities provided by these UAVs helped Israel neutralize the Syrian air defenses in the 1982 Lebanon War without losing any pilots. They also used drones to perform reconnaissance on the enemy's position, jam

Figure 1-29. Concept of a future uncrewed combat aerial fighter.

30. For a summary of the teams doing research and development work in this field, see: "Record Oil Spill Response Research Projects," Bureau of Safety and Environmental Enforcement, US Department of the Interior, https://www.bsee.gov/site-page/remote-sensing-and-surveillance.

communications, and act as decoys. In the late 1980s, Iran also deployed drones in the Iran–Iraq War, the first time an armed drone was used in war.

Uncrewed combat aerial vehicles (UCAVs), also known as combat drones, have no onboard pilot and usually carry aircraft ordnance such as missiles. They are usually under real-time human control, with varying levels of autonomy. Because the equipment used by human pilots (cockpit, armor, ejection seat, flight controls, pressure, and oxygen controls) is not needed, the weight and size of the UCAVs (Figure 1-30) are reduced compared to those of human-crewed aircraft.

The military use of drones has increased during the past 25 years. Drone missions grew from 50 strikes under President George W. Bush to more than 400 under President Barack Obama. Americans started to wonder if remote-controlled killing should be a permanent fixture of national policy as the media showed that civilians had become collateral damage. In 2022, thousands of drones from a number of nations (Turkey, Syria, United States, etc.) were being used in the undeclared war between Russia and Ukraine. They served to spot enemy positions, launch missiles, and direct artillery fire, and they have satellite communications capability. International policies or regulations on the use of drones do not exist. No one has agreed on whether using drones on the territory of another nation is an act of war, and if it is, who can authorize it?

Some or most attacks are carried out in secret and are not circumscribed by any international law. This is inconsistent because if US military forces were to attack another nation openly, it could be considered an act of war. However, the intelligence agencies operate in a netherworld where the same rules do not apply. As you read this book, the

Figure 1-30. A US Air Force MQ-9 Reaper, a UCAV, during a training mission.

Source: "MQ-9 Reaper - 090609-F-0000M-777.JPG," by Paul Ridgeway, US Air Force, August 7, 2008, Public Domain, http://www.af.mil/News/Photos.aspx?igphoto=2000398487.

drone strikes carried out by the CIA in Ukraine, Pakistan, Yemen, Afghanistan, Iran, and Somalia are probably continuing. Congress should create a legal framework for defining the authorization process for drone strikes, and the international community should decide that a declaration of war is required before such military attacks are used.

Today, no such rules exist. Drone strikes are part of targeted killing campaigns. Determining the total number of combatant and noncombatant civilian deaths is nearly impossible, although the Bureau of Investigative Journalism (BIJ) periodically reports such estimates. A poll conducted in the United States in 2013 asked registered voters if they "approve or disapprove of the US military using drones to carry out attacks abroad on people and other targets if they are deemed to be threats to the United States." The results indicated that some 75% approved. Outside the United States, there is widespread opposition to drone killings. In a 2014 survey of 44 nations asking about using drones to target extremists in countries such as Pakistan, Iran, Iraq, Libya, Yemen, and Somalia, majorities or pluralities in 39 countries were opposed. The United States, Kenya, and Israel were the only countries where at least half the public supported drone strikes, and Israel had the highest pro-drone attitude, with 65% approving of their use.

We all know that the AI war has already begun, as some 50 nations are developing robot soldiers that no longer need human direction. Although such troops are not superintelligent yet, they can already think for themselves. In short, this process is out of control and can lead to disaster. A particularly dangerous development is the expansion of uncrewed military activity into space and equipping these AI-controlled machines with nuclear arms.

AI applications such as drones are spreading in an uncontrolled manner, and this is prompting some serious legal and ethical questions. In military applications, AI can guide autonomous vehicles and drones to swarm targets thousands of miles away, and these drones often are unable to tell friend from foe. The civilian applications of drones are also uncontrolled, and 90% of airports are unprepared to protect against them. No air security or electronic fences are available, and most drones are not provided with integral GPS controls to prevent them from entering restricted areas.

To get this process under control, international regulations that prohibit the design and use of machines that can take lives without human involvement must be developed and enforced. It is essential that the direction of AI development is kept under control and that AI is prevented from being used in morally repugnant applications.

1.7 Opportunities and Risks

The development of AI must be controlled. AI must be developed as a tool that will serve humankind to make life better, easier, and happier. In other words, we must control the direction in which AI applications evolve to ensure that AI will be used as a helpful tool and that it will not be allowed to do whatever it *can* but only what it *should*.

This is important from both safety and cultural points of view. Today, neither aspect is under control. From a safety perspective, we have no cyber defense systems that would effectively protect our communications, power distribution, water supply, transportation, financial, or industrial sectors. The primitive cyber defense that exists consists of scattered efforts, but there are no standardized, multilayer national or global security products or strategies in place. Even the need for such uniform and standardized cybersecurity systems is just beginning to be recognized.

It is important to recognize the cultural impact of out-of-control AI development. If we do not rein it in, it can divert human evolution, and even scarier, it can gradually turn the coming generations from "homo sapiens" (wise man) into "homo clicksters" (Figure 1-31).

In theory, computers can emulate the biological brain and, therefore, they can also emulate or even exceed some aspects of human intelligence. On the other hand, computers will never have a conscience or moral or ethical values; their actions will always depend on the algorithms loaded into them or on algorithms that direct them to self-learn their environment. In other words, they will remain dual-purpose tools that can help us to a better future or lead us into disaster.

Figure 1-31. The generation bringing in the age of clicking.

Comparing the evolution of AI to the nature of industrial processes, we can also note that the response times of the two processes are dramatically different. In industry, when we increase the heat input into an energy system, for example, the system responds in minutes because its time constant is short and its capacity is relatively small. On the other hand, the time constants of cultural processes are long and, once triggered, are not easily reversed. Therefore, if we do not establish a "cell-phone-resistant" cultural environment now, the impact can affect generations. We need to motivate children to stop staring at the screens on their phones and to discuss ideas with their friends, play outside, read books, and listen to music; in short, we should encourage them to have a life experience akin to that of previous generations. If this is not done, cultural devolution can start. Using process control language, if a process, such as the cultural impact of the Internet, is left uncontrolled, it can destroy culture as we know it.

1.7.1 Promising AI Applications

Thus far, most AI applications in use belong to the category of "weak AI," where the particular algorithm can perform only one task at a time. For example, it might play chess, load the dishwasher, get a beer, feed the cat, and clean and declutter the house (if you leave and come back 6 hours later). It may even provide companionship for the elderly, make sure that they take their pills, and help them remember their grandchildren's names. Such tasks are endless and are multiplying rapidly.

AI has already assisted breakthroughs in health care, scientific discoveries, improved weather forecasting, safer and more efficient transportation, personalized education, and improved manufacturing and agriculture, as well as the evaluation of complex and multivariable processes such as GW. In no particular order—and without attempting completeness[31]—I next describe a few examples of some existing or developing AI applications that are useful. As some of these applications have already been discussed in this chapter, I will not elaborate further.

1.7.1.1 Voice and Face Recognition

In the area of hearing, voice recognition is evolving and being used for many purposes, including security. Voice analysis on the phone can determine not only who is calling but also such complex and critical facts as whether the caller is having a heart attack.

IBM introduced its first speech recognition machine in 1962. Voice recognition seeks to identify an individual user's voice, whereas speech recognition focuses on translating

31. For a more complete discussion of the many applications of AI, see: "Applications," National Artificial Intelligence Initiative Office, accessed March 5, 2022, https://www.ai.gov/strategic-pillars/applications/.

a verbal format into text. Voice recognition is the cycle that empowers a software package to perceive and react to verbally expressed words and then change them into a format that a machine understands. Such software packages as Apple's Siri, Google's Alexa, and Google Dictate can transcribe dictated words into text. This allows one to essentially type with one's voice, speaking emails and documents into existence by activating the microphone option on a device.

Automatic language translation has also significantly improved. Another hearing-related application is lip reading. This dual-purpose application can help individuals with speech impairments to communicate; it can also be used as a tool for surveillance or spying.

One type of biometric software employs facial recognition, whereby the features that make a face unique are stored in a database. In this case, if a person's features match the stored data, all the information concerning that person becomes available for any purpose. Facial recognition has been used successfully in airport security, criminal investigations, and forensic services. Because the images used in such systems can be recorded using surveillance cameras without the knowledge of the people being photographed, such techniques can be both useful and unethical.

1.7.1.2 Safety

The AI applications that are advancing the fastest are safety-related surveillance systems (Figure 1-32). Such applications can include observation from a distance using electronic equipment (such as closed-circuit television cameras) or the interception of electronically transmitted information (such as Internet traffic or phone calls). Surveillance is a dual-purpose application because it can be useful not only for intelligence gathering or investigation by law enforcement but also for illegal purposes by criminal organizations, foreign governments, and others. For example, face recognition and voice recognition enable authoritarian governments to carry out surveillance

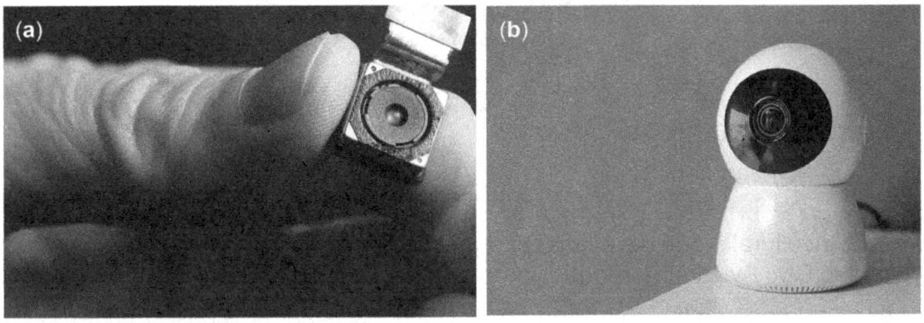

Figure 1-32. AI security cameras.

to classify potential enemies or target propaganda and misinformation for maximum effect.

Secretly surveilling or gathering personal information is a violation of privacy, and as such, it is often opposed by civil libertarian groups. However, its legal limitations are not yet fully defined. Today, as with so many other AI applications, the evolution of surveillance technology is also out of control.

1.7.1.3 Agricultural Applications

Agricultural drones (Figure 1-33) are used to monitor crop growth and increase crop production. Their sensors have digital imaging capabilities that provide farmers with improved information and allow them to inspect their fields from the sky. Agriculture drones can, for example, examine plant leaves, diagnose a disease, and suggest treatment. The bird's-eye view they provide can reveal problems related to irrigation, soil variation, and pest or fungal infestations. Multispectral images obtained from near-IR and visual cameras can help farmers decide issues such as which fungicide to use and how much to apply.

1.7.1.4 Insurance Applications

AI has an almost unlimited range of applications in all types of office work. Consider the insurance industry as an example. AI can build deep-learning models to evaluate customer risk profiles more accurately than humans and thus provide optimal insurance prices. In the area of document processing, AI systems can handle low-skill, repetitive tasks that are prone to errors. Similarly, simple queries can be handled by

Figure 1-33. An agricultural drone developed by Agridrones Solutions Israel that can carry up to a 25 kg load, certified by the Civil Aviation Authority of Israel.

Source: "4X-UHJ Agridrones d.jpg" by Agridrones Solutions Israel is licensed under CC BY-SA 4.0, https://en.wikipedia.org/wiki/Agricultural_drone.

chatbots, which can also be used for intelligent call routing to forward customers to specific agents based on their needs. In addition, angry customers can be directed to more experienced call agents, and voice analytics can help determine whether a customer is lying when submitting a claim.

New devices and software packages are becoming available to assist in reaching these goals. For example, smartphone sensors are being used to help insurance companies distinguish between "risky" and "safe" driving habits. This distinction is based on measuring the acceleration, braking, turning, and other habits of the drivers.

1.7.1.5 Schools and Teaching

In the schools of the future, grading, taking attendance, or teaching languages can all be assisted by AI. AI could also provide a personalized level of education and take on some of the teaching roles by providing students with basic information. At the same time, teachers could move into the role of classroom facilitator or learning motivator. It is likely that the next generation of students will grow up with an AI companion that knows their personal and school history. Therefore, it will know each student's strengths and weaknesses. AI will help teach and study both the students and the teacher. It could help the teacher by automating grading not only of multiple-choice materials but also of open-response assignments, for example, by "clustering" similar student responses, thereby allowing the teacher to comment only once on each cluster.

AI can also report on whether the explanation of a particular question was not well understood and needs to be rethought. It is hard to say where this process will lead because AI that utilizes a deep-learning software architecture is just beginning to enter the field of education. As with most others, this application also includes risks because it can easily be used by students to copy material from the Internet or by dictatorships to brainwash people, as is already done by the Chinese in their "reeducation" of the Uighurs.

1.7.1.6 Companion and Home Robots

Not too long ago, people believed that mobile robots would soon permit older people to live independently at home, with robots acting as partners and assistants. However, it turned out that getting around a house is difficult, so home companion robots are no longer on the near horizon. Specialized robots can pick things up from the floor but do so relatively slowly. Similarly, a robot can make a bed, but again slowly. Therefore, even special robots custom-designed for single tasks cannot yet perform at human speed. Despite these difficulties, some companies believe that the aging of people will be the catalyst for developing more sophisticated companion robots.

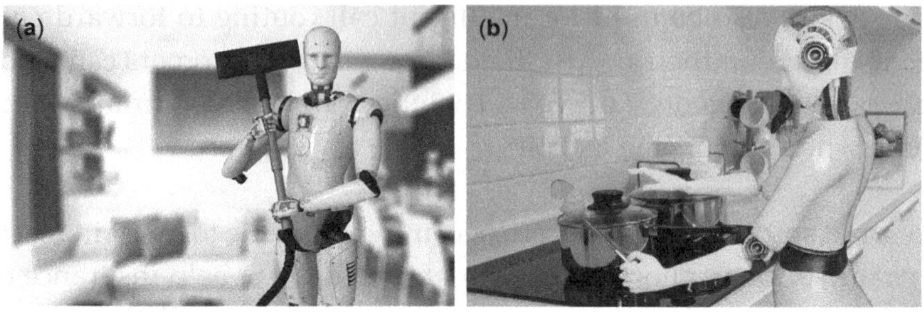

Figure 1-34. Single-task robots.[32]

The development of home robots has turned out to be more difficult than expected. Although such robots are capable of performing ordinary household tasks such as getting a cup of coffee, loading and unloading the dishwasher, and vacuuming (Figure 1-34), the present prototypes can handle such tasks only in highly controlled environments.

The designers of self-driving cars have figured out ways to navigate in unknown environments and "self-create" maps to reach destinations. However, in a home with densely packed objects and moving humans, software applications similar to those in self-driving cars are insufficient. Nevertheless, Toyota, Alphabet's Google X research laboratory, and Amazon have active home robotics research efforts in progress. They believe that the demographic shifts of aging societies will eventually provide a large market for such products, but they also realize that it will take decades before such products are available at affordable prices.

1.7.2 Harmful AI Applications

Because AI is a dual-purpose tool, it can also do harm in the wrong hands. It can do harm not only as a new weapon of war, as discussed earlier, but also as a tool like the "Big Brother," described by George Orwell, who watches us all. Here, I mention some areas in which AI must be regulated and must not be allowed to evolve freely.

1.7.2.1 Surveillance versus Privacy

Facebook and Google (and who knows who else) can track our every move; analyze the implications of our every search on the Internet; weigh the nature of our web browsing activities; and assess every photo we view, every video we watch, and all the

---

32. For additional information, see the following websites: Mariella Moon, "Robot home can anticipate and cater to your needs," Engadget website, March 11, 2016, accessed November 1, 2022, https://www.engadget.com/2016/03/11/brain-of-things-robot-homes/. Glenda Bartosh, "Deluxe helpers on the home front," Pique Newsmagazine website, September 6, 2017, accessed November 1, 2022, https://www.piquenewsmagazine.com/whistler/deluxe-helpers-on-the-home-front/Content?oid=4318761.

information we post. Based on the accumulated data, they can place us into various groups and categories. The lists of people in each of these groups can then be marketed, sort of like "human trading." These Big Brothers might be spying only to make money, but when a list is sold, they no longer have any control over how it will be used, whether the purpose of the buyer is harmless, such as a pet food company marketing to dog owners in Connecticut, or harmful, such as a radical group spreading disinformation to undermine democracy or authoritarian governments using smart spyware that enables machine learning to classify potential enemies.

1.7.2.2 Disinformation

AI can be used to spread disinformation and propaganda to manipulate public opinion. This process is currently uncontrolled, and if allowed to become out of control, it can have harmful social consequences. The algorithms that protect against inciting hate or spreading disinformation are still in a primitive state, operating only at the keyword level. This means that, in their attempt to filter out disinformation, Facebook and others only look at statistical trends in word usage. These algorithms cannot identify and block ideas, and even if they could, there are no clear legal definitions of which types of lies are protected by freedom-of-speech laws and which are not. In fact, the more lies we hear without consequences, the less upset we get when we hear another one, and this can lead to the acceptance of the evolution of a truthless culture.

This age of "alternative facts" can destroy the integrity of our culture, as well as our trust in the integrity of our leaders or the reliability of the information reported in the media. It would be desirable to set the allowable limits of lying and resolve where free speech ends and punishable speech begins. By the way, the same applies to the role of money in elections because using it to influence elections or other decisions is not a form of free speech; it is bribery. Today, it is also possible to create believable pictures, videos, and animated movies that can circulate through the Internet and spread false information.

Clearly, the use of AI to spread disinformation is not the fault of the AI. It is the fault of those who use it for that purpose and society as a whole, which—up to now—has allowed AI to do what it *can* and has not restricted it to do only what it *should*. AI can serve valuable purposes in robotics, entertainment, education, and similar applications, but being a dual-purpose tool, it can also cause harm. Protection against disinformation is both a technical challenge and an ethical and legal one. In addition, when dictatorships use AI tools to brainwash whole societies, justify criminal acts, or attack the very foundations of democracy, there are few nonviolent means of protection against them.

1.7.2.3 Computer Activity Monitoring

A very controversial area of AI applications is the monitoring of computer activity, including accessing data stored on hard drives and data transferred over computer networks such as the Internet. This monitoring can become widespread and can be carried out covertly, as nearly all Internet traffic can be monitored. As with all dual-purpose AI applications, this one can be useful because it allows governments to maintain social control, recognize and monitor threats, and prevent and investigate criminal activity. On the other hand, because most governments have an unprecedented ability to monitor the activities of their citizens and influence them as they wish, AI technology can also be harmful.

AI gives governments new methods of preserving power.[33] Surveillance powered by AI allows despots to automate the monitoring and tracking of their opposition in ways that are far less intrusive than traditional surveillance. These digital tools enable authoritarian regimes to cast a wider net than they could with human-dependent methods, and they can do so using far fewer resources. No software program needs to be purchased to monitor people's text messages, read their social media posts, or track their movements. Moreover, once citizens learn to assume that all of these things are happening, they alter their behavior without the regime resorting to physical repression.

1.7.2.4 Employment

The increased presence of robots in the workplace is resulting in the displacement of human workers, which will continue in the coming decades. Some believe that, eventually, nearly half of the currently held jobs (from welding in factories to agriculture) will be taken over by robots. These robots are single-task machines using weak AI (also called "applied AI" or "narrow AI"), and although they can replace workers, they need to be designed and maintained, which also requires a workforce. Therefore, robotization does not necessarily cause unemployment, but it definitely results in a need for reeducation as it reduces the number of blue-collar jobs and increases the need for white-collar workers. Robots do not necessarily create unemployment; after all, they are only tools that make human work easier as long as humans know how to use them.

Others believe that, although robots displace workers, technological innovation creates more new industries and jobs. I believe transforming our energy industry into a green economy will create many more jobs than automation destroys.

---

33. "The Digital Dictators. How Technology Strengthens Autocracy," Global Committee for the Rule of Law, March 10, 2020, accessed November 1, 2022, http://globalcommitteefortheruleoflaw.org/the-digital-dictators-how-technology-strengthens-autocracy/.

1.7.2.5 Workplace Ethics

The goal of automation is usually to optimize the operation and maximize the efficiency of a controlled process. AI systems, such as those developed at Amazon to track worker productivity and fire them if their performance does not live up to expectations, are unacceptable because they dehumanize the workplace. Black boxes can be useful as one of the tools of human decision-making, but they should never be allowed to become the decision-makers themselves. For example, résumés can be screened by computers, but the hiring decision must always be made at a hiring interview. This is a sensitive area because algorithmic bias can lead to unfair outcomes if AI is used for processes such as credit rating or hiring.

Process control theory has shown that the time constants of changes and their corrections must be similar. This is not the case in some AI applications. For example, changes in workplace practices are slow, while the change AI creates is much too fast for stable control. It is true that the spread of robots motivates innovation, establishes new industries, and increases the demand for a better-educated workforce, but preparing such a workforce takes time and requires a much-improved education system for the next generation. It seems to me that AI and computer science, in general, should become a basic component of school curricula, just as mathematics is today.

1.7.2.6 Cyberattacks

Today, cyberattacks are exploding because the so-called bad guys are just as smart as the good guys. The bad ones are working on clever ways of hacking, and the good ones are trying to protect against hacking attacks by designing even more clever anti-hacking software.

The goals of cyberattacks can range from simply criminal (e.g., ransomware) to sabotage, espionage, or even war. Previous cyberattack programs include Stuxnet, NotPetya, and WannaCry; by now, they are too numerous to list. Some argue that this trend might lead to nonlethal wars that could eliminate all direct human casualties, as in the substantial damage caused to Iran's nuclear program by the Stuxnet attack. My view is that we should not wait to see where cyberattack capabilities lead. We should not wait for a cyber–Pearl Harbor before setting up an International Cyber Control Agency to establish multinational agreements, develop treaties, and motivate unwritten agreements to govern cyberactivity. This can be done just as the International Atomic Energy Agency has done for nuclear technology. Naturally, this will not happen unless such international institutions are created that are more powerful than nation-states, which is unlikely to occur for centuries, if ever.

1.8 Conclusions

AI has opened a new chapter in human history, one that is probably just as important as or even more important than the discovery of electricity. I say even more important because electricity changed only our physical environment, whereas AI has the potential to change our cultural environment as well, thereby influencing human evolution itself. Today, the development of this new tool is completely out of control, and although it could take us into a better future, if left uncontrolled, it could also lead us onto the slippery slope of destruction.

My experience in process control tells me that leaving a poorly understood process uncontrolled can lead to disaster. On the other hand, it also tells me that controlling any process requires understanding its characteristics and knowing its desirable state (set point) and the road that can lead to reaching that destination. In the case of the evolution of AI, we should also consider the interactions of AI development with other parallel processes. These other loops include the control of GW because AI and GW can reinforce or conflict with each other. AI can tremendously boost global education by explaining the causes of wildfires, loss of polar ice, hurricane-caused storm surges, and so on. At the same time, the jobs created by the transition to a carbon-free, green-energy economy can also help solve the unemployment created by automation.

As I noted earlier, the evolution of this process is out of control; consequently, it is doing all it *can* rather than doing only what it should. If left uncontrolled, AI can destroy our culture and civilization.

As described in this chapter, the effects of AI have also entered into our private lives because AI can be used for manipulative advertising and for the development of racially biased algorithms to apply to the allocation of public assistance services or hiring, for example. Protection is needed against such activities as collecting private information and using group-targeted telephone calls and email campaigns to obtain credit card information or passwords.

Therefore, nationwide or even global regulations are required to forbid some aspects of the use of AI. Early steps are beginning to be taken, particularly in the European Union. In short, the time has arrived for an effort akin to the Geneva Convention to set the legal limits for the allowable use and applications of AI.

Yes, AI is a useful tool. We can delegate many of our routine mental functions to it, but it is also a double-edged sword, and we must stay in control of it. For example, on the one hand, we must not allow the elimination of steering wheels and brakes from our cars, but on the other hand, we must bear in mind that human operators can also fail, just as computers do. Therefore, neither should be allowed to automatically "overrule" the other; AI should select and implement the safer of the two instructions if they conflict.

Finally, but very importantly, we must be aware that our cultural environment must also be protected. We must create global institutions, regulations, and enforcement structures that will protect us from passively submitting to Big Brother or from sinking into an age where the relevance of facts or truth is destroyed. In short, I hope that the contents of this book will help to get AI under control so that it can mark the beginning of a better future for all of humanity.

On the one hand, we should remember what Stephen Hawking said:

"The rise of powerful AI will be either the best or the worst thing ever to happen to humanity. We do not yet know which."

On the other hand, we should also remember what Vladimir Putin, President of Russia declared:

"Whoever becomes the leader in this sphere will become the ruler of the world."

2

Global Warming Processes

Table of Contents

Global warming (GW) is caused by Earth receiving more heat than it loses. Compared to the reference global average surface air temperature of 14.0°C (57.2°F),[1] which is the average of the 1951 to 1980 period, it warmed by about 1.2°C (2.2°F) by 2022. During the last century, the global average temperature increased by 7% (from about 14.0°C [57.2°F] to about 15.2°C [59.4°F]). For a while after World War II, the difference (anomaly) between the preindustrial average temperature and the actual temperature remained stable for a while, but when air pollution started to rise, it too started to increase. When the United States passed the Clean Air Act (CAA), a federal air quality law in 1963, Congress was focused on air pollution and did not realize the danger of global warming (GW). The CAA was amended many times and in some areas it was very successful. For example, the amendments in 1990 committed the United States to a reduction in sulfur emissions and have been particularly successful in fighting acid rain. A similarly successful effort was made to protect the ozone layer, which was being depleted by refrigerant and aerosol emissions. In 1987, 197 nations agreed to phasing out chlorofluorocarbons (CFCs), an effort which is likely to succeed in closing the ozone hole over the Antarctic. In 2022, GW was rising at a rate of about 0.2°C (0.36°F) per decade.

---

1. "World of Change: Global Temperatures," NASA Earth Observatory website, accessed March 8, 2022, https://earthobservatory.nasa.gov/world-of-change/decadaltemp.php.

GW can be stopped by either reducing heat input or increasing heat removal. Achieving either is difficult because the planet's heat balance is the combined total of many complex and poorly understood subprocesses.

Although warming by a degree or so does not sound like much, it actually is a lot because of the gigantic mass of the planet. Just keep in mind that a decrease of just 1°C to 2°C (1.8°F to 3.6°F) was all that it took to plunge Earth into the Little Ice Age[2] and that a 5°C (9°F) drop was enough to bury a large part of North America under a towering mass of ice 20,000 years ago.

To restabilize the planet's temperature, the heat added must be reduced, the heat leaving must be increased, or both. First I describe how to increase the amount of heat leaving.

The amount of radiation reflected back into outer space can be improved by increasing the *albedo* (reflectivity) of the planet's surface by making the human "footprint" lighter. This can be achieved by increasing the albedo of our roofs, roads, and agriculture. Assuming that humankind can increase the approximately 30% reflectivity of Earth's surface (an albedo of 0.3) to about 32%, that would remove enough heat to eliminate GW. Unfortunately, this potential is not being exploited, and in fact, Earth's albedo is dropping because highly reflective snow and ice are melting and are being replaced by highly heat-absorbing water and land. In addition, this process is self-accelerating because melting increases GW and increased GW further increases melting.

GW is caused by the sum of two main subprocesses, one occurring in the oceans and the other occurring in the atmosphere. These subprocesses respond at different speeds because the mass of the oceans is over a hundred times greater than that of the atmosphere. As a result, atmospheric processes respond more quickly to an upset in the planet's heat balance than oceans. Consequently, in some locations, the atmosphere has already warmed by some 4°C to 5°C (7°F to 9°F), which is causing melting in the Arctic. At the same time, other areas are experiencing severe droughts, heavy rains, wildfires, heat waves, water shortages, hurricanes and flooding causing coastal water surges, and wildlife destruction. In contrast, changes in oceanic processes, such as increases in thermal expansion and CO_2 degassing or alterations in ocean currents, El Niño cycles, or the Pacific Decadal Oscillation (PDO), are just beginning.

In addition to GW being a complex process, the cooling method is a difficult task. This is because human behavior must be changed to increase cooling. Changing human

2. "Global Temperatures," NASA Earth Observatory website, accessed December 1, 2022, https://earthobservatory.nasa.gov/world-of-change/decadaltemp.php.

behavior is slow and difficult because almost the only manipulation people respond to is financial self-interest. As of 2022, strong financial motivation for changing behavior to combat climate change does not yet exist.

In this and the following two chapters, I discuss the GW process, including its many subprocesses. In Chapter 3, I discuss the various potential sources of available green energy. In Chapter 4, I describe the design and control of the various methods of storing and transporting green energy.

GW is a controversial topic because it conflicts with the interests of the fossil-fuel and nuclear industries and because different groups favor different approaches. For example, the utility industries favor investing in updating the aging centralized grid systems, whereas others favor investing in local electricity distribution systems, which are less expensive, faster to build, more efficient, and better suited for using intermittent energy sources (such as solar and wind).

The information and analysis presented here are from the perspective of a process control engineer who is approaching the analysis of this nonindustrial climate change process as he would that of an industrial process. The analysis focuses on the dynamic and interacting nature of the GW process and aims to predict the dates by which its various symptoms are likely to evolve fully, particularly focusing on the tipping points at which marked accelerations are likely to occur. The behavior of the GW process is evaluated from two viewpoints: (1) assuming that it remains out of control and (2) that the required steps will be taken to convert our energy economy to a carbon-free one by 2050.

2.1 Introduction

The carbon footprint of the 7.9 billion humans on this planet has caused the release of enormous amounts of greenhouse gases (GHGs), mostly carbon, into the air and oceans. About half of the emitted carbon stays in the air and warms the planet through the greenhouse effect, and the other half is dissolved in the oceans, causing acidification, and is absorbed by vegetation. The main cause of carbon emissions is the burning of fossil fuels, but there are many others, some of which are not yet active and will be triggered as GW continues,[3] such as deforestation, wildfires, loss of agricultural production, oceans degassing, and melting permafrost.

---

3. The role that burning fossil fuels plays in GW is common knowledge, whereas several other factors also contribute to the positive feedback that further increases GW. Among these factors are permafrost melting (permafrost contains twice as much carbon as the atmosphere plus large quantities of methane), ocean degassing, destruction of rain forests (and corals, which some refer to as the "rain forests of the sea"), animal waste, and traditional methods of tilling (some estimates indicate that 50 GTC has been removed from the soil and emitted into the air). Most GW models consider only the carbon released by burning fossil fuel and often disregard the other factors that I consider in my analysis of the GW process.

From the beginning of the industrial age (about the 1880s) until now, the CO_2 concentration in the atmosphere has risen from about 280 ppm to nearly 415 ppm, which corresponds to an increase in the weight of carbon in the atmosphere from 600 to 880 gigatons of carbon (GTC). The situation is worse because annual carbon emissions are still increasing. In 2018, the yearly carbon emissions amounted to 8.75 GT. This corresponds to total carbon dioxide equivalent (CO_2e) emissions of about 32.1 GT (the weight of 100,000 Empire State Buildings). This also means that about 3.5 kg (18 lb) of CO_2 is emitted daily by every man, woman, and child on this planet. In 2021, the emissions of CO_2e amounted to about 36 $GTCO_2e$ (about 10 GTC), as its concentration increased by 2.75 ppm from the previous year. From a control perspective, this data describes a process that is not only out of control but is in a runaway state.

Consider the atmosphere as a tank and the CO_2 concentration as the level (415 ppm) in that tank. If the goal were to bring this concentration down to where it was before the industrial age (280 ppm), a process control engineer would immediately close the inlet and open the outlet valve on that tank, thereby removing the excess carbon until the level had dropped to 280 ppm. From that point on, the engineer would control the system so that inflows and outflows remain equal.

This is what we would do in industry, but the atmosphere is not a tank and we do not (yet) know how to remove CO_2 from it. As the CO_2 concentration in the air increases, the oceans dissolve more of it; however, this dissolution process is balanced to some extent by degassing because the warmer the water, the less CO_2 it can dissolve. Moreover, acidification of the oceans kills the coral reefs that protect the seashore[4] and serve to absorb CO_2 (similar to the rain forests). The net effect is an increase in the CO_2 concentration in the air.

This analysis sounds logical, except that it does not consider the time constants of the two processes. CO_2 dissolution is relatively fast; ocean warming is slower. In other words, in the future, when the CO_2 concentration in the air has already stabilized (dissolution is no longer increasing), warming of the water, and therefore degassing, will continue. If this proves to be the case, then as the temperature rises, the oceans will start releasing more and more CO_2 in a positive feedback cycle. (An increase in CO_2 causes the ocean temperature to rise; a rise in temperature reduces the solubility of CO_2; the reduced solubility increases degassing, which, in turn, increases the CO_2 even further; and the cycle continues.)

---

4. According to research at Stanford University, coral reefs provide protection from storms and rising sea levels. By breaking up waves, coral reefs protect an estimated 200 million people from natural disasters and rising sea levels.

Figure 2-1. According to some estimates, 90% of the coral reefs will die by 2100.

As shown in this and the coming chapters, there are positive feedback effects in the GW process that can push the process to a tipping point. In addition, for example, the acidification of the oceans also decreases the oxygen content of the waters, which has already fallen by 2% in the past 50 years. The condition of our oceans is also harmed by the accumulation of human waste, including plastics. According to some estimates, by 2030, there will be more plastic[5] in the oceans than fish. Some estimates of the damage caused by plastics are even worse.

The dissolution of CO_2 causes acidification of the oceans. Today, the pH of the oceans is the lowest it has been in the past 2 million years (pH 8.1), and it is expected to drop further and reach pH 7.8 by 2100. (Lower pH corresponds to higher acidity.) This acidification harms ocean life and devastates coral reefs (Figure 2-1) because carbonic acid is formed when CO_2 dissolves in the ocean's water, impeding the production of the calcium carbonate needed to build the coral exoskeleton (shell). About 25% of the coral reefs have already died, and that percentage is estimated to reach 90% by 2100.[6] Just in the past 3 years, one-fifth of all corals have died, and scientists report that we have killed nearly half of all coral reefs during the industrial age. Coral bleaching[7] also occurs because of increased solar irradiation and chemical pollution (e.g., about 14,000 metric tons of sunscreen are deposited in the oceans yearly).

---

5. This is particularly unfortunate because we have the technology to produce biodegradable plastics from agricultural products.

6. "Coral Reefs," Kids Discover Online website, accessed March 8, 2022, https://online.kidsdiscover.com/unit/ecosystems/topic/coral-reefs?ReturnUrl=/unit/ecosystems/topic/coral-reef.

7. "When corals are stressed by changes in conditions such as temperature, light, or nutrients, they expel the symbiotic algae living in their tissues, causing them to turn completely white." They are not dead and can recover but they are more susceptible to disease. "What is coral bleaching?," NOAA website, accessed December 7, 2022, https://oceanservice.noaa.gov/facts/coral_bleach.html.

As discussed in more detail in Section 2.2.5, "Tipping Points," as GW increases, it is likely to reach and pass several tipping points that could cause the rate of temperature rise to increase sharply. Some of these are as follows:

1. The point at which the ocean's CO_2 solubility drops, causing a reduction in absorption and the release of immense quantities of CO_2
2. When the global albedo (reflectivity) drops because the highly reflective ice and snow melt and low-reflectivity ocean water and permafrost surfaces replace them
3. When permafrost starts to melt and release CO_2 and immense amounts of methane, which has a greenhouse effect that is 25 times greater than that of CO_2
4. When (if) the lower clouds reflect more energy back to Earth (greenhouse effect) than the upper clouds reflect back into space

2.2 Cycle of Life on Earth

Earth is the only planet we know of that has an atmosphere containing more than a trace amount of oxygen. Earth's atmosphere protects the planet from meteorites and from becoming an ice ball. In addition, the oxygen in Earth's atmosphere appears to be one of the keys to life itself. Scientists concluded that oxygen was first made on Earth about 2.7 billion years ago by symbiotic cyanobacteria (blue-green algae). These microbes conduct photosynthesis, using sunshine, water, and CO_2 to produce carbohydrates and oxygen.

The other key factor in sustaining life is Earth's magnetic field,[8] which reaches above the ionosphere and keeps the atmosphere in place by protecting the planet from the charged particles of the solar wind and cosmic rays that would otherwise strip away the upper atmosphere, including the ozone layer that protects the Earth from harmful ultraviolet (UV) radiation. The atmosphere also thermally insulates the planet, preventing its surface temperature from dropping to extreme lows during the night or rising to extreme highs during the day, as occurs on the Moon. The temperature is maintained in a life-supporting range, which guarantees that water remains present in all three states (solid, liquid, and vapor). In addition, in this ingenious design, the global temperature is probably also stabilized by the effect that a warmer body reflects more heat into outer space than a cooler body. However, a contradictory argument also

8. Earth's molten and rotating iron core generates a magnetic field that is maintained over astronomical time scales. This magnetic field is essential to keeping the atmosphere in place and protecting our planet from harmful UV and X-ray solar radiation.

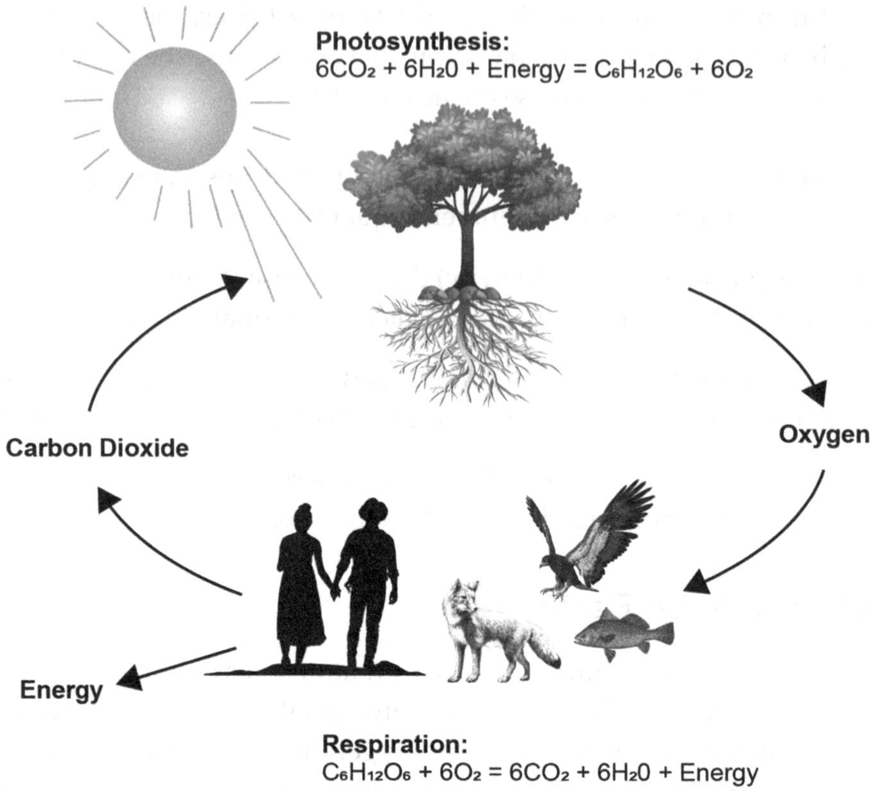

Figure 2-2. The cycle of life on Earth.

exists, which says that ice melts as the planet warms, causing the planet's albedo (heat reflection) to drop. My view that the first effect is stronger is supported by the fact that, during the past 1 million years, Earth's average temperature has never dropped below 5°C or increased above 20°C.

Next, let us consider how the whole cycle of life operates on our planet. An oversimplified summary of this cycle can be stated as follows: Plants use energy from the Sun to combine carbon from the CO_2 in the air with hydrogen from the water in the ground to produce carbohydrates (glucose, cellulose, etc.) while releasing oxygen back into the air (photosynthesis shown at the top of Figure 2-2). Animals and humans inhale oxygen and eat the food (carbohydrates), which they oxidize to obtain energy while exhaling the combustion products CO_2 and water (respiration shown at the bottom of Figure 2-2).

For millions of years, the vegetation on Earth "inhaled" the CO_2 the animals exhaled, and despite ice ages, meteorite impacts, volcanic eruptions, and so on, the CO_2 concentration in the air remained stable, between 180 and 280 ppm. For the past million years, it has never exceeded 280 ppm! Then came the Industrial Revolution, when humans started to burn the fossil-fuel deposits on Earth and the CO_2 concentration of the air started to rise, increasing the carbon content of the air by some 280 GTC (1030 $GTCO_2$).

Today, it is over 410 ppm (corresponding to about 880 GTC [3,230 $GTCO_2e$] in the atmosphere) and still rising. Over the past 20 years, it increased at a rate of about 2.0 ppm/year; in 2019, it increased at an even higher rate of 2.75 ppm/year. The rise slowed during the pandemic, but by 2022 the rate was increasing again.

2.2.1 Process Characteristics

One of the goals of this book is to analyze the nature of GW using only factual data that are scientifically verified. Naturally, such data is available only to document the trends of the past, therefore, the future behavior of the GW process can only be estimated.

I am only an engineer, not a climate scientist, so I have only a simple overall view of the relationship between insolation (the rate at which solar energy is received in a given area, usually given in units of W/m^2) and the changes in insolation as the Earth's distance from the Sun varies due to the elliptical shape of its path around the Sun and the changes in the angle of Earth's axis of rotation. (Milankovitch cycles). Insolation varies with the distance of the Earth to the Sun, and that distance varies with the tilt of the Earth's axis of rotation relative to the orbital plane, the motion of the axis through time, and the elliptical shape of Earth's orbit. The interactions and feedback effects among these movements result in ice ages every 100,000 years or so (Figure 2-3).

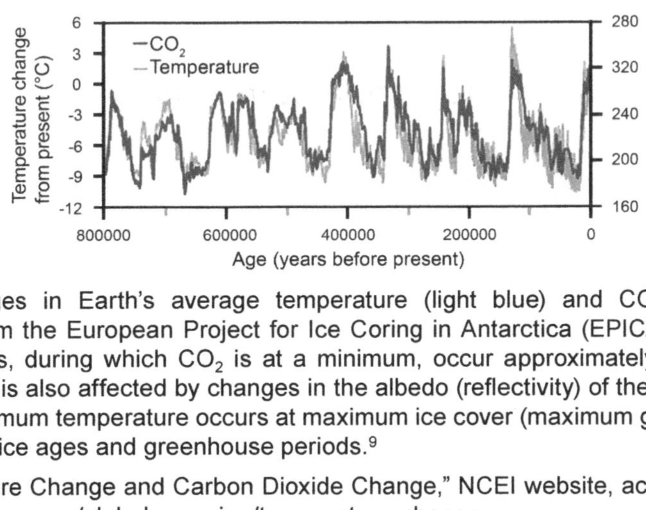

Figure 2-3. Changes in Earth's average temperature (light blue) and CO_2 concentration (dark blue) measured from the European Project for Ice Coring in Antarctica (EPICA) Dome C ice core in Antarctica. Ice ages, during which CO_2 is at a minimum, occur approximately every 100,000 years. Global temperature is also affected by changes in the albedo (reflectivity) of the planet due to changes in its ice cover. Minimum temperature occurs at maximum ice cover (maximum glacification), so climate alternates between ice ages and greenhouse periods.[9]

Source: "Temperature Change and Carbon Dioxide Change," NCEI website, accessed March 8, 2022, https://www.ncdc.noaa.gov/global-warming/temperature-change.

9. Source data is from the following studies: J. Jouzel, V. Masson-Delmotte, O. Cattani, G. Dreyfus, S. Falourd, G. Hoffmann, B. Minster, J. Nouet, J. M. Barnola, J. Chappellaz, H. Fischer, J. C. Gallet, S. Johnsen, M. Leuenberger, L. Loulergue, D. Luethi, H. Oerter, F. Parrenin, G. Raisbeck, D. Raynaud, A. Schilt, J. Schwander, E. Selmo, R. Souchez, R. Spahni, B. Stauffer, J. P. Steffensen, B. Stenni, T. F. Stocker, J. L. Tison, M. Werner, and E. W. Wolff, "Orbital and Millennial Antarctic Climate Variability over the Past 800,000 Years," *Science* 317, no. 5839 (August 10, 2007): 793–797. D. Lüthi, M. Le Floch, B. Bereiter, T. Blunier, J. M. Barnola, U. Siegenthaler, D. Raynaud, J. Jouzel, H. Fischer, K. Kawamura, and T. F. Stocker, "High-resolution carbon dioxide concentration record 650,000–800,000 years before present," *Nature* 453 (May 15, 2008): 379–382.

The details are complicated, but the bottom line is that we are currently nearing the end of the last cycle (Figure 2-4); the global temperature is rising. After previous ice ages, this rise has never exceeded about 3°C (5.4°F), and as the figure shows, as of 2020, the rise has already passed 1°C (1.8°F).

GW is a lag-dominant process because the ocean and land masses of the planet are immense, and so are their heat storage capacity and inertia. Still, the principles that define its behavior are very similar to those of any other process, including industrial ones. Therefore, the analysis of the GW process will show not only that its time constants are long but also that the consequences of these slowly evolving conditions are vast and hard to stop because of the mass and inertia of the planet, and the accumulated CO_2 stays in the atmosphere for a long time. In fact, the global temperature will continue to rise even after we no longer increase the amount of carbon in the atmosphere because the previously accumulated GHGs are still there.

The nature of this process differs from the nature of industrial heat-balance processes because both its "set point" and its "manipulated variable" are determined by the same factor; they are up to humankind. The set point (target) is our collective decision on the acceptable maximum GW, and the manipulated variable is the required change in humankind's behavior to reach that desired goal.

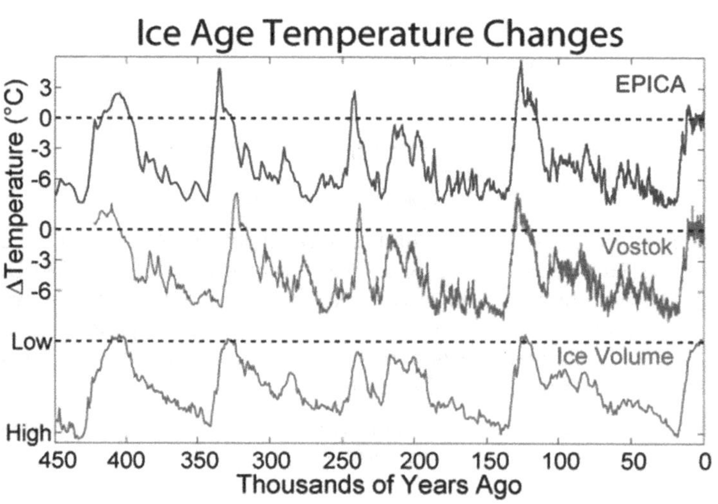

Figure 2-4. Global temperature rise measured in ice cores taken by the European Project for Ice Coring in Antarctica (EPICA) (blue) and at Vostok Station (green), both in East Antarctica, showing that the maximum GW was about 3°C (5.4°F) in the past 450,000 years.

Source: "Ice Age Temperature.png" by Robert A. Rohde is licensed under CC BY-SA 3.0, https://commons.wikimedia.org/w/index.php?curid=466270, accessed March 8, 2022.

Controlling industrial processes is easy because the measurements are accurate, the set points are undisputed, and the manipulated and controlled variables are different. In the case of GW, the measurements (information concerning the actual global conditions) can be misunderstood or falsified, and the set points of the model-based control (MBC) algorithms (the targets for the upper limit of GW) are debated. Until now, the net effect of all of these factors has been overall inaction because the uninformed and confused humankind remained idle.

On one hand, no one among the 7.9 billion of us wants a more polluted and hotter planet, and everyone wants cheap and abundant energy supplies to support our modern lifestyle. This is the conflict that opens the door to confusion and inaction. The confusion arises because, in the case of "GW denialists" (those who refuse to accept the facts about GW, even though they are supported by the majority of scientific and historical evidence) might say that we should not worry about warming because it was "only" 0.12°C (0.22°F) per decade during the past 50 years. An average person hearing that number tends to agree that it sounds insignificant. On the other hand, if that same change is stated in terms of heat energy that "equals the heat release of five Hiroshima bombs every second,"[10] or if our emission was described as equaling the weight of five Empire State Buildings every hour, the average person would consider it significant.

It should also be noted that, during the 5 years from 2015 to 2020, the rate of temperature increase reached 0.2°C/decade (0.36°F/decade) and is continuing to rise. The total temperature increase that occurred during the industrial age (also called *climate forcing* or *climate anomaly*) has already exceeded 1.0°C (2.0°F)[11] (Figure 2-5).

We can better visualize the GW process by considering the similarities between controlling it and controlling the temperature in a hot water tank. For both processes, we can determine their dead times (the time period before they start responding), their lag times (the time period it takes for the temperature to start to respond to a change in the heat input), and their gains (the rate at which the systems respond after the dead

---

10. Ron Brackett, "World's Oceans Warming at Rate Equal to Five Hiroshima Bombs Dropping Every Second," The Weather Channel website, January 14, 2020, accessed December 1, 2022, https://weather.com/science/environment/news/2020-01-14-world-oceans-warming-five-hiroshima-bombs-a-second.

11. "Global Temperature," NASA Global Climate Change website, accessed March 8, 2022, https://climate.nasa.gov/vital-signs/global-temperature/. "State of the Climate Summary Information," National Centers for Environmental Information website, December 2014, accessed December 1, 2022, https://www.ncdc.noaa.gov/sotc/summary-info/global/201412.

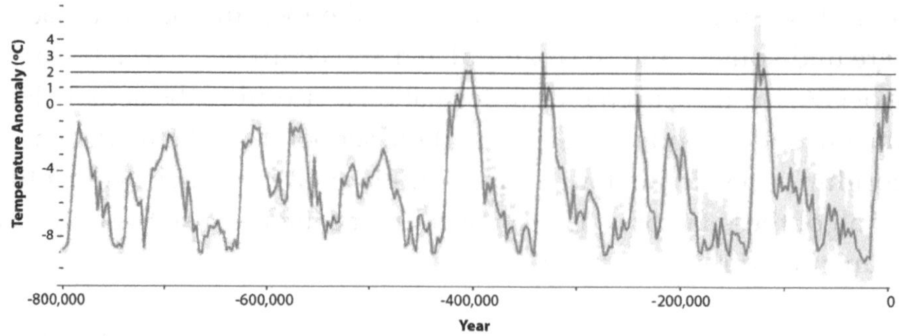

Figure 2-5. NASA prepared this 800,000-year record of the global temperature anomaly by analyzing air bubbles trapped in glacial ice.

Source: "How is Today's Warming Different from the Past?," NASA Earth Observatory, published June 3, 2010, https://earthobservatory.nasa.gov/features/GlobalWarming/page3.php.

time has passed). We can also determine all their other dynamic characteristics. For example, we can measure their temperature responses to changes in the thickness of the thermal insulation of the tank.

We can regard GHGs as the insulation of our planet. GHGs reflect solar energy that would otherwise escape into space back to Earth. Some estimates suggest that if there were no GHGs in the air, the global temperature would be about –18°C (0.0°F), and if the atmosphere consisted of only GHGs, the temperature could exceed 400°C (752°F). During the past century, global temperature has increased by only about 1.2°C (2.36°F); today,[12] it is still rising.[13] The analysis in this book will show that if the present inaction continues, the GW rise[14] will reach or exceed the 1.5°C (2.7°F) limit by 2030.

Because the atmosphere already contains 880 GTC (3,230 $GTCO_2$) and our emissions add nearly 10 GTC (37 $GTCO_2$) each year, the planet's temperature will continue to increase even after we reduce our emissions to net zero. The temperature will not start dropping until we become carbon-neutral and start removing the CO_2 in the atmosphere or start cooling the planet in some other way. To put the yearly carbon emissions

---

12. The size of the temperature rise depends on how the reference line is selected. If the reference is taken as the average temperature between 1880 and 1889, then (according to the NASA report published in the *New York Times* on February 7, 2019) the rise in 2016 reached 1.2°C, and the rise in 2018 was about 1.1°C.

13. Kendra Pierre-Louis, "Ocean Warming Is Accelerating Faster Than Thought, New Research Finds," *New York Times*, January 10, 2019, https://www.nytimes.com/2019/01/10/climate/ocean-warming-climate-change.html.

14. Scientific calculations indicate that, if all fossil fuels on Earth were burned, the temperature rise would reach 8.3°C (17°F) by the year 2300. Such a rise has occurred only once, some 65 million years ago, when, because of the impact of an asteroid in the Yucatán Peninsula in Mexico, the CO_2 concentration in the air increased sharply, and with it, the global temperature increased by some 10°C (18°F), causing the destruction of 96% of all species on the planet (including the dinosaurs).

of about 10 GTC into perspective, note that, for the current global population of 7.9 billion, this corresponds to daily carbon emissions per person of 3.3 kg (12 kg of CO_2).

2.2.2 Material Balance

To return the planet to preindustrial conditions—using the hot water tank analogy—we would need not only to stop increasing the thickness of the thermal insulation on the tank every year (equivalent to yearly emissions) but also remove all of the "insulation" that accumulated during the previous century.

Increasing the outflow from the tank (removing carbon from the air) is difficult but not impossible. For example, trees can do that temporarily because the CO_2 they absorb during growth is released when they are cut down some decades later. Therefore, tree planting is actually a carbon-neutral process in the long run. We could reduce the global temperature by reflecting more solar radiation into space, growing plants in salt water, or storing CO_2 underground, but such efforts have not been attempted at scale.

We can also reduce our carbon emissions (the inflow into the tank). For example, we can convert our throw-away economy,[15] improve public transportation, increase the use of electric cars, better insulate our homes, use smart thermostats and appliances, minimize animal products in our diet, convert to no-till farming,[16] and introduce a carbon tax (not a cap-and-trade system, but an actual tax). Unfortunately, few of these steps have been undertaken. Some people also consider direct carbon capture from the air,[17] but this does not seem to be realistic because the CO_2 concentration in the air is not high enough for underground storage to be practical. Instead of storing CO_2, a more realistic method is to reduce CO_2 emissions at their source (i.e., the power plant), where the concentration is high, as suggested by the Nobel Prize-winning chemist George Oláh.[18]

The only approach that would have a real impact on reducing carbon emissions is to stop burning fossil fuels and convert to using carbon-free "green" energy sources.

---

15. By this term, I am referring not only to the plastic bags and packaging or food waste but also to the overall manufacturing philosophy of industry, which is based on making products that have short life spans and are hard or expensive to recycle. This "throw it away and buy a new one" attitude applies to everything from mobile phones to washing machines.

16. No-till farming (also called zero tillage or direct tilling) is a way of growing crops or managing pastures from year to year without disturbing the soil through tillage. Reduced tillage increases the amount and variety of life in and on the soil. The most powerful benefit of no-tillage farming is the improvement in the biological fertility of the soil, which makes it more resilient.

17. "World's largest plant capturing carbon from air starts in Iceland," Reuters website, September 13, 2021, accessed December 1, 2022, https://www.reuters.com/business/environment/worlds-largest-plant-capturing-carbon-air-starts-iceland-2021-09-08/.

18. Kevin Bullis, "The Methanol Economy," MIT Technology Review, March 9, 2006, accessed December 1, 2022, https://www.technologyreview.com/2006/03/09/229450/the-methanol-economy/.

This can make the planet carbon-neutral, a state in which the CO_2 concentration in the atmosphere (the thermal insulation of the tank) is no longer increasing. However, even that will not remove the accumulated excess carbon that is already in the atmosphere. To do that, we need new technologies.

In other words, even after we stop burning fossil fuels, the planet will *not* start cooling; it will only reach a new steady-state temperature, one that is higher than the one that existed prior to the industrial age. This will be the case even if the Paris Agreement goals of limiting the temperature rise to 1.5°C (2.7°F) (or at most 2.0°C [3.6°F]) and the United Nations (UN) target of cutting emissions by 7.6% per year (about 0.8 GTC [3 GTCO2e] per year) by 2030 are both met, and if the European Union and American goals of reaching carbon neutrality by 2050 are met. (For a summary of GW-related targets and goals, see Chapter 6.)

2.2.3 Global Heat Balance

I begin my description of the overall GW process by describing the heat balance and the heat-transfer processes at work on our planet. Our planet is surrounded by space, which is at a temperature of –270°C (–454°F). Earth is protected from that cold by an atmosphere, the greenhouse effect of which (much like our clothing in the winter) protects Earth from excessive cooling (Figure 2-6). If we did not have an atmosphere, Earth's surface temperature would be similar to that of the Moon. When sunlight shines, the temperature on the Moon's surface can reach +127°C (+260°F), and when the Sun goes down, the temperature can dip to –173°C (–280°F).

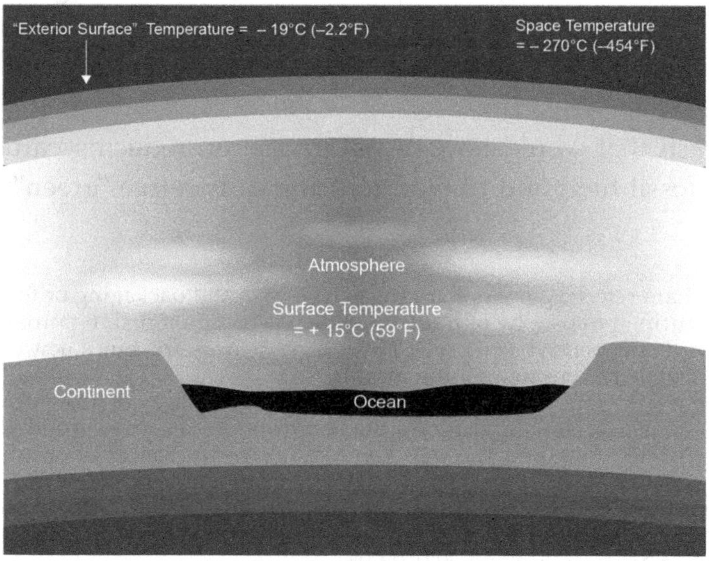

Figure 2-6. Space surrounding Earth is very cold, yet the surface of our planet is at a livable average temperature because of the warming effect of its atmosphere.

2.2.4 Trends in Temperature Rise

So far, we have seen that GW is potentially a runaway process. Now, let us discuss its interactive nature. The nature of this interaction is such that an increasing CO_2 concentration increases the global temperature, and an increasing temperature further increases CO_2 concentration through such processes as wildfires and permafrost melting. Another form of interaction has to do with the fact that the more stress humankind applies to its environment, the more inhospitable the environment becomes. Some call this phenomenon the interactive feedback effect.

During the past two centuries, the global population increased 7- to 8-fold, and the per capita energy consumption rose about 5-fold. The production of goods increased 20-fold (and, without recycling, these goods end up in the environment as garbage). Because of these increases, we spend 10% of our gross domestic product (GDP) on energy, about 80% of which comes from fossil-fuel sources. Mostly, this is coal, but we are also burning 12.5 GT of oil (having a heating capacity of 3.5×10^{17} Btu) and a rapidly increasing quantity of natural gas every year. Figure 2-7 depicts some temperature rise predictions for the coming decades depending on our actions regarding GW. (Figure 6-10 in Chapter 6 provides similar data.)

Converting to using green energy sources can stop GW, and we have abundant solar energy. The average solar energy received on Earth is about 5 to 6 million quads per year (1 Q = 10^{15} Btu), whereas our yearly global energy consumption is about 600 Q. In other words, our total energy consumption corresponds to about 0.01% of the yearly global insolation. This much solar energy can be collected on a small percentage of the area of the Sahara or by covering just the roofs of our buildings and our parking lots with solar panels.

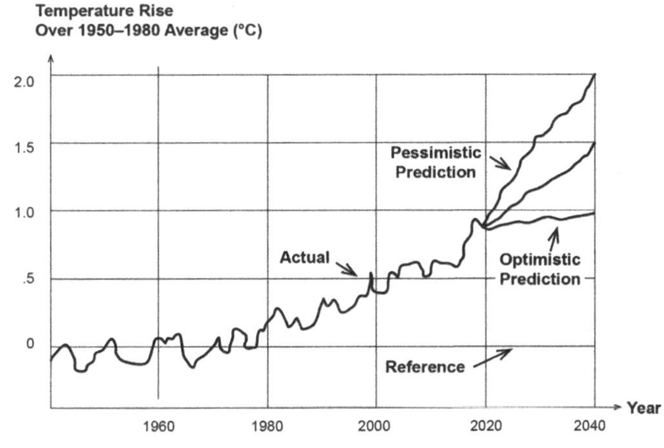

Figure 2-7. GW in the coming decades will depend on our actions today.

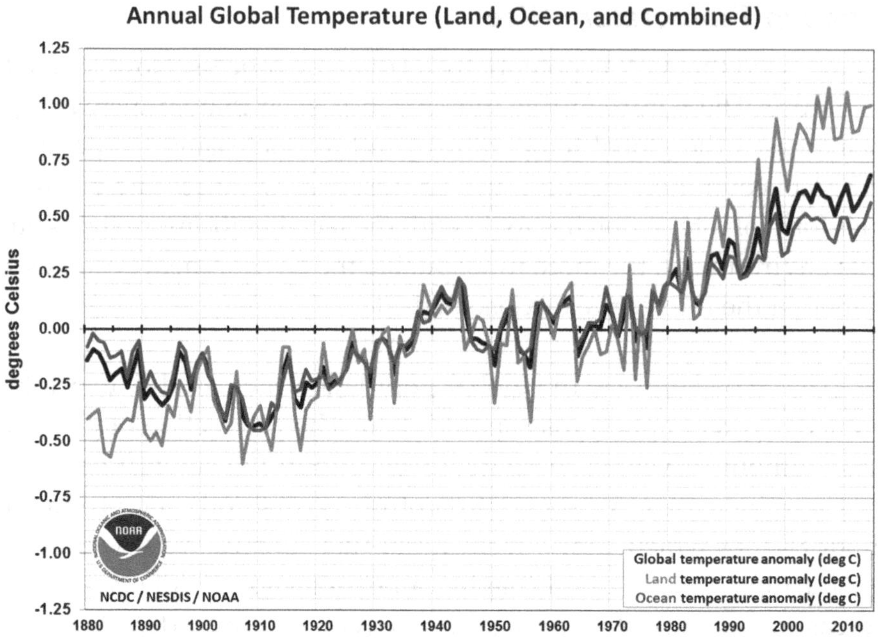

Figure 2-8. Increase in the average temperature anomaly of Earth between 1880 and 2010: (black) entire globe, (tan) land area, and (blue) oceans. During the past decade, GW increased even faster.

Source: "Global Climate Report – Annual 2014: Various Global Temperature Time Series, updated through 2014," NOAA/NESDIS/NASA, accessed March 8, 2022, https://www.ncdc.noaa.gov/sotc/global/2014/13/supplemental/page-4.

As GHGs such as methane (CH_4), nitrous oxide (N_2O), water vapor (H_2O), and carbon dioxide (CO_2) have accumulated in the atmosphere, they have reflected more and more energy back to Earth, thereby increasing the temperature of the surface of the planet (see Figure 2-8). The resulting temperature rise is not uniform. It varies by location and is generally higher on land and lower on the surface of the oceans because of the mass of the oceans.

Some 93% of the extra solar energy received by our planet due to the greenhouse effect has been absorbed by the oceans, 3% by the continents, 3% by the ice at the poles, and 1% by the atmosphere. Yet, because of their gigantic mass, the oceans have warmed more slowly; over the past half-century, they have absorbed nearly 100 times more heat than the atmosphere.

Naturally, switching from fossil fuels to green energy would reduce GW by reducing the greenhouse effect. We do not know the precise time constants, gains, and reaction rates of this heat-transfer process, but we do know that these time constants are long and the reactions are slow in such large-capacity, lag-dominant processes because they have large dead times and delayed responses. We also know that positive feedback from other processes can accelerate with time. For example, the dynamics of this process will change dramatically as the ice cover melts at the poles. However, according to climate scientists, complete melting has never happened during the past million years.

2.2.5 Tipping Points

The *feedback effect* is very familiar to process control engineers, but it is not always well understood by others. A positive feedback effect is when a change in variable A causes a change in variable B, and this change in variable B triggers a further change in variable A, and so on. In GW, when the rise in temperature reaches certain points called *tipping points*, feedback processes are triggered that initiate feedback effects that further increase the rate of temperature rise.

Climate drivers of significance during the industrial era include those associated with anthropogenic activity (caused directly by human activities) and, to a lesser extent, those of natural origin, such as changes in solar irradiance, volcanic eruptions, and the El Niño–Southern Oscillation. Human activities that affect Earth's climate generally result in changes in the concentrations of GHGs and small airborne particles (aerosols) in the atmosphere and changes in the reflectivity of Earth's surface. The increase in radiative forcing (RF, the imbalance between the heat received and the heat leaving Earth) due to these activities has far exceeded the relatively small net increase due to natural factors, which include changes in the amount of energy received from the Sun and the cooling effect of volcanic eruptions.

Some of the tipping points that are approaching are self-accelerating, and the sizes of their impacts approach the effects of human emissions. Such processes are triggered by the following phenomena:

- Ice melting reduces Earth's albedo by uncovering land and water surfaces.
- Permafrost melting releases GHGs, including methane.
- Ocean warming lowers CO_2 solubility, causing degassing emissions.
- Ocean evaporation increases humidity (by 7%/1°C [4%/1.8°F]), which increases the greenhouse effect and reduces Earth's albedo.

If we stabilize the anthropogenic climate drivers that are currently increasing the RF of the atmosphere, Earth's energy balance and climate will eventually reach a new steady state, an equilibrium in which equal amounts of energy enter and leave the planet. When and at what global temperature this will occur is up to us.

In the remainder of this section, I describe in more detail some of the tipping points in the GW process. Some scientists predict that reaching any of these tipping points can possibly cause the GW process to become irreversible, whereas others believe that they will only cause a faster increase in temperature until a self-stabilized higher value is reached. I am not knowledgeable enough to quantify the combined consequences of the positive feedback effects that occur at the different tipping points. However, I can say with a fair amount of certainty that the various prediction models tend to underestimate both the size and the speed at which tipping points are approaching.

2.2.5.1 Albedo Effect

The fight against GW can be fought on two fronts, but as of now, we are fighting on only one. That is, we are trying to reduce heating, but we are doing nothing about increasing the cooling of the planet. Cooling would occur if we increased the amount of heat leaving the planet by increasing Earth's reflection of the solar radiation back into space. It can be achieved by increasing the albedo of the planet.

The name albedo comes from Latin and means "whiteness." The albedo value expresses the portion of the incoming radiation that is reflected back into space. The albedo of a completely white surface is 1.0 (100% reflection), and that of a completely black surface is 0.0 (0% reflection, total absorption). Today, Earth's albedo is about 0.30, meaning that the average reflection from the total surface in a year is about 30% of the incoming radiation. That amount of reflection can be increased. The reflection of asphalt, for example, is 5% to 10%, whereas that of fresh snow is about 80%. Therefore, we can cool the planet if, for example, we increase the whiteness of the asphalt we use.

Let us take a look at the numbers. As shown in Figure 2-9, the total incoming shortwave radiation received from the Sun is 340 W/m^2. Of that, 29% to 30% is reflected back into space. Of the amount reflected back into space, most (22% to 23%) is reflected by the atmosphere and the clouds, whereas 7% is reflected by the land and ocean surfaces. As of now, we have no means to affect the size of the reflection by the oceans or the atmosphere, but we could change the albedo of Earth's land surface. Specifically, we could increase the current 7% (23.8 W/m^2) reflected back into space by increasing the albedo of the land surface, and by doing so, we could cool the Earth.

Today, GW is actually decreasing Earth's albedo. This is because the rising temperature melts snow and ice as the globe is warming, and their surfaces are replaced by water or land. Therefore, the reflection from these surfaces drops from 80% for fresh snow to less than 10% for water or wet soil. The melting reduces reflection, increases heat absorption, and causes the global temperature to rise.

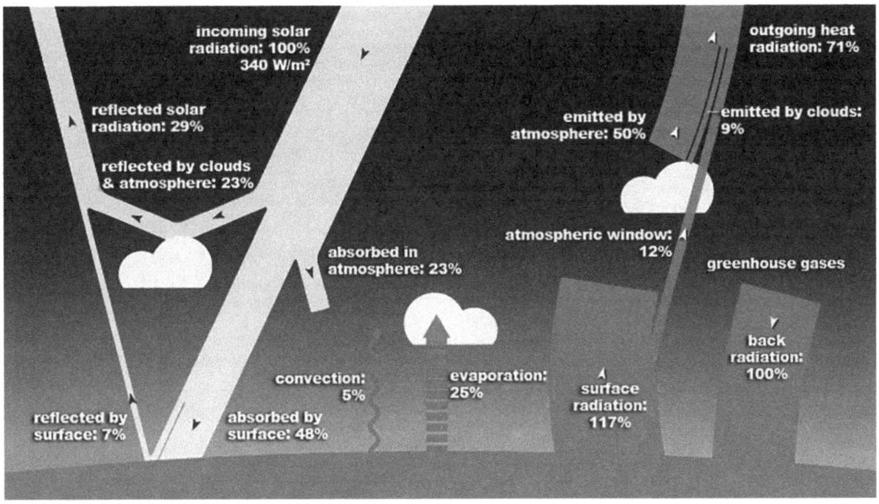

Figure 2-9. By increasing the heat that is reflected back into space (by increasing the albedo, the whitening of the surface of the planet), we could cool Earth.

Source: "Diagram showing the Earth's energy budget, which includes the greenhouse effect (NASA).png" by Robert Simmon, from "Climate and Earth's Energy Budget" by R. Lindsey, NASA Earth Observatory, Public Domain, January 14, 2009, https://commons.wikimedia.org/w/index.php?curid=22474641.

Note that GW in other locations can temporarily increase reflection. For example, increased temperatures also result in the drying of the sand in the deserts of the tropics, which increases their reflection from about 10% to 25% until the sands are completely dry.

Let us do a crude calculation to get a sense of the size of the impact of changing the albedo. Today, the total land reflection of the planet is about 7% (23.8 W/m^2), and humanity is present on about 10% of Earth's surface. Therefore, if we want to generate a cooling effect of 1.0 W/m^2 (from 23.8 to 24.8 W/m^2), we have to increase the reflection from the populated areas by 10 W/m^2 or to about 33.8 W/m^2. Considering that the present albedo of the planet is about 0.3 (23.8 W/m^2), increasing that value to 33.8 W/m^2 would require an increase in the albedo of the human-populated area by about 13% (from 30% to 43%). It seems to me that increasing the reflectivity (whiteness) of the surfaces of our roofs, roads, parking lots, and agricultural land (by planting lighter-colored crops) could accomplish most of that change in the global albedo. The advantage of this method of reducing GW is that it is relatively inexpensive and can be done relatively quickly compared to eliminating or reducing carbon emissions, which will take many decades. Another advantage is that the total population of the planet can participate in this effort, which also creates jobs.

Now, consider the tipping point that can be caused by a continued drop in albedo. The Arctic is warming three to four times faster than the rest of the world. The highly reflective (up to 80%) snow and ice are melting and are being replaced by heat-absorbing

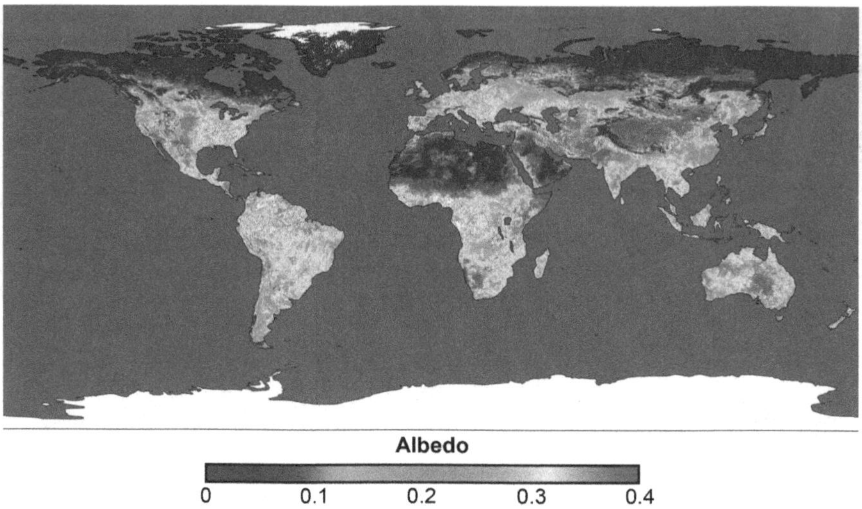

Figure 2-10. Values of the albedo (reflectivity) of the various surfaces on the planet according to the NASA Earth Observatory. Here, 0 means that no solar radiation is reflected back to outer space, and 0.4 means that 40% of the solar energy is reflected back by the ice and snow surfaces in the north and the dry sand in the Sahara. (See color images after page 306.)

Source: "Global Albedo," NASA Global Observatory, July 10, 2002, https://earthobservatory.nasa.gov/images/2599/global-albedo.

surfaces (ocean water and permafrost) that reflect only a small portion (less than 10%) of the incoming solar radiation. This causes a feedback effect because, as melting causes more solar energy to be absorbed, the temperature rises causing more ice to melt which further increases the amount of heat absorbed, and the cycle continues. The Arctic Ocean covers about 3% of the surface area of the planet, and if all of its ice cover melted, its albedo would drop from 0.8 to 0.1 (Table 6-3). This would reduce the overall reflection of the planet from the present 30% to about 29%, which could increase the present GW by more than 50%.

As mentioned earlier, the albedo of the oceans is less than 0.1 (10% reflection), whereas that of the fresh snow is up to 0.8 and that of sea ice varies from approximately 0.35 to 0.5.[19] Therefore, the albedo of Arctic surfaces would drop by at least 0.3 if the highly reflective polar and glacier ice melted and its surface was replaced by radiation-absorbing water or permafrost having an albedo of about 0.1. If this occurred, the heat input of the planet would rise, and the feedback cycle would accelerate. It seems to me that this effect is already working because the albedo is already dropping in the Arctic, and the Arctic is warming much more rapidly than the rest of the planet. This rise in GW will continue as long as the ice is melting on the planet, and for that reason, it could be approaching a tipping point (Figure 2-10).

19. "Thermodynamics: Albedo," National Snow and Ice Data Center website, last updated April 3, 2020, https://nsidc.org/cryosphere/seaice/processes/albedo.html.

It seems to me that one of the least expensive and fastest ways to cool the planet is by increasing its albedo. It would take only education and financial motivation, for example, to reduce the use of tilling in agriculture or to increase the use of cover crops, which, in addition to being more reflective, will also reduce carbon emissions and serve to retain both moisture and nutrients. Naturally, making our roofs and roads lighter in color[20] and any other steps that make the human presence on Earth more reflective, or whiter, serves the same purpose. In addition, doing this would provide some of the jobs needed to employ the parts of the workforce displaced by robotization.

2.2.5.2 Permafrost Effect

Permafrost is permanently frozen organic material below the surface of frozen land. Permafrost regions cover about 22 million square kilometers (about 24% of the exposed land surface) of the Northern Hemisphere[21] (Figures 2-11 and 2-12). Soils of the northern permafrost region contain about 1600 GT of organic carbon, nearly twice as much as currently contained in the atmosphere.[22]

Upon melting, the upper layers of the frozen land (permafrost) release CO_2, and the lower layers release methane in enormous quantities, where methane is a GHG that is

Figure 2-11. Permafrost currently covers about 24% of the land area of the Northern Hemisphere, which is about 3% of the land surface area of the planet.

20. The *New York Times* reported on September 25, 2022, that according to the DOE, if solar collectors collected solar energy on all American roofs and parking lots, the collected energy would meet 80% of the nation's electricity needs.

21. "SOTC: Permafrost and Frozen Ground," National Snow and Ice Data Center website, April 4, 2017, accessed March 11, 2022, https://nsidc.org/cryosphere/sotc/permafrost.html.

22. T. Schuur, "Permafrost and the Global Carbon Cycle," November 22, 2019, accessed March 11, 2022, https://arctic.noaa.gov/Report-Card/Report-Card-2019/ArtMID/7916/ArticleID/844/Permafrost-and-the-Global-Carbon-Cycle.

25 times more powerful[23] than CO_2. Some scientists estimate that the amount of carbon in the permafrost is greater than that contained in all of our fossil-fuel deposits. The effect of its melting would be greater than burning all of our fossil-fuel reserves, but such estimates are little more than guesses.

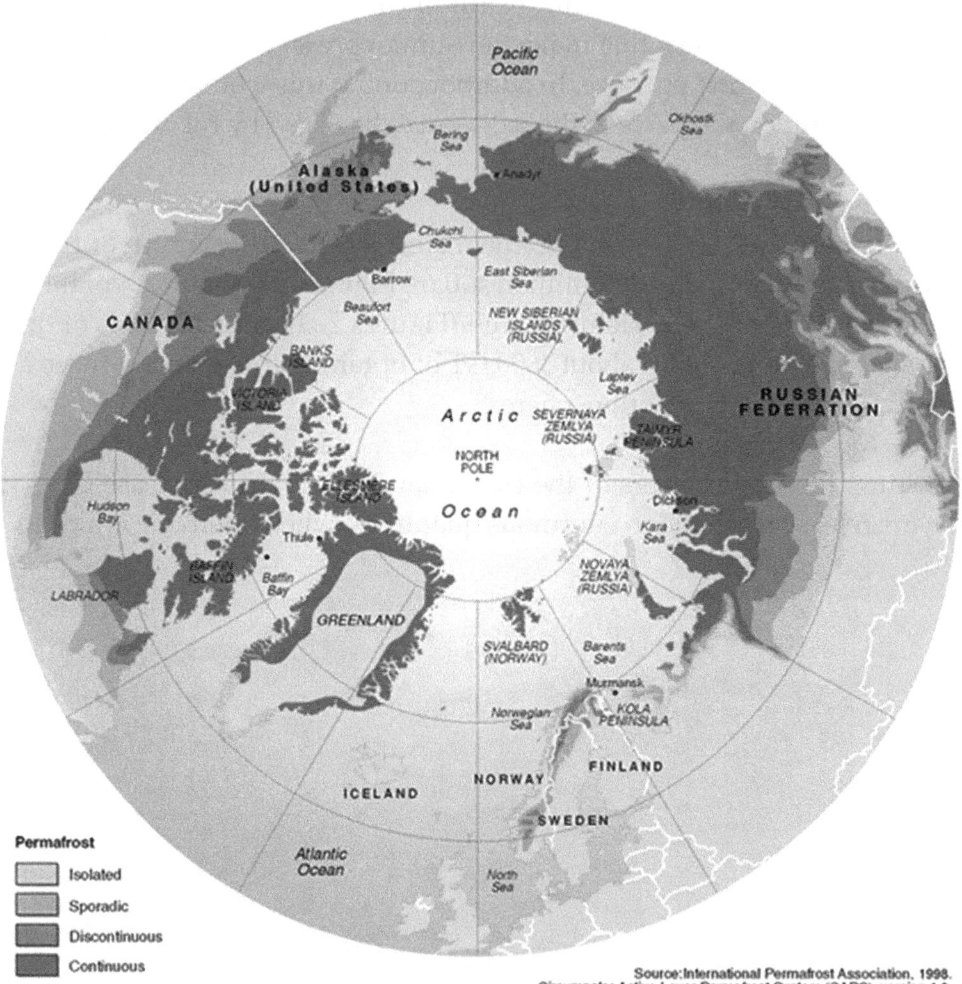

Figure 2-12. Permafrost covers about 24% of the land area in the Northern Hemisphere. According to the National Snow and Ice Data Center, it contains over 1500 GTC (5500 $GTCO_2$).

Source: "State of the Cryosphere: Permafrost and Frozen Ground," NASA National Snow and Ice Data Center, last updated April 4, 2017, https://nsidc.org/cryosphere/sotc/permafrost.html.[24]

---

23. "Overview of Greenhouse Gases," US EPA website, accessed March 8, 2022, https://www.epa.gov/ghgemissions/overview-greenhouse-gases. Methane's lifetime in the atmosphere is much shorter than that of carbon dioxide, but it is more efficient at trapping radiation than CO_2. Pound for pound, the comparative impact of CH_4 is 25 times greater than that of CO_2 over a 100-year period.

24. For additional information, see: Brian Resnick, "Melting permafrost in the Arctic is unlocking diseases and warping the landscape," *Vox*, updated Nov 15, 2019, accessed December 1, 2022, https://www.vox.com/2017/9/6/16062174/permafrost-melting.

The melting permafrost also triggers a feedback effect because, as it melts, it releases its GHG content, which causes more warming that, in turn, causes further melting. The permafrost has already started melting. This is demonstrated by the self-ignited subterranean gas explosions in Siberia, where the accumulated high-pressure gases explode as they find their way to the surface. The latest explosion occurred at Crater 17 on the Yamal Peninsula in the Russian Arctic, where the gas produced by partially decayed organic matter under the thawing layer broke through, leaving a 100-meter-deep crater.

In process control, such self-intensifying lag-dominant processes are usually modeled to cover a time period of about five times their lag time. As with the albedo effect, this permafrost effect has already started, and our GW projection models (including the UN representative concentration pathways—RCPs) should take it into account (Figure 2-12).

2.2.5.3 Degassing of the Oceans

Under normal conditions, the oceans absorb about 30% of all CO_2 emitted into the atmosphere. As the planet warms, degassing (increase in CO_2 release) occurs because the solubility of CO_2 in water decreases as the temperature increases. As the solubility of the water drops, the oceans absorb less CO_2 and release immense amounts of carbon into the atmosphere. This further increases the concentration of GHGs, so the temperature rises even higher. Some scientists estimate that this degassing can add 600 GTC (2,200 $GTCO_2e$) to the air per degree Celsius of temperature rise. If that value is accurate, each degree of GW would add twice the amount of carbon as was added throughout the industrial age.

I should mention that the increase in the GHG concentration in the air has a balancing effect on the positive feedback cycle. This is because, as the GHG concentration in the air rises, the amount of these gases absorbed by the oceans also rises. Because of this negative feedback, the net effect is somewhat debatable.

Our damage to the oceans also involves acidification. Normally, seawater is slightly basic (today, its pH value is about 8.1), and ocean acidification involves a shift toward neutrality (pH 7.0). Today, the pH of the oceans is the lowest it has been in a million years, and it is still falling. It is estimated that it will reach a pH of 7.8 by 2100. This has already caused a 30% loss of coral reefs, and this loss is estimated to reach 90% by 2100. This topic is discussed in more detail in Section 2.5.2.

Estimates of the damage to the oceans are highly unreliable. One study, for example, suggested that, by 2030, there will be more plastic in the oceans than fish, and a newspaper reported that the total weight of plastics in the world will exceed the weight of all ocean and land animals by 2050.

2.2.5.4 Water Cycle and Cloud Formation Effects

It is estimated that the volume of water in the atmosphere at any one time is about 3,100 cubic miles (mi^3) or 12,900 cubic kilometers (km^3), which is only about 0.001% of the total water on Earth. This quantity is rising because as global temperature rises, evaporation from the land and oceans increases. When warm air from the oceans is moved by the wind (or hurricanes) over the colder coastal land areas, the vapor condenses into clouds until the water falls and the rain completes the water cycle. About 90% of water in the atmosphere is produced by evaporation from water bodies, while the other 10% comes from transpiration from plants.

Because water vapor is a GHG, rising water vapor concentrations in the air also increase evaporation, further increasing the water vapor concentration in the atmosphere and, in turn, increasing GW.

Water vapor contributes about half of the Earth's greenhouse effect, yet water vapor does not control Earth's temperature; instead, the concentration of water vapor in the air is controlled by the temperature. This is because each 1°C (1.8°F) increase in temperature increases the water-vapor-holding capacity of the air by 7%. Today that rate of increase is about 1% to 2% per decade, and as the water vapor concentration increases, GW increases further.

The lifetime of water vapor in the atmosphere is about 2 weeks, whereas CO_2 remains in the atmosphere for centuries. Water vapor accounts for 36% to 66% of the greenhouse effect when the sky is clear and between 66% and 85% when it is cloudy. If the planet is at a constant temperature, humidity remains approximately constant because precipitation occurs when the air cools and evaporation occurs when the planet warms. Therefore, this water cycle normally does not contribute to GW.

When an increase in GHG concentration causes the planet's temperature to rise, the cycle of evaporation and condensation is intensified by a positive feedback loop because warmer air can carry more water vapor. When the water vapor concentration rises, it increases the greenhouse effect, which further increases GW, and the cycle continues as Earth warms. For example (according to one source[25]), if the CO_2 concentration doubled by rising to 560 ppm from the preindustrial level of 280 ppm, the water vapor would add another 1°C (1.8°F) to the global temperature.

---

25. See: "Feedback mechanisms and climate sensitivity," *Britannica*, accessed March 8, 2022, https://www.britannica.com/science/global-warming/Feedback-mechanisms-and-climate-sensitivity.

Clouds carry solar energy from the warm tropics to other parts of the globe. As GW accelerates evaporation, adding more water vapor to the air, the increase in water vapor strengthens the greenhouse effect. However, more water vapor may also lead to the formation of more clouds, which might help cool (shade) the planet by increasing the reflection of solar energy back into space.

There is agreement among scientists that GW increases the evaporation of the oceans and, therefore, increases the water circulation rate. It would seem logical that a higher water circulation rate would be accompanied by higher cloud formation. On this point, there is no agreement. Therefore, further research is needed to resolve whether extra cloud formation contributes to GW. The majority view is that GW increases as cloud formation increases. In that case, increased cloud formation also has a positive feedback effect on GW. Figure 2-13 illustrates some of the effects of clouds on the global heat balance.

As with the influence of the oceans on GW, the water cycle (the water content of the air) also has two consequences. Specifically, it increases the evaporation of bodies of water, which, in addition to increasing the strength of hurricanes and flooding, also increases cloud formation. Clouds cause two effects: In the upper regions, they reflect

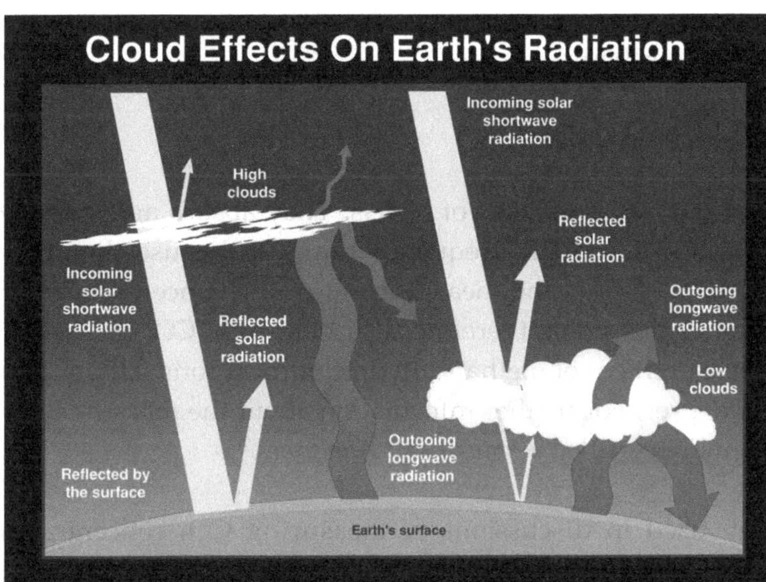

Figure 2-13. Clouds generate a greenhouse effect by reflecting some of the longwave radiation (thermal IR radiation) that reaches them from the surface of the Earth. The high clouds in the upper regions reflect incoming solar radiation back into space. The net effect changes with the type of clouds because low and bright clouds are cool by day (warm at night), whereas thin and high clouds (because of their low albedo) are warm both day and night.

Source: "Cloud Effects on Earth's Radiation," NASA Visible Earth, published April 18, 2000, https://visibleearth.nasa.gov/view.php?id=54219.

more incoming solar energy back into space, causing cooling. At the same time, lower clouds reflect more radiation back to Earth, which causes warming through the greenhouse effect.

I tried to calculate the sizes of these conflicting effects but could not do so. However, it does seem that these effects more or less cancel out because the total reflection from the land surface seems to be relatively stable. Nevertheless, because these are lag-dominant processes, the balance might change over time. Clouds make the world a warmer place because much of the energy that would otherwise escape into space is captured in the atmosphere; without clouds, Earth's average surface temperature would climb. Most climate models predict that clouds will amplify GW slightly. However, direct observational evidence is still limited. Human GHG emissions and clouds are the most significant sources of uncertainty in predicting the amount of future GW.

It is also clear that the gain of this process (the rate of increase in GW) will soar when the planet's albedo drops. This is another case in which reliable data is unavailable. We know that an increase in GW causes the albedo to change in two directions. Melting ice and snow reduces the albedo, whereas drying sand increases it. I believe that the effect of the melting at the poles and the melting of the glaciers is greater and, therefore, that the total effect will be a drop in the global albedo, so the resulting tipping point could soon be reached.

2.3 Greenhouse Gases

The atmosphere contains a number of GHGs. They absorb and emit radiant energy within the thermal infrared (IR) frequency range and cause the greenhouse effect (Figure 2-13). A discussion of the measurement of the concentrations of all GHGs is beyond the scope of this book; therefore, I discuss only CO_2 measurement in detail and cover the measurement of methane, hydrofluorocarbons (HFCs), and water vapor to a much lesser degree. For further information about the measurement details of all GHGs, see the handbook titled *Analysis and Analyzers.*[26]

The terminology used in discussions of measuring CO_2 concentrations results in much confusion. This is because there should be a distinction between the terms CO_2 and CO_2 equivalents (CO_2e) when stating atmospheric concentrations of GHGs. The meanings of these terms are not clearly understood, which results in confusion when discussing future mitigation targets in units of parts per million. This is for several reasons.

26. Béla G. Lipták and Kriszta Venczel, eds., *Instrument and Automation Engineers' Handbook: Analysis and Analyzers*, vol. II, 5th ed. (Boca Raton: CRC Press, 2016).

The total greenhouse effect is the sum of the effects of several gases present in different proportions in the atmosphere. Several gases are positive GHGs because they increase the greenhouse effect. Their typical proportions in the air are as follows: CO_2, 75%; methanol, 14%; nitrogen oxide, 8%; others, 3%. Ozone, on the other hand, is a negative GHG because it decreases the greenhouse effect by blocking solar radiation from entering the atmosphere.

In addition, the percentages of GHGs cannot be added together because some of the gases are more powerful than others. For example, the GW potential of a ton of methane is over 25 times that of a ton of CO_2. In addition to these effects, one should consider the length of time each of the various emitted gases remains in the atmosphere. For example, CO_2 can persist for a century or longer,[27] and methane remains for 10 to 20 years,[28] whereas aerosols disappear almost immediately. Another contributing factor is the role of water vapor (Figure 2-14), which, on one hand, has the highest total greenhouse effect but, on the other hand, varies greatly with the weather.

For all these reasons, when talking about the GHG concentration in the air, some people think in terms of actual measurements of CO_2 concentrations in parts per million,

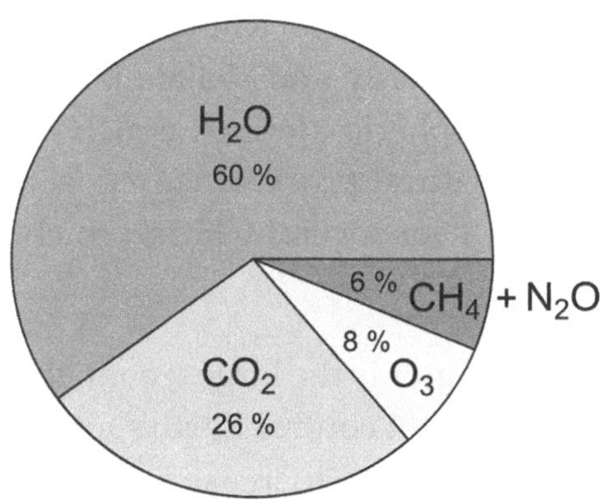

Figure 2-14. Contributions of the most important GHGs to the greenhouse effect.[29]

Source: Eni Generalic, "Global Warming and Mankind," EniG. Periodic Table of the Elements website, January 22, 2020, Figure 6, accessed September 14, 2020, https://www.periodni.com/global_warming_and_mankind.html.

---

27. The lifetime of CO2 in the air is difficult to determine because several processes remove it from the air. About 75% dissolves in the ocean over a period of 20 to 200 years. The rest is removed by slower processes that take up to several hundred or even thousands of years, including chemical weathering and rock formation. This means that once it is in the atmosphere, CO2 can continue to affect the climate for many centuries.

28. Methane is removed by chemical reactions and usually persists for about 12 years. Therefore, while methane is a much more potent greenhouse gas than CO2, its effect is short-lived.

29. For more information, see: J. T. Kiehl and Kevin E. Trenberth, "Earth's Annual Global Mean Energy Budget," *Bulletin of the American Meteorological Society* 78, no. 2 (February 1997): 197–208, https://doi.org/10.1175/1520-0477(1997)078<0197:EAGMEB>2.0.CO;2.

whereas others attempt to determine the CO_2e value instead. These numbers are not the same. For example, if one adds together the positive forcing effects of all of the individual GHGs and disregards water vapor and aerosols, today's CO_2e value calculates to about 560 ppm, whereas the present measured value of the CO_2 concentration is about 415 ppm.

So how do I discuss GHG concentrations in this book? I decided to use the concentration that our satellites measure (today, about 415 ppm), which is just CO_2, but I note when a value reflects the total CO_2 equivalent, in which case I use the notation CO_2e.

Water vapor is usually not mentioned as GHG, but it certainly is one. It is omitted from the list of GHGs because its concentration is increased by GW and not the other way around. In fact, water vapor makes the largest contribution to the greenhouse effect (Figure 2-14). It is always present in the atmosphere in the form of clouds or humidity. The evaporation of the oceans increases as the oceans warm, so the amount of water vapor in the atmosphere is increasing. Each 1°C (1.8°F) rise in GW results in a 7% increase in the water vapor concentration in the air before condensation. Water vapor thus provides a positive feedback to GW because, as the temperature of the oceans rises and evaporation increases, the concentration and, therefore, the greenhouse effect of water vapor also rises.

The strength of the greenhouse effect of a GHG is determined by two of its characteristics. One is its ability to absorb energy and radiate it back to Earth (called *radiative efficiency*). The second is its atmospheric lifetime, which measures how long the gas stays in the atmosphere before natural processes (e.g., chemical reactions) remove it. Table 2-1 lists these characteristics for several GHGs, which quantify the strengths of their global warming potentials (GWPs).

As mentioned previously, water vapor is the largest contributor to Earth's greenhouse effect. However, water vapor does not control Earth's temperature; it is controlled by Earth's temperature. If there had been no increase in the amounts of noncondensable GHGs (such as carbon dioxide) in the air, the amount of water vapor in the atmosphere would not have changed. On the other hand, the addition of noncondensable gases to the atmosphere causes the temperature to increase, leading to an increase in the amount of water vapor, which further increases the temperature (positive feedback effect). The warming due to the increase in the concentration of noncondensable gases causes more water vapor to enter the atmosphere, and this feedback increases the effects of the initial RF of the GHGs.[30]

---

30. Marshall Shepherd, "Water Vapor Vs Carbon Dioxide: Which 'Wins' In Climate Warming?," Forbes website, June 20, 2016, https://www.forbes.com/sites/marshallshepherd/2016/06/20/water-vapor-vs-carbon-dioxide-which-wins-in-climate-warming/?sh=7847a03a3238.

Table 2-1. GWPs and atmospheric lifetimes of several GHGs.

Greenhouse Gas	Chemical Formula	GWP[a]	Atmospheric Lifetime (years)
Carbon dioxide	CO_2	1	Over 100
Methane	CH_4	25	10 to 20
Nitrous oxide	N_2O	265	121
Chlorofluorocarbon-12 (CFC-12)	CCl_2F_2	10,200	100
Hydrofluorocarbon-23 (HFC-23)	CHF_3	12,400	222
Sulfur hexafluoride	SF_6	23,500	3200
Nitrogen trifluoride	NF_3	16,100	500

[a]Reported for a 100-year time horizon.

Second only to water vapor, CO_2 makes the next greatest contribution to the greenhouse effect. The main source of CO_2 emissions is the burning of fossil fuels. About half of these emissions are absorbed by vegetation and dissolve in the oceans, causing acidification. Some of the absorbed CO_2 is released back into the atmosphere when the oceans warm and their solubility decreases (degassing). Organisms in the sea can also convert dissolved CO_2 into calcium carbonate, an ingredient of limestone rock, in which carbon can be stored for millennia. However, the rock-forming process is very slow. Overall, the time CO_2 can linger in the atmosphere for over a century. Therefore, even if we stopped emitting CO_2 today, our past CO_2 emissions would continue warming the planet for a long time.

Like the oceans, growing vegetation also removes CO_2; however, when a plant dies and rots, it releases the same amount of CO_2 as it absorbed during its growth, so the net effect of plants over time is nearly zero. Actually, when wood is burned, its carbon emission and pollution effects are worse than that of fossil fuels.

Methane, the main component of natural gas, is the next most powerful GHG. In addition to being emitted by the natural gas industry, it is also released from a host of biological sources, including rice production, animal manure, cow digestion, and the breakdown of waste in landfills. Methane accounts for about 10% of US GHG emissions. This gas is much better at trapping heat than CO_2 (some 25 to 30 times better), but it does not remain in the atmosphere as long as CO_2 (about a century). It is broken down in about a decade as it reacts with hydroxyl radicals (neutrally charged OH ions made from bound atoms of oxygen and hydrogen).

Reducing methane emissions associated with human activity by 50% could mitigate GW by 0.2°C during the next 30 years. At Stanford University in 2022, researchers found that oil and gas-producing regions emit several times greater quantities of methane than previously estimated. They used airborne sensors to detect methane leaks from individual oil and gas production facilities, and they found that more than 9% of all methane produced in some regions leaked into the atmosphere. This is several-fold higher than the Environmental Protection Agency (EPA) estimates (1.4%) and well above those in the published literature. The recently started "natural gas war" and damaging pipelines worsen this situation.

Another powerful group of GHGs is composed of chemical coolants called HFCs that have 1000 times the heat-trapping punch of CO_2. In October 2016, delegates from 197 nations meeting at a conference in Kigali, Rwanda, set hard targets for the global phaseout of HFCs. HFCs contribute an estimated 11.5% to the present-day effects of GHGs on climate and climate change.[31]

2.3.1 CO_2 Measurements

CO_2 (and other gases that consist of two or more dissimilar atoms) absorbs IR radiation in a characteristic, unique manner. Water vapor, methane, CO_2, and carbon monoxide (CO) are all gases that can be measured with IR sensors. Therefore, IR detectors are the most widely used CO_2 analyzers. They are stable and highly selective to CO_2. They also have a long lifetime, and because the measured gas does not contact the sensor, they can withstand high humidity, dust, dirt, and other harsh conditions.

A precise knowledge of CO_2 concentration in the atmosphere is necessary because CO_2 plays a key role (Figure 2-15) in GW. Instruments used to measure atmospheric CO_2 concentrations must be accurate to 1 ppm or even better.

In the past, CO_2 concentration was often measured by taking grab samples from the air near Earth's surface. These instruments used calibration standards consisting of mixtures of CO_2 and nitrogen and achieved accuracies of ±0.2 ppm. This error at normal atmospheric CO_2 concentrations was equivalent to ±0.06% full-scale accuracy.

In addition to laboratory grab-sample measurements, continuous industrial CO_2 measurements were made using portable or permanently installed nondispersive infrared (NDIR) CO_2 analyzers. The portable units are usually battery-operated, and their ambient samples are received by a combination of diffusion and convection effects in the sensor head without any pumps or filtering. These units have digital displays, one or two alarm settings, and analog or digital output signals. If gas filter correlation

31. "Chlorofluorocarbons (CFCs) and hydrofluorocarbons (HFCs)," Minnesota Pollution Control Agency website, accessed March 8, 2022, https://www.pca.state.mn.us/air/chlorofluorocarbons-cfcs-and-hydrofluorocarbons-hfcs.

(GFC) is combined with single-beam, dual-wavelength designs, the analyzer is virtually immune to the obstruction of the optics. This, in turn, prevents drift and reduces the frequency of recalibration.

For accurate atmospheric measurements, pulsing laser detectors employing light detection and ranging (LiDAR[32]) are used. These operate at wavelengths slightly greater than 2 μm, where CO_2 is highly absorbing, but H_2O (water vapor) is not (Figure 2-15). The goal of wavelength selection is to minimize the overlap between the absorption bands of water vapor and CO_2 because measurement errors can occur if the IR bands overlap.

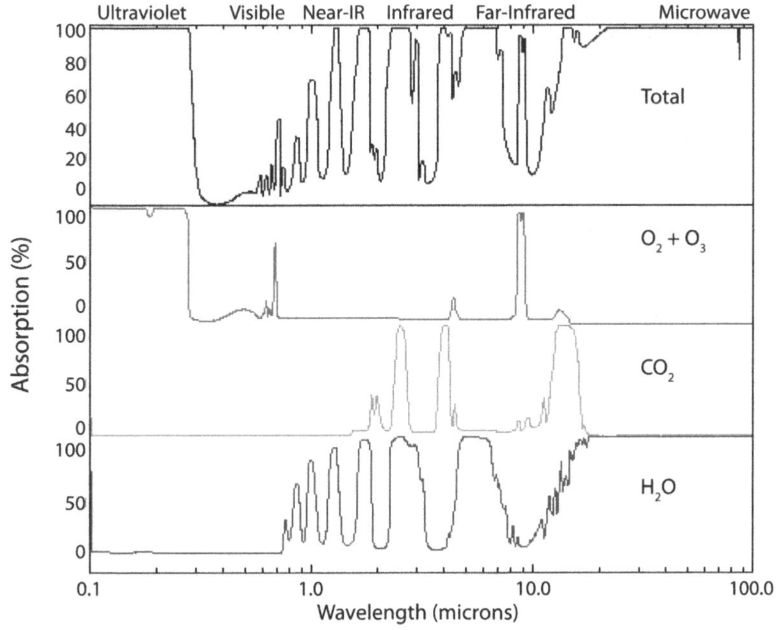

Figure 2-15. CO_2 has three large absorption bands in the IR region at about 2.7, 4.3, and 15 μm. Water has several absorption bands in the IR range and some well into the microwave region. One wavelength frequently used in CO_2 measurements is slightly greater than 2 μm, where CO_2 is highly absorbing, but H_2O is not.

Source: T. J. Nelson, "Cold Facts on Global Warming," *randombio Blog*, published March 17, 2003, last updated January 1, 2011, Figure 4, http://www.randombio.com/co2.html.[33]

32. Sometimes also called *light imaging, detection, and ranging*. Originally, it was called a portmanteau of light and radar. It is sometimes called *3D laser scanning*, a special combination of 3D scanning and laser scanning.

33. Original data from the following sources: José P. Peixoto and Abraham H. Oort, *Physics of Climate* (New York: Springer, 1992). Gary E. Thomas and Knut Stamnes, *Radiative Transfer in the Atmosphere and Ocean* (Cambridge, UK: Cambridge University Press, 1999).

Figure 2-16. Artist's conception of OCO-2, the second successful high-precision (better than 0.3%) CO_2 observing satellite.

Source: "Orbiting Carbon Observatory," NASA/JPL, December 9, 2008, https://www.nasa.gov/mission_pages/oco/multimedia/1.html.

Today, the use of grab-sample-type sensors is mainly limited to industrial applications. The Total Carbon Column Observing Network (TCCON) was formed in 2004 to monitor atmospheric CO_2 concentrations, and by 2015, it included 23 monitoring stations worldwide. TCCON is a global network of measurement stations that measure the amounts of CO_2, methane, carbon monoxide, nitrous oxide, and other trace gases in Earth's atmosphere.

TCCON's goal is to investigate the flow (or flux) of carbon between the atmosphere, land, and ocean (the so-called carbon budget or carbon cycle). This is achieved by measuring the atmospheric mass of carbon (the airborne fraction). TCCON measurements have improved the scientific community's understanding of the carbon cycle and urban GHG emissions. TCCON can also assist in validating measurements made by airplane-, drone-, or satellite-mounted systems. It does that by providing independent readings to compare with the space-based measurements when the satellite or airplane is flying over the TCCON site locations. For example, the TCCON provides the primary measurement validation data set for the Orbiting Carbon Observatory-2 (OCO-2) mission, and TCCON has also been used to validate other space-based measurements of CO_2 (Figure 2-16).

2.3.2 NASA's Role in CO_2 Measurements

In 2004, ground-based total column measurements mostly replaced surface-only CO_2 monitoring, and later, they were supplemented by airplane, drone, and, eventually, satellite monitoring. Space-based measurements of CO_2 help to answer questions about Earth's carbon cycle; however, remotely sensing trace gases has several challenges. Most techniques rely on observing IR light reflected off Earth's surface. Changes in surface albedo[34] and viewing angles can affect measurements. Clouds can also interfere with accurate measurements, so platforms may include instruments to measure and compensate for clouds.

The first satellite mission was the Interferometric Monitor for Greenhouse Gases (IMG) on board the ADEOS I satellite in 1996. This mission lasted less than a year. Since then, additional space-based measurements have been implemented, including those from two high-precision (better than 0.3% or 1 ppm accuracy[35]) satellites, namely, the Greenhouse Gases Observing Satellite (GOSAT) and OCO-2.

Two decades of efforts by NASA in developing LiDAR and a 2 μm transmitter resulted in a new capability for remotely measuring CO_2 levels in Earth's atmosphere. Specifically, NASA developed a pulsed, integrated path differential absorption (IPDA) LiDAR instrument that incorporates high-energy, double-pulse lasers with high repetition rates. The laser pulses can be tuned and locked near the 2.05 μm wavelength, which is an ideal spectral region for CO_2 sensing. Separated by 150 μs, the first pulse is tuned to a high CO_2 absorption wavelength, and the second pulse is tuned to a low absorption wavelength. When the pulses are aimed at a hard target, the difference between the return signals correlates with the average amount of CO_2 in the column between the instrument and the target.

NASA's OCO-2 satellite, launched in 2014, used passive monitoring, which means that the CO_2 measurements are made by observing the gas's effect on sunlight. As sunlight passes through the atmosphere, CO_2 and other molecules absorb light at distinct frequencies. The pattern of absorption can be as unique as a fingerprint. The intensity of the sunlight reflected from Earth's surface, which is partially absorbed by CO_2 and oxygen, is recorded by the satellite and used to calculate how much CO_2 there is between the satellite and the surface. Measurements at different altitudes make it possible to monitor the global distribution of CO_2 and its increase over time (Figure 2-17).

---

34. Albedo (whiteness) is a dimensionless measure of a body's reflectiveness of solar radiation. As mentioned in the text, a black body absorbs all radiation and has an albedo of 0.0 (0%), whereas a body that reflects all incoming radiation has an albedo of 1.0 (100%). The albedo of fresh snow is about 80%; ice, 35%; forest and dark soil, 10%; and water (including ocean water), about 7%.

35. Carbon dioxide measured at NOAA's Mauna Loa Atmospheric Baseline Observatory peaked for 2022 at 420.99 parts per million in May, an increase of 1.8 parts per million over 2021, pushing the atmosphere further into territory not seen for millions of years.

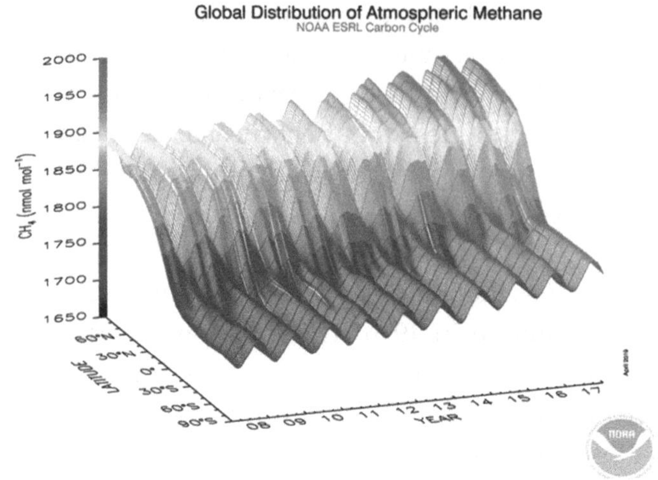

Figure 2-17. CO_2 concentrations in the marine boundary layer at different latitudes between the North and South Poles measured between 2008 and 2017. (See color images after page 306.)

Source: "Global Distribution of Atmospheric Carbon Dioxide," National Oceanic and Atmospheric Administration (NOAA), October 2018, https://gml.noaa.gov/ccgg/gallery/figures/co2_surface_color.png.[36]

The key limitation of passive measurements is that they can only operate in daylight, whereas it is also necessary to gain a total, 24-hour understanding of the CO_2 cycle. It is important to know what is happening at night because plants are inactive at night or in the winter. Their respiration also varies at high latitudes, including at the poles, where there is a complete absence of sunlight in the winter.

In contrast to passive monitoring systems, which work with reflected sunlight, the Anthropogenic Carbon Emissions System (ACES) team works with active monitors, where the lasers send continuous-wave beams down to Earth. ACES is a monitoring package installed on an airplane that flies at about 30,000 ft. It has five lasers that can focus on the ocean's surface and uses LiDAR to measure CO_2. Unlike the passive OCO-2 satellite, ACES provides active monitoring. Its pulsed LiDAR beam concentrates all the laser energy into very short pulses, about one-millionth of a second each. The main accomplishment of this system is the generation of high laser powers at precisely controlled frequencies, which is needed for precise CO_2 measurements of better than 1.0 ppm precision (at 410 ppm, that accuracy is ±0.24%).

36. For further information, see: "Carbon dioxide in Earth's atmosphere," Wikipedia, accessed October 27, 2022, https://en.wikipedia.org/wiki/Carbon_dioxide_in_Earth%27s_atmosphere.

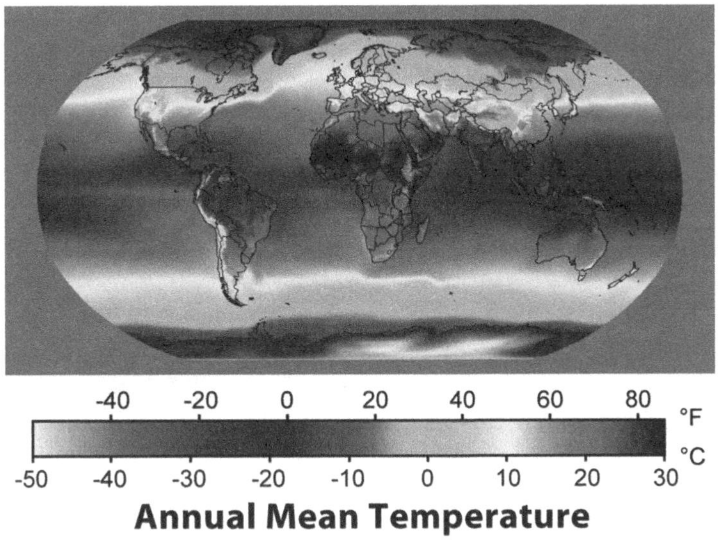

Figure 2-18. Annual mean temperatures around the planet. (See color images after page 306.)

Source: "Annual Average Temperature Map.jpg" by Robert A. Rohde, February 15, 2008, is licensed under CC BY-SA 3.0, https://commons.wikimedia.org/w/index.php?curid=3558400.

2.4 Oceanic Processes

2.4.1 Ocean Temperature and Density

As already noted multiple times, we have to understand a process to be able to influence or control it. The most important aspect is to understand the dynamic behavior and interactions of the subprocesses, which then combine into the "personality" of the total process.

Basically, the climate is the result of a heat-balance process in which heat input is received from the Sun. This input is a variable. It varies not only with the seasons but also with the location (longitude and latitude[37]) of the area and the composition of the atmosphere (Figure 2-18).

Tropical oceans receive a lot of direct overhead sunlight for much of the year, so their water is warm. On the other hand, summer is the only time polar regions receive sunlight; even then, it is never from directly overhead, so water in these places tends

---

37. The Equator has a latitude of 0°, the North Pole has a latitude of 90° North (written as 90° N or +90°), and the South Pole has a latitude of 90° South (written as 90° S or −90°). The latitude of an arbitrary point is the angle between the equatorial plane and the normal to the surface at that point, where the normal to the surface of a sphere is along the radius vector. The Arctic is the area north of the Arctic Circle at 66° N; the Antarctic is the region to the south of 66° S.

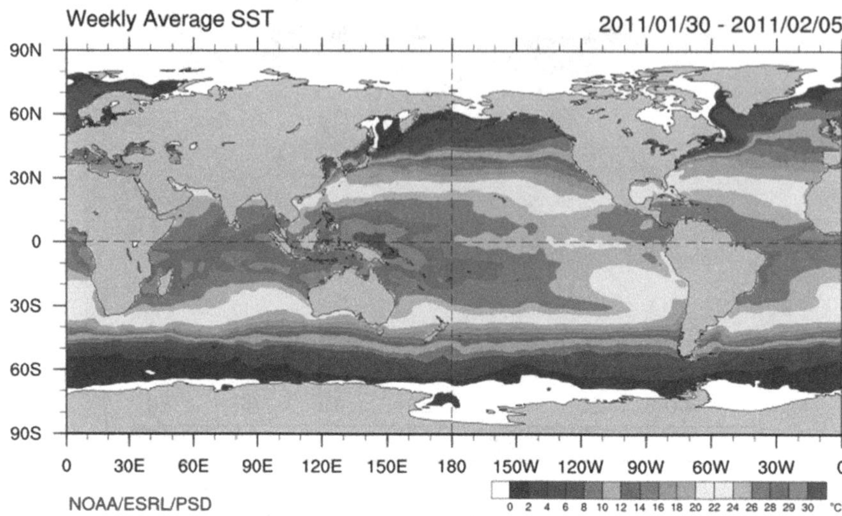

Figure 2-19. SSTs during the first 5 days of February 2011. (See color images after page 306.)
Source: "weeklysst.gif" by Earth System Research Laboratory Physical Sciences Division, February 5, 2011, Public Domain, https://commons.wikimedia.org/w/index.php?curid=13262187.

to be cold. The amount of sunlight that hits the temperate regions (between the tropics and the poles) varies between summer and winter. The variation in the amount of solar energy absorbed means that the ocean surface can vary in temperature from a high of about 28°C (82.4°F) in the tropics to a very low –2°C (28.4°F) near the poles. Figure 2-19 shows the sea surface temperatures (SSTs) on the planet during the first 5 days of February 2011.

The oceans play an important role in absorbing most of the energy received from the Sun (about 90%), thereby reducing the speed of GW. Water warms more slowly than air and also holds more heat. This is because land has a smaller "heat capacity" than water, which means it needs less heat to raise its temperature. The volumetric heat capacity of water is some 3,400[38] times higher than that of dry air (and about 1000 times higher than that of wet air); therefore, it takes much more energy to raise the temperature of water than of the same volume of air. This has a dual effect because, although the increasing GHG concentration in the air increases GW, the immense heat absorption capacity of the oceans slows GW. Naturally, although the oceans absorb the excess heat as it is received, the absorbed heat energy will eventually be (slowly) released (Figure 2-19).

38. "Table of specific heat capacities," Wikipedia, accessed March 8, 2022, https://en.wikipedia.org/wiki/Table_of_specific_heat_capacities. Ilja, March 22, 2016 (11:29 p.m.), answer to Andrew Lancaster, "Water vs. Air: Heat Absorption/Capacity," March 22, 2016 (10:11 p.m.), *Physics Stack Exchange*, https://physics.stackexchange.com/questions/244989/water-vs-air-heat-absorption-capacity.

At a temperature of 4°C (39.2°F), salinity causes the ocean water density to increase to about 1.0278 g/cm^3. Ocean temperatures range from 2°C to 28°C (35.6°F to 82.4°F), but they can be even higher near hydrothermal vents[39] or closer to land. Salinity is usually 35 PSU (practical salinity unit), but it can range from 28 to 41 PSU (1 PSU = one gram of salt per liter of water). Ocean temperature also varies from the surface to the bottom (Figure 2-20), giving a vertical structure to most of the ocean. There is an upper layer of water, up to 200 m (656 ft) deep, that is warmed by the Sun and has the same temperature throughout. Below that is a layer called the *thermocline*, which can extend to 1000 m (3280 ft) in places and in which the temperature slowly cools with depth. Below the thermocline layer, the temperature of the deep ocean (80% of the ocean) is the same down to the bottom.

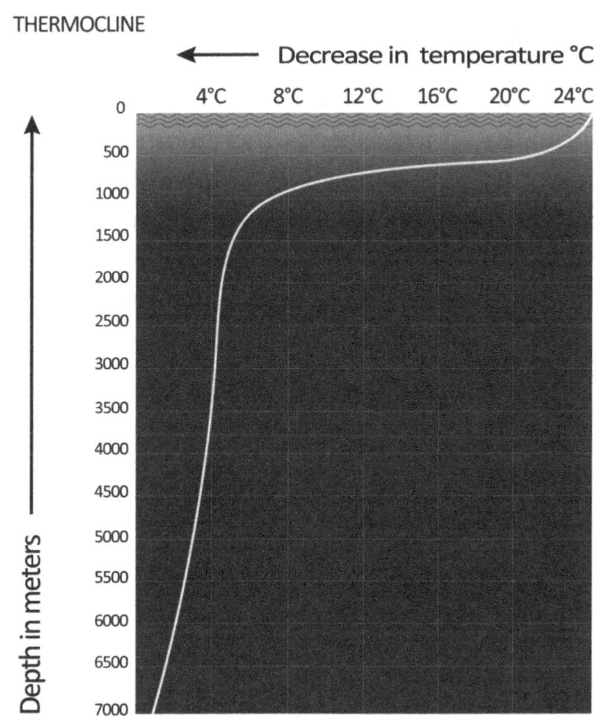

Figure 2-20. Water gets colder and denser (heavier) with increasing ocean depth.[40]

Source: "THERMOCLINE.png" by Praveenron, February 8, 2014, is licensed under CC BY-SA 3.0, accessed March 13, 2022, https://commons.wikimedia.org/w/index.php?curid=31055923.

39. Hydrothermal vents are found in places on Earth where two tectonic plates move away or toward one another. In these locations the cold seawater is heated by hot magma and reemerges to form the vents, where it can reach temperatures of over 700°F. Hot seawater in hydrothermal vents does not boil because of the extreme pressure at the depths (1,500 feet) where the vents are formed. These vents represent potential energy sources, but to date have not been exploited.

40. For further information, see: "Thermocline," Wikipedia, accessed March 8, 2022, https://en.wikipedia.org/wiki/Thermocline.

Cold water has a higher density than warm water. Water gets colder with depth because cold, salty ocean water sinks to the bottom of the ocean basins, below the less dense warmer water near the surface. The sinking and transport of cold, salty water at depth combined with the wind-driven flow of warm water at the surface creates a complex pattern of ocean circulation called the *global conveyor belt.*

Temperature and density share an inverse relationship. As temperature increases, density decreases; conversely, when water temperature decreases, its density increases, but only to a point. At a temperature of 4°C (39.2°F), pure water reaches its maximum density. If cooled further, it becomes less dense, which explains why when water freezes at 0°C (32°F) and ice is formed, it floats. Salinity and density share a positive relationship. As the amount of salt in the water (salinity) increases, density also increases. Salinity, and therefore density, can decrease from the addition of fresh water as a result of the melting of polar ice (e.g., in Greenland).

Because of the immense mass of Earth, GW is very slow. Today, Earth receives a total of 5 million quads annually, which exceeds the amount radiated back into space by 1.6 W/m^2 (24,000 Q/year). Of this difference, 93% is absorbed by the oceans (Figure 2-21). This results in a yearly sea temperature rise of about 0.015°C/year (0.027°F/year). This very small increase in the oceans' temperatures is caused by the absorption of immense quantities of heat. The average ocean surface temperature today is about 17°C (62.6°F), which varies with season, location, and so on. Between 1900 and 2000, the ocean surface temperature increased by some 0.7°C (1.26°F), and today, it is rising at a rate of about 0.015°C per year.

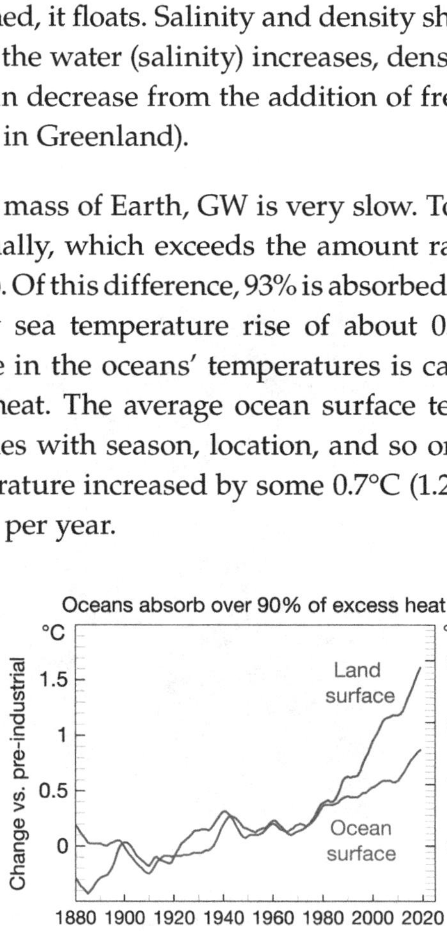

Figure 2-21. NASA data shows that land temperature increased faster than sea surface temperatures. The baseline for these curves is the preindustrial average temperature.[41]

Source: "Land vs. Ocean Temperature" by Efbrazil, March 2, 2020, is licensed under CC BY-SA 4.0, accessed March 13, 2022, https://commons.wikimedia.org/w/index.php?curid=87627678.

41. For additional information, see: "Sea surface temperature," Wikipedia, accessed March 8, 2022, https://en.wikipedia.org/wiki/Sea_surface_temperature.

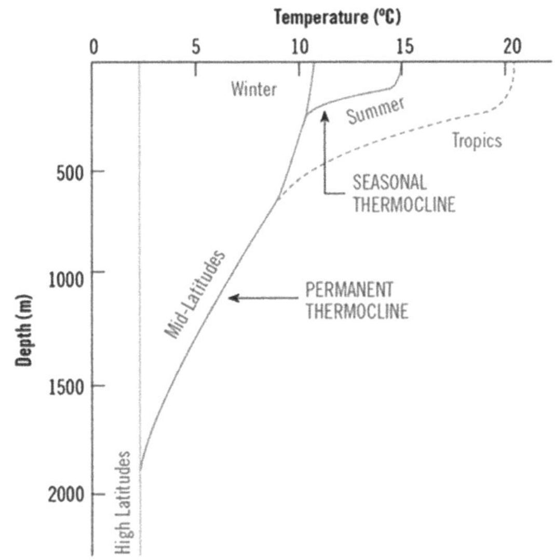

Figure 2-22. Seasonal variations in the rate at which water gets colder with increasing depth.

Source: "ThermoclineSeasonDepth.png" by Welcome1To1The1Jungle, July 24, 2014, is licensed under CC BY 3.0, https://commons.wikimedia.org/w/index.php?curid=48498086. Figure with fonts and line colors modified courtesy of *Control Global*. Béla Lipták, "Can we reliably measure the rate of the rise of global warming at 0.01°C/yr?," June 3, 2020, Figure 2, https://www.controlglobal.com/articles/2020/can-we-reliably-measure-the-rate-of-rise-of-global-warming-at-0-01cyr/.

Our ability to detect such small temperature increases improved with the introduction of the Argo floats, 3000 of which are deployed in the ice-free areas of the world's oceans, profiling temperature and salinity down to a depth of about 2000 m. These floats dramatically reduce the time and money required to gather such data by research ships using traditional over-the-side methods of ocean observations. In addition, the data gathered by the floats is accurate to about ±0.002°C.

The amount of imbalance between the incoming and outgoing energy is called RF, and according to the *Fourth Assessment Report* (AR4) of the UN's Intergovernmental Panel on Climate Change (IPCC), its current level is assumed to be 1.6 W/m^2 (with a range of uncertainty from 0.6 to 2.4 W/m^2).

Figure 2-20 shows that the water temperature decreases as ocean depth increases, but not linearly. At about 450 m (1500 ft), there is a seasonal thermocline in which the temperature starts dropping quickly (Figure 2-22). This layer separates the mixed and unmixed regions of ocean water.

The most accurate temperature measurements are made with a four-wire resistance temperature detector (RTD) design using a class AA platinum sensor. Such RTDs can provide 0.01°C (0.02°F) accuracy. Note that, in some cases, vendor specifications state only the resolution and give the impression that resolution equals accuracy. This is incorrect. (An easy way to remember the difference between accuracy and resolution is to think of a clock, which has a resolution of one minute—the equal distance between the 60 markings—whereas its accuracy can vary.)

In the past, average ocean surface temperatures were monitored by taking samples or attaching thermometers to buoys or ships. Radiosondes followed, and starting in the 1950s, satellites measuring microwave emissions were used. Other technologies, such as those leveraging global positioning system (GPS) satellite signals, have also been tried, but they have limited recording length, and their accuracy has not been clearly established.

The basic operating principle of one of the most successful devices for measuring ocean temperature profiles correlates density change with temperature. Fish use this method to rise or sink while swimming. The density of an object is the ratio of its weight to its volume. If the density of a floating object decreases, it will rise, and vice versa. Fish change their weight-to-volume ratios, too. The weight of a fish is constant, so when its volume decreases, so does its density. Therefore, when a fish's swim bladder is pressurized, its volume rises, and its buoyancy also rises (its density drops). Conversely, buoyancy decreases when the bladder is deflated and the fish sinks. This allows a fish to change its own density and sink or rise.

Measuring the increase in the surface temperature of the oceans does not represent the total heat absorbed because ocean temperature changes with depth. Therefore, a more accurate reading is obtained if the average temperature is measured along with the depth profile. Today, the Argo program is doing just that: measuring temperature and salinity profiles from the surface to a depth of 2000 m (6562 ft) at almost 3000 to 4000 points around the globe (Figure 2-23). The project was conceived and designed by US academics and scientists at the National Oceanic and Atmospheric Administration (NOAA) to obtain a uniform and systematic method for measuring the physical state of the upper layers of oceans and to place that information into climate computer models in near real-time. (The name Argo comes from the name of the ship in the tale of Jason and the Argonauts from Greek mythology.)

Argo floats use the flexible bladder method to rise and sink to measure the heat content and temperature profile of the ocean at different elevations. At its tip, an Argo float contains a bladder filled with an oil that is less dense than water. By varying the bladder volume, the Argo floats use density changes to drive their movements. If the density of an object in the ocean drops below that of the water, the object will rise, whereas if its density increases above that of water, it will sink. An Argo float (Figure 2-24) does the same. It uses flexible bladders to change its volume (just as fish do) to sink or rise. The hydraulic

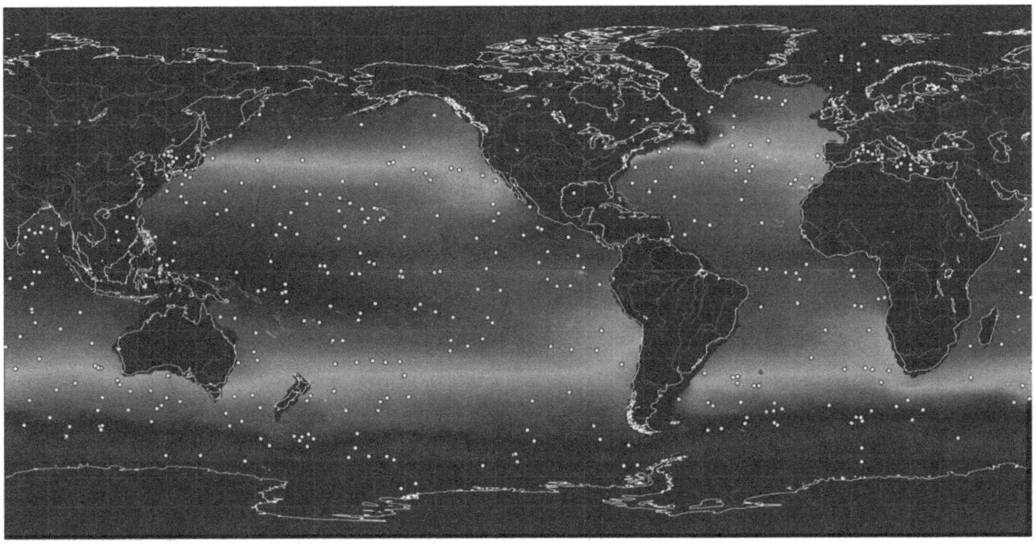

Figure 2-23. As of October 2020, nearly 4000 Argo floats (yellow dots) had been installed around the world. The background colors indicate ocean temperatures (red, warmest; blue, coldest). (See color images after page 306.)

Source: "Argo's Status: Implementation Status," Argo Program website, accessed September 16, 2020, https://argo.ucsd.edu/about/status/.

bladder at the tip of the float and a pneumatic bladder on the side vary the float's density. When the density rises, the float sinks, and when it decreases, the float rises.

Nearly 4000 Argo floats are deployed in the ice-free areas of the world's oceans, and they profile the temperature and salinity down to a depth of about 2000 m. These floats dramatically reduce the time and cost of gathering such data by research ships using traditional over-the-side methods of ocean observations. Moreover, the data gathered is claimed to be accurate to about ±0.002°C (±0.004°F). The reasoning behind the claimed increase in accuracy is that when one has thousands of measurements of the same value, some of the sensor errors cancel out, and a more accurate average reading results. We will never know if this logic is valid because we will never have a more accurate thermometer to check this one.

The Argo floats are battery-powered autonomous units that spend most of their life drifting at a "parking depth" of about 1000 m for about 9 days during each measurement cycle. They are stabilized by automatically matching their density with that of the water around them (Figure 2-25). At 10-day intervals, they automatically reduce their own volume, decreasing the volume of their bladders by pumping some of the fluid from them and thereby increasing their density so that they will sink to a depth of 2000 m. The floats then automatically reduce their density by increasing their volume (expanding the bladder) and rise to the surface over about a 6-hour period, during which they measure the temperature and salinity profiles of the water. When they

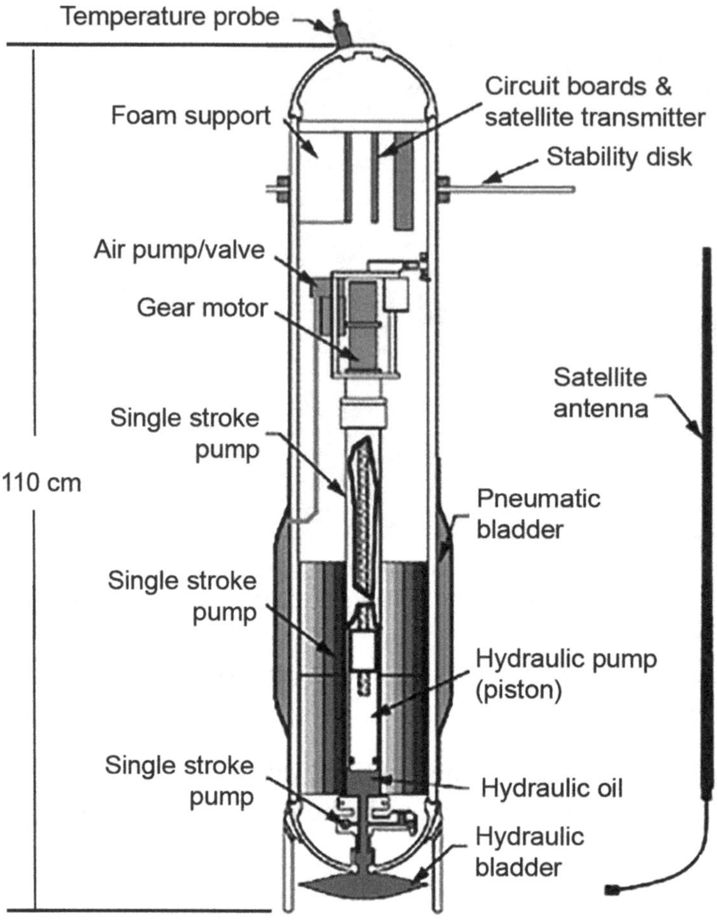

Figure 2-24. Schematic of an Argo SOLO-II float.

Source: Reprinted with permission. "Float Design Schematic," Physical Oceanography Division, Atlantic Oceanographic and Meteorological Laboratory, NOAA, accessed August 16, 2021, https://www.aoml.noaa.gov/phod/argo/webpage_sections/doc/argo_intro_float_design.php.

surface, the floats transmit the profiles they collected to satellites. After about 20 minutes on the surface, the bladders deflate, the floats sink, and the cycle is repeated. The floats are designed to make about 150 such cycles annually.

Using this method, the total heat content of the oceans is measured, and by detecting the rate at which the heat content rises, the rate of GW is accurately determined.

2.4.2 Acidification and Corals

The increased concentration of CO_2 in the air causes acidification of the slightly alkaline oceans. The ocean's pH has already dropped to 8.1 (its lowest value in 2 million years), and it is projected that it could drop to 7.8 by 2100. This acidification harms ocean life and devastates the coral reefs (Figure 2-26) because carbonic acid is formed by the dissolution of CO_2 in the ocean's water, and this carbonic acid impedes the

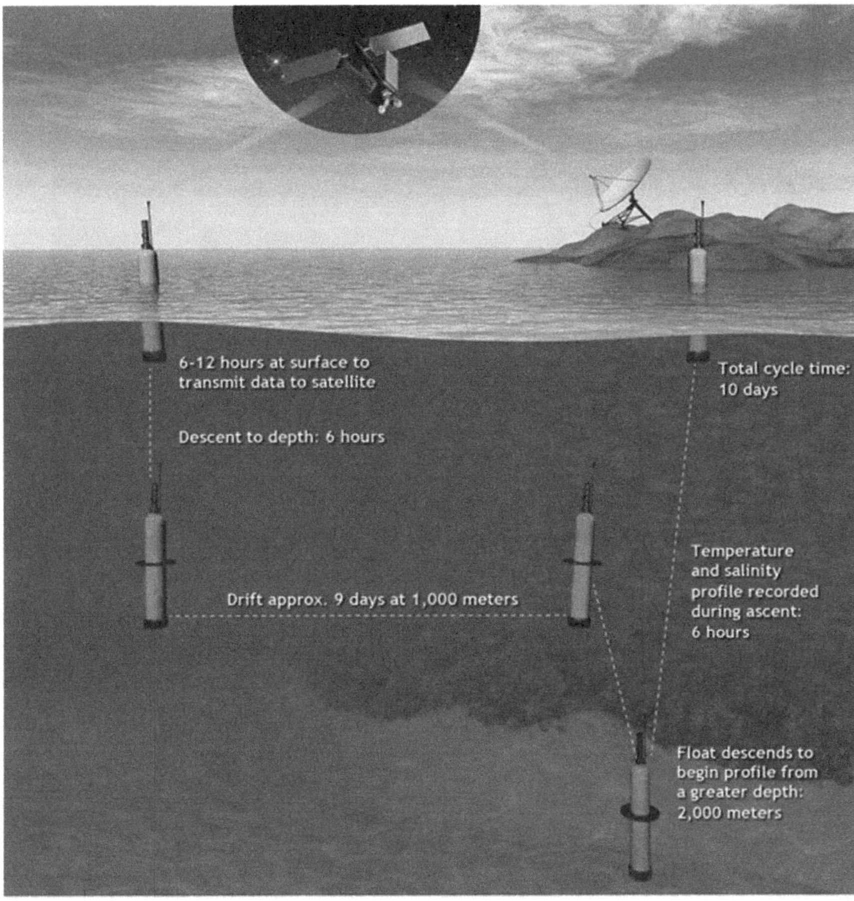

Figure 2-25. Ten-day data collection cycle of an Argo float.

Source: Dale C. S. Destin, "The Argo revolution," NOAA, December 5, 2014, https://www.climate.gov/news-features/features/argo-revolution.

production of the calcium carbonate needed to build the coral exoskeleton. In just the past 3 years, one-fifth of all corals have died. In addition, coral bleaching also occurs because of increased solar irradiation and increased warming. Coral provides a home for 25% of the marine fish[42] and, therefore, a large portion of food for people. One might note that overfishing of the oceans is also in progress. For example, China has already reduced its local fish population to the point that Chinese ships are now fishing in distant oceans.

---

42. "Hidden beneath the ocean waters, coral reefs teem with life. Coral reefs support more species than any other marine environment and rival rainforests in their biodiversity. Countless numbers of creatures rely on coral reefs for their survival. These important habitats are threatened by a range of human activities. Many of the world's reefs have already been destroyed or severely damaged by an increasing array of threats, including pollution, unsustainable fishing practices, and global climate change. However, we can still protect and preserve our remaining reefs if we act now. NOAA is leading U.S. efforts to study and conserve these precious resources for future generations." *Source:* "Coral Reefs: Rainforests of the Sea," National Ocean Service website, accessed March 8, 2022, https://oceanservice.noaa.gov/ocean/corals/welcome.html.

Figure 2-26. As the oceans become more acidic, corals' ability to form skeletons through calcification is inhibited, and they die.

Coral's other important role is to reduce coastal flooding and erosion, although they could dissipate as much as 97% of incident wave energy. Yet, as of this writing, there are no effective coral reef restoration projects in progress that would serve such coastal flood risk reduction.

Some refer to coral reefs as "rainforests of the sea," but their impact on climate change is not yet resolved. This is because CO_2 is absorbed by photosynthesis but is released by calcification. On the other hand, there is no debate about their role in helping us to breathe. Although coral reefs cover only 0.0025% of the ocean floor, they produce half of the oxygen in the atmosphere.

Clearly, reducing the concentration of GHGs in the atmosphere will ultimately aid in recovering coral reefs. The climate-induced weakness of these ecosystems makes them vulnerable to other dangers, such as chemical pollution by many substances, including sunscreen. As an example of chemical pollution, some 14,000 metric tons of sunscreen are deposited in the oceans each year.[43]

Process control technology plays an important role in monitoring the condition of coral reefs from several thousand feet overhead. Cameras detect colors outside the range of human vision that describe the chemistry of the corals. Bleaching, which can lead to

---

43. It has been estimated that 14,000 T of sunscreen goes in the world's oceans every year. This is causing genetic damage to coral polyps and other marine animals, making coral susceptible to disease, causing mortality in developing young corals, and increasing the severity of coral bleaching. *Source:* "Protecting Florida Keys Coral Reefs by Banning Oxybenzone!," Surfrider Foundation website, February 5, 2019, accessed December 1, 2022, https://www.surfrider.org/campaigns/protecting-our-florida-keys-coral-reefs-by-banning-oxybenzone.

death, is precipitated by a change in the chemistry that the cameras can see. Imaging spectroscopy sensors and LiDAR instruments emit 500,000 pulses of light per second and record the echoes, revealing the locations and three-dimensional shapes of every object beneath the plane. High-resolution cameras collect data at 4 cm resolution. Back in the laboratory, deep-learning algorithms are used to determine whether a coral is alive or dead and whether it is bleached (and therefore at risk of death) or on its way to becoming so. In some cases, they can even determine which species it is.

In remote mapping, each device has a different geometry and measurement pattern that must be aligned to make an accurate map; otherwise, the result would look like a 3-D movie viewed without the glasses. An inertial motion unit (built by Northrop Grumman for use on military warheads and so sensitive that its location must be reported to the US State Department to ensure the technology is handled properly) aligns the images. Thereby, progress is being made toward building a spectral-chemical library for all 830 known species of coral.

2.4.3 Global Water Cycle

The interactions among global subprocesses are mainly caused by heat transfer between the land, the oceans, and the atmosphere, where "heat conveyors" cause the movement of water and air. The response time of atmospheric processes is short (they are fast processes); their heat capacity and inertia are both low. In contrast, the response of oceanic processes is extremely slow (long time constants), whereas their heat capacity and inertia are both enormous.

The distribution of water among land, ocean, and atmosphere plays an important role in determining the water movements in the oceans (currents) and the air movements in the atmosphere (winds, rain, hurricanes), which determine the climate. The total quantity of water on the planet is relatively constant, but its distribution among land, ocean, and atmosphere is not, nor is the ratio among its three phases (solid, liquid, and vapor). This is because, as these phases change, conversions among them (melting, vaporization, condensation, freezing) have different dynamics and cause interactions.

The melting of ice reduces the quantity of water on land and increases the quantity of water in the oceans; melting also cools the receiving waters. This cooling affects the ocean currents. The evaporation of water moves it from the oceans into the atmosphere and takes heat from the oceans into the atmosphere. The energy received by the atmosphere heats the air causing hurricanes, rainstorms, and other weather events as the water vapor condenses (or, in the winter and at higher elevations, freezes) when it is moved by the winds over land. The condensed water feeds the rivers; the frozen water (snow) temporarily accumulates on the land and forms glaciers. The rate of this water circulation and the resulting changes in the movements of both air and water (winds and ocean currents) interact and do not stabilize until their driving force (the heat input from the Sun) stops increasing.

2.4.4 Centrifugal, Gravitational, and Coriolis Forces

Earth's rotation causes centrifugal and Coriolis forces that determine the movement of the air and water. The centrifugal and Coriolis effects are caused by Earth's shape and rotation about its own axis. Earth rotates eastward, which, as viewed from the North Pole, is counterclockwise (from west to east). Thus, as Earth's surface moves[44] eastward, the oceans and the atmosphere lag behind, which, when observed from the North Pole, is clockwise.

The winds and ocean currents usually move east to west. However, the direction is affected by other forces, such as thermal forces, because the temperature difference between the tropics and the polar regions also causes air movements as the wet and warm (low-density) air rises near the Equator and the cold and dry (high-density) air drops at the poles. There are also other influences, such as the effects of landmasses and the tidal acceleration caused by gravitational interactions with the Moon.

Because a point near the Equator, where Earth's diameter is largest, moves (rotates) faster than a point near the poles, anything that moves on Earth's surface (including water and air) is affected by Coriolis forces. One can visualize this phenomenon by imagining a person standing near the Equator and throwing a ball (a projectile in Figure 2-27) in a straight line to a friend standing farther north. Because the thrower is moving faster than the friend, the ball will land to the right of the friend. The same occurs with the wind and ocean currents. The flows that are moved from east to west by the rotation of the planet are bent to the right (clockwise) by the Coriolis force in the Northern Hemisphere and to the left (counterclockwise) in the Southern Hemisphere (Figure 2-27).

The gravitational, centrifugal, and Coriolis forces[45] are relatively constant, but the thermal forces (generated by the "heat engine" of GW) are changing. The winds and ocean currents convey the heat. The ocean currents have more influence on the climate because the mass of the oceans is more than 1000 times greater than that of the atmosphere, and because of the ocean mass, it takes longer for the effects to become

44. Because Earth's circumference at the Equator is 24,901 miles, a spot on the Equator rotates at approximately 1037 mph or 1670 km/h. At the North and South Poles, the speed is effectively zero.

45. The Coriolis effect is named after the French mathematician Gaspard Gustave de Coriolis (1792–1843) who studied the transfer of energy in rotating systems only two centuries ago. This shows how recent our understanding of the various processes on the planet is.

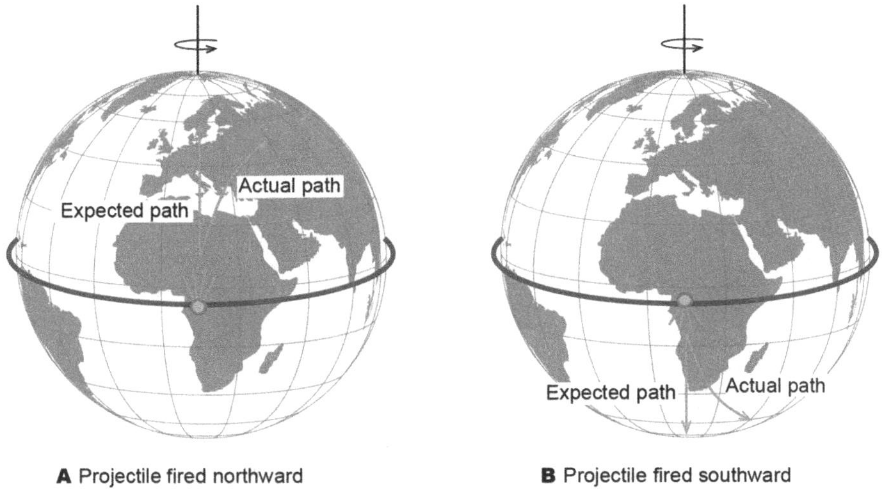

Figure 2-27. The Coriolis force causes air and water currents to bend toward the east. (See color images after page 306.)

evident. Therefore, the atmospheric effects of GW are faster and are already notice-able: dry regions are becoming drier, which is causing wildfires, and wet regions are becoming wetter because of more rain and storm surges caused by hurricanes, both of which cause flooding. These symptoms will increase with time as long as GW acceler-ates them.

In the following section, I discuss some of these interacting subprocesses in a little more detail, starting with the oceanic ones.

2.4.5 Ocean Currents and the Gulf Stream

Temperature and salinity affect the density of water, resulting in gravitational forces moving the water vertically up and down through the ocean layers. The centrifugal and Coriolis forces move the water horizontally. The great heat conveyor belt (GHCB) transports thermal energy from the Equator to the north (Figure 2-28).

Three forces combine to drive the GHCB. The first is the Coriolis effect, which is a consequence of the globe's diameter being larger near the Equator than elsewhere and of the resulting faster movement of the planet's surface near the Equator. The second is that the surface of the Pacific Ocean is 0.2 m (8 in) higher than that of the Atlantic Ocean. This is due to the water being less dense on average on the Pacific side because the temperature and salinity levels cause warmer water to expand. Such sea level dif-ferences are common when short sections of land divide ocean basins. The GHCB's third driving force is the difference in the salinity and temperature of the waters near the Equator and near Greenland in the north.

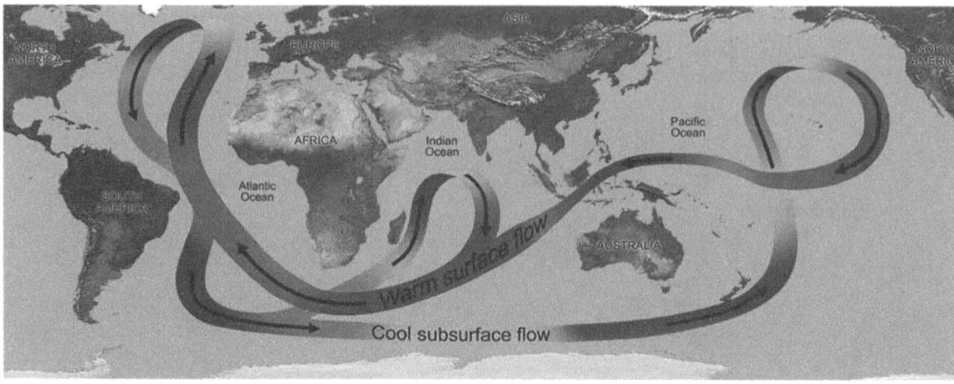

Figure 2-28. The dominant current, called the *great ocean conveyor belt* or the *great heat conveyor belt* (GHCB), moves ocean water around the globe. Warm water on the surface (red) moves northwest (from the Pacific to the Atlantic Ocean). Near Greenland, the cold, salty water (blue) sinks to the bottom of the ocean and returns southeast from the Atlantic back to the Pacific Ocean. (See color images after page 306.)

Source: "Why does the ocean get colder at depth?," National Ocean Service website, accessed March 8, 2022, https://oceanservice.noaa.gov/facts/coldocean.html.

Consequently, the oceans' currents operate as follows: As warm water has relatively low salinity (and is thus lighter), it travels east to west (due to the wind and the rotation of the globe) on the surface of the Atlantic. Some of it evaporates, and this evaporation increases its salinity, making it heavier. As this water continues north, because of the Coriolis effect and because of the landmass of South America blocking its way, cold winds from the land start to cool it, making it even heavier. Finally, near the coast of Greenland, it sinks to the bottom of the Atlantic and turns back. Far larger than all rivers of the planet combined, this gigantic underwater river travels south, becoming warmer and less dense. Completing the loop, as the water's temperature rises and its salinity decreases, its density decreases, it rises to the surface in the Pacific Ocean, and it starts once again to move west.

When the water carried by the GHCB passes through the various seas and oceans, the geography of the landmasses and other effects tend to break it up into subcurrents. One of these is the Gulf Stream, which, on the way north, heats the East Coast of the United States and Europe (Figure 2-29).

As the ice on Greenland melts, more fresh water pours into the ocean, which further reduces its density by lowering its salinity. Both of these effects contribute to reducing the density of the northern end of the Gulf Stream. As its density drops, the Gulf Stream has less of a driving force to sink to the bottom of the ocean and return south. As a result, it slows.

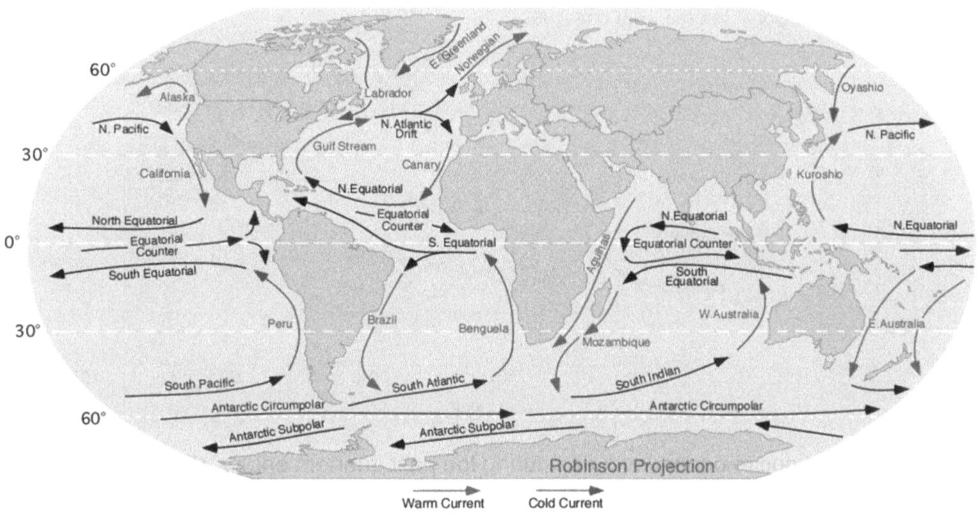

Figure 2-29. Schematic of oceanic currents,[46] where red is warm and black is cold. In both hemispheres, warming currents (red) move heat toward the poles. The Gulf Stream is one current that warms the East Coast of the United States and Europe. Some believe that it is slowing down because the ice on Greenland is melting as a result of GW, reducing the warming of the East Coast. (See color images after page 306.)

Source: "Corrientes-oceanicas.png" by Michael Pidwirny, August 10, 2007, Public Domain, https://commons.wikimedia.org/w/index.php?curid=37108971.

2.4.6 Ocean Levels and the Albedo Effect

The media often give the impression that ocean levels are rising more rapidly than they really are (Figure 2-30). The fact is that the yearly rise of 3 to 4 mm (~5 mm in 2022), although immense given the size of the oceans, is a less immediate concern than some of the other consequences of climate change. We should realize that the floods experienced today are not the consequence of rising sea levels but of hurricanes driving the surface waters toward the shores (storm surges) and the increased water circulation on the planet caused by the increased evaporation of the oceans. These floods threaten the coasts because they can contaminate soil and groundwater with salt.

On the other hand, ocean levels are projected to rise by some 0.3 m (1.0 ft) by the end of 2022 and by about 1.0 m (more than 3 ft) by the end of the century. When combined with tide and wind effects, this will cause substantial flooding in many areas. It is projected that, by 2050, flooding will cause some 25 million people to find themselves

46. The current loops (large systems of rotating ocean currents, or gyres) are as follows: Viking gyre = E. Greenland + Labrador + Norwegian, Columbus gyre = Gulf stream + N. Atlantic Drift + Canary + N. Equatorial, Navigator gyre = S. Equatorial + Brazil + South Atlantic + Benguela, Majid gyre = S. Indian + S. Equatorial + Mozambique + W. Australia, Aloet gyre = Alaska + N. Pacific + Oyashio, Turtle gyre = Kuroshio + N. Equatorial + California, and Heyerdahl gyre = Peru + S. Pacific + E. Australia. The gyres existing at the poles are the Polar bear, Storkerson, and Melville gyres at the North Pole and the Penguin gyre at the South Pole.

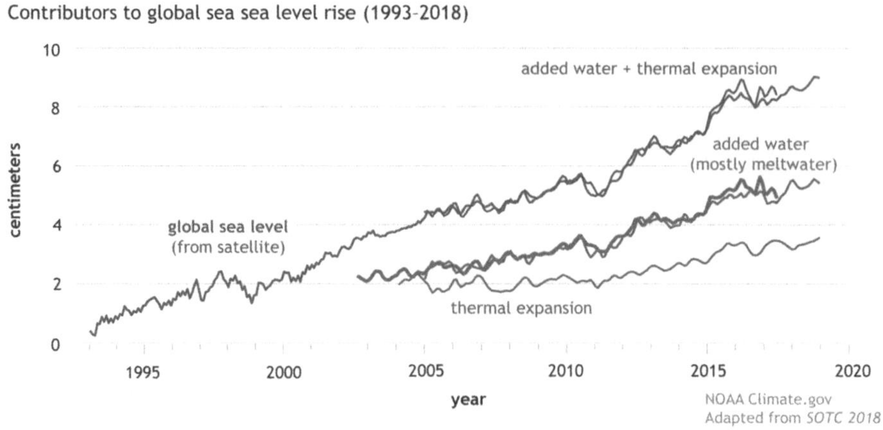

Figure 2-30. Contributions to ocean level rise during the past quarter century. (See color images after page 306.)

Source: Rebecca Lindsey, "Climate Change: Global Sea Level," Climate.gov website, January 25, 2021, accessed March 13, 2022, https://www.climate.gov/news-features/understanding-climate/climate-change-global-sea-level.

living below the high-tide line in coastal cities. Specifically, if GW accelerates, each 1°C (1.8°F) increase in temperature will result in a sea level rise of 2.3 m (7.5 ft) (equivalent to 1.3 m [4.2 ft] per 1°F). Moreover, if all of the ice on Greenland melted, the sea level rise would be 7.7 m (25.6 ft). If all of the ice on Antarctica melted, the sea level rise would be 61 m (210 ft), and if all of the ice on the planet melted, the rise would be 66 m (220 ft). For the foreseeable future, it seems that such an extreme amount of melting is unlikely to occur, except possibly in Greenland.

Unfortunately, people often focus on addressing the symptoms instead of the cause. One example is the expensive proposal to defend Europe against rising sea levels[47] by building one dam between Scotland and Norway and another between England and France (Figure 2-31). Because the North Sea is up to about 1000 ft (300 m) deep, build-ing these dams would consume the total concrete capacity of the planet and eventually turn the North Sea into a freshwater lake, thereby destroying all of the saltwater life forms in it. Also, it would interfere with the Gulf Stream, which would have serious climate consequences.

Naturally, I hope that such things will not happen and that we will stop the rise of ocean levels in time by converting to the use of green fuels. The approach of attempting to mitigate the symptoms of GW is not limited to ocean level rise; it extends to all

47. For more information, see: "Researchers: 'Only a dam across the North Sea helps against projected sea rises'," Water News Europe website, February 16, 2020, https://www.waternewseurope.com/researchers-only-a-dam-across-the-north-sea-would-help-against-projected-sea-rises/.

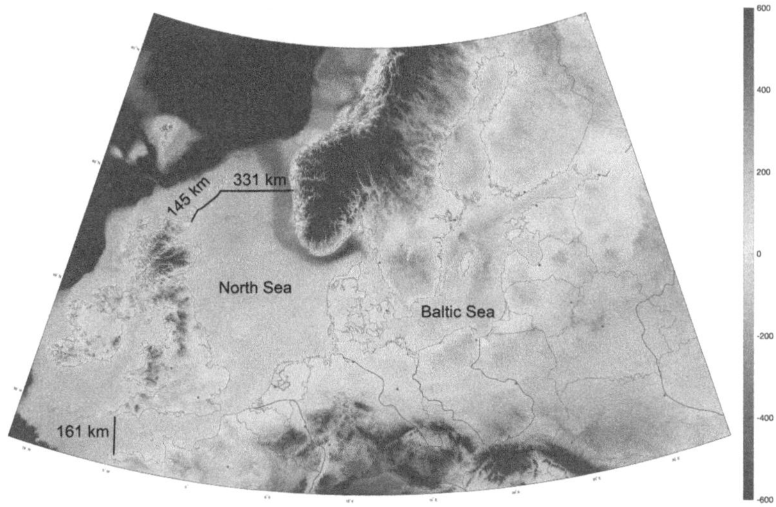

Figure 2-31. A 300-m-deep dam system between England and France and between Scotland and Norway (black) would potentially cost $1 trillion and turn the North Sea into a freshwater lake.

Source: Sjoerd Groeskamp and Joakim Kjellsson, "NEED: The Northern European Enclosure Dam for if climate change mitigation fails," Bulletin of the American Meteorological Society 101, no. 7 (July 2020): E1174–E1189, https://doi.org/10.1175/BAMS-D-19-0145.1. Licensed under CC BY 4.0.

cases in which we are attempting to live with the symptoms of climate change instead of eliminating its causes.

Let us consider why the North Sea level is rising and how and why melting occurs. Both of Earth's polar regions are cold, primarily because they receive far less solar radiation than the tropics and midlatitudes. At both poles, the Sun never rises more than 23.5° above the horizon, and both locations experience 6 months of continuous darkness. Moreover, most of the sunlight that does shine on the polar regions is reflected back into the atmosphere by the bright white surface of the snow.

The South Pole is much colder than the North Pole because it is covered by a thick ice sheet, which sits on a continent. The ice sheet thickness at the South Pole is roughly 2750 m (about 9000 ft) tall, and its top surface is close to a mile and a half above sea level. Antarctica is by far the highest continent on Earth. In comparison, the North Pole rests in the middle of the Arctic Ocean, where the surface of the floating ice is only a foot or so above the surrounding sea. The Arctic Ocean also acts as an effective heat reservoir, warming the cold atmosphere in the winter and drawing heat from the atmosphere in the summer.

The best-known consequence of GW is the rise of ocean levels due to the melting of polar ice. Yet, if one examines the melting process from a process control perspective, it

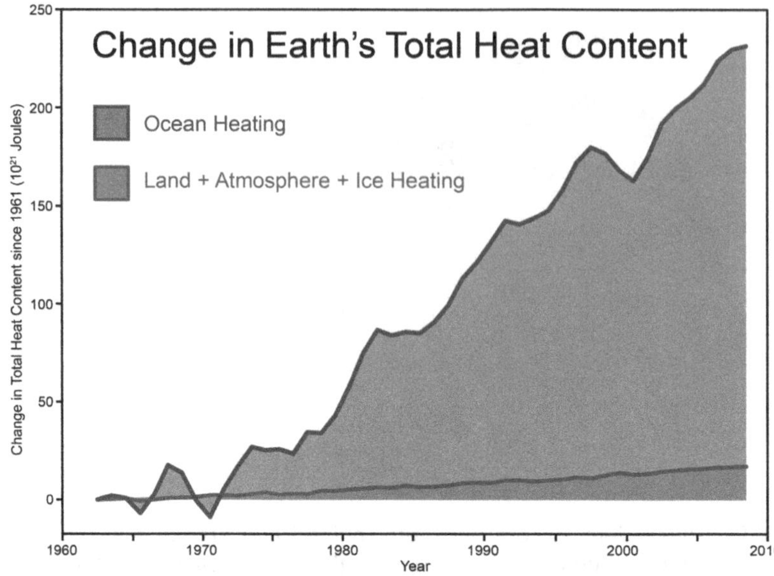

Figure 2-32. Global heat accumulation: Some 90% of the GW energy accumulates in the oceans. By 2010, GW caused the accumulation of more than 2.0×10^{23} J (2.0×10^5 Q), and by the time of this writing, this accumulation had increased by another 25%. (See color images after page 306.)

Source: "Total Heat Content (2011 update)" by Skeptical Science is licensed under CC BY 3.0, accessed March 8, 2022, https://static.skepticalscience.com/graphics/Total_Heat_Content_2011.jpg.

becomes obvious that ocean levels rise very slowly and that many other consequences of climate change occur much more rapidly (e.g., migration of species from the tropics caused by increasing temperatures and drought in many areas).

The slowness of sea level rise is caused by the immense surface area (362 million square kilometers) and the mass of the oceans. This enormous mass keeps the rise of ocean temperatures below that of land areas under 0.2°C (0.36°F) per decade despite the huge increase in solar heating. The data in Figure 2-32 shows that the heat content of the oceans is much greater and rises much faster than that of the atmosphere and land areas.

To date, about one-third of the increase in ocean levels has been due to thermal expansion, and one-third has been caused by the melting of polar and glacier ice, as the ocean has been getting progressively warmer. The ocean levels have increased only 7 to 10 cm (3 to 4 in) since 1993 (Figure 2-33), and according to the Center on Global Energy Policy, it is expected to rise by about another 1 m (3.3 ft) by the end of this century.

I do not intend to minimize the seriousness of this rise! The consequences of such a relatively small rise can still be disastrous. For example, an increase of just 0.6 m (2 ft) would flood the towns and villages of 30 to 80 million people. Still, 2100 is a long time

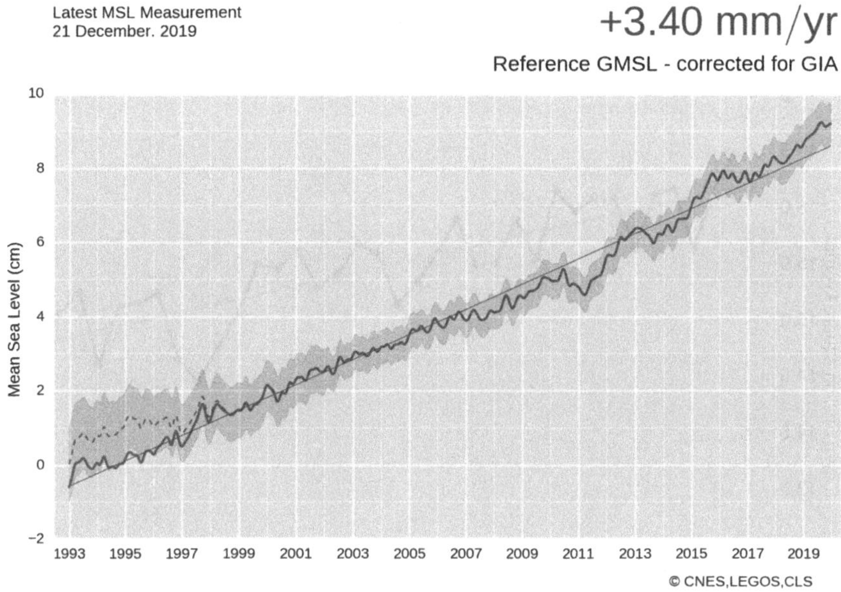

Figure 2-33. The rate at which the sea level increased between 1993 and 2019. In 2019, it was about 5 mm (0.2 in) per year. Abbreviations: GIA, glacial isostatic adjustment (also called postglacial rebound); GMSL, global mean sea level.

Source: Reprinted with permission. "Mean Sea Level from Altimetry – 1993–2019," Aviso website, accessed March 8, 2022, https://www.aviso.altimetry.fr/en/data/products/ocean-indicators-products/mean-sea-level.html.

from now, and hopefully, we will correct the situation by then. As of today, the sea level is rising by about 3 to 5 mm (0.12 to 0.20 in) per year, however, the main reason for flooding is not the rise of ocean levels but the flooding due to rainstorms and hurricanes.

The melting of floating sea ice affects global temperature but not sea level, whereas melting ice on land affects both. The total ice volume on the planet is about 35 million cubic kilometers. Of this total volume, the land ice is about 33 million cubic kilometers (about 2.5 million cubic kilometers on Greenland, 30 million cubic kilometers in Antarctica, and 1 to 2 million cubic kilometers elsewhere in the planet's glaciers).

Most studies agree that the loss of ice mass is approximately 1% per decade, whereas the decrease in the area that the ice covers is about 10% per decade. Figure 2-34 shows the trend in the Arctic for the month of March from 1979 to 2014. The area of the Arctic Ocean is 14 million square kilometers (5.4 million square miles); the area covered by ice changes with the season, reaching its maximum in February and its minimum in September. The minimum coverage in the Arctic occurred in September 2012, when it reached 3.41 million square kilometers (1.32 million square miles). All studies agree that the rate of melting is increasing. This rate is the highest in the Arctic waters (13% per decade). In the summer of 2022, the area of the Arctic Sea ice tied the tenth lowest

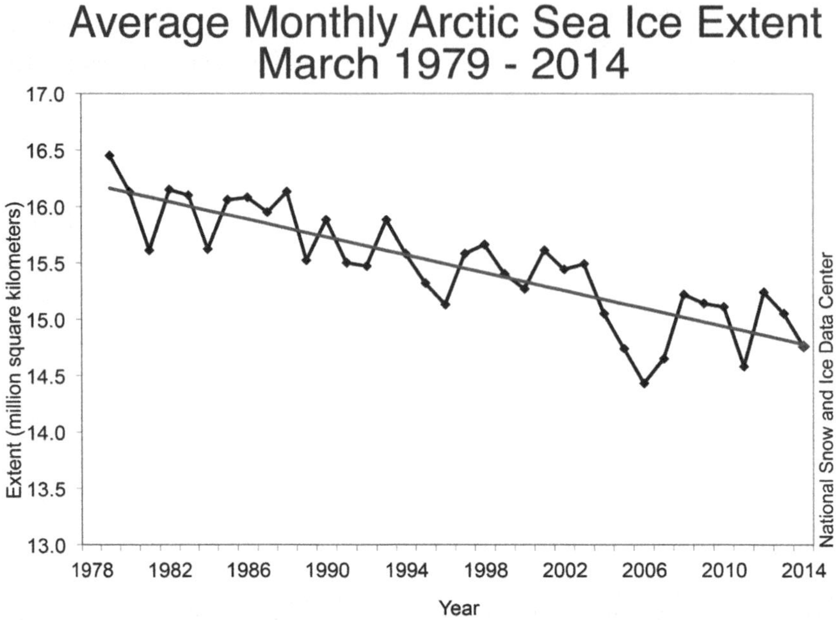

Figure 2-34. The trend in the area of ice cover in the Arctic in the month of March from 1979 to 2014 shows a decline of 2.6% per decade relative to the 1981–2010 average.

Source: "Arctic sea ice at fifth lowest annual maximum," National Snow and Ice Data Center website, April 2, 2014, http://nsidc.org/arcticseaicenews/2014/04/arctic-sea-ice-at-fifth-lowest-annual-maximum/.

area on record. According to some projections, if the present rate of GW continues, the Arctic Ocean could be ice-free in the summer between 2030 and 2040.

Let us consider the rate at which the ocean level is rising[48] and how much cooling is provided by the melting of this ice. As of this writing, the total yearly melting on the planet is about 1000 GT/year (or, in terms of volume, 1090 km^3/year). According to NASA,[49] each millimeter of ocean level rise is a result of the melting of 364 GT of ice. If the melting rate is 1000 GT per year and the area of the oceans is 3.6×10^8 km^2, then

---

48. Melt data:
    - Melt in Antarctica, 219 GT/year ("200 Billion Tonnes of Antarctic Ice Is Melting into the Sea Each Year and It's Expected to Increase," IFLScience website, accessed March 8, 2022, https://www.iflscience.com/environment/200-billion-tonnes-of-antarctic-ice-is-melting-into-the-sea-each-year-and-its-expected-to-increase).
    - Melt of all the glaciers in 2019, 390 GT (Doyle Rice, "Good-bye glaciers: 390 billion tons of snow and ice melts each year as globe warms," *USA Today*, April 8, 2019, https://www.usatoday.com/story/news/nation/2019/04/08/glaciers-melting-ice-melts-due-global-warming-sea-levels-rise/3405569002/).
    - Melt in Greenland in 2019, 200 GT (Bob Berwyn, "Greenland's Nearing a Climate Tipping Point. How Long Warming Lasts Will Decide Its Fate, Study Says," Inside Climate News website, December 23, 2019, https://insideclimatenews.org/news/23122019/greenland-ice-sheet-climate-tipping-point-temperature-duration-sea-level-rise-pnas-study).
49. "Understanding Sea Level," NASA Sea Level Change website, https://sealevel.nasa.gov/understanding-sea-level/global-sea-level/ice-melt.

this melting corresponds to an ocean level rise of about 3 mm per year. This is consistent with the actual measurements at the time of this writing (5 to 6 mm/year [0.20 to 0.24 in year]) because about half of the ocean level rise is due to the thermal expansion of ocean waters.

Now we come to the amount of cooling provided by melting. It takes 2.3×10^5 Btu (2.3×10^{-10} Q) to melt one ton of ice, so the cooling effect of melting 1000 GT of ice is 230 Q. Compared to the yearly RF of our planet, the amount of excess heat that is causing GW (28,000 Q), this cooling effect is relatively negligible.

The total volume of ice on the planet is about 33 million cubic kilometers, so melting all of this ice would require about 6 million to 9 million quads of heat. We can also approach this calculation in another way: If 1 mm of ocean level rise is caused by the melting of 364 GT of ice, and if it is assumed that half of that rise comes from thermal expansion (with the other half coming from melting ice: glaciers, 21%; Greenland ice sheet, 15%; and Antarctic ice sheet, 8%) and that the present rate of yearly ocean level rise is 5 to 6 mm (0.20 to 0.24 in), then the weight of ice that melts yearly is about 910 GT. The heat required to melt this quantity is $[(9.1 \times 10^{11}\ \text{T}) \times (0.23 \times 10^6\ \text{Btu/T})]/(10^{15}\ \text{Btu/Q}) = 210$ Q. Therefore, if this calculation is correct, it seems that the cooling effect of the melting ice on the planet is insignificant compared to the heat excess causing GW. Although the cooling effect of the melting ice is negligible, the heating effect resulting from the melting of the ice cover on land and oceans is significant because it increases the absorption and reduces the reflection of solar energy back into space.

If we consider the longer range, between 1900 and 2016, the sea level rose by 16 to 21 cm (6.3 to 8.3 in). More precise data from satellite radar[50] measurements reveal an accelerating sea level rise of 7.5 cm (3.0 in) from 1993 to 2017, indicating a trend of roughly 30 cm (12 in) per century. This acceleration is due primarily to human-caused[51] GW,[52] which is driving the thermal expansion[53] of seawater and the melting of land-based ice sheets[54] and glaciers.[55] Between 1993 and 2018, the thermal expansion of the oceans contributed 42% to sea level rise; the melting of temperate glaciers, 21%; the melting of

50. "Radar," Wikipedia, accessed November 1, 2022, https://en.wikipedia.org/wiki/Radar.
51. "Attribution of recent climate change," Wikipedia, accessed November 1, 2022, https://en.wikipedia.org/wiki/Attribution_of_recent_climate_change.
52. "Climate Change," Wikipedia, accessed October 27, 2022, https://en.wikipedia.org/wiki/Climate_change.
53. "Thermal expansion," Wikipedia, accessed November 1, 2022, https://en.wikipedia.org/wiki/Thermal_expansion.
54. "Ice sheet," Wikipedia, accessed November 1, 2022, https://en.wikipedia.org/wiki/Ice_sheet.
55. "Glacier," Wikipedia, accessed November 1, 2022, https://en.wikipedia.org/wiki/Glacier.

Greenland's ice sheet,[56] 15%; and the melting of the Antarctic[57] ice sheet, 8%. Climate scientists expect the rate to increase further during the twenty-first century.

As climate research[58] into determining past and present sea levels improved, computer model[59] projections also became more accurate. For example, in 2007, the IPCC projected a high-end estimate for sea level rise of 60 cm (2 ft) through 2099,[60] but their 2014 report raised the high-end estimate to about 90 cm (3 ft). A number of later studies have concluded that the global sea level rise by the end of the century will reach 200 to 270 cm (6.6 to 8.9 ft). A conservative estimate of the long-term projections is that each degree Celsius of temperature rise triggers a sea level rise of approximately 2.3 m (4.2 ft/°F).

Albedo commonly refers to the whiteness of a surface, with 0 meaning black and 1 meaning white. A value of 0 means the surface is a perfect absorber, absorbing all incoming energy, and a value of 1.0 means the surface is a perfect reflector, reflecting all incoming energy. The Arctic Ocean covers about 1% of Earth's surface. The melting of that sea-covering ice results in substantial heating because ice acts like a mirror, reflecting solar energy back into space. In contrast, uncovered ocean and land surfaces absorb solar energy. This occurs because, as more and more ice cover disappears from the surface of the Arctic Ocean (6.5×10^6 km^2), the surface changes from a highly reflective ice surface into a heat-absorbing land and water surface. This is called the *ice-albedo feedback effect*. In addition to the albedo effect, the melting permafrost heats the planet because it releases large amounts of GHGs, including methane, that further contribute to GW.

The ocean reflects only 7% of the incoming solar radiation and absorbs the rest, whereas sea ice reflects 35% to 65% of the incoming energy. Therefore, sea ice absorbs less solar energy than water and keeps the surface cooler. Snow has an even higher albedo than sea ice, so thick sea ice covered with fresh snow reflects as much as 80% of the incoming solar radiation. (This albedo of 0.8 applies only to fresh snow because as its surface turns to ice, its albedo drops.) The snow insulates the sea ice, maintaining cold temperatures and delaying ice melt in the summer. After the snow does begin to melt, the resulting shallow melt ponds deepen, and the surface albedo can drop to 0.15. As a result, melt ponds cause higher energy absorption and more rapid ice melting.

---

56. "Greenland," Wikipedia, accessed November 1, 2022, https://en.wikipedia.org/wiki/Greenland.
57. "Antarctica," Wikipedia, accessed November 1, 2022, https://en.wikipedia.org/wiki/Antarctica.
58. "Climatology," Wikipedia, accessed November 1, 2022, https://en.wikipedia.org/wiki/Climatology.
59. "Climate model," Wikipedia, accessed November 1, 2022, https://en.wikipedia.org/wiki/Climate_model.
60. "Sea Level Rise," Wikipedia, accessed October 27, 2022, https://en.wikipedia.org/wiki/Sea_level_rise.

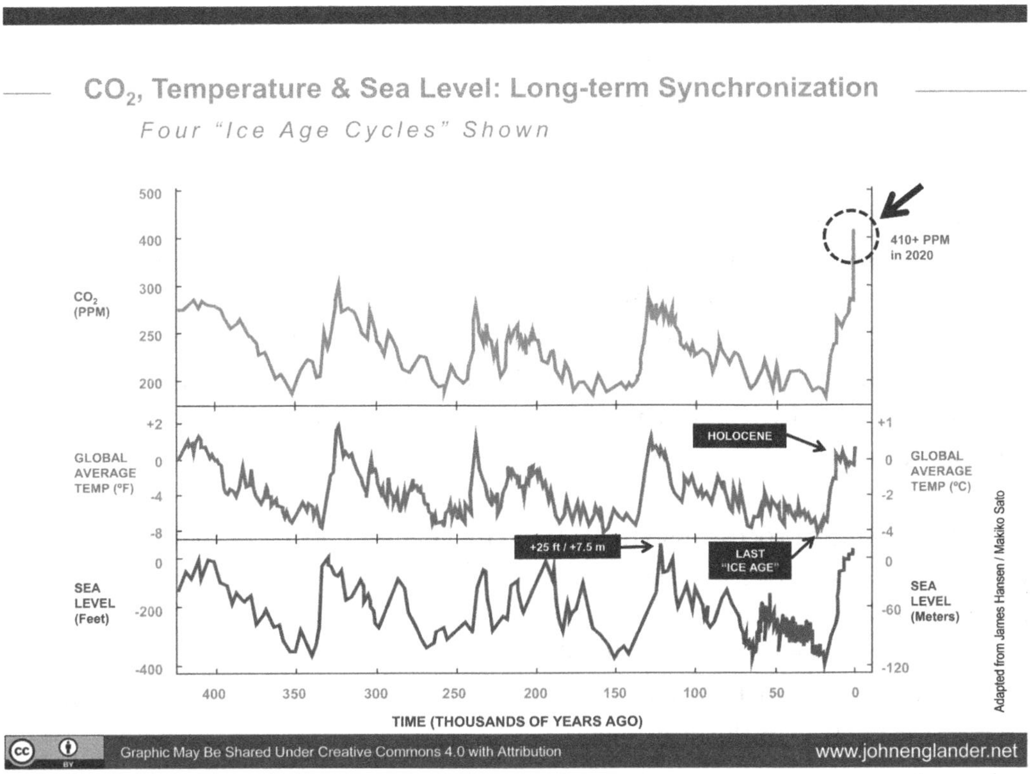

Figure 2-35. During the past 420,000 years, sea level variations followed the changes in atmospheric CO_2 concentration and global temperature. It is logical to assume that this will continue.

Source: Courtesy of John Englander, http://www.johnenglander.net.

A cycle evolves in which the melting snow and ice causes the albedo to drop, so heat absorption increases, which further increases melting. In addition, melting permafrost indirectly increases the temperature because this permanently frozen soil stores massive amounts of methane and carbon. As a result of warming, permafrost releases stored methane and CO_2—powerful heat-trapping gases—further increasing heating.

The total warming at the poles is due to changes in clouds, water vapor, surface albedo, atmospheric temperature, and season, with greater warming occurring in the summer than in the winter. The combination of the heating effects of the albedo and permafrost is much greater than the cooling effect of the melting ice. Therefore, the overall process of ice melting at the poles or in glaciers heats the planet. For all of the listed reasons, ocean level rise is a low-gain, time-delayed process, whereas the atmospheric greenhouse effect of the rise in CO_2 concentration is a fast process (Figure 2-35).

Even if emissions stopped today, a substantial additional rise in sea level is inevitable because the sea level is rising in a doubly delayed process. It is delayed twice because the heat first has to accumulate in the ocean before the ocean temperature starts rising, and this accumulation starts only after the concentration of atmospheric GHGs increases. Consequently, the ocean level rises very slowly. In addition, the water produced by melting is distributed over an immense area (3.62×10^8 km^2, 72% of the planet's surface).

So today, the total yearly rise in the level of the oceans is only about 5 mm (0.2 in), half of which is caused by thermal expansion. On the other hand, process control indicates that, after this time delay has passed, the melting process will accelerate and, in some areas, it might become irreversible. This can occur in the Arctic Ocean when the melting due to rising ocean temperature is further accelerated by the shrinking of the area of the reflective ice, which causes more heat absorption and further emissions of GHGs (including significant amounts of methane).

This doubly delayed process also has a large inertia, and once the melting rate reaches a tipping point, it can become hard to stop. Theoretically, if all land ice on our planet melted in thousands of years, the level of the 362 million square kilometers of oceans would rise by some 69 m (226 ft). In 2019, the rate of melting in the Arctic exceeded 500 GT per year, and most models agree that, at this rate, in a decade or two, the Arctic could become ice-free in the summer.

Greenland is warming more than twice as fast as the rest of the planet. If the 2.5 million cubic kilometers of ice on Greenland melts, the ocean levels will rise by some 7 m (23 ft). Greenland has been ice-free in the past, and most models predict that it could become ice-free again in about 1000 years.

In the Antarctic, less than 250 GT of ice melts in a year (3000 GT melted in 25 years).

All of these numbers are probably confusing. What might be easier to remember is that the yearly rise in sea level today is only about 5 mm (0.2 in) and that each additional 1.0°C (1.8°F) increase in global temperature results in a sea level rise of approximately 0.3 m (about 12 in). In other words, most flooding today is caused by storm surges driven by hurricanes and increased water circulation (heavy rains), not by rising ocean levels.

Compared to ocean level rise, other effects of GW are more substantial. One of these effects involves the increased evaporation of the oceans and the drying out of land near the Equator. Consequences include increased hurricane formation and heavy rain in the wet areas due to the increased rate of water circulation, making dry areas drier and wet areas wetter, along with causing drought and wildfires in the dry areas. Another effect of the land becoming drier is the lowering of groundwater levels. Such a loss of water impacts agriculture and the availability of drinking water. Evidence of

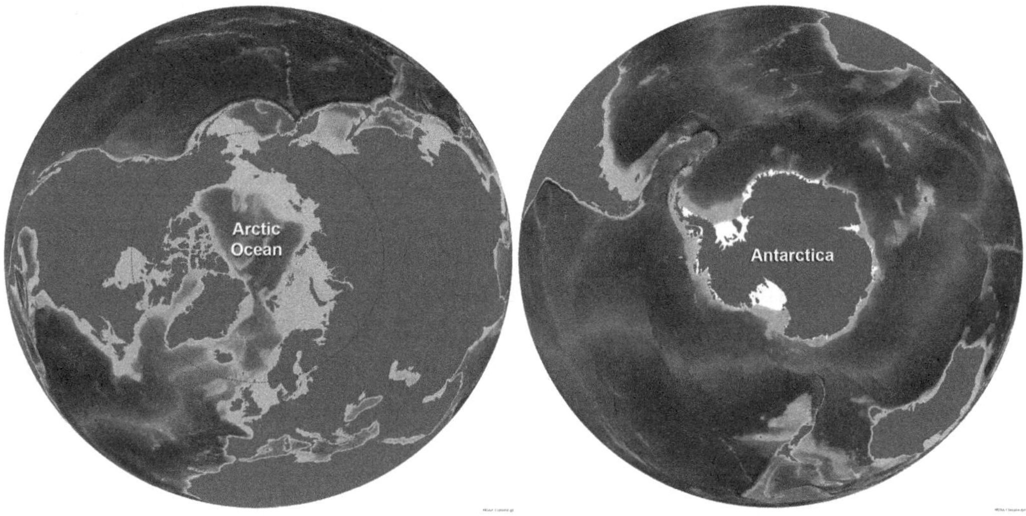

Figure 2-36. The Arctic and Antarctic regions are geographic opposites, not just because they sit on opposite ends of the globe. They also have opposite land–sea arrangements. In the Arctic, there is an ocean surrounded by continents, whereas the Antarctic is a continent surrounded by oceans.

Source: Michon Scott and Rebecca Lindsey, "Polar Opposites: the Arctic and Antarctic," NOAA, July 10, 2011, accessed March 14, 2022, https://www.climate.gov/news-features/understanding-climate/polar-opposites-arctic-and-antarctic.

this effect of GW can be seen in dried-out riverbeds and the fact that the Sea of Galilee is beginning to look like the Dead Sea. What is not as obvious is the process going on underground: As the groundwater level drops, the salt water of the oceans moves in. The consequences of this are many and cannot be solved by desalination of the salt water, although huge desalination facilities have been built in some areas (such as in Israel), and they do help a little.

2.4.7 Arctic and Antarctic Ice Measurements

Process control plays an important role in the remote measurement of the areas, thicknesses, and other properties of Arctic, Antarctic, and glacier ice formations (Figure 2-36), including detecting the rate at which they are melting. Many instruments are available for these measurements, but their detailed description is beyond the scope of this book.[61] The most common ones are the Atmospheric Infrared Sounder (AIRS) and OCO-2.[62]

---

61. For a full description of the various measurement devices and sensors, see: Béla G. Lipták and Kriszta Venczel, eds., Instrument and Automation Engineers' Handbook, 5th ed. (Boca Raton: CRC Press, 2016).

62. For detailed information on these systems, see the following websites: "AIRS," JPL, https://airs.jpl.nasa.gov/. "Orbiting Carbon Observatory-2," JPL, accessed March 8, 2022, https://ocov2.jpl.nasa.gov/.

In the Arctic, NASA's Operation IceBridge studies the effects of polar ice on Earth's climate. Researchers use highly sophisticated airborne instruments on retrofitted aircraft to measure annual changes in the thickness and movements of the ice. Also, as already described, floats from the Argo program can detect ocean salinity and temperatures at depths of about 2000 m (6500 ft) and communicate the readings back to satellites.

Although it can be difficult to measure ice melt, a new NASA satellite (ICESat-2—Ice, Cloud, and Land Elevation Satellite—launched in September 2018) is expected to help scientists gather enough data to gauge future changes in the Greenland and Antarctic ice sheets and estimate their annual heights to accuracies within the width of a pencil.[63]

2.4.7.1 Laser Altimetry

By the 1990s, laser altimetry[64] from aircraft revealed that ice sheets were thinning rapidly enough on Greenland's coastal margins to create an imbalance as outlet glaciers lost ice mass to the ocean.[65] Later surveys using NASA's Airborne Topographic Mapper (ATM) showed that the thinning rates had increased.

2.4.7.2 Operation IceBridge

The largest airborne survey of polar ice is known as Operation IceBridge, a NASA mission designed to fill the gap between ICESat, which ceased functioning in 2010, and the launch of ICESat-2 in September 2018. Researchers used data from these yearly surveys to reveal that the rate of ice discharge from the two fastest glaciers draining the Greenland and Antarctic ice sheets was increasing. The study also relied on data from NASA's ATM and the University of Kansas's ice-depth sounder, both of which

---

63. Thorsten Markus, Tom Neumann, Anthony Martino, Waleed Abdalati, Kelly Brunt, Beata Csatho, Sinead Farrell, Helen Fricker, Alex Gardner, David Harding, Michael Jasinski, Ron Kwok, Lori Magruder, Dan Lubin, Scott Luthcke, James Morison, Ross Nelson, Amy Neuenschwander, Stephen Palm, Sorin Popescu, C.K. Shum, Bob E. Schutz, Benjamin Smith, Yuekui Yang, and Jay Zwally, "The Ice, Cloud, and Land Elevation Satellite-2 (ICESat-2): Science requirements, concept, and implementation," *Remote Sensing of Environment*, 190 (2017): 260–273, accessed March 14, 2022, https://icesat-2.gsfc.nasa.gov/sites/default/files/multimedia/Markus_etal_RSE_2017.pdf.

64. Beata M. Csatho, Anton F. Schenk, Cornelis J. van der Veen, Gregory Babonis, Kyle Duncan, Soroush Rezvanbehbahani, Michiel R. van den Broeke, Sebastian B. Simonsen, Sudhagar Nagarajan, and Jan H. van Angelen, "Laser Altimetry Reveals Complex Pattern of Greenland Ice Sheet Dynamics," *Proceedings of the National Academy of Sciences of the United States of America* 111, no. 52 (December 30, 2014): 18478–18483, https://doi.org/10.1073/pnas.1411680112.

65. W. Abdalati, W. Krabill, E. Frederick, S. Manizade, C. Martin, J. Sonntag, R. Swift, R. Thomas, and W. Wright, "Outlet Glacier and Margin Elevation Changes: Near-Coastal Thinning of the Greenland Ice Sheet," *Journal of Geophysical Research* 106, no. 33 (December 1, 2001): 33729–33742, https://doi.org/10.1029/2001JD900192.

have made almost yearly surveys of the Jakobshavn Glacier in Greenland since 1991 and of the Pine Island Glacier in Antarctica since 2002.

2.4.7.3 Radar Interferometry

Satellite synthetic-aperture radar interferometry techniques measure the speed at which ice streams move, as well as the position of the grounding line, which separates ice grounded over bedrock from ice floating on the ocean. One example of the power of this technique is the measurement of the retreat of Antarctic glaciers. Researchers using synthetic-aperture radar data from European remote sensing satellites ERS-1 and ERS-2 from 1992 to 2011 found that the Pine Island, Thwaites, Haynes, and Smith-Kohler glaciers in West Antarctica are rapidly retreating.

The differences among sets of radar altimetry measurements for the same time intervals—the "interference" of interferometry—enabled the researchers to compensate for the tracking drift caused by the loss of gyroscopes in ERS-2. The study produced color-coded maps of the velocity of recession of the glaciers, which showed the movement of the glacial grounding lines. The authors concluded that "this sector of West Antarctica is undergoing a marine ice sheet instability that will significantly contribute to sea level rise in decades to centuries to come"[66] because the bed on which the ice rests is retrograde—that is, it deepens going inland, toward the south. The same interferometric techniques were used to map the seaward flow of networks of glaciers across the entire continent from 2007 to 2009.

2.4.7.4 Paleo Observations

As with ancient sea levels, the collapse and regrowth of ice sheets over the ages can be measured using paleontological proxies. One intriguing study examined coral terraces that had been thrust some 20 to 140 m (66 to 459 ft) above present-day sea level in Papua New Guinea because of tectonic uplift. The researchers linked the episodic rise and fall of sea level between 75,000 and 25,000 years ago to alternating warm and cold periods during the last ice age, lasting roughly millennia but transitioning on the scale of decades. In addition, these bouts of warming and cooling could be read in the patterns of coral growth on the uplifted terraces. The resulting sea level changes were attributed to the partial collapses and later regrowth of the Northern Hemisphere ice sheets, causing sea levels to rise and then fall. The vertical position of coral on the terraces told the story: During periods of falling relative sea level, coral growth was spotty, whereas rising sea levels allowed the coral to build more robust colonies at higher elevations.

---

66. Michon Scott and Rebecca Lindsey, "Polar Opposites: the Arctic and Antarctic," July 10, 2011, accessed March 14, 2022, https://drupal-www.climate.woc.noaa.gov/news-features/understanding-climate/polar-opposites-arctic-and-antarctic.

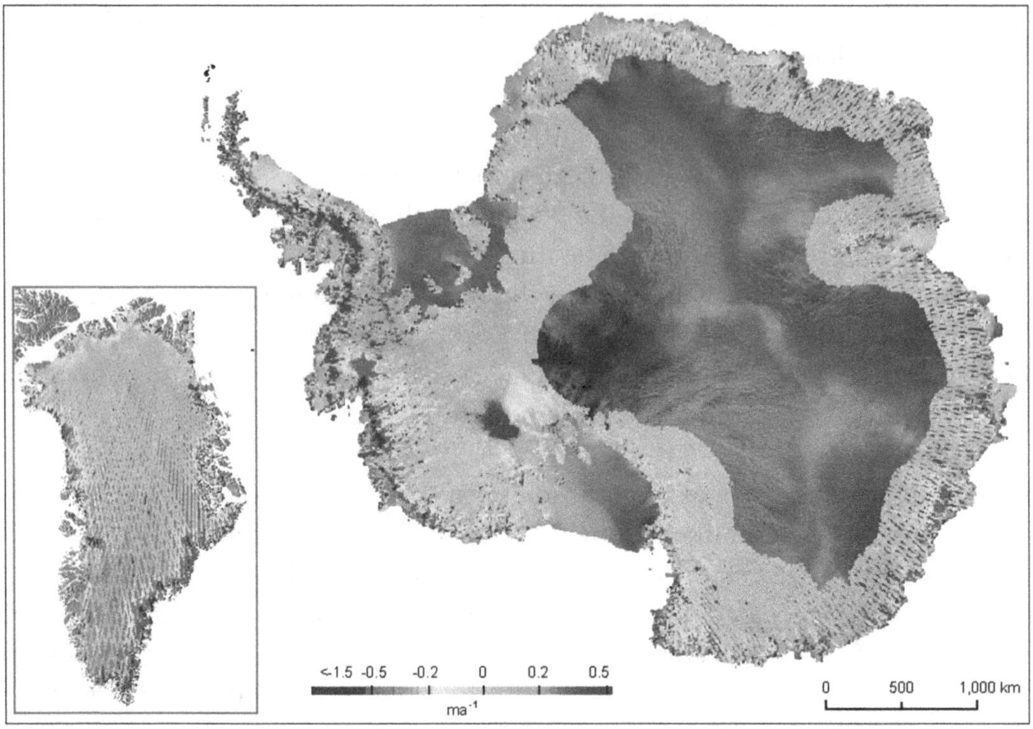

Figure 2-37. Satellite altimetry photograph from NASA's ICESat reveals areas of dynamic thinning (red) in Antarctica (*main image*) and Greenland (*bottom left*). The colors indicate ice loss (red) and gain (blue) in meters. (See color images after page 306.)

Source: "Fast-Flowing Glaciers Thin Greenland and Antarctic Ice Sheets," NASA, October 3, 2009, https://www.nasa.gov/images/content/389001main_ICESatThinningLead_226x160.jpg.

2.4.7.5 Satellite Altimetry

The first attempt to measure ice sheets using satellite radar altimetry (Figure 2-37) took place in the early 1980s, with observations of only limited parts of Greenland and Antarctica. Since then, however, altimetry technology has become an important means of determining ice sheet [67]and glacial mass balance—that is, the ledger of gains and losses in ice mass. Although the primary measurements are by laser altimetry (NASA's ICESat and ICESat-2 missions), which has high accuracy and a very small footprint, radar altimetry is also used, starting with ESA's ERS-2 (1995) and Envisat (2002) missions and continuing with the 2010 ESA CryoSat-2 mission, optimized for ice work.

---

67. Ice sheet elevation has long been observed by radar altimetry, and the recent successful launch of CryoSat-2 represents "a great improvement in these observations [because] it addresses many of the limitations of polar monitoring with previous systems, which were primarily targeted at the oceans." H. D. Pritchard, S. B. Luthcke, and A. H. Fleming, "Understanding ice-sheet mass balance: Progress in satellite altimetry and gravimetry," *Journal of Glaciology*, 56 (2010): 1151–1161, accessed March 14, 2022, doi:10.3189/002214311796406194.

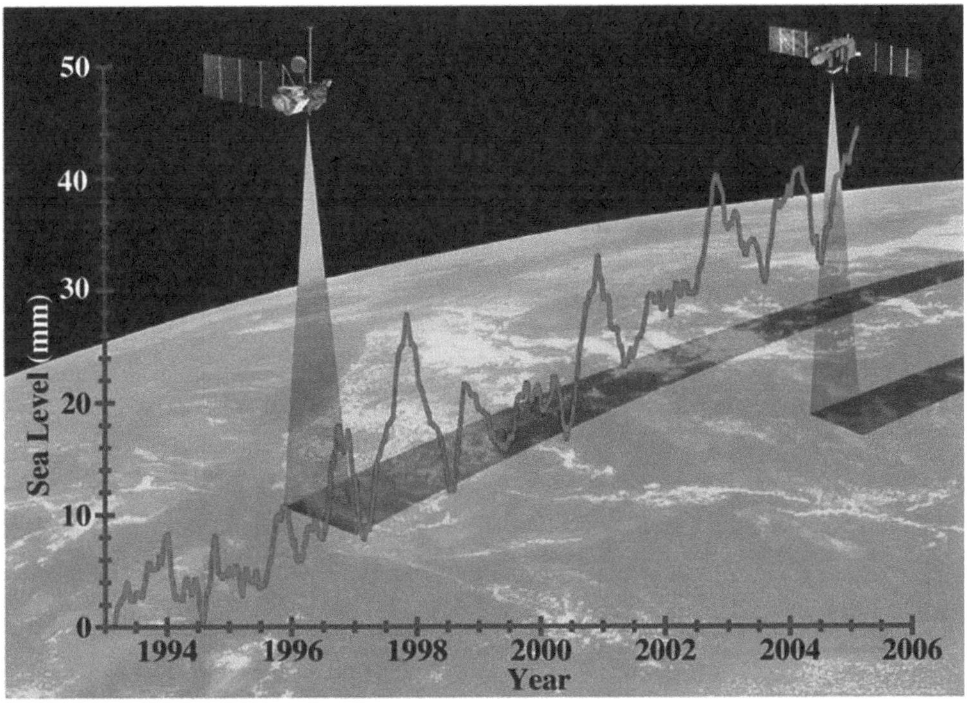

Figure 2-38. Sea level monitoring data resulting from satellite gravimetry tracking. Sea Level (mm) refers to the sea level rise.

Source: "JPL Sea Level Satellite Measurement Reports," NASA/JPL, last updated April 2013, http://ossfoundation.us/projects/environment/global-warming/sea-level-rise.

Although ICESat-2's laser beams can pass through some wispy clouds, other clouds block the beam, reflecting light back to the satellite. Cloudy days are not necessarily bad news, however, as atmospheric scientists can gather information on the cloud heights, compositions, densities, and more to advance the field of cloud physics. With the satellite's ability to detect elevation changes on such a fine scale, scientists expect that the data will lead to unexpected discoveries.

The advent of gravimetric measurements with the twin Gravity Recovery and Climate Experiment (GRACE) satellites in 2002, along with the more recent deployment of floating Argo sensors, opened the way to achieving closure of the sea level budget. GRACE measures changes in water mass, including terrestrial storage in the form of groundwater, rivers, snow, and ice and mass changes within the ocean itself, as well as the movement of water between land and ocean (Figure 2-38).

Early attempts did not achieve closure of the sea level budget, leading to concerns about possible instrument drift. However, more recent efforts have led to reports

Figure 2-39. The Arctic region (surrounded by eight countries) could become ice-free by 2030\ to 2040. *Source:* "Political Map of the Arctic.pdf," US State Department, circa 2014, accessed March 14, 2022, https://en.wikipedia.org/wiki/File:Political_Map_of_the_Arctic.pdf.

of closure for more extended periods, including a NOAA report covering the years 2005 to 2013.[68]

2.4.8 Ice Cover of the Arctic and Greenland

The name Arctic came from the Greek word "arctos," which translates as "bear." "The Arctic: A Delicate Icy Ecosystem,"[69] article on the SciTechDaily website provides an excellent description of the area and its inhabitants.

> The Arctic consists of the Arctic Ocean, adjacent seas, and parts of Alaska (United States), Canada, Finland, Greenland (Denmark), Iceland, Norway, Russia, and Sweden [see Figure 2-39]. Land within the Arctic region has seasonally varying snow and ice cover, with predominantly treeless permafrost (permanently frozen underground ice) containing tundra. Arctic seas contain seasonal sea ice in many places.

---

68. Eric Leuliette, *The Budget of Recent Global Sea Level Rise, 2005–2013* (Washington, DC: National Oceanic and Atmospheric Administration, September 30, 2014).

69. European Space Agency (ESA), "The Arctic: A Delicate Icy Ecosystem," SciTechDaily website, accessed December 1, 2022, October 7, 2020, https://scitechdaily.com/the-arctic-a-delicate-icy-ecosystem-video/.

> The Arctic region is a unique area among Earth's ecosystems. The cultures in the region and the Arctic Indigenous peoples have adapted to its cold and extreme conditions. Life in the Arctic includes zooplankton and phytoplankton, fish and marine mammals, birds, land animals, plants, and human societies.

What makes the South Pole so much colder than the North Pole is that it sits on top of a very thick ice sheet, which sits on a continent, whereas the Arctic ice is close to (floating in) the warming ocean. The main reason the Antarctic surface is colder than the Arctic is that it is farther from the warming oceans. As a result, the signs of change due to GW are absent or inconclusive in the Antarctic.

The Arctic Ocean is warming more than three times faster than the global average of the oceans. This is because, whereas the cooling effect of ice melting is negligible, the heating effect of the shrinking ice cover is significant. The Arctic Ocean covers about 1% of Earth's surface, and when that surface is covered with ice, most of the solar radiation reaching it is reflected back into space. The loss of sea ice results in substantial heating because, whereas ice acts like a mirror, the uncovered ocean and land surfaces absorb about 80% of the incoming solar radiation. This occurs because, as the ice cover disappears, the surface changes from a highly reflective one into a heat-absorbing land and water surface. As mentioned previously, this is called the ice-albedo feedback effect.

The Arctic is melting much faster than expected. According to a report from the Arctic Council's Arctic Monitoring and Assessment Programme,[70] it could even be ice-free in the summers by the late 2030s. Previous studies had forecast an ice-free North Pole in the summer by midcentury. Since the late 1970s, the Arctic has lost an average area of 20,800 mi^2 of sea ice per year, and today, about 300 to 500 GT of ice is lost yearly. This is nearly five times the melting rate in 2000 (Figure 2-40).

According to Wikipedia:[71]

> The Arctic ice pack undergoes a regular seasonal cycle in which ice melts in spring and summer, reaches a minimum around mid-September, [and] then increases during fall and winter. Summer ice cover in the Arctic is about 50% of [the] winter cover. Some of the ice survives from one year to the next. Currently, 28% of Arctic basin sea ice is multiyear ice, thicker than seasonal ice: up to 3 to 4 m (9.8 to 13.1 ft) thick over large areas, with ridges up to 20 m (65.6 ft) thick. [In addition to] the regular seasonal cycle, there has been an underlying trend of declining sea ice in the Arctic in recent decades.

---

70. The Arctic Monitoring and Assessment Programme "provid[es] reliable information on the status of, and threats to, the Arctic environment, and provid[es] scientific advice on actions to be taken in order to support Arctic governments in their efforts to take remedial and preventive actions relating to climate change." Arctic Monitoring and Assessment Programme home page, accessed March 14, 2022, https://www.amap.no/.

71. "Arctic ice pack," Wikipedia, accessed March 8, 2022, https://en.wikipedia.org/wiki/Arctic_ice_pack.

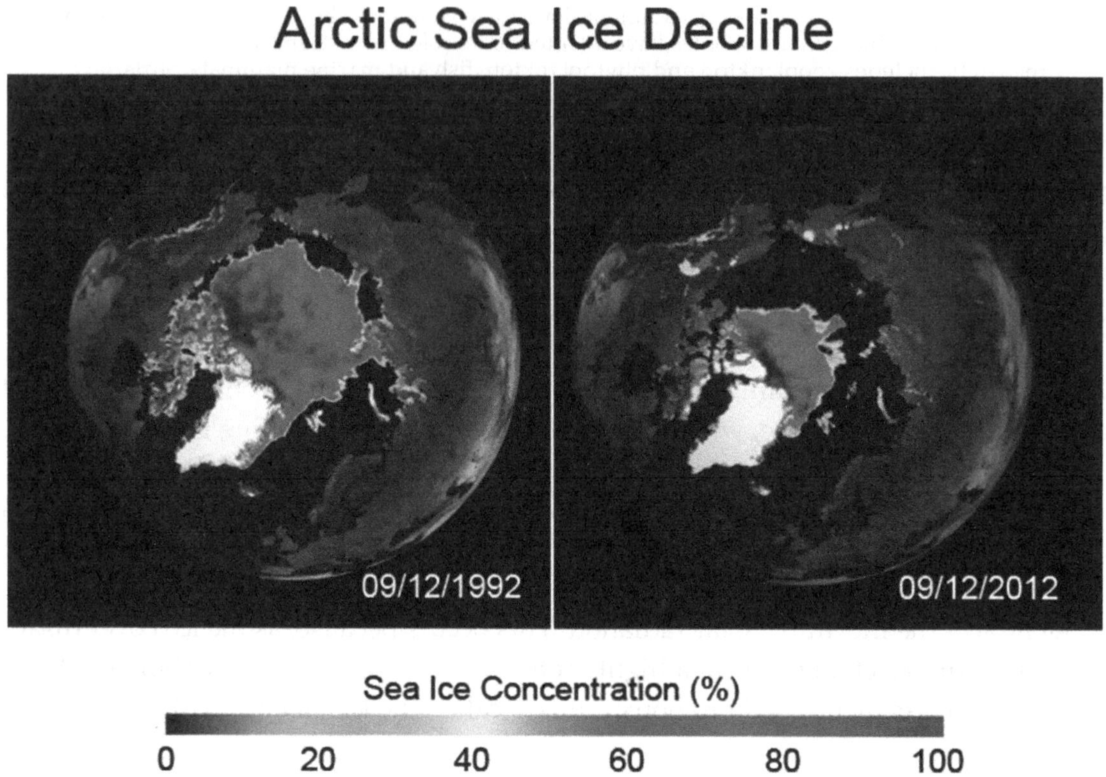

Figure 2-40. The area of high concentration of sea ice (the amount of ice in the seawater in September) in the Arctic decreased by 50% between 1992 and 2012. This reduction (3 million square kilometers or 1.2 million square miles) is nearly one-third of the area of the contiguous United States. (See color images after page 306.)

Source: Reprinted with permission. "Arctic Sea Ice Decline," The Cryosphere Today web page, Department of Atmospheric Sciences, the University of Illinois at Urbana-Champaign, November 10, 2013, https://data.globalchange.gov/report/nca3/chapter/appendix-climate-science-supplement/ igure/ arctic-sea-ice-decline. Data from the National Snow and Ice Data Center.

The sea ice loss reported in the IPCC *Fourth Assessment Report* (AR4) states that GHG forcing is largely, but not wholly, responsible for the decline in Arctic Sea ice extent. A study from 2011 suggested that internal variability enhanced the GHG-forced sea ice decline over the past few decades. In 2020, NASA found that the summer Arctic Sea ice extent is shrinking by 12.6% per decade. It has been established that the region is the warmest it has been in 40,000 years, and that the Arctic-wide melt season has lengthened at a rate of 5 days per decade (from 1979 to 2013), dominated by a later autumn freeze. According to other sources, the melting sea ice accelerates the acidification of the Arctic Ocean and affects ocean currents.

The effects of GW in the Arctic include rising temperatures, loss of sea ice (as shown in Figure 2-41), and melting of the Greenland ice sheet. Potential methane release from the

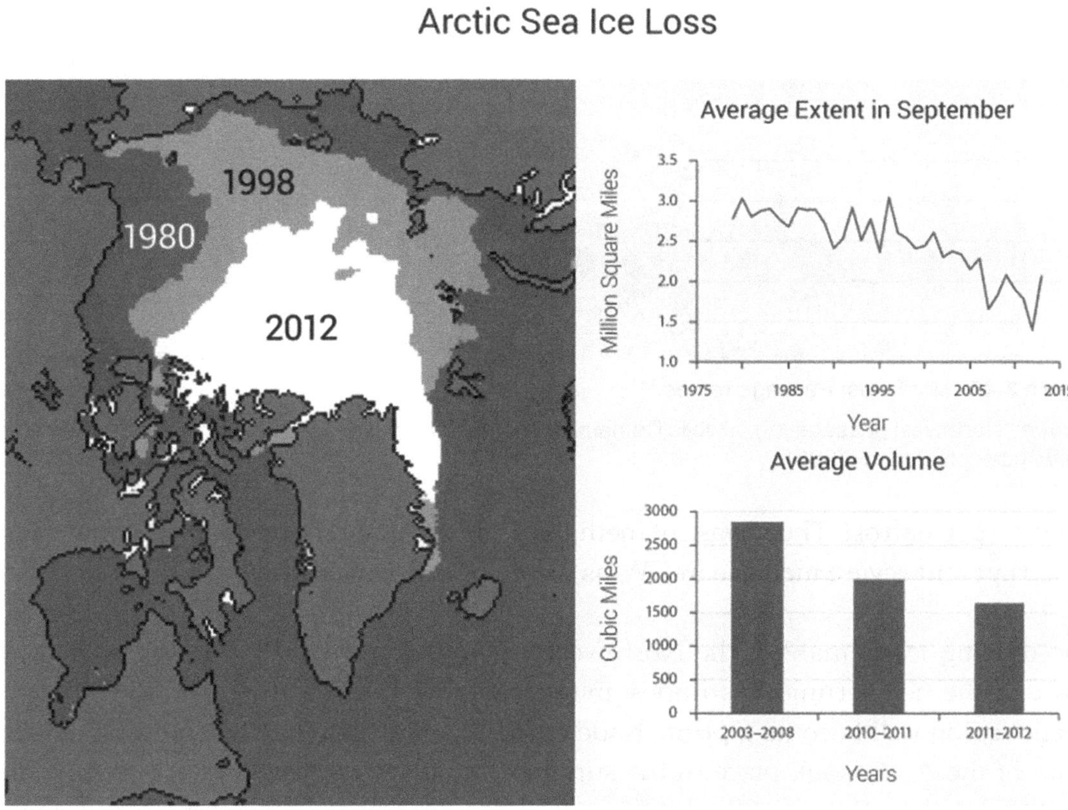

Figure 2-41. The area of the ice cover of Greenland is shrinking.

Source: "Decline in Arctic Sea Ice Extent," National Snow and Ice Data Center, November 13, 2013, https://nca2014.globalchange.gov/sites/report/files/images/web-large/CS_arctic-sea-ice-loss_V2.png.

region, especially through the thawing of permafrost and methane clathrates,[72] is also a concern. Arctic warming is a leading indicator of GW. The melting of Greenland's ice sheet is also linked to polar amplification.[73]

Climate models predict much greater warming in the Arctic than the global average, resulting in significant international attention to the region. In particular, there are concerns that Arctic shrinkage, a consequence of the melting of glaciers and other ice in Greenland, could soon contribute to rising sea levels worldwide. The current Arctic warming is leading to fears of ancient methane and carbon being released from

---

72. Methane clathrates (also called methane hydrates, hydromethane, methane ice, fire ice, natural gas hydrates, and gas hydrates) are solid clathrate hydrates in which large amounts of methane are trapped within the crystal structure of water, forming a solid similar to ice.

73. Polar amplification describes the fact that, for any change in Earth's net radiation balance (e.g., greenhouse intensification), the change in temperature near the poles tends to be greater than the planetary average change in temperature.

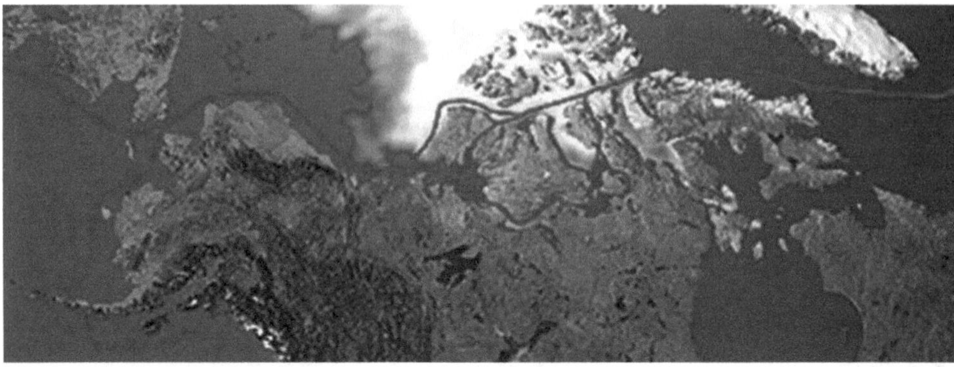

Figure 2-42. Northwest Passage routes.[76]

Source: "Northwest passage.jpg," Public Domain, accessed March 8, 2022, https://commons.wikimedia.org/w/index.php?curid=76051.

thawing permafrost. The release of methane and CO_2 stored in permafrost could cause an abrupt and severe increase in GW, as they are both potent GHGs.

The melting ice is making the Northwest Passage (Figure 2-42), the shipping route through the northernmost latitudes, more navigable, raising the possibility that the Arctic region will become a prime trade route. One harbinger of the opening navigability of the Arctic took place in the summer of 2016 when the Crystal Serenity successfully navigated the Northwest Passage, a first for a large cruise ship. In addition, it is believed that the Arctic seabed may contain substantial oil fields that may become accessible.

The Northwest Passage is the sea route between the Atlantic and Pacific oceans that runs through the Arctic Ocean, along the northern coast of North America via waterways through the Canadian Arctic Archipelago.[74] The eastern route along the Arctic coasts of Norway and Siberia is called the Northeast Passage.[75]

2.4.9 The Titanic

The iceberg that sank the RMS *Titanic*[77] was calved on the Greenland coast in the Arctic Ocean, and the currents took it to the Atlantic Ocean. Of the 15,000 to 30,000

---

74. "Arctic Archipelago," Wikipedia, accessed December 8, 2022, https://en.wikipedia.org/wiki/Arctic_Archipelago.
75. "Northeast Passage," Wikipedia, accessed December 8, 2022, https://en.wikipedia.org/wiki/Northeast_Passage.
76. For further information, see: "Northwest Passage," Wikipedia, accessed March 8, 2022, https://en.wikipedia.org/wiki/Northwest_Passage.
77. For a detailed description of this accident, see: Alasdair Wilkins, "What Happened to the Iceberg That Sank the *Titanic*?," *WIRED Magazine*, April 16, 2012, https://www.wired.com/2012/04/titanic-iceberg-history/.

Figure 2-43. The iceberg that the Titanic probably struck.

Source: "How Large Was the Iceberg That Sank the Titanic," US Department of Homeland Security, US Coast Guard, last updated June 19, 2009, https://www.navcen.uscg.gov/how-large-was-iceberg-sank-titanic.

icebergs calved by the Greenland glaciers each year, probably about 1% make it to the Atlantic. On April 15, 1912, this iceberg was some 5000 miles (8046 km) south of the Arctic Circle.

The photograph in Figure 2-43 was taken by the chief steward of the German ocean liner *SS Prinz Adalbert*, which on April 15 was sailing through the North Atlantic mere miles from where the *Titanic* had sunk early that morning. The steward was not even on the lookout for icebergs; he spotted a streak of red paint along one iceberg's base, which most likely meant that a ship had collided with it within the past 12 hours.

The iceberg likely was born some 3000 years ago from the snowfall on the western coast of Greenland sometime about 1000 BCE. After a few months, this snow was turned into a more compacted form called *firn*, which then, over subsequent decades, was compressed into dense ice by the weight of newer snow on top of it. The resulting huge mass of ice, or glacier, was slowly forced farther westward toward the sea. When it finally reached the coast of the Arctic Ocean, the lapping tides broke off chunks of the ice, and icebergs were calved from the glacier. In this way, the old snow, some 30 centuries later, entered the ocean as an iceberg.

The water temperature on the night the *Titanic* sank was thought to be about –2°C (28°F), just below freezing. (This temperature is close to the freezing point of salt water.) The average life expectancy of an iceberg in the North Atlantic is only about 2 to 3 years from calving. This means the iceberg the *Titanic* struck broke off from Greenland in 1910 or 1911 and was gone forever by 1912 or 1913.

2.5 Ice on Land

Talking about GW the other day, a friend of mine said that ice on Earth is the "air conditioner" of the planet because as it melts, it cools the Earth in the same way ice cools a gin and tonic. As I have already discussed and will discuss in more detail, the opposite is the case. The melting of ice substantially increases GW through the albedo effect. Table 2-2 summarizes the locations and quantities of ice on the planet and their melting rates.

2.5.1 Antarctica

On a planet where every square meter is claimed by some nation (often by more than one) and where nations spend much of their national wealth to move borders (or to build walls), the continent of Antarctica has escaped that fate (Figure 2-44).

Antarctica is governed internationally through the Antarctic Treaty System, which was signed in 1959 by the 12 countries whose scientists were active in and around Antarctica at the time. Among the original signatories of the Antarctic Treaty were seven countries—Argentina, Australia, Chile, France, New Zealand, Norway, and the United Kingdom—with territorial claims for parts of Antarctica, some overlapping. Some parties recognize territorial claims, and others maintain that they reserve the right to make a claim, yet there is peace. All positions are explicitly protected in Article IV of the Antarctic Treaty, which has, to date, successfully preserved the *status quo.*

Table 2-2. Locations, weights, volumes, areas, and melting rates of ice on our planet.

Location	Weight (PT)	Volume (10^6 km^3)	Area (10^6 km^2) (% of Earth's Surface)	Melting Rate in 2019 (GTy)
Antarctica	28	30	15 [3%]	200 to 300
Greenland	2.6	2.8	2.2 [0.4%]	200[78] to 300
Other Glaciers	1.0	1.0	16 [3.1%]	350
Arctic Ocean	Floating ice	Thin ice	14 [2.8%]	500
Total	32	35	51 [10%]	>1000[79]

78. Carolyn Gramling, "The Arctic is burning and Greenland is melting, thanks to record heat," *Science News*, August 2, 2019, accessed March 14, 2022, https://www.sciencenews.org/article/arctic-burning-greenland-melting-thanks-record-heat.

79. Bob Berwyn, "Greenland's Nearing a Climate Tipping Point. How Long Warming Lasts Will Decide Its Fate, Study Says," Inside Climate News website, December 23, 2019, accessed December 8, 2022, https://insideclimatenews.org/news/23122019/greenland-ice-sheet-climate-tipping-point-temperature-duration-sea-level-rise-pnas-study.

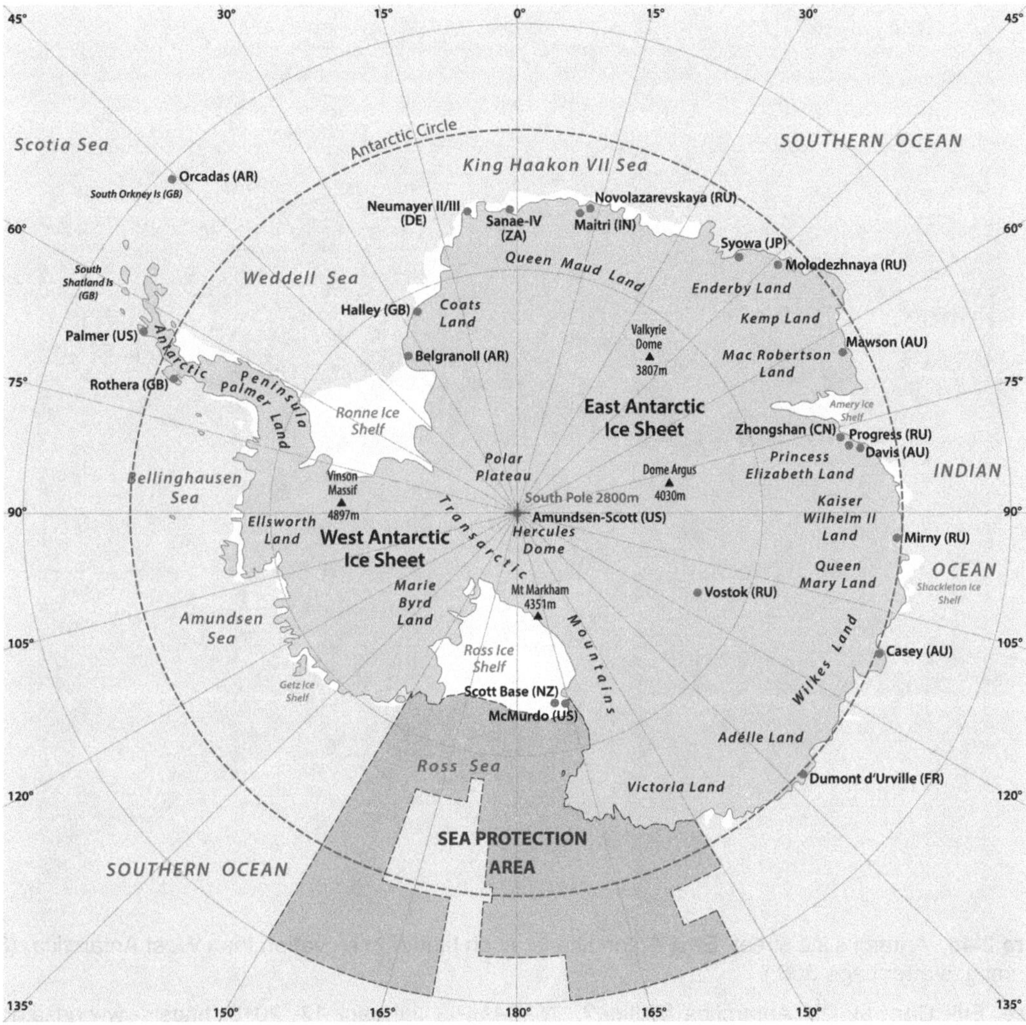

Figure 2-44. Map of the continent of Antarctica.

West Antarctica is much lower than East Antarctica, and the retreat of its glaciers (Figure 2-45) is being accelerated by ice shelf collapse. Ice shelves are the parts of a glacier that extend past the grounding line toward the ocean; they are the most vulnerable to warming seas. A long-standing theory in glaciology is that ice shelves tend to buttress (support the end walls of) glaciers, with their mass slowing ice movement toward the sea. This was confirmed by the spectacular collapse of the Rhode Island-sized Larsen B shelf along the eastern edge of the Antarctic Peninsula in 2002.

The amount of ice on Antarctica affects the level of the oceans. Antarctica is technically a desert—averaging 166 mm (6.53 in) of precipitation per year—but snow that falls on it tends to stay and become packed ever tighter. Antarctica is also the highest, driest, coldest, and brightest of the seven continents. It is larger than the combined

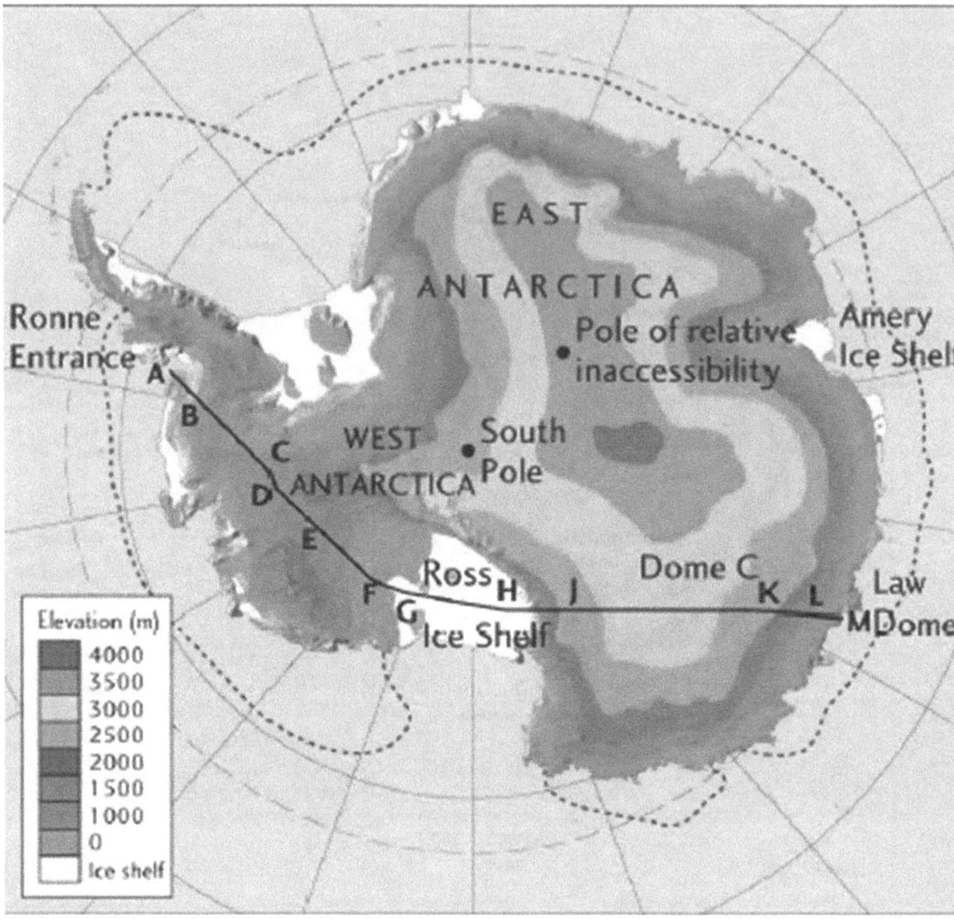

Figure 2-45. Antarctic ice sheet. East Antarctica is much higher in elevation than West Antarctica. (See color images after page 306.)

Source: Erik Conway, "Is Antarctica Melting?," NASA/JPL, January 12, 2010, https://www.nasa.gov/topics/earth/features/20100108_Is_Antarctica_Melting.html.

area of the United States and Mexico and is almost completely covered by a thick layer of ice that averages more than one mile in thickness and is nearly three miles thick in some places (Figure 2-46). This ice accumulated over millions of years through snowfall. Currently, the Antarctic ice sheet contains 90% of the volume of the ice on Earth, and if it melted, it would raise the sea levels worldwide by more than 61 m (200 ft). The eastern ice sheet is terrestrial, mainly in mountainous regions and above sea level, whereas the western ice sheet is marine, with the base of the ice sheet lying below sea level. The maximum ice thickness exceeds 4 km (2.5 miles) and is reached in the inte-rior regions of eastern Antarctica, where the surface elevation is high. In these areas, the temperature is lower than elsewhere.

As glaciers in West Antarctica have melted, heavy snowfalls in East Antarctica and the interior of West Antarctica have led to an increase in ice sheets. Although this seems counterintuitive, the accumulation of snow in the Antarctic is, in fact, a sign of GW. The warmer the temperatures, the more moisture there is in the air. In turn, the more moisture there is, the more snowfall there will be.

A combination of processes contributes to ice sheet flows, some occurring within the ice mass and others at the boundary between the ice and its base. Plastic creep[80] contributes significantly to the flow in the interior regions such as Greenland and Antarctica.

2.5.1.1 Temperature

The Antarctic is colder than the Arctic and is warming more slowly. Even though the North Pole and the South Pole get the same amount of sunlight, the South Pole is much colder than the North Pole. The average summer and winter temperatures at the North

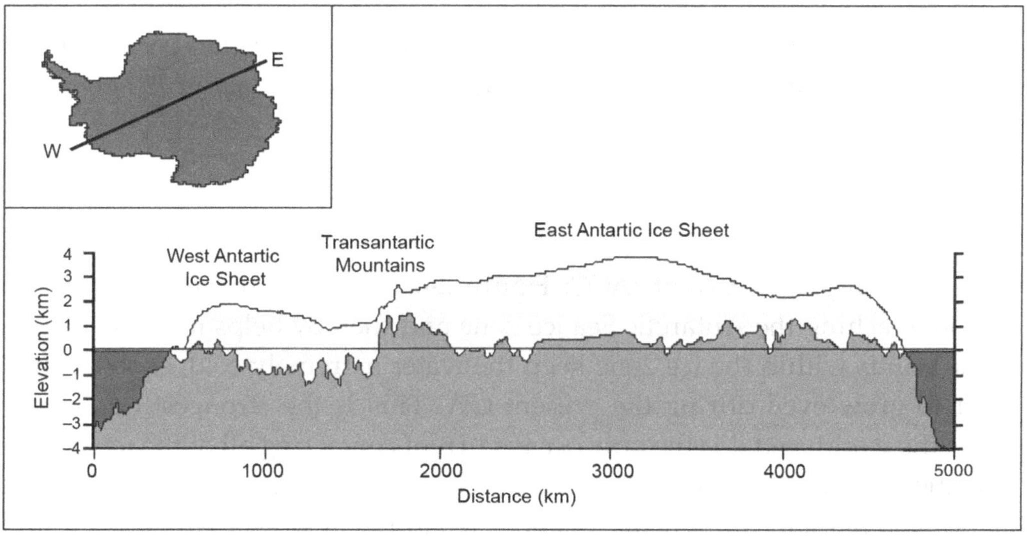

Figure 2-46. The cross-sectional profile of the Antarctic ice sheet.[81] Inset: Locations of profile end points.
Source: Reprinted with permission. "Geological and Geophysical Setting," in *Exploration of Antarctic Subglacial Aquatic Environments: Environmental and Scientific Stewardship*, (Washington, DC: The National Academies Press, 2007): Chapter 2, Figure 2.3, https://www.nap.edu/read/11886/chapter/4.

80. This is because the viscosity of the ice decreases with increasing stress. Ice creep is non-Newtonian (its viscosity is not constant, it varies with stress).

81. Based on BEDMAP bed topography from Lythe and Vaughan (2001) and surface topography from Liu et al. (1999): Matthew B. Lythe and David G. Vaughan, "BEDMAP: A New Ice Thickness and Subglacial Topographic Model of Antarctica," *Journal of Geophysical Research B: Solid Earth* 106, no. B6 (June 10, 2001): 11335–11351. Hongxing Liu, Kenneth C. Jezek, and Biyan Li, "Development of an Antarctic digital elevation model by integrating cartographic and remotely sensed data: A geographic information system-based approach," *Journal of Geophysical Research B: Solid Earth* 104, no. B10 (October 10, 1999): 23199–23213.

Pole are 0°C (32°F) and –40°C (–40°F), respectively, whereas the corresponding temperatures at the South Pole are –28°C (–18°F) and –60°C (–76°F). One of the reasons for the difference is that the Arctic is ocean surrounded by land, whereas the Antarctic is land surrounded by ocean. The ocean under the floating Arctic ice is warmer than the ice; therefore, it can melt the ice in the summer. Antarctica is dry and high. In Antarctica, there is land under the ice and snow, not ocean. Also, Antarctica has mountains where the temperature drops as elevation increases.

The ice in West Antarctica is shrinking because the ocean under the ice is warming. Some researchers believe that, even if the GW limits specified in the Paris Agreement were met, the melting in West Antarctica could be irreversible. With an average elevation of about 2500 m, less warm air is transported from the lower latitudes. As the melting in West Antarctica shows, when the altitude of the ice is reduced, the ice becomes more prone to melting.

The Antarctic's climate is warming, but not as fast as the Arctic's because it is less affected by the warming ocean. The melting rate in Antarctica (about 250 gigatons per year—GTy) is about half the rate in the Arctic (500 GTy).

2.5.1.2 Antarctic Circumpolar Current

Another factor that keeps the South Pole colder than the North Pole is the mighty Antarctic Circumpolar Current (ACC; Figure 2-47), which prevents warmer ocean water from reaching the Antarctic Sea ice zone and thereby helps to isolate the continent. The winds within the ice zone keep the water extremely cold, enabling the sea ice cover to grow even during the present GW. This is the strongest ocean current system in the world, and it is the only ocean current contacting all of the major oceans: the Atlantic, the Indian, and the Pacific. It flows clockwise from west to east around Antarctica and has a mean flow rate of about 100 to 150 Sv (1 sverdrup [Sv] = 1 million cubic meters per second), making it the largest ocean current in the world. It is circumpolar because there is no landmass to block its way around Antarctica. It also keeps the warm ocean waters away from Antarctica, enabling the continent to maintain its huge ice sheet.

The ACC (Figure 2-47) is well known to sailors because it flows from west to east speeding up travel in that direction; however, it makes sailing from east to west extremely difficult. The eastward momentum caused by the winds drives water parcels outward from Earth's axis of rotation (i.e., northward) as a result of the Coriolis force. This force causes moving objects on Earth's surface to be deflected to the left in the Southern Hemisphere (with respect to the direction of travel). The horizontal deflection effect is greater near the poles because the effective rotation rate around a local vertical axis

Figure 2-47. The ACC, also called the *West Wind Drift*, contributes to keeping the continent of Antarctica isolated from the warmer waters outside the current's circle.

Source: Reprinted with permission. Ceridwen Fraser, "Antarctica May Not Be as Isolated as We Thought, and That's a Worry," *The Conversation*, May 26, 2016, http://theconversation.com/antarctica-may-not-be-as-isolated-as-we-thought-and-l-a-worry-59969.

is highest there, and the effect decreases to zero at the Equator. Rather than flowing directly from high-pressure to low-pressure areas, as they would in a nonrotating system, winds and currents tend to flow to the left in the Southern Hemisphere. I discussed the Coriolis effect in relation to Figure 2-27 and will discuss it in more detail in Section 2.6.1, "Hurricanes (Typhoons) and Storm-Surge-Caused Flooding."

2.5.1.3 Ocean Levels

A study published in *Nature* reported that ice loss from Antarctica caused sea levels to rise by 7.6 mm (0.29 in) from 1992 to 2017,[82] with two-fifths of this increase occurring after 2012. (Over the past two decades, global sea levels have risen by approximately

---

82. The IMBIE Team, "Mass balance of the Antarctic Ice Sheet from 1992 to 2017," *Nature* 558 (2018):219–222, accessed March 14, 2022, https://www.nature.com/articles/s41586-018-0179-y/.

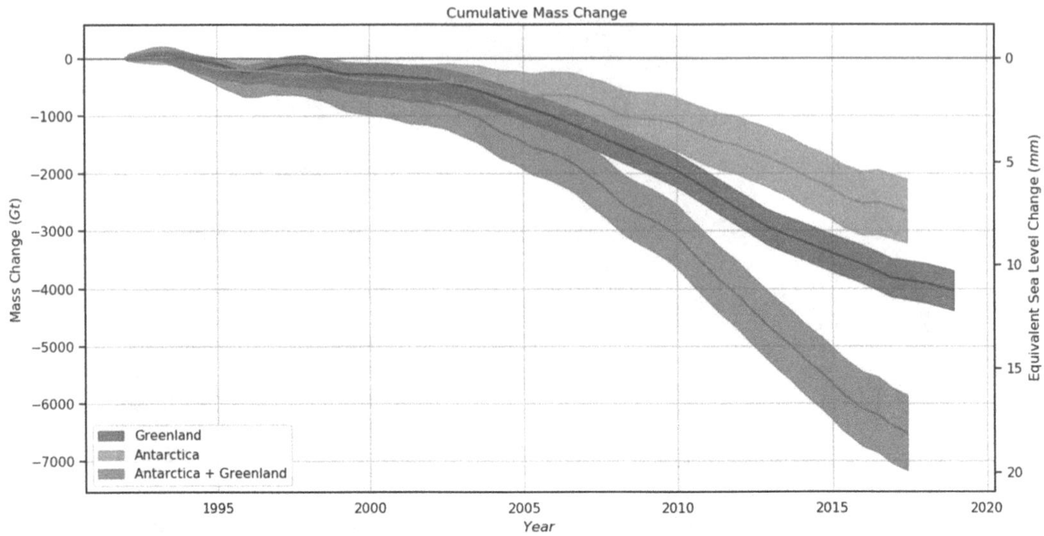

Figure 2-48. The amount of ice loss and resulting sea level change caused by the melting across Antarctica (yellow), Greenland (blue), and if these losses are combined (green). (See color images after page 306.)

Source: Reprinted with permission from Andrew Shepherd.

3.2 mm/year.) According to some sources, Antarctic ice melt is now driving a sea level rise of 0.6 mm/year—suggesting that the melting rate has increased substantially during the past 5 years.

Figure 2-48 shows the total amount of ice loss from Antarctica and Greenland from 1992 to 2017.

It sounds insignificant when one hears that the melting of ice in the Antarctic is causing a yearly increase in the global average sea level of only about 0.6 mm. However, if one hears (as some researchers suggest) that enough Antarctic ice has melted to cover Texas with 4 m (13 ft) of water, it sounds significant indeed! The various models today predict a sea level rise of 0.3 to 1.8 m (1 to 6 ft) by the end of this century. In other words, there is an awful lot of ice at the South Pole, and if all of the Antarctic ice melted, sea levels would rise by more than 61 m (200 ft). From a process control perspective, this is a typical high-capacity, low-gain process. If we continue our present rate of GHG emissions, by 2500, the sea level will rise by 13.6 m (45 ft). Equally importantly, even if we become carbon-neutral (stop increasing the carbon content of the atmosphere completely), melting will continue, and the sea level will rise—because of the amount of GHG already in the air—only much more slowly.

Another unpredictable factor is that the West Antarctic ice sheet could collapse, and we have no idea if this will happen. We know that the relatively warm ocean has been melting these sheets from below for decades, and consistent ice loss has been observed over the past 20 years. This ice sheet is sitting on land that is below sea level. This means that the ocean currents can accelerate its melting. The ocean has already begun breaking off chunks of the ice, eroding it from below. Such "melting from below" is causing huge parts of the ice sheet to collapse into the ocean. For example, in January 2020, when the Florida-sized Thwaites Glacier collapsed, the water temperature was measured as 2°C (35°F) at the grounding line at the base of the ice.

2.5.2 Glaciers

The melting of the 2000 glaciers on the planet (350 GTy) adds more water to the oceans than the melting on Antarctica (250 GTy). From a process control perspective, it is important to understand that glaciers are not stagnant; they are in continuous movement, and, therefore, they are simultaneously subject to the effects of both the melting and the movement. Glaciers contain 90% of all the ice on the planet and supply most rivers with water. They are located on Greenland in the Arctic and on Antarctica; and between latitudes 35° N and 35° S, they occur only in the Himalayas, Andes, Rocky Mountains, and a few high mountains elsewhere. They cover about 10% of Earth's land surface. Glaciers create unique landforms similar to riverbeds.

A glacier usually originates when a landform called a *cirque*, a typically armchair-shaped geological formation (such as a depression between mountains), collects the snow that falls into it, and the snow is compressed by its own weight. Glaciers form only on land and are distinct from the much thinner sea ice and lake ice that form on the surfaces of bodies of water. Glacier formation begins when the rate of snow accumulation exceeds the rate of losses (ablation[83]). The snow undergoes repeated freezing and thawing cycles, changing into granular ice called firn. Under the pressure of the layers of ice and snow above it, this granular ice fuses into denser and denser firn. Over a period of years, layers of firn undergo further compaction and become glacial ice, which is constantly moving under its own weight. The yearly rate of glacial ice melting is approaching 1000 GT/year.

Glaciers store water in the form of ice during the colder seasons and release it later as meltwater. At high altitudes and in Antarctic environments, the summer temperature is often insufficient to release meltwater. The mass of glaciers is affected by long-term climatic changes; therefore, glaciers are both sensitive indicators of climate change

83. "In glaciology and meteorology, ablation—the opposite of accumulation—refers to all processes that remove snow, ice, or water from a glacier or snowfield. Ablation refers to the melting of snow or ice that runs off the glacier, evaporation, sublimation, calving, or erosive removal of snow by wind. Air temperature is typically the dominant control of ablation, with precipitation exercising secondary control. "Ablation," Wikipedia, accessed August 29, 2022, https://en.wikipedia.org/wiki/Ablation#Glaciology.

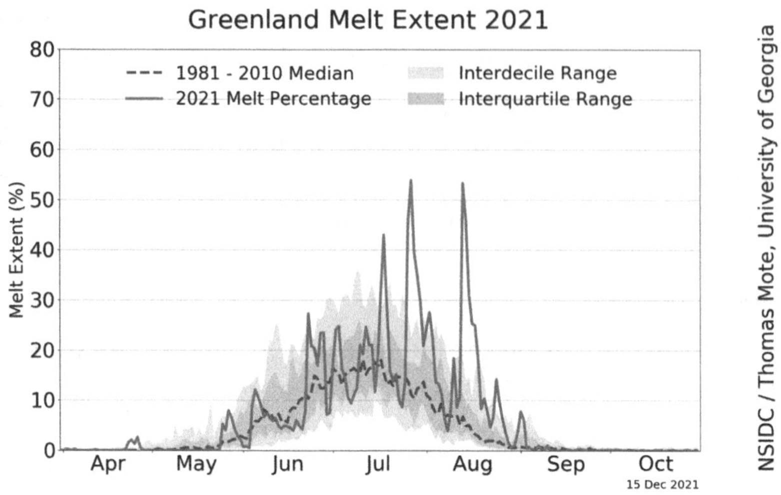

Figure 2-49. Percentages of melting by month for the years 1981 (blue) and 2018 (red) showing that, in Greenland, almost all glacial melting occurs between May and September. (See color images after page 306.)

Source: "Greenland Melt Extent 2018," National Snow & Ice Data Center, accessed March 15, 2022, http://nsidc.org/greenland-today/.

and causes of sea level changes. Glaciers are blue because the air bubbles, which give a white color to ice, are squeezed out of glacial ice.

In addition to rising sea levels, melting glaciers are also causing the land of floating islands to rise. For example, scientists say that Iceland is rising as GW melts the island's ice caps.[84] Geologists have long known that, as glaciers melt and become lighter, the surface of islands rebounds upward as the weight of the ice decreases. For example, Iceland has risen by about 0.5 m (20 in) since 1930, and it is projected to rise by nearly 2 m (6 ft) by the end of this century.

2.5.3 Greenland

The temperature on Greenland is increasing about twice as fast as the temperature of the rest of the planet. Greenland has made the largest contribution to global sea level rise, and its contribution has been increasing every decade. Its melt grew from an equivalent global sea level rise of 0.09 mm/year between 1992 and 2001 to a rise of 0.59 mm per year between 2002 and 2011.[85] The estimated total loss of ice per year is about 500 GT. This loss occurs during the summer months (Figure 2-49) and is due in

84. For more information, see: Charles Q. Choi, "Why Iceland Is Rising So Fast," Live Science website, January 2015, https://www.livescience.com/49751-iceland-rising-melting-ice.html.

85. I. Velicogna, T. C. Sutterley, and M. R. van den Broeke, "Regional Acceleration in Ice Mass Loss from Greenland and Antarctica Using GRACE Time-Variable Gravity Data," *Geophysical Research Letters* 41, no. 22 (November 2014): 8130–8137, https://doi.org/10.1002/2014GL061052.

part to changes in precipitation or surface melting and in part to changes in glacial discharge.

It takes about 3000 years from the time snow falls on Greenland to the time the resulting ice enters the Arctic Ocean. In the case of Greenland, the frozen water within the glaciers is slowly forced farther westward toward the sea. When it finally reaches the coast of the Arctic Ocean, the lapping tides break off chunks of ice, and icebergs are calved from the glacier some 30 centuries after their source water was first deposited. As mentioned in Section 2.4.9, "The *Titanic*," the iceberg that sank the *Titanic* was calved off Greenland.

The calving process[86] is a result of small cracks and fractures in the ice growing into larger crevasses. It can be triggered at the glacier terminus by stretching due to uneven ice flow rates, force imbalances, ocean undercutting, or buoyant forces. This occurs if a glacier's surface thins below the level needed for flotation. In that case, the end of the glacier will become buoyant and lift off the land underneath. As the glacial surface continues to become thinner, buoyancy increases pushing the end of the glacier up and away from the land and causing crevasses. Finally, calving occurs, resulting in large icebergs.

Greenland has many glaciers, all of which are shrinking in size (Figure 2-50). Greenland is the world's largest island, and its ice sheet is about a mile thick. If all the snow and ice on Greenland completely melted, the global average sea level would rise by about 7 m (about 23 ft).

2.5.4 Other Glaciers

There are thousands of glaciers in addition to those in Antarctica and Greenland,. The total ice mass of the Arctic, Greenland, and other ice deposits amounts to more than 1000 GT. Aside from rain, they provide the main water supply for our rivers. Their melting is already in progress, causing flooding and uncovering solar-heat-absorbing land surfaces. They feed the rivers with fresh water, and once they are all melted, their loss will threaten the water supplies of billions of people.

2.5.4.1 Himalayan Glaciers

While the global media's attention is on Greenland, Antarctica, and the Arctic, it is important to note that the Himalayan glaciers will also face a catastrophic meltdown this century if there is no immediate effort to reduce carbon emissions. The Himalayan

86. For further reading, see: Jacob Bendle, "Calving of freshwater glaciers," AntarcticGlaciers.org website, updated July 27, 2020, accessed March 15, 2022, http://www.antarcticglaciers.org/glacier-processes/glacial-lakes/calving-of-freshwater-glaciers/.

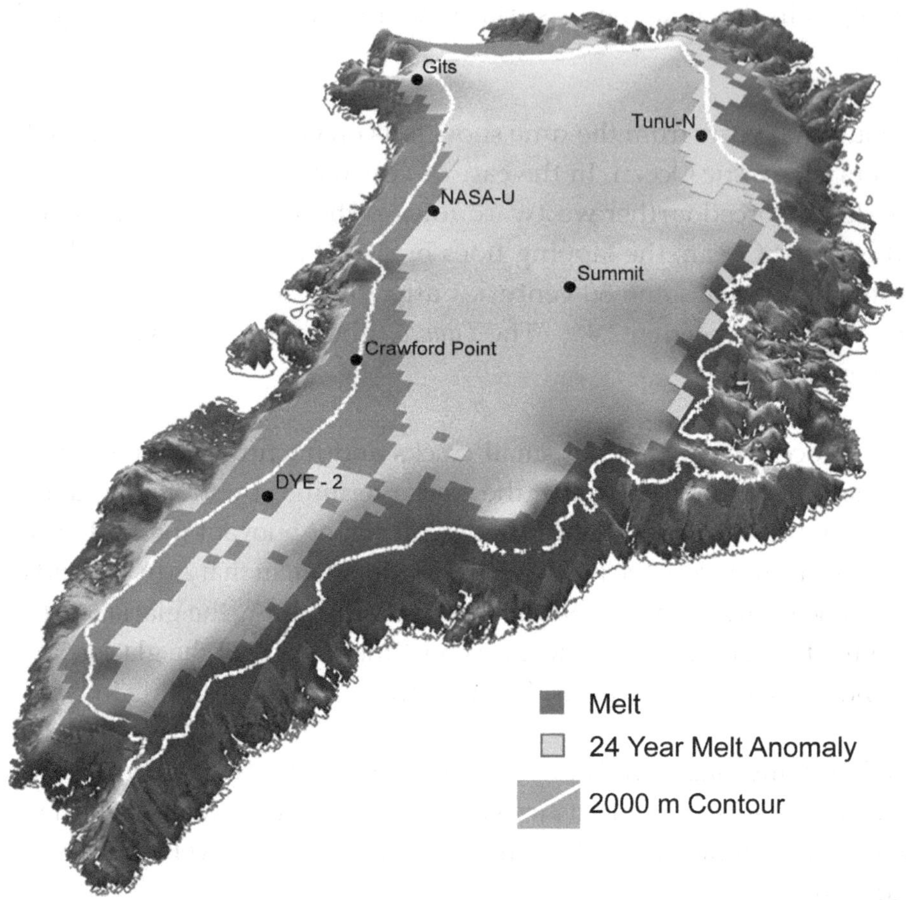

Figure 2-50. Computer graphic showing the 2002 extent of melting of Greenland's ice sheet in pink.

Source: Reprinted with permission. Evelyne Yohe, "Vanishing Ice," National Snow and Ice Data Center, last updated July 28, 2020, https://earthdata.nasa.gov/learn/sensing-our-planet/vanishing-ice. Image courtesy of Konrad Steffen and Russell Huff, CIRES, the University of Colorado at Boulder.

glaciers have been losing almost 0.5 m (about 20 in) of ice each year since the start of this century, and this rate has doubled in recent years. Spanning 2000 km (1243 miles) and harboring some 600 GT of ice, the Himalayan glaciers supply over one billion people with water for irrigation, hydropower, and drinking. Recently, the glaciers have lost about 8 GT of water per year, potentially threatening water supplies for hundreds of millions of people across parts of Asia. In addition, this will cause flooding and mass population movement until all of the ice has melted. After that, hundreds of millions of people, currently depending on the Indus, Ganges, Yangtze, and Mekong rivers for their water supplies, will experience massive water shortages.

In September 2022, one-third of Pakistan was flooded. The flood was caused by a 10-fold amplification of the monsoon rainfall by GW and melting glaciers that caused

Figure 2-51. Trift glacier at the Susten pass in the Swiss Alps.
Source: "Susten.jpg" by Johnw, September 26, 2004, Public Domain, https://upload.wikimedia.org/wikipedia/commons/5/50/Susten.jpg.

the overflow of the Indus River. The immense destruction and the death of over 1100 people resulted from climate change, caused mainly by other nations.

According to an article from the Inside Climate News website,[87] various measurements indicate that the glaciers in the Central and Eastern Himalayas are shrinking 40 cm per year. Some parts are receding by up to 30 m per year, while the peaks are warming between 0.3°C and 0.7°C faster than the global average. This makes it probable that the Himalayan mountains will lose more than one-third of their ice by the end of the century. This "loss of Himalayan ice would have devastating consequences for 1.6 billion people living in the mountains and downstream countries. Climate models show that summer flow in the Indus, Ganges and Brahmaputra Rivers and their

87. Inside Climate News website, February 5, 2019, accessed December 1, 2022, https://insideclimatenews.org/news/05022019/everest-himalayan-mountains-climate-change-glacier-ice-loss-water-crisis-icimod-global-warming-report/#:~:text=Himalayan%20peaks%20are%20warming%20between,the%20mountains%20and%20downstream%20countries.

snow-fed tributaries will rise as the glaciers melt, but will then start decreasing as the ice disappears."[88]

2.5.4.2 Swiss Alps

Switzerland has more glaciers (Figure 2-51) than any other Alpine country, and it generates 60% of its electricity from hydropower. The Swiss Alps are warming faster than the rest of the world; the 1500 glaciers of the country are all retreating, and some are expected to disappear within this century. The retreating glaciers usually feed lakes that can be used as reservoirs for hydroelectric plants, but excessive melting can flood these reservoirs, and once all of the ice melts, the plants will become useless.

2.6 Atmospheric Processes

The mass of the oceans is almost 300 times greater than the mass of the atmosphere. Therefore, the response of atmospheric processes is much faster than that of oceanic processes. This is why we are already experiencing atmospheric consequences of GW (e.g., hurricanes, storm-surge-caused flooding, drought, wildfires); oceanic consequences (e.g., sea level rise, degassing) will come much later.

Winds are caused by pressure differences. According to Archimedes' principle, because warmer air has a lower density, it tends to rise through the surrounding denser, cooler air. Many aspects of the weather are a consequence of this effect. If the ocean surface is warmer than the nearby land, the ocean air will rise, and colder, denser air from the land will blow in to take its place. Conversely, if the land's surface is warmer than the ocean's, the wind will blow from the ocean to the land.

The pressure differences between the atmospheric belts around the globe are due to the Coriolis effect, which is caused by the faster movement of Earth's surface (due to the larger diameter of the globe) near the Equator than toward the poles. For this reason, the air masses moving toward the pole tend to move east in the Northern Hemisphere and west in the Southern Hemisphere.

In addition to causing hurricanes, the increased moisture content of the air can cause gigantic rains. In 2021, in Germany and elsewhere in Europe, 18 to 20 cm (7 to 8 in) of rain fell (more than ever in the past 1000 years), killing hundreds of people. The reason for such events is the increased humidity and a decrease in wind speed, resulting in

88. For more information about the overall situation, refer to the detailed report titled *The Hindu Kush Himalaya Assessment: Mountains, Climate Change, Sustainability, and People,* compiled by the International Center for Integrated Mountain Development (ICIMOD). https://lib.icimod.org/record/34383.

the rain staying longer. This slowing effect is caused by the warming in the Arctic, which reduces the temperature difference that moves the wind to the north. When these gigantic rains are combined with glacier melting, the result can be floods such as the one that covered one-third of Pakistan and killed 1100 people in 2022.

2.6.1 Hurricanes (Typhoons) and Storm-Surge-Caused Flooding

Although some scientists believe there is enough evidence to say that climate change caused by human activities is the reason for the increase in the number and strength of hurricanes during recent hurricane seasons, other researchers emphasize that climate change is not the only cause.

Most agree that the intensity, frequency, and duration of North Atlantic hurricanes and the frequency of the strongest hurricanes have increased since the early 1980s. In 2022, the hurricane that caused destruction in Puerto Rico and Florida has shown that hurricane intensity and rainfall are increasing as the climate continues to warm the planet.

Hurricanes are usually categorized based on their maximum sustained wind speeds according to the Saffir-Simpson wind scale[89] (Table 2-3). For example, a Category 1 hurricane has wind velocities of at least 74 mph (119 km/h, 33 m/s, 64 kn).[90] The highest classification in the Saffir-Simpson scale is Category 5, which has sustained wind velocities of more than 156 mph (251 km/h, 70 m/s, 136 kn). Overall, this classification system can indicate the potential damage and flooding that a hurricane will cause upon landfall.

The wind patterns called hurricanes in the North Atlantic and central and eastern North Pacific are known in the Northwest Pacific as typhoons; both are more generally classified as tropical cyclones. They are all initiated by hot spots and pressure depressions that evolve on the surface of warm oceans and supply the energy (fuel) for hurricanes. For hot spots to cause air to rise and start forming hurricanes, the ocean's water has to be warmer than 26.7°C (80°F) to a depth of about 61 m (200 ft).

The air density above a hot spot is low for two reasons: First, air density decreases as its temperature rises. Second, higher temperatures promote greater water evaporation

---

89. The Saffir-Simpson Hurricane Wind Scale is a 1 to 5 rating based on a hurricane's maximum sustained wind speed that is used to estimate potential property damage. Hurricanes of Category 3 and higher are considered major hurricanes because of their potential for significant loss of life and damage. Category 1 and 2 storms are still dangerous, however, and require preventive measures.

90. The units of measure referenced are km/h, kilometers per hour; m/s, meters per second; and kn, knots.

Table 2-3. The Saffir–Simpson Hurricane Wind Scale is used to define hurricane categories according to their maximum sustained wind speeds.

Tropical Cyclone Categories			
Category	**Wind Speed (mph)**	**Damage at Landfall**	**Storm Surge (feet)**
1	74 to 95	**Minimal** While wind speeds are quicker than the speed a cheetah can run, there is not much damage to property.	4 to 5
2	96 to 110	**Moderate** Wind speeds are as fast or faster than a baseball pitcher's fastball. This category cyclone may break windows and destroy trees.	6 to 8
3	111 to 129	**Extensive** This wind speed is similar, or close, to the serving speed of many professional tennis players. This category cyclone can break windows and doors.	9 to 12
4	130 to 156	**Extreme** With winds quicker than the world's fastest rollercoaster, there is much damage to property in this category cyclone.	13 to 18
5	157 or higher	**Catastrophic** This category cyclone causes the most damage. With wind speeds similar to the speed of some high-speed trains, it destroys buildings.	19+

Source: "Hurricane Dynamics," My NASA Data, accessed March 8, 2022, https://mynasadata.larc.nasa.gov/basic-page/hurricane-dynamics.

at the ocean's surface, and the increased moisture content decreases the air's density. Above a hot spot, the hot and moist air rises, causing a chimney effect that reduces the pressure at the ocean surface to below atmospheric. This pressure differential ranges from –1 to –4 inHg (–25.4 to –101 mmHg, –33.9 to –135 mbar).[91] As this pressure depression pulls in more hot and moist air from the ocean's surface outside the hot spot, the hurricane grows stronger and stronger.

As the air at the center of the hurricane rises, the Coriolis force causes it to start spinning. North of the Equator, surface winds flow inward trying to fill in the low-pressure region above the hot spot and are deflected to the right by the Coriolis force, setting up a counterclockwise rotation. In the Southern Hemisphere, the inflowing winds are deflected to the left; therefore, a clockwise rotation is started. This circulatory motion is typical of all cyclones and hurricanes. A cyclone or tropical storm becomes a

91. The units of measure are inches of mercury (inHg) and millimeters of mercury (mmHg).

Figure 2-52. Hot spots on the ocean surface cause the hot, humid air to rise, and the resulting pressure depression pulls in the colder and denser air at the bottom as the Coriolis effect rotates the whole body of air. This is the process by which hurricanes form.[92]

Source: "Fran September 4 1996 1700Z.png" by NOAA (storm image) and NASA (land image), September 4, 1996, Public Domain, https://commons.wikimedia.org/w/index.php?curid=66833980.

Category 1 hurricane (Table 2-3) when its spinning velocity reaches 74 mph (119 km/h). Hurricane velocities of up to 200 mph (320 km/h) have been recorded.

As the spinning moist, warm air of a hurricane rises (Figure 2-52) into the cooler, upper regions of the atmosphere, it cools. The moisture in the air condenses, and great quantities of energy are released into the atmosphere. The latent heat released by condensation provides the energy that accelerates this powerful dynamic process. This energy drives the convection cycle, which increases the hurricane's lethal spinning velocity. The energy content and the hurricane velocity continue to increase as long as the hurricane is above warm water.

Hurricanes move in the direction of the prevailing trade winds. At the outer perimeter of the hurricane, the condensed moisture returns to the ocean as rain. The temperature difference between the core and the perimeter of the hurricane is about 8.3°C (15°F).

---

92. For additional information, see: "Hurricane Fran," Wikipedia, accessed March 8, 2022, https://en.wikipedia.org/wiki/Hurricane_Fran.

Once the hurricane moves over land, its energy supply (the hot, moist air from the ocean's surface) is cut off, its "engine" stops, and it dies.

So, how does GW make hurricanes and tornadoes[93] more powerful and more destructive? It is because the increasing temperature of the ocean's surface causes more evaporation, which, in turn, decreases the already very low pressure at the core of the hurricane. In other words, the higher the ocean surface temperature, the more frequent and powerful the hurricanes become.

GW also influences the formation of thunderstorms and tornadoes. Computer modeling[94] suggests that two atmospheric processes contribute to this influence. One is the convective available potential energy (CAPE) created by the warming of the atmosphere at lower elevations, which causes warm, moist air to rise to higher altitudes. The other process that contributes to tornado formation is vertical wind shear, caused by variations in the wind speed with height.

CAPE[95] is the buoyancy of a parcel of air or the amount of energy that the air parcel would possess if lifted a certain distance vertically through the atmosphere. It is a form of fluid instability found in thermally stratified atmospheres where a colder fluid lies atop a warmer one. Thunderstorms are created when an air mass that is less dense than the surrounding air (its buoyant force is greater than its weight) rises, creating vertically developed clouds due to the rising motion. When CAPE is high, vertical wind shear is more likely to be high. This means that the total frequency and severity of thunderstorms grow as a result of increasing GW.

Yet this view is not supported by all predictive models (Figure 2-53). Some suggest that the relationship between SSTs and the power dissipation index (PDI) of hurricanes[96] is strong and project that the destructive potential of hurricanes will increase by roughly a 300% change in PDI by 2100 (Figure 2-53a). Alternative statistical relation-

---

93. NASA scientists have developed a new climate model that indicates that the most violent severe storms and tornadoes may become more common as Earth's climate warms. See: "Global Warming Will Bring Violent Storms and Tornadoes, NASA Predicts," Science Daily website, August 31, 2007, accessed December 1, 2022, https://www.sciencedaily.com/releases/2007/08/070830105911.htm.

94. See: Ker Than, "Global Warming to Spawn More Severe U.S. Thunderstorms: Study," *National Geographic*, September 24, 2013, accessed December 1, 2022, https://www.nationalgeographic.com/science/article/130924-global-warming-climate-change-thunderstorms-tornadoes.

95. CAPE is a term that is used frequently during severe-weather season. It provides a way to quantify the amount of energy available in the atmosphere to develop and strengthen thunderstorms. Thunderstorms are fueled by the buoyancy of air parcels—basketball-sized volumes of air—in the atmosphere whose properties can be analyzed and tracked. Jack Sillin, "What Is CAPE?," blog.weather.us website, July 5, 2017, accessed December 1, 2022, https://blog.weather.us/what-is-cape/.

96. PDI, as an aggregate of storm intensity, frequency, and duration, is a way to express the destructiveness of a hurricane, and it provides a measure of total hurricane power over a hurricane season. PDI for a tropical cyclone is defined as "the sum of the maximum one-minute sustained wind speed cubed, at six-hourly intervals, for all periods when the cyclone is at least tropical storm strength." For additional information, see: "Climate Change Indicators: Tropical Cyclone Activity," US EPA website, accessed December 5, 2022, https://www.epa.gov/climate-indicators/climate-change-indicators-tropical-cyclone-activity.

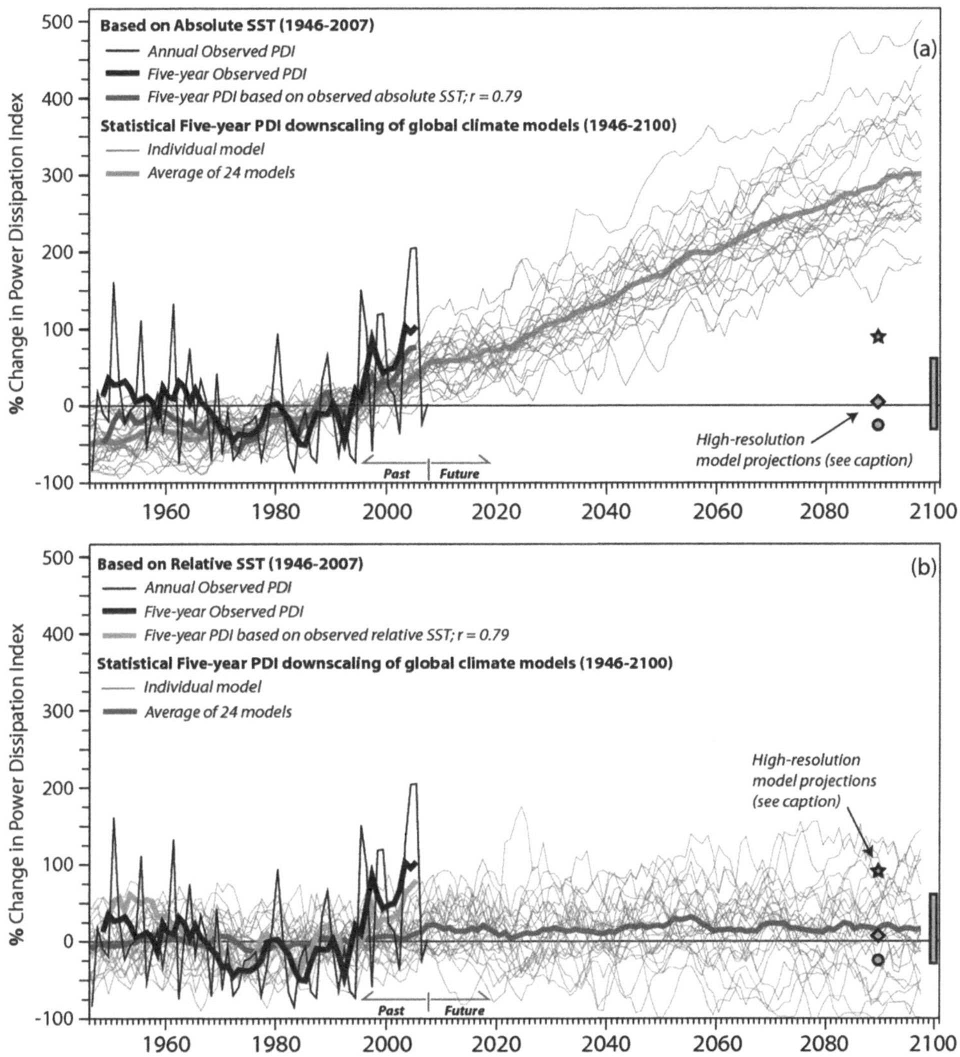

Figure 2-53. Past and extrapolated variations in the PDI of the Atlantic Ocean basin as calculated from the absolute (a.) and relative (b.) SSTs. PDI is a way to quantify the destructiveness of hurricanes, and SST is the surface temperature of the sea. According to the orange line in panel 2-53a, as the absolute value of the SST rises, so does the PDI. On the other hand, according to the blue curve in panel 2-53b, the PDI does not change much when the relative SST (compared with the temperatures in other basins) is considered. (See color images after page 306.)

Source: Gabriel A. Vecchi, Kyle L. Swanson, and Brian J. Soden, "Whither Hurricane Activity?," Science 322, no. 5902 (October 31, 2008): 687–689, https://www.gfdl.noaa.gov/wp-content/uploads/files/user_files/gav/publications/vss_08_diverge.pdf.

ships between the PDI and the relative SST (Figure 2-53b) imply only modest future long-term trends of Atlantic hurricane activity (PDI) with greenhouse warming.

Storm surge is one of the most damaging consequences of hurricanes.[97] According to an article on hurricane effects on The Weather Channel's website:

97. "Hurricane Damages and Effects," The Weather Channel website, August 24, 2015, accessed December 1, 2022, https://weather.com/safety/hurricane/news/hurricane-damages-effects-20120330.

> Storm surge is a rapid rise in the level of water that moves onto land as the eye of the storm makes landfall. Generally speaking, the stronger the hurricane, the greater the storm surge.
>
> As a hurricane approaches the coast, its winds drive water toward the shore. Once the edge of the storm reaches the shallow waters of the continental shelf, water piles up. Winds of hurricane strength force the water onto shore.
>
> At first, the water level climbs slowly, but as the eye of the storm approaches, the water rises rapidly. Wave after wave hits the coast as tons of moving water hammer away at any structure on the coastline. A cubic yard of water weighs about 1,700 pounds [587 kg/m^3].
>
> The surge is greater if a hurricane's track is perpendicular to the coastline, allowing the surge to build higher. The storm surge is also greater if the storm affects a bay or if it makes landfall at high tide. The greatest storm surge occurs to the right of where the eye makes landfall.

2.6.2 Polar Vortex

Because the atmosphere is held in place by gravity, it should rotate with the Earth, and there should not be any air circulation unless some force upsets the atmosphere's equilibrium. The source of this force is the Sun, which fuels the winds by creating temperature differences, which, in turn, cause pressure differences that make the air move from high- to low-pressure areas. Wind speed and direction are determined by the pressure gradient between neighboring high- and low-pressure areas. In the Northern Hemisphere (because of the Coriolis force caused by Earth's rotation), winds circulate clockwise around areas of high pressure and counterclockwise around regions of lower pressure.

Over hot spots, the warm, humid air rises and sucks in drier and colder (therefore denser) air at the bottom, and the Coriolis force (in the Northern Hemisphere) rotates the evolving cyclone counterclockwise over this hot spot. Over cold spots, the opposite occurs. In those locations, the cold, dry air sinks instead of rising, creating a high-pressure zone and clockwise circulation.

Above the poles, in the upper regions of the stratosphere[98] (at the poles, the stratosphere starts at about 7 km [4.3 mi] and reaches as high as 50 km [30 mi]), the amount of solar energy received is low (decreasing to none at all in the winter). The low angle of the arriving solar radiation results in very low temperatures, which, in turn, leads to dense and relatively high-pressure air. In the stratosphere, the temperature rises with altitude. Below the stratosphere, in the higher regions of the troposphere, the temperature decreases with increasing altitude. This reversal is caused by the ozone layer, which absorbs the UV light of the Sun (Figure 2-54) and prevents it from entering and heating the troposphere. A temperature increase contributes to a decrease in the pressure in the stratosphere; therefore, the vortex force swirls the air counterclockwise

---

98. Béla Lipták, "Understanding the polar vortex process," *Control Magazine*, May 24, 2019, accessed December 1, 2022, https://www.controlglobal.com/articles/2019/understanding-the-polar-vortex-process/.

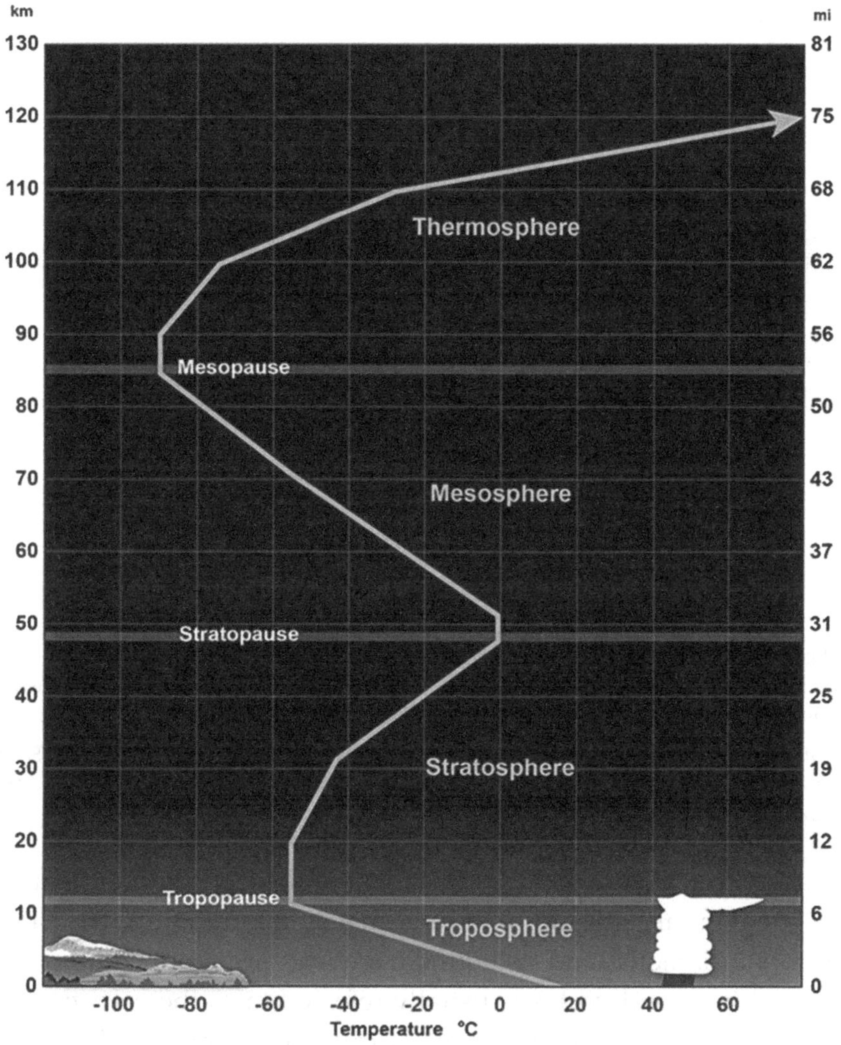

Figure 2-54. At the planet's surface, the average temperature is somewhat less than 20°C (68°F). As elevation rises to the top of the troposphere (at about 10 km [6.2 mi]), the average temperature gradually drops to about –60°C (–76°F). When the altitude reaches the ozone layer,[99] which is at the border between the troposphere and the stratosphere (i.e., about 10 km [6.2 mi]), the temperature starts rising again and reaches 0°C (32°F) at the border between the stratosphere and the mesosphere at an elevation of about 50 km (31 mi).

Source: "Average temperature profile for the lower layers of the atmosphere," National Weather Service, NOAA, accessed March 8, 2022, https://www.weather.gov/jetstream/layers.

and inward toward the pole. This effect helps to keep this cold and swirling body of air near the poles in a phenomenon called the *polar vortex*.

99. The ozone layer acts as a shield protecting Earth and its inhabitants from harmful UV radiation. It prevents the radiation from penetrating an organism's (plant, animal, and human) protective layers, such as skin, and damaging the organism's DNA molecules.

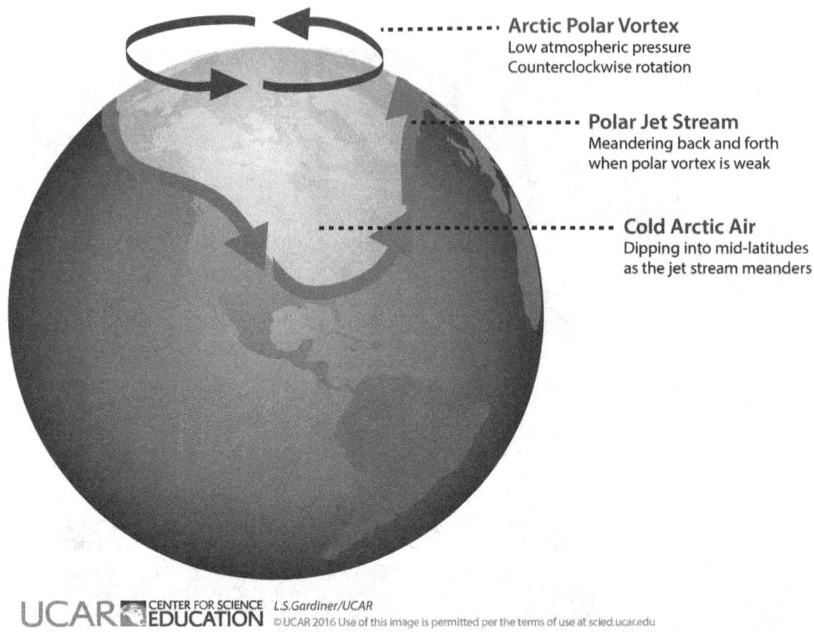

Figure 2-55. Illustration of the stratospheric (dark blue) and tropospheric (light blue) polar vortices. The rotation of both is counterclockwise (west to east). When GW weakens the polar jet stream (light blue), that results in breaking it up and causes the formation of meanders that dip down to the south.

Source: Reprinted with permission. L. S. Gardiner, "Why the Polar Vortex Keeps Breaking out of the Arctic," University Corporation for Atmospheric Research, 2016, https://scied.ucar.edu/blog/why-polar-vortex-keeps-breaking-out-arctic.

Below the stratospheric polar vortex, which forms only in the winter, is another vortex in the troposphere (at an altitude of approximately 8 km [5 mi]), which operates all year. It is larger in diameter, and its boundary winds are more commonly called the *polar jet stream*. This second vortex is usually referred to as the *polar vortex*. The large difference in temperature between the air of this polar vortex and the air farther to the south in the midlatitudes drives the polar jet stream. The interface between the cold, dry air mass nearer the pole and the warm, moist air farther to the south defines the location of the polar front. The polar front is centered at roughly 60° N latitude (Figure 2-55), and the wind it creates is usually called the *polar jet*, which also rotates counterclockwise (west to east).

A polar vortex strengthens in the winter and weakens in the summer because its strength depends on the temperature difference between the midlatitudes and the poles. When the vortex is strong, it is a clearly defined single vortex, and the Arctic air is well contained within the vortex because of the rotation's high velocity. Under such conditions, the diameter of the path of the polar jet stream is stable, and it forms a boundary between the very cold Arctic air and the warmer air over the midlatitudes. In the winter, this path is near the border between the United States and Canada.

GW destabilizes and upsets this weather pattern because the Arctic is warming at least *twice as fast as the rest of the planet.* This warming occurs when the highly reflective ice surface on the Arctic Ocean is replaced by water which absorbs much more solar heat. This is the reason for the fast temperature rise called the albedo effect.[100]

The consequence of this warming is the development of relatively warm spots on the ocean's surface, which warms the air in those areas. This makes the temperature difference between the Arctic and the midlatitude air temperatures less distinct, which causes the polar jet stream to weaken. When weaker, it can break into two or more vortices, and when very weak, the flow of Arctic air becomes even more disorganized. Under such conditions, the polar jet stream sends cold Arctic air to the south, bringing rapid and sharp temperature drops. In this case, the polar jet stream meanders instead of making a beeline around the planet. The meandering causes the polar vortex to wobble so that the cold, southbound Arctic air can extend over the United States, Europe, or Asia.

In this way, the rising warm air caused by the hot spot's albedo effect tends to break up the stable polar vortex (Figure 2-56) and generate fluctuations, reversals, and "daughter vortices" (also called *kinks* or *wobbles*). These daughter vortices can bring extremely cold air to parts of the United States that are typically well south of the jet stream.

2.7 GW Effects Already Being Felt

Compared to the Earth's land and oceans, its atmosphere has a small mass, shorter time constants, and faster reactions. For this reason, the atmospheric processes caused by GW have already started, whereas the oceanic ones are just beginning. The effects of GW are greater in the Northern Hemisphere than in the Southern Hemisphere because the land is warming faster than the oceans, and the proportion of the land area is greater in the north (Figure 2-57). Oceans warm about half as fast as land because most of the heat received by the oceans leaves through evaporation and because the atmosphere's water-holding capacity increases by about 7% for every 1.0°C (1.8°F) of temperature rise, which contributes to increased evaporation and, therefore, more heat removal.

As water circulation rises, wet places get wetter (Northern Europe, eastern North and South America, and northern and central Asia) because they receive more rain, and dry places (such as large areas of the Southern Hemisphere, including major portions of Australia, Africa, and South America) get drier due to receiving less rain. The number of days with very heavy precipitation has increased by 58% over the past 50 years in the northeastern United States and by 27% in the Midwest.

In dry areas, droughts result in decreased levels of rivers, lakes, and groundwater and reduced soil moisture in agricultural areas. Precipitation has already declined in the

100. The albedo effect (reflectiveness) of white snow is 70% to 90%, whereas that of water is about 10%.

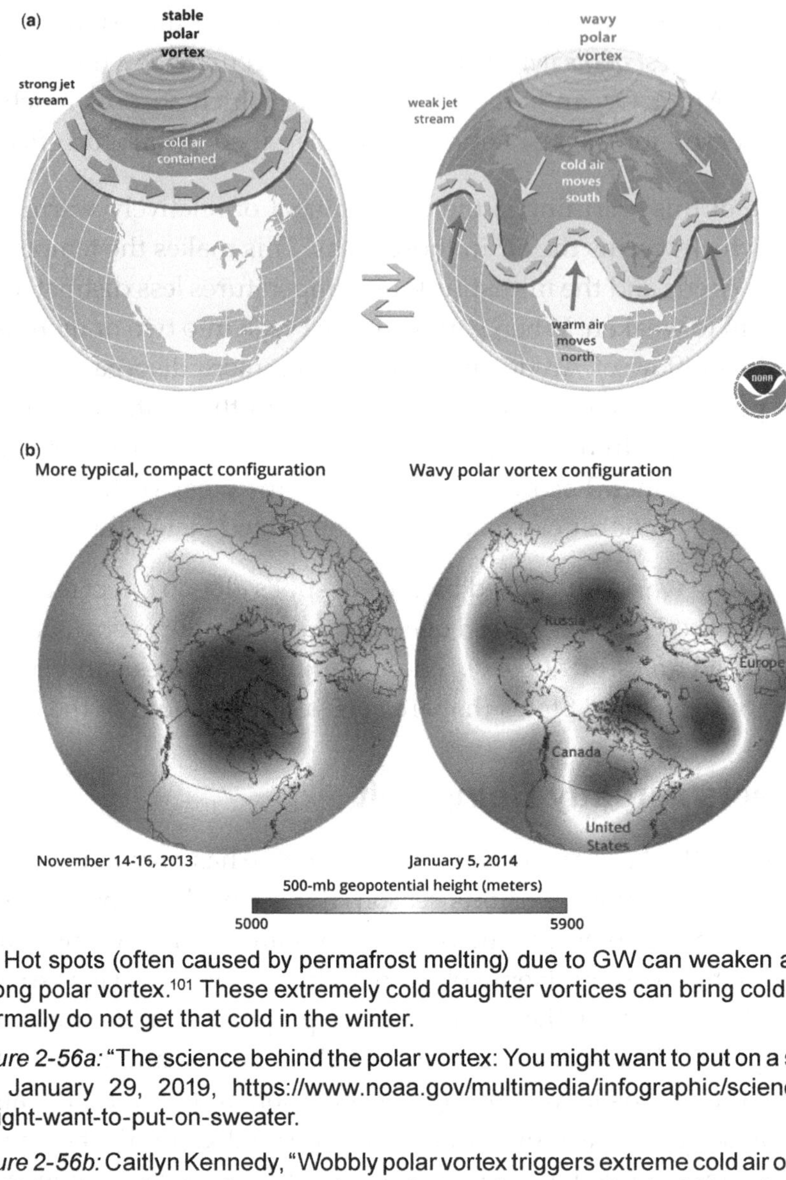

Figure 2-56. Hot spots (often caused by permafrost melting) due to GW can weaken and break up the otherwise strong polar vortex.[101] These extremely cold daughter vortices can bring cold air southward to areas that normally do not get that cold in the winter.

Source of Figure 2-56a: "The science behind the polar vortex: You might want to put on a sweater," NOAA, last updated January 29, 2019, https://www.noaa.gov/multimedia/infographic/science-behind-polar-vortex-you-might-want-to-put-on-sweater.

Source of Figure 2-56b: Caitlyn Kennedy, "Wobbly polar vortex triggers extreme cold air outbreak," NOAA, January 8, 2014, https://www.climate.gov/news-features/event-tracker/wobbly-polar-vortex-triggers-extreme-cold-air-outbreak.

tropics and subtropics, including Southern Africa, the Sahel region of Africa, Southern Asia, the Mediterranean, Australia, and the southwestern United States. Even areas that remain relatively wet, such as the tropics, can experience long, dry conditions between extreme precipitation events. Scientists expect that the water resources in drier areas will further decline by an additional 30% by midcentury.

---

101. For additional information, see: Marlo Garnsworthy, "This Week in Ice: Dec. 24, 2017 – Jan 6, 2018," Polar Bird Blog, January 6, 2018, accessed December 1, 2022, https://wordybirdsci.com/2018/01/06/this-week-in-ice-dec-24-2017-jan-6-2018/. "Polar vortex," Wikipedia, accessed December 1, 2022, https://en.wikipedia.org/wiki/Polar_vortex.

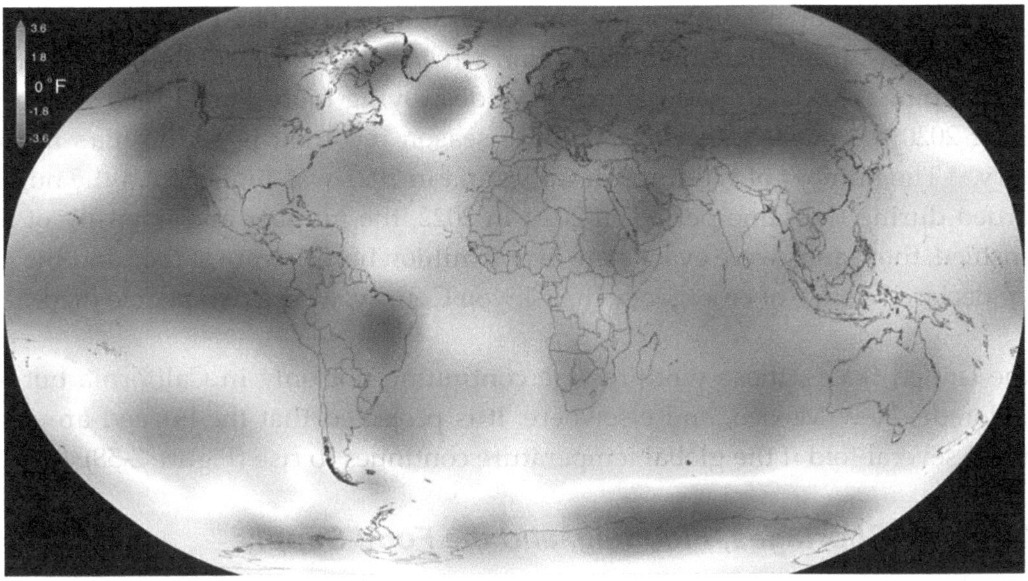

Figure 2-57. The warming of the globe is not uniform. It is greater in the Northern Hemisphere because the land area is greater in the north and land warms faster than oceans. The northern part of the Atlantic Ocean presents an exception because the melting ice from Greenland creates a relatively colder area that also cools the northeastern United States. (See color images after page 306.)

Source: "NASA, NOAA Analyses Reveal Record-Shattering Global Warm Temperatures in 2015," Scientific Visualization Studio/Goddard Space Flight Center, January 20, 2016, https://www.nasa.gov/press-release/nasa-noaa-analyses-reveal-record-shattering-global-warm-temperatures-in-2015.

2.7.1 Wildfires

Precipitation has substantially declined, and dry seasons have lengthened in many areas including the US Southwest. Today, wildfires in the United States (similar to Australia and elsewhere) burn more than twice the area they did in 1970, and the average wildfire season is 78 days longer. For much of the western United States, some projections suggest that an additional 1°C (1.8°F) temperature rise could increase the area of wildfires by 600%.

In 2020, Colorado had its second largest fire on record.[102] More than 1 million acres were torched in Oregon, and California had its worst fire year ever. In total, across the United States, 10 million acres were burned (40,500 km^2, or 15,625 mi^2). According to the National Interagency Fire Center,[103] from January 1 to August 31, 2020, there were 39,829 wildfires compared with 33,559 wildfires during the same months in 2019,

---

102. Kasha Patel, "East Troublesome Fire Spreads to the Rockies," NASA Earth Observatory website, accessed December 1, 2022, https://earthobservatory.nasa.gov/images/147452/east-troublesome-fire-spreads-to-the-rockies.

103. "National Report of Wildland Fires and Acres Burned by State," National Interagency Fire Center, accessed December 1, 2022, https://www.predictiveservices.nifc.gov/intelligence/2020_statssumm/fires_acres20.pdf.

an increase of about 19%. As of the end of 2020, nearly 10,000 fires had burned in California, consuming more than 4.2 million acres, more than 4% of the state's roughly 100 million acres of land, and damaging or destroying more than 10,000 structures, making 2020 the most destructive wildfire season recorded in California's modern history.[104] The number of wildfires in California in 2021 was higher than the number recorded during the same period in 2020. In 2022, the number and intensity of fires intensified; thousands were evacuated, over 2 million homes were at risk, and the governor declared a state of emergency in Siskiyou County, where two people died.

In the United States, these wildfires are continuing not only in California but also in Colorado, New Mexico, and elsewhere. It is projected that the burned areas will increase several-fold if the global temperature continues to rise (Figure 2-58).

2.7.2 Water Shortages, Flooding, and the Food Supply

It was already discussed that increased water circulation causes an increase in cloud formation, which contributes to increases in the greenhouse effect and GW. In addition, the warming of the oceans causes more evaporation, which not only fuels hurricanes and typhoons (discussed in detail in Section 2.6.1) but also brings large amounts of moisture into the air. When this moist air arrives over land in temperate regions, it cools, condenses, and causes rain and storm surge flooding. Moreover, as shown in Table 2-3, storm surges driven by the wind can cause devastating damage as the water hits structures on the shore. The process that causes flooding in the temperate zones of the planet also causes drying and water shortages in some of the tropical zones. Both GW-caused water shortages and overpopulation, which further increase the water use for agricultural irrigation and the per capita water use, can result in billions of people running out of drinking water. In some cases, the groundwater levels drop, and damage is caused by the ocean salt water that replaces it. In addition, the water shortages result in the shrinking of the total global food supply.

As the temperature in tropical zones rises, the evaporation of water increases, causing a lack of rain and drying of the soil. Water shortages already exist in large parts of the world. Countries that are home to one-fourth of the global population face the risk of running out of water. According to the World Resources Institute, from India to Iran to Botswana, some 17 countries worldwide are currently under extreme water stress (see Figure 2-59).

---

104. "2020 Incident Archive," Cal Fire website, accessed March 8, 2022, https://www.fire.ca.gov/incidents/2020/.

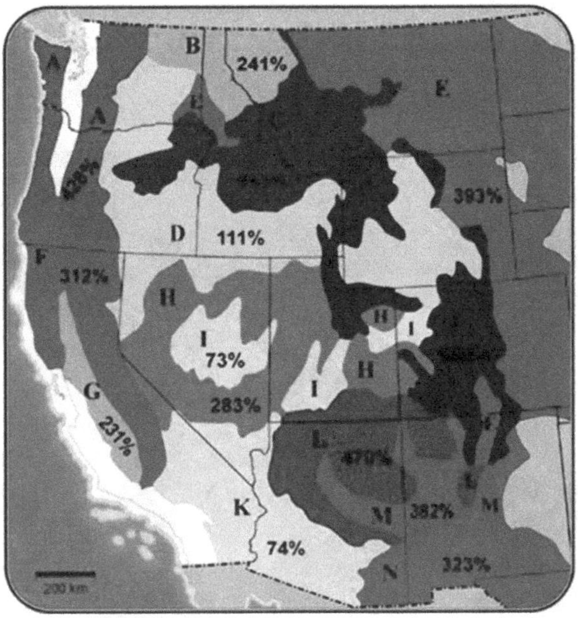

Legend

- A – Cascade Mixed Forest
- B – Northern Rocky Mt. Forest
- C – Middle Rocky Mt. Steppe-Forest
- D – Intermountain Semidesert
- E – Great Plains-Palouse Dry Steppe
- F – Sierra Steppe-Mixed Forest
- G – California Dry Steppe
- H – Intermountain Semidesert/Desert
- I – Nev.-Utah Mountains-Semidesert
- J – South Rocky Mt. Steppe-Forest
- K – American Semidesert and Desert
- L – Colorado Plateau Semidesert
- M – Ariz.-New Mex. Mts. Semidesert
- N – Chihuahuan Semidesert

Figure 2-58. Projected percentage increases in the probability of wildfire in the western United States by 2070 to 2090 compared to the reference period of 1961 to 1990 if the global temperature further increases by another 1°C (1.8°F).

Source: Climate Stabilization Targets: Emissions, Concentrations, and Impacts over Decades to Millennia (Washington, DC: National Research Council, 2011), 180, Figure 5.8, https://www.nap.edu/read/12877/chapter/11#180.

According to the World Resources Institute, water withdrawal has more than doubled globally since the 1960s, and the rise in this demand shows no signs of slowing. In the regions shown in Figure 2-59, irrigated agriculture, industries, and municipalities withdraw more than 80% of their available water supply. Forty-four countries—home to one-third of the world—face high levels of stress, where, on average, more than 40% of the available supply is withdrawn. Even in other parts of the world, communities intermittently face "Day Zero" in their water supplies.

Drought also impacts agriculture because it limits the amount of water available for irrigation and lowers groundwater levels. The UN Intergovernmental Panel is constantly warning that shortages in the food supply will develop as a lack of water and an increase in erosion reduces food yields or as desertification destroys life itself.

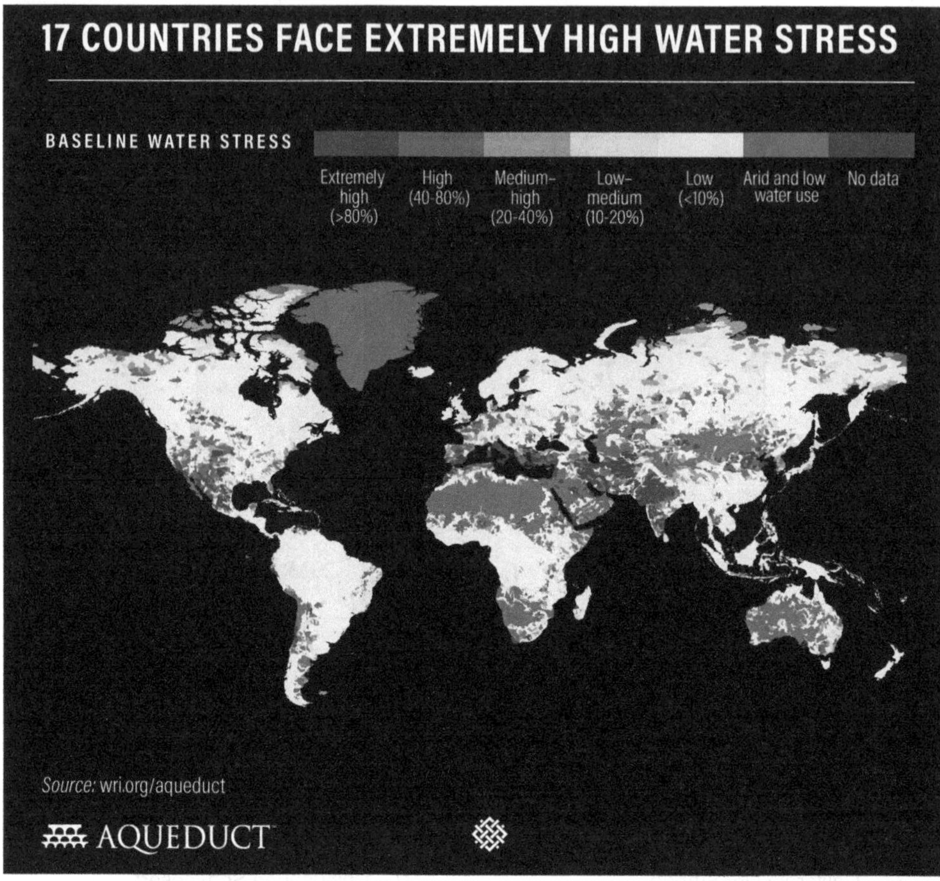

Figure 2-59. According to the World Resources Institute, about one-quarter of the world's population in the indicated regions is experiencing extreme water shortages. (See color images after page 306.)

Source: Reprinted with permission. Rutgers Willem Hofste, Paul Reig, and Leah Schleifer, "17 Countries, Home to One-Quarter of the World's Population, Face Extremely High Water Stress," Aqueduct, World Resources Institute website, August 6, 2019, https://www.wri.org/blog/2019/08/17-countries-home-one-quarter-world-population-face-extremely-high-water-stress.

It is essential to educate farmers. Best practices indicate that, after crops are harvested, farmers should use cover crops to return nitrogen and carbon to the soil by nourishing the fungi and microbes that pull CO_2 and nitrogen out of the air and draw them down to the root zones of the soil. In addition, farmers should minimize tillage because it increases drying, removes carbon from where it is needed, and becomes a source of carbon emissions into the air.

2.7.3 Indirect Symptoms: Panama Canal Example

There are hundreds of indirect symptoms of GW, too many to be listed. Water shortages, flooding, and other water-related phenomena are well-reported and generally understood. Other effects are not. As an example, I describe one significant effect that is usually not reported but is already occurring all around the world. This is the effect of GW on waterways. Here, I use the Panama Canal situation as an example not only because the canal itself is an important waterway and its operation depends on the availability of water but also to illustrate how process control can mitigate such seemingly unsolvable problems.

Dry areas of the planet are getting drier, and wet areas are getting wetter. In regions where the dry and rainy seasons alternate, the dry seasons are getting longer. This change in weather has many consequences, some of them unexpected, such as the impact on the operation of the Panama Canal.

The Panama Canal revolutionized trade and travel in the Western Hemisphere by cutting travel time between the US West Coast and Europe in half. One of the effects of this cut in travel distance is a tremendous saving in CO_2 emissions. Yearly, 20,000 ships (50 per day) carry 500 million tons of goods through the canal. In 2022, this represented 6% of global trade or 70% of the trade between Europe and the United States. Each time a ship transits the canal, about 50 million gallons of fresh water are lost through the locks to the oceans. The drought caused by GW endangers the canal operation because of the shortening of the rainy season and the consequences of hurricanes in the area.

The Panama Canal is a good illustration of two aspects of the indirect effects of GW. One is that the GW effects are not recognized as being the cause. The second is that process control theory and analysis can be applied to these processes, and through their application, we can make operations safer and more efficient. I believe that illustrating this approach with a specific example is helpful.

Until the twentieth century, the shipping distance between the east and west coasts of the United States was 13,000 miles. Construction of the Panama Canal reduced that distance to 5200 miles. Today, some 600 million tons of goods are transported through that canal each year on nearly 15,000 ships, corresponding to about 25% of the planet's seaborne containerized cargo.

The sea level Suez Canal between the Red and the Mediterranean seas, built by French engineer Ferdinand de Lesseps, successfully opened in 1869. In 1880, de Lesseps started work on the Panama Canal. He made the mistake of trying to build a 51-mile, sea-level canal, which would have required cutting through mountains of rock, such as the Culebra Mountain, in a tropical rainforest region. After digging up some 60 million cubic meters of earth and rock, and after many landslides, yellow fever epidemics, and other accidents that killed up to 22,000 workers, the project went bankrupt in 1889, wiping out the savings of some 800,000 investors. de Lesseps and his son were convicted of misappropriating funds and of knowing that some 2500 newspapers and magazines and some 150 deputies were paid off to promote financing the project.

In 1904, the United States purchased the remaining assets of the French project and forced the independence of Panama. A treaty was then signed giving the rights to the United States to build, indefinitely administer, and defend the Panama Canal Zone. This agreement has sometimes been misinterpreted as the "99-year lease," which it was not.

That same year, the construction project, which President Theodore Roosevelt strongly supported, was restarted. Thanks to the chief engineer, John Frank Stevens, the main error made by the French (the goal of building the canal at sea level) was eliminated. After some 5600 deaths, mosquito-spread diseases were nearly eliminated through the efforts of William C. Gorgas, the chief sanitation engineer. The project was concluded 2 years ahead of schedule in 1914.

Panama never entirely accepted that the Canal Zone would be administered by the United States, and the tension between the two nations increased until relations broke in 1964. In 1979, President Carter signed a treaty promising that the ownership of the canal would be transferred to Panama in 20 years. A complicated period followed during which the United States first supported the former CIA-informer-turned-dictator Manuel Noriega and then, in 1998, turned against him and invaded Panama. Still, the treaty signed by President Carter in 1979 was respected, and on December 31, 1999, the canal was turned over to Panama.

In 2006, Panama decided by a national referendum to expand the canal. This expansion added two sets of locks (one at the Pacific end and one at the Atlantic end) and widened the old canal to allow traffic in three lanes. The construction of the expanded waterway cost $5.25 billion, involved 37,600 workers, and used 44 million cubic meters of concrete and 192,000 metric tons of steel.

The canal was inaugurated in 2016. Today, ships large enough to carry 13,000 containers can pass through it in both directions (the old canal could only handle ships carrying 3000 containers). The average passage time through the canal is about 24 hours, which includes about 16 hours of waiting time at the locks. During the canal's first 20 months of operation, three ships crossed the expanded canal, mainly carrying Chinese goods to the United States. The per-ship cost of each passage varies widely and can reach $1 million. Because the sizes of container-cargo ships are constantly increasing, it is already projected that the larger locks that serve the new expansion, completed in 2016, will become too small by 2030.

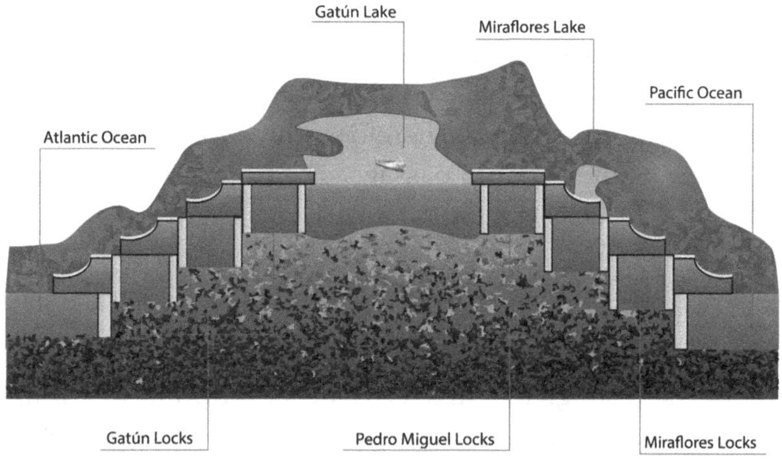

Figure 2-60. Process control can save 66% of the water needed to keep the Panama Canal operational during the dry season.

Source: Courtesy of *Control* magazine. Béla Lipták, "Controlling the Panama Canal," January 23, 2019, Figure 1, https://www.controlglobal.com/articles/2019/controlling-the-panama-canal/.

The Panama Canal locks use the water of two lakes (Gatún and Miraflores) for operation (Figure 2-60). As ships start to travel from one ocean to the other, they enter the first lock, in which the water level at the time of entering is the same as that of the ocean because its gate is open on the ocean side. Once the ship is inside the lock chamber, the ocean-side gate is closed, and water from the lake above is admitted to raise the water level and the ship. This is repeated in the following locks until the ship reaches the level of the lake. At the other end of the canal, the opposite process occurs as the ship is lowered three steps to the ocean level. The energy supply of this operation is the gravity flow of the water from the lakes above.

Gatún Lake is one of the world's largest artificial lakes, created to serve as a water reservoir for the canal. It is filled by the Chagres River and rainwater runoff from the surrounding rainforests. The water collected during the rainy season must be sufficient to operate the canal during the dry season. Specifically, 26 million gallons of water are needed for the passage of each ship. When the canal was expanded, the number of ships, their size, and the size of the locks increased, which naturally required more water. In anticipation of this increased water demand, the capacity of the lakes was also increased by increasing their water levels.

When analyzing this process, a process control engineer will first consider the goals of the application (the set points of the control loops) and then identify the variables that can be manipulated to achieve those goals (manipulated variables). The next step is to review the potential sources of upsets.

As with other processes today, operating the canal will also be affected by rising global temperature because this rise increases the length of the dry season in the tropics. This means that the quantity of water collected in the artificial lakes during the rainy season must last longer than in the past. Therefore, one of the goals has to be to reduce the water requirement of the per-ship passage. The means of achieving this goal (the available manipulated variables) include:

1. Maximizing water savings by introducing the use of water-saving basins.
2. Adjusting the passage fees to motivate the use of optimum timing and distribution of the passage of ships. (Whenever we want to manipulate human behavior, the most effective manipulated variable is the pocketbook.)
3. Considering the use of adjustable-length lock chambers.

In option 1, when a ship is lowered, the drained water is sequentially sent to water basins at elevations where it can serve as the water source to lock chambers for lifting an entering ship instead of being drained into the ocean. This reduces the water usage of the canal by 60%, but only if, during a particular time period, the number of ships moving in each direction is the same, the ships move at the same speeds, and their water displacements (sizes and weights) are close to being equal. Process control algorithms can easily calculate how to reach this balance—human operators cannot—and doing so is important because, otherwise, the water in the storage basins will run out on the side where the traffic is higher.

Process control algorithms can only suggest the most water-saving operation, but they cannot force the ships to follow that recommendation. The best motivation for the ships to follow that recommendation is to increase the cost of passage (manipulated variable) if they do not.

If GW increases further, these steps will eventually become insufficient. At that point, the options will change to using pumped ocean water to provide hydraulic lifts, converting the canal into a sea level passage, or replacing this canal with a sea level canal in Nicaragua.

2.8 Conclusions

For a long time, human evolution was guided mostly by Nature. Today, we are beginning to become the masters of our own evolution. In Chapter 1, "Controlling AI," I addressed the out-of-control AI development that is changing our cultural environment.

In this chapter, I described GW, which will determine our physical environment in the future. I have shown that both processes are out of control, even though both are controllable. In the coming chapters, I describe how to control these processes and show what will happen if we do not. We have reached a fork on the road of human evolution, and for our children's sake, we must take the right turn and prevent the destruction of our cultural and physical environments.

In this chapter, I first showed that GW is taking place and that its causes are largely anthropogenic. After that, I analyzed the personality of this heat-transfer process and found it to be a self-stabilizing one. I quantified the cause of GW by showing that the Earth receives more heat from the Sun than it radiates back into space. I also calculated that (as is the case for all warm objects in cold surroundings) as the temperature of the globe rises, so does its heat loss to outer space. Once the global temperature rises to the point where the cooling effect of this heat loss equals the heat gain caused by the excess solar energy retained, the global temperature will stop increasing, and the heat balance of the planet will stabilize.

This sounds like a happy conclusion; it suggests that all we need to do is wait and GW will stop rising on its own. Unfortunately, this is not the case because if we continue to burn fossil fuels, the planet's temperature will self-stabilize, but at a global temperature that will destroy our civilization. In other words, Nature will eventually restabilize this heat-balance process, but the steady-state temperature reached will be too high for human survival.

The main difference between, say, controlling the temperature of a nuclear reactor and controlling GW is that when a reactor overheats, we increase the speed of the cooling-water pumps and remove the excess heat. In the case of the GW process, the "pump" is human behavior, and while increasing the pump speed is easy, changing human behavior is not. The most effective way to change human behavior is through our pocketbooks. The methods include tax credits, carbon taxes, and cap and grade methods, but in my view, market forces are the most effective tools. Therefore, we must make the conversion to a carbon-free green energy economy profitable. Using the above industrial parallel, what flows in the GW "cooling pump" is money, and its flow rate can be increased only by raising the cost of fossil fuels or making green energy cheaper.

The option of increasing the cost of fossil fuels through taxation or because of energy wars should be avoided. Therefore, it is essential that books like this one prove that modern technology—assisted by process control—can make green energy less expensive, and it already has in many cases. For example, in California, "gridless" homes are spreading because the electricity they generate costs less. This is because the fuel for green electricity is free, and the generated electricity does not need to be transported. Today storage can be provided by batteries, which can be replaced with solar-hydrogen storage when it becomes available and connecting to the grid will no longer be needed.

Progress has also been made in building larger-scale green power plants. For example, the Randolph Solar Project in Charlotte County, VA, will operate on approximately 3000 acres and generate as much electricity as a nuclear reactor (800 MW).

While we have started developing the conversion technology, progress is practically nonexistent, and emissions and GW are still increasing. I found in this analysis that we are running out of time. We must, and we can, eliminate our GHG emissions in about two decades. I have shown that most currently used future projection models are overly optimistic because they disregard the positive feedback effects that will be triggered as we reach the many tipping points (permafrost melting, ocean degassing, etc.) as GW rises. I have also shown that GW will continue to increase even after we reach a zero-carbon energy economy because the already-accumulated GHGs will continue to warm the planet.

This warming effect can only be eliminated by increasing the albedo (reflectivity) of the planet or by emitting no carbon so that the oceans and plants can start removing the carbon that is already in the air. This would require converting to a hydrogen-fueled energy economy in two decades, which is impossible.

I have also shown that the Paris carbon emissions reduction goals will not be met, nor will the goal of limiting the rise in global temperature to 1.5°C (2.7°F) above preindustrial levels. In fact, even if we were able to stay below the 1.5°C (2.7°F), we would still see flooded coastlines, intensified droughts, wildfires, hurricanes, debilitated industries, and water shortages harming agriculture and natural ecosystems, all of which would cause a crippling wave of poverty-fueled mass migrations. This assessment is not just my conclusion; it is also the view of the United Nations. According to a report by the UN's IPCC, if we reach that temperature limit, the damage will cost some $54 trillion.[105]

---

105. Wu, "12-Year Ultimatum."

Today, most climate control goals are expressed in terms of target dates. The language used in defining these dates is essential and often misunderstood. Reaching net-zero conditions does not mean that carbon emissions are stopped, but only that the amount of carbon added to the atmosphere no longer exceeds the amount that land plants and the oceans remove. Also, the public should understand that even cutting emissions to net zero does not reduce the quantity of carbon already accumulated in the air; it only reduces the rate at which its concentration is rising. Therefore, achieving net zero will not decrease GW; it will only reduce the rate at which it is increasing.

I am optimistic about the future because we know what needs to be done, and we have the tools and know-how to do it. I am optimistic because books like this one show that the key argument against using carbon-free green energy—namely that it is only intermittently available—is false, and the problem of converting it to continuously available hydrogen has been resolved. I am encouraged because the younger generations are beginning to understand this, as was shown in March 2019 when an estimated 1.6 million students from 125 countries protested the present inaction. I am also optimistic because moving from the fossil to the hydrogen age will stop energy wars, and the jobs it creates will trigger an economic boom not seen since the end of WW2. Most importantly, now that the free market forces are starting to support our efforts, the cost of green energy is dropping and the cost of fossil fuels is rising. Finally, I know that we will stop using fossil fuel not because we have exhausted the supply or because we got tired of fighting wars for it but because we will follow the example of our Stone Age ancestors who switched to bronze, not because they ran out of stone but because bronze was better.

3
Green Energy Sources

Table of Contents

Before discussing green energy sources, I would like to give an example of the value of "green design" in general. In September 2022, hurricane Ian, a Category 4 hurricane, devastated Southwest Florida. Yet the residents of Babcock Ranch—an innovative community north of Fort Myers, where the homes are built using green design—remained untouched. It was not flooded and did not lose electricity, water, or the Internet.

Babcock Ranch calls itself "America's first solar-powered town." Its nearby solar array—made up of 700,000 individual panels—generates more electricity than the neighborhood uses in a state where most electricity elsewhere is generated by burning natural gas. Besides the energy-related advantages of the "green town" design, it should be noted that the town did not suffer from hurricane damage due to its resilient and sustainable design. The buildings are surrounded by "flood gardens" for stormwater attenuation. The homes have no basements, are built one foot above flood elevation, and have hip roofs and hurricane-proof windows; all utilities are routed underground. This shows that the carbon-free energy economy of the future can not only help to stop global warming but in many cases, the design resulting from it can also improve the safety of communities from climate-related damage.

In the previous chapter, I discussed the "personalities" (dynamic characteristics) of the various climate change subprocesses in order to understand their dynamics, sizes, and speeds at which they contribute to the overall process of global warming (GW). In this chapter, I describe the types and dynamics of carbon-free, green energy sources.

We saw that some processes warm and some cool the planet, some provide negative feedback (reduce the effects of GW), whereas others provide positive feedback (accelerate the effects of GW). For example, the melting ice cover on the oceans and the melting permafrost have large, self-accelerating, positive feedback on GW. According to

NASA, these characteristics have caused the Arctic to be the fastest-warming region on the planet. The amount of warming varies with both the location and the season. For example, the Siberian heat wave in the spring of 2020 caused the Arctic Ocean ice melt season to start early; the Arctic temperatures were 8°C to 10°C (14°F to 18°F) warmer than average as the ice coverage kept declining.

To control any process, we must fully understand it, and only then can we determine what control system should be used to control it. We must select the sensors needed to measure the existing conditions and compare those conditions with the desired conditions or set points. If there is a difference between the two, that is the error we have to eliminate. These values become the inputs of the model-based control (MBC) algorithm, which generates the output signal that manipulates the process. Naturally, to select the correct algorithm, we must understand the personality of the controlled process as well as the dynamics, time constants, time lags, capacities, and other characteristics of the green energy sources.

Therefore, in this chapter, I identify the nature of several green energy sources. I do not describe the associated sensors and control algorithms needed to utilize them because this information is already available in the literature.[1]

According to the International Energy Agency (IEA), the total global energy consumption was 418 EJ (450 Q) in 2021. That number is probably low today because of the pandemic and the jump in energy costs caused by the war in Ukraine, which probably caused a temporary drop in consumption that will not last long. Therefore, I assume that today's consumption is about 600 Q.

Figure 3-1 provides historical data and projections on total energy consumption and the relative sizes of different energy sources. Interested readers can find a large amount of data and projections in the sources listed in the footnotes.[2,3]

According to Figure 3-1, both the total energy use and the use of fossil fuels are projected to keep rising. Based on my study of the subject, I believe this projection is probably correct. The 2022 war in Ukraine, however, changes the details of the projections

---

1. Béla G. Lipták, ed., *Instrument and Automation Engineer's Handbook,* 5th ed. (Boca Raton, FL: CRC Press), 2016.

2. *Advancing the Global Renewable Energy Transition: Highlights of the REN21 Renewables 2018 Global Status Report in Perspective* (Paris: REN21 Secretariat, June 2018), https://www.ren21.net/wp-content/uploads/2019/05/GSR2018_Highlights_English.pdf.

3. *International Energy Outlook 2017* (Washington, DC: US Energy Information Administration [EIA], September 14, 2017), https://www.eia.gov/outlooks/ieo/pdf/0484(2017).pdf.

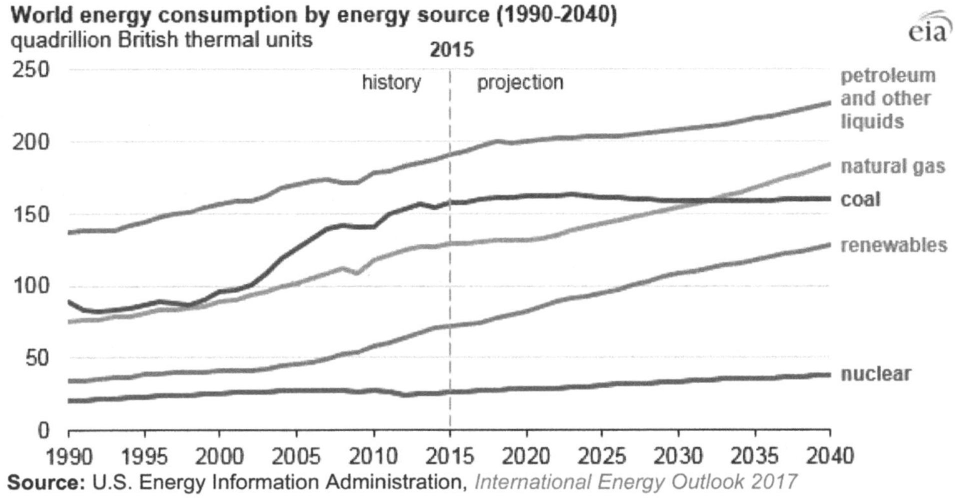

Figure 3-1. Global energy consumption by energy source over the 1990 to 2040 time period according to the US Energy Information Administration (EIA).

Source: "EIA projects 48% increase in world energy consumption by 2040," Today in Energy website, May 12, 2016, https://www.eia.gov/todayinenergy/detail.php?id=26212.

because reduced supplies and increased fossil-fuel prices will temporarily increase the profitability of these industries (particularly for natural gas). Still, in the long run, their high cost will put them at a competitive disadvantage relative to green energy sources. It seems that, in the long term, the use of green energy sources will rise, and the use of fossil fuels (particularly the burning of coal) will drop.

As to what the future holds, your guess is as good as mine. The fact is that not even the IEA has issued a single projection. They issued four World Energy Outlook scenarios in 2021. A summary of the four scenarios is given below:

1. The **Stated Policies Scenario** (STEPS) is based on the assumption that policies issued in 2021 will be met and concludes that in that case, GW will exceed 2.6°C above preindustrial levels by 2100 and will still be rising.
2. The **Announced Pledges Scenario** (APS) assumes that all climate commitments will be met in full and on time and concludes that in that case, GW will rise to around 2.1°C by 2100 and will continue to increase.
3. The **Sustainable Development Scenario** (SDS) assumes that in addition to #1, a surge in clean energy policies and investment will occur, and advanced economies will reach net zero emissions by 2050. In that case, it concludes that GW will peak at 1.7°C by 2050 and could decline to 1.5°C by 2100. The energy

supply will be 55% renewable and electricity generation will be 58% renewable by 2050.

4. The **Net Zero Emissions by 2050 Scenario** (NZE) assumes that net zero CO_2 emissions will be reached locally by 2050. In that case, it estimates that GW will peak at 1.7°C by 2050 and decline to 1.4°C by 2100. Half of the global energy consumption will be in the form of electricity, and 90% of that half will be from green sources (70% by wind and solar photovoltaic—PV) by 2050.

According to a new IEA report, "clean energy investment could be on course to exceed $2 trillion per year by 2030, an increase of over 50% compared to today."[4]

3.1 Green Energy

Optimists say that 100 years from now, our concerns about GW and nuclear proliferation will have long been solved and forgotten. In contrast, pessimists say that, by the end of the century, the collapse of human civilization will be in progress, and we will go the way of the dinosaurs. I am among the optimists, and the main reason for writing this book is because it is important for all to understand that we know how to overcome our problems and direct the transition to a sustainable and safe future.

According to the IEA's *Global Energy Review 2020:*[5]

> Renewable energy has so far been the energy source most resilient to COVID-19 lockdown measures. ... In Q1 [the first quarter of] 2020, global use of renewable energy in all sectors increased by about 1.5% relative to Q1 2019. Renewable electricity generation increased by almost 3%, mainly because of new wind and solar PV [photovoltaic] projects completed over the past year. ... Despite supply chain disruptions that have paused or delayed activity in several key regions, the expansion of solar, wind, and hydro power is expected to help renewable electricity generation to rise by nearly 5% in 2020. This growth is smaller than anticipated before the COVID-19 crisis, however. A faster recovery would have a minimal impact on renewable energy production, though it would enable more new renewables-based projects to be completed. If recovery is slower, renewable energy would still increase, making renewables the energy source the most resilient to the current COVID-19 crisis.

In 2019, renewable electricity generation[6] increased as more than 100 GW of solar photovoltaic (PV) projects and about 60 GW of wind power projects were completed. In

4. Anmar Frangoul, "Energy security, not climate change, is driving clean energy investment, IEA chief says," published October 27 2022, updated October 27 2022, https://www.cnbc.com/2022/10/27/clean-energy-investment-may-hit-2-trillion-a-year-by-2030-iea.html.

5. IEA (International Energy Agency), "Renewables," in *Global Energy Review 2020* (Paris: IEA Publications, July 2020), 34–39, https://www.iea.org/reports/global-energy-review-2020/renewables.

6. "Renewable energy," Wikipedia, accessed November 1, 2022, https://en.wikipedia.org/wiki/Renewable_energy.

2020, the share of renewables in global electricity generation jumped to nearly 28%, and the share of solar PV and wind power reached 9% of renewable energy generation. The report forecasted that the world's *total* renewable-based power capacity would grow by 50% between 2019 and 2024. This increase of 1200 GW would be equivalent to the total current power capacity of the United States. Solar PV is expected to account for 60% of the rise. The share of renewables in global power generation is expected to rise to 30% by 2024.

In May 2022, the IEA reported that annual renewable capacity additions broke a new record in 2021, increasing 6% to almost 295 GW. Despite the continuation of pandemic-driven supply chain challenges and construction delays, an 8% further increase is expected in 2022, reaching almost 320 GW. The IEA also noted that the Russian invasion of Ukraine is sending shock waves through energy and agriculture markets, and the impact of this cannot be predicted, but it can result in an unprecedented global energy crisis.

I know process control can play a major role in converting from our dependence on fossil and nuclear energy technology to the age of using green energy. In this chapter, I discuss these clean energy sources, focusing primarily on solar energy, but I also briefly discuss hydraulic, wind, geothermal, and other renewable energy sources. I discuss nuclear energy, not because I recommend its use but because it is also a zero-carbon energy source. Within that discussion, I also explain why I believe its use should be abandoned.

It is important to address the terminology related to renewable fuels because it is often misused. Biofuel is a fuel that is produced through contemporary biological processes, such as agriculture and anaerobic digestion. If the source (the biomatter) can regrow quickly, the resulting fuel can be described as a "carbon-neutral renewable" energy source. In other words, biofuels are not "carbon-free" energy sources, as are solar, wind, geothermal, and hydraulic sources. They are only *carbon-neutral* or *net-zero* carbon sources because they emit carbon dioxide (CO_2) when burned, and it takes time for them to regrow and consume the same amount of carbon as they emitted upon burning. Tree growth and wood decomposition represent a short-term carbon cycle because while growing trees convert carbon dioxide into biomass, the decomposing trees release carbon dioxide back into the atmosphere. Decaying forest wood and the decomposition of wood and paper products in landfills releases CO_2 and methane in large quantities, and burning wood can be more harmful than burning fossil fuels.

I differentiate between "carbon-free" and "carbon-neutral or renewable" energy sources. When I use the second term, I am talking about energy sources that release GHG emissions in the same amount as they absorb years later. When I call an energy source "carbon-free," I mean a source that does not release carbon at all. Finally, when I want to refer to both categories, I use the term *green energy* sources. Some people refer to the burning of biomass as *traditional renewable* or *biomass renewable* sources while referring to *carbon-free* sources as *modern renewable* (Figure 3-2).

Another aspect of well-intentioned tree planting that should be mentioned is that one should not plant just any tree anywhere because that can upset the natural ecosystem.

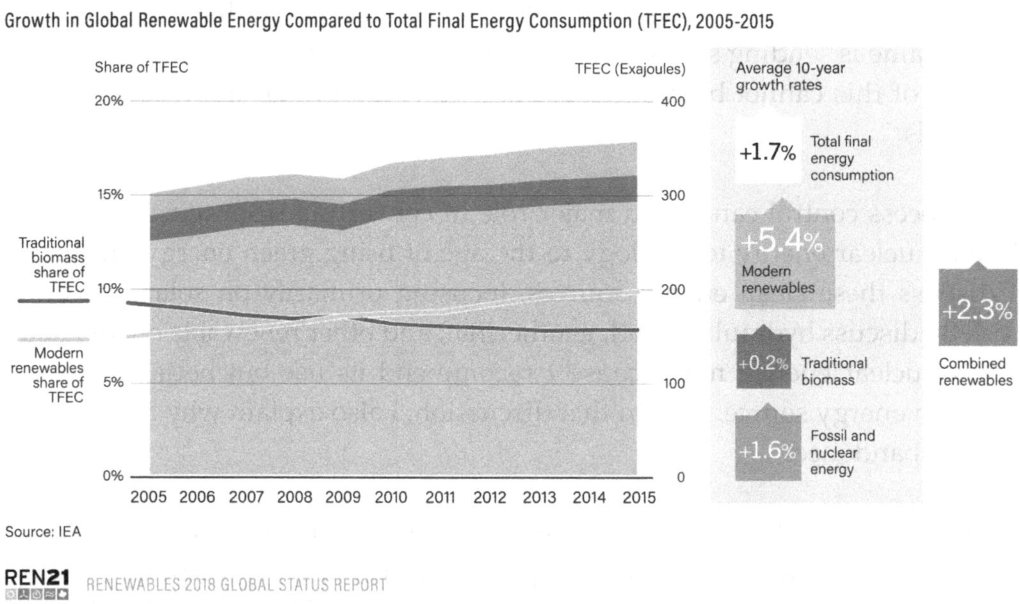

Figure 3-2. From 2005 to 2015, the total final energy consumption (TFEC) increased at a rate of about 2% per year, whereas the rate of increase of modern energy sources (solar, wind, etc.) was more than 5% per year. After 2015, this trend accelerated.[7] (See color images after page 306.)

Source: Reprinted with permission from *Advancing the Global Renewable Energy Transition: Highlights of the REN21 Renewables 2018 Global Status Report in Perspective* (Paris: Renewable Energy Policy Network, June 2018), https://www.ren21.net/wp-content/uploads/2019/05/GSR2018_Highlights_English.pdf.[7]

---

7. "Solar PV was the star performer for the second year in a row, with newly installed capacity increasing by about 33% (at least 98 gigawatts, GW) over the record-setting additions in 2016. Approximately 402 GW of solar PV was up and running worldwide by the end of 2017. About 52 GW of wind power was added globally in 2017, bringing the cumulative total installed capacity to around 539 GW—an increase of nearly 11% over 2016. The global offshore wind market grew 30%." *Advancing the Global Renewable Energy Transition: Highlights of the REN21 Renewables 2018 Global Status Report in Perspective* (Paris: Renewable Energy Policy Network, June 2018), https://www.ren21.net/wp-content/uploads/2019/05/GSR2018_Highlights_English.pdf.

It is a much better practice to fence in an area around the forest and let nature expand the forest without human interference.

Similarly, the category labeled "Modern renewables" in Figure 3-2 is not truly modern because hydraulic energy sources are not modern, and the other sources (solar, wind, and geothermal) are renewable, *inexhaustible zero-carbon* sources. I emphasize that this imprecise use of terminology is a problem because not understanding the difference between carbon-neutral and zero-carbon sources confuses not only the public but also the people writing the climate treaties and setting renewable energy goals. I know it is hard to change the language in common use, but it is still desirable to distinguish these categories and clearly understand that a zero-carbon economy is our ultimate goal.

In addition to using these terms correctly, I try to discuss global energy trends without using too many numbers because the sizes of the numbers, when dealing with global energy, are so large that they are incomprehensible. Therefore, I mention only a few numbers and use them to show the relationships between them to help you understand the relationships and trends. The unit of energy I use most often is the quad, Q, so named because it is equal to one quadrillion British thermal units (1 Q = 10^{15} Btu).[8] Each quad of energy is equal to the energy produced by 33 nuclear power plants in a year, the energy content of 10,000 supertankers of oil, 400,000 rail cars of coal, or 28 billion cubic meters of natural gas.

The distribution of global energy sources in 2019 was as follows:[9]

- Oil = 33.06%
- Coal/peat/shale = 27.04%
- Natural gas = 24.23%
- Hydroelectricity = 6.45%
- Nuclear = 4.27%
- Wind = 2.18%

---

8. One quad is one quadrillion British thermal units (1 Q = 10^{15} Btu), where 1 Btu is the energy needed to increase the temperature of 1 lb (about 454 g) of water by 1°F (0.56°C). Conversions to some of the other large energy units used in this chapter are as follows: 1 Q = 25.24 MTOE (megatons of oil equivalent) = 0.3 PWh = 1.05 EJ. See Chapter 6, "Data Summary," for more information on units and conversions.

9. "Energy consumption by source, World," Our World in Data website, accessed March 12, 2022, https://ourworldindata.org/grapher/energy-consumption-by-source-and-region?country=~OWID_WRL

- Solar = 1.11%
- Other green sources = 1.0%

These percentages remained the same in 2020 (below), but after that, oil use dropped and the use of natural gas and green energy sources increased.

- Oil = 31.2%
- Coal = 27.2%
- Natural gas = 24.7%
- Hydro = 6.9%
- Nuclear = 4.3%
- Others = 5.7%

In 2021, the IEA estimated the total global energy consumption as 440 Q (416 EJ) and projected that it will rise to 820 Q by 2040. As shown in Figure 3-3, in the past few years, both the total energy use and the fossil energy use (except for coal) continued to rise.

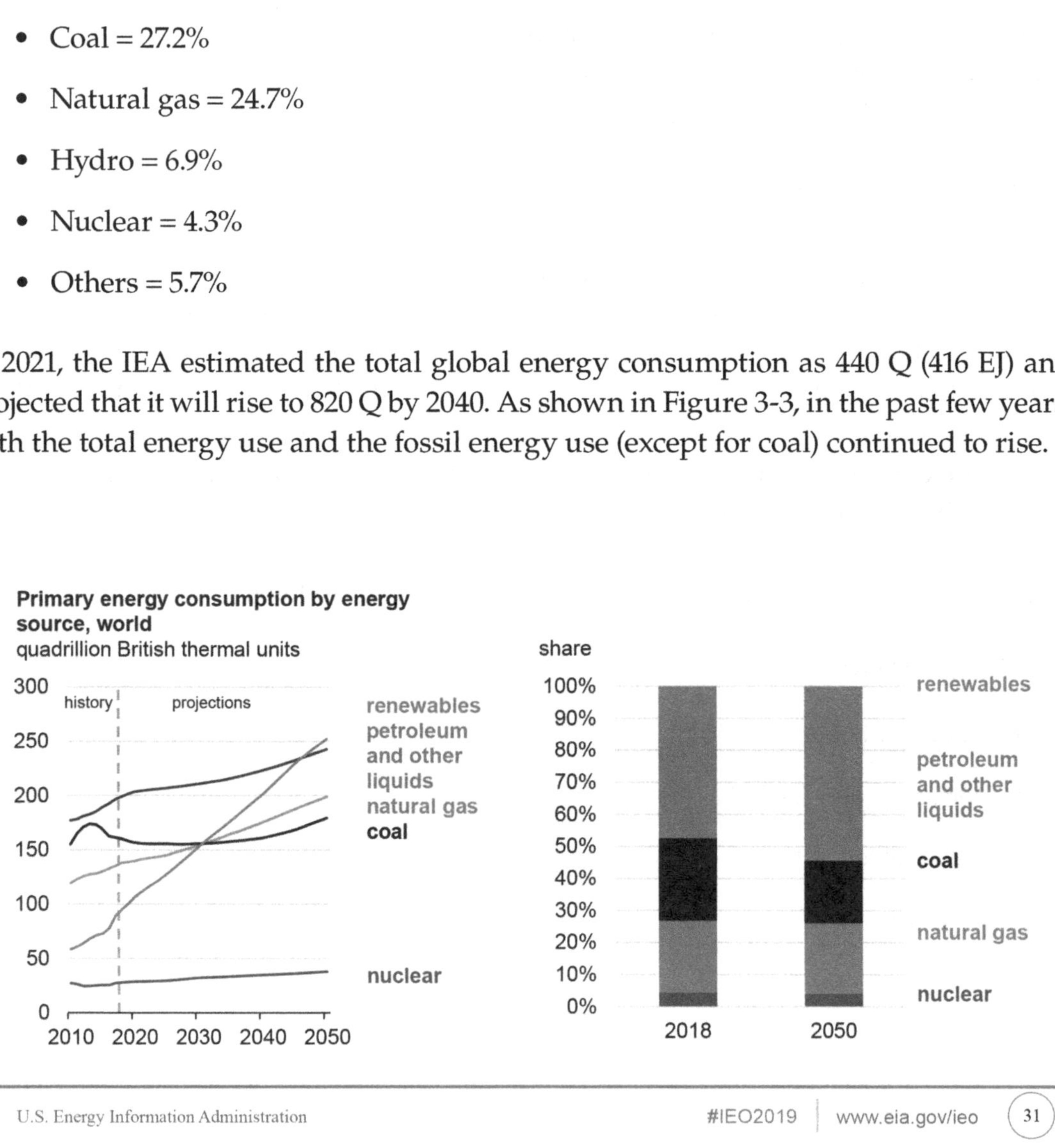

Figure 3-3. Projection of global primary energy consumption between 2010 and 2050 in (quads) (1Q = 10^{15}Btu). (See color images after page 306.)

Source: "EIA projects nearly 50% increase in world energy usage by 2050, led by growth in Asia," Today in Energy Blog, September 24, 2019, https://www.eia.gov/todayinenergy/detail.php?id=41433.

Global electricity consumption is about 15% of the total energy use.[10] The IEA estimated[11] that, by 2040, the total consumption will reach 820 Q[12] and by 2050, it will reach 920 Q. By these dates, green energy sources will become the largest energy source (Figure 3-3). One should note that China is an exception to the global trend because, as of this writing, its carbon emissions are greater than those of the United States, the European Union, and India combined, and they are projected to keep increasing for decades.

So, what have we learned from all of these projections? Well, very little actually! The most important conclusion is that they sharply disagree. Therefore, such projections have little value and only confuse the public.

From process control, we know that *lag-dominant* processes, such as GW, show their true personalities (their likely long-term behavior) only after their lag time has passed. Naturally, the future behavior of the GW process will depend highly on whether fossil fuels continue to be used. This means that all future projections that go much beyond the present have little value. In other words, process control theory reinforces the view that dramatic action is needed now. The other feature that all of these figures have in common—and I hope they are all wrong—is that they predict the continuation of fossil energy use or even its domination for decades. These projections show that it will take intensive education and taxation to prevent this from happening. Financial and public pressure is needed to change the present trend and to overcome the misinformation campaign of the powerful fossil-fuel lobby because my analysis indicates that we have a little more than a decade to prevent a climate disaster.

Today, the global share of renewable energy is only about 15% to 20%, and of that, less than one-half is carbon-free (solar, wind, geothermal, hydraulic); the other half is traditional, meaning the burning of wood and biomass. Some institutions (such as the EIA) project that, by 2040, the global use of carbon-free sources will still only be 15%. Naturally, I hope that they will turn out to be wrong, but this will not occur without a gigantic joint global effort.

---

10. This is equivalent to an energy content of 2100 MTOE/year or 24,500 TWh/year. The contributions of various fuels to the generation of this electricity are as follows: coal/peat, 40.8%; natural gas, 21.6%; hydraulic energy, 16.4%; nuclear power, 10.6%; renewables such as biomass, solar, and wind, 6.3%; and oil, 4.3%.

11. *"EIA projects world energy consumption will increase 56% by 2040," Today in Energy, US EIA, July 25, 2013*, https://www.eia.gov/todayinenergy/detail.php?id=12251.

12. Some estimates suggest that the total fossil-fuel deposits of the planet amount to 75,000 Q. I have no idea where this number came from, but I do know that, just as our ancestors in the Stone Age did not switch to using bronze because they ran out of stone, we should also leave the fossil-fuel deposits where they are and switch to hydrogen.

One cannot blame the fossil-fuel companies for not wanting to leave about $40 trillion worth of their resources in the ground, nor can one blame our political leaders for not rejecting campaign contributions from the fossil-fuel and nuclear lobbies. Yet, one must still urge the general public to learn the scientific information, such as the data in this book, disregard misinformation, and act now! We must not allow the GW process to become a batch-type process.[13] We know how to stabilize it, and we have the tools needed to do that. All we need is the will.

3.1.1 Green Energy Sources

In this chapter, I discuss solar energy in detail because, in my view, it is the primary energy source for our future energy supply. In the next chapter, I focus on the technology for storing and transporting solar energy. Although other renewable energy sources (hydraulic, wind, geothermal, etc.) are also valuable, I discuss them to a lesser extent because, in this transition period, their contributions are less critical. Finally, I discuss nuclear energy not because I recommend its use but only because it is carbon-free; otherwise, in my view, it is too dangerous and should be phased out.

Table 3-1 includes data from three reports showing the shares of global energy consumption for different energy sources that provided our total energy supply in 2013, 2015, and 2017. Note that green energy sources are broken down by type in only the 2017 data.

Table 3-1. Percentages of global energy consumption in 2013, 2015, and 2017.

Type	2013[a]	2015[b]	2017[c]
Total fossil	78.4%	78.7%	79.6%
Nuclear	2.6%	2.2%	2.2%
Total green energy	19%	19.2%	18.2%[d]

[a]"Total World Energy Consumption by Source (2013)" by Delphi234 is licensed under CC0 1.0, https://upload.wikimedia.org/wikipedia/commons/b/b6/Total_World_Energy_Consumption_by_Source_2013.png.
[b]John Constable, "The Global Status of Renewable Energy: Taking Stock," The Global Warming Policy Forum, December 6, 2017, https://www.thegwpf.com/the-global-status-of-renewable-energy-taking-stock/.
[c]José Eustáquio Diniz Alves, "O longo caminho da transição energética," EcoDebate, September 3, 2018, https://www.ecodebate.com.br/2018/09/03/o-longo-caminho-da-transicao-energetica-artigo-de-jose-eustaquio-diniz-alves/.
[d]Of this 18.2%, carbon-neutral biomass provided 12.1%; hydropower, 3.1%; and zero-carbon sources (solar, wind, etc.), 2.3%. From the data, we can see that essentially no progress was made during a period of 4 years (2013 to 2017). In fact, the use of fossil fuels increased slightly, and the total use of green energy sources dropped slightly.

---

13. A batch process is one that has a beginning and an end.

In 2020, the global yearly green energy consumption[14] amounted to about 8300 TWh (28.3 Q) and came from the following sources:

- Biofuels = 1000 TWh
- Hydroelectric = 4200 TWh
- Wind = 1500 TWh
- Solar = 900 TWh
- Other = 700 TWh

The total global green electricity generation capacity in 2020 amounted to about 3000 GW, and the sources were as follows:

- Biofuels = 600 GW
- Hydropower = 1100 GW
- Solar = 780 GW
- Wind = 518 GW

In 2020, green energy sources supplied 28% of the total electricity consumption. For a detailed breakdown of the global sources of electricity generation in 2019, see *Table 6-4* in Chapter 6.

Note that the year 2020 presented somewhat of an anomaly. In response to the exceptional circumstances stemming from the coronavirus pandemic, the IEA's annual *Global Energy Review* for 2020[15] reported that countries in total lockdown experienced an average 25% decline in energy demand; however, the total effect on global consumption was minimal.

In 2020, in terms of specific fuels, global coal demand fell by 9% compared with the first quarter of 2019. Oil demand fell by nearly 5% and gas demand fell by about 2%, whereas green energy sources were the only category that posted a growth. Electricity demand was depressed by 20%. Demand fell for all other sources of electricity as well, including nuclear power.

14. "Renewables," in *Global Energy Review 2020.*
15. IEA (International Energy Agency), *Global Energy Review 2020* (Paris: IEA Publications, April 2020), https://www.iea.org/reports/global-energy-review-2020.

For the full year of 2020, oil demand fell by 9%; coal demand fell by 8%, mainly because electricity demand was at least 5% lower; and gas and nuclear energy demands were lower because of the reduced electricity demand. Only the demand for renewables increased.

That year, global CO_2 emissions declined by 8%, falling to the 2010 level. Such a year-on-year reduction was the largest ever, six times larger than the previous record reduction of 0.4 GT in 2009—caused by the global financial crisis—and twice as large as the combined total of all previous reductions since the end of World War II. As with previous crises, the rebound from 2021 to 2022 was also substantial.

Figure 3-4 illustrates the prevailing terminology chaos in discussions of the different types of emissions. As already discussed, it is misleading to apply the term *renewable* to all sources that do not increase the net carbon content of the air. We should clearly distinguish *zero-carbon* fuels, which emit no carbon at all, from *carbon-neutral* ones, which emit carbon when burned but later absorb the same amount when they regrow.

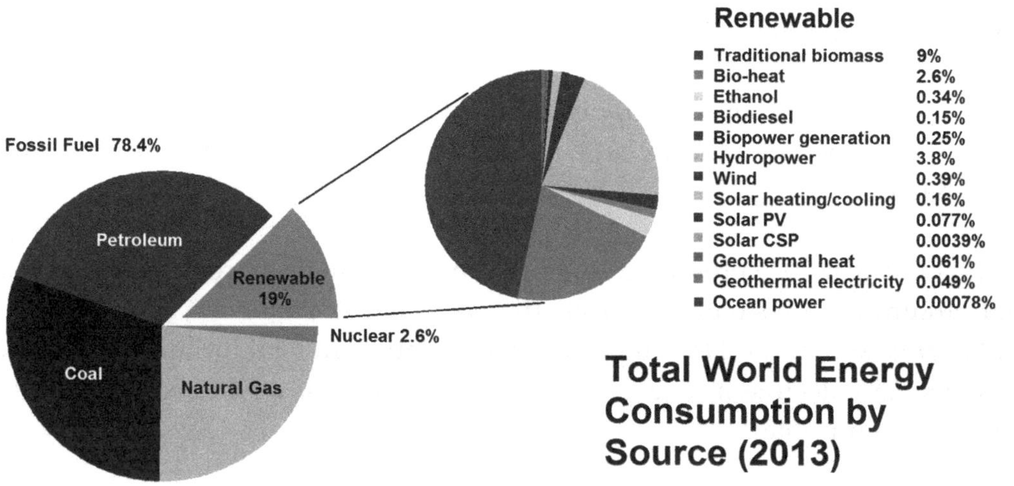

Figure 3-4. Illustration of the confusing terminology used to describe global renewable energy consumption in 2013.[16]

Source: "Total World Energy Consumption by Source 2013.png" by Delphi234, June 24, 2014, is licensed under CC0 1.0, https://commons.wikimedia.org/wiki/File:Total_World_Energy_Consumption_by_Source_2013.png.

16. For more recent data, see: *International Energy Outlook 2020* (US Energy Information Administration, October 14, 2020), www.eia.gov/ieo. For 2018 data, see: IEA (International Energy Agency), *Renewables 2018: Analysis and Forecasts to 2023* (Paris: IEA Publications, October 2018), https://www.iea.org/renewables2018/.

Figure 3-4 illustrates this problem, as it lists both carbon-free and carbon-neutral sources as *renewable*. I emphasize the importance of using the correct terminology because an increase in the share of zero-carbon energy sources means progress, whereas an increase in the share of carbon-neutral sources does not. I also emphasize the need for accurate terminology because inaccuracy opens the door to political footballs, where politicians can imply progress when, in fact, they are just using different names for the same thing. For example, adding the word *net* to the proposed 50% cut in emissions might lead the public to believe that we plan to cut our total emissions in half when, in fact, *net* means that we plan to cut in half only the amount by which the GHG emissions exceed their absorption by plants and the ocean.

This lack of distinguishing between the two green fuel categories is wrong because it implies progress even in cases when nothing or little has been done. When we read that the total share of *renewable* energy sources has increased to 19%, we think that we have made progress, but when we read a source, such as the *BP Statistical Review of World Energy 2019*,[17] more carefully, we see that the renewable and zero-carbon categories were combined.

Energy sources that emit CO_2 should not be combined with the ones that do not. If we distinguish these two categories, we see that less than half of the sources included in the 19% in Figure 3-4 (solar, wind, geothermal, hydraulic) are truly *zero-carbon*, whereas biofuels (all biomass, bioheat, biopower, ethanol, etc.) are only *carbon-neutral*. When burning biofuels, similar GHG emissions occur as with fossil fuels, and only if and when they are replanted will they absorb the same amount of carbon as they emitted when burned.

3.1.2 The IEA Position

To emphasize the above points, I quote here from the IEA's 2018 report discussing the projected growth of the renewable energy industry for the 5 years between 2018 and 2023.[18] There are more recent reports available, but I have selected this one to allow the reader to compare these 2017 projections with the actual trends in the past few years. We should note that, in 2018, even this important institution did not distinguish between zero-carbon and carbon-neutral renewable energy sources.

> Modern bioenergy is the overlooked giant within renewable energy. Modern bioenergy (excluding the traditional use of biomass) was responsible for half of all renewable energy

17. *BP Statistical Review of World Energy 2019*, 68th ed. (London, bp p.l.c., June 2019), https://www.bp.com/content/dam/bp/business-sites/en/global/corporate/pdfs/energy-economics/statistical-review/bp-stats-review-2019-full-report.pdf.
18. *Renewables 2018*, IEA.

> consumed in 2017—it provided four times the contribution of solar photovoltaic (PV) and wind combined. Most modern bioenergy is used in final energy consumption to deliver heat in buildings and for industry.
>
> Bioenergy is the largest source of growth in renewable consumption over the period 2018 to 2023. Bioenergy—as solid, liquid, or gaseous fuels—will account for 30% of the growth in renewable consumption in this period. This is a result of the considerable use of bioenergy in heat and transport. Other renewables have less penetration in these two sectors, which account for 80% of total final energy consumption.
>
> In 2023, bioenergy will remain the predominant source of renewable energy, although its share of total renewable energy declines from 50%, in 2017, to 46% as the expansion of both solar PV and wind accelerates in the electricity sector.

Below I quote many numbers concerning the use of green energy sources. I mention these numbers to show that the use of green energy is rising, not because one would be expected to remember the individual numbers. According to this report, the share of renewables was expected to grow by one-fifth in the next 5 years and reach 12.4% by 2020. The report also estimated that renewables would have the fastest growth in the electricity sector, providing almost 30% of the power demand in 2023, up from 24% in 2017. Renewables were forecast to provide 70% of the growth in global electricity generation during this period, led by solar PV and followed by wind, hydropower, and bioenergy. Hydropower was forecast to remain the largest renewable source, meeting 16% of global electricity demand by 2023, followed by wind (6%), solar PV (4%), and bioenergy (3%). In addition, according to the report, the power and heat sectors—which include building heating—would account for the largest overall share of using renewables in meeting energy demand in 2023. They expected that renewable heat consumption would increase by 20% over the forecast period and reach a share of 12% of the heating sector demand by 2023.

3.1.3 Relative Costs

The data in this section is based on the conditions that existed prior to the war in Ukraine. One should keep in mind that the predictions and estimates given here will be affected by inflation and by the outcome of the war. On the other hand, we also know that the incentives of the 2022 Inflation Reduction Act in the United States could quintuple the solar generating capacity of the nation by 2025. However, according to an evaluation by Princeton University, solar production in the United States needs to grow 20-fold to fully decarbonize the economy by 2050. An example of a project that will help meet that goal is the Randolph Solar Project in Charlotte County, VA, which is scheduled to start up in 2025. The project will produce as much electricity as a nuclear reactor (800 MW) and is estimated to cost "only" $1.5 billion (less than half of the cost and one-fifth of the construction time of a nuclear power plant with the same production capabilities).

Before discussing the technical and control aspects of the various renewable energy sources, I will provide a brief summary of their costs and competitiveness. I use so-called units of levelized cost of energy (LCOE) when comparing the different technologies. The LCOE represents the total cost to build and operate a new power plant over its life divided into equal annual payments and amortized over the expected annual electricity generation of the expected life period. It reflects all the costs including initial capital, return on investment, continuous operation, fuel, and maintenance, as well as the time required to build a plant and its expected lifetime.

LCOE does not include any government or state incentives or transportation and distribution costs. These costs often double the production cost of electricity. For example, in the case of my home (in Connecticut, United States), the cost is $0.22/kWh for electricity from the grid but only $0.118/kWh for the electricity from the solar collectors on the roof. The reason for the difference is that the solar collectors that generate electricity are installed at the location where the electricity is used; therefore, avoiding the costs of transportation and maintenance of the grid. In general, solar installation costs have been falling, as shown in Figure 3-5. This figure is used only to show the speed at which solar electricity costs dropped during the past decade.

The market value of renewable energy systems has exceeded $1 trillion providing10 million jobs, and both of these values are growing. Some of the numbers I quote come from the January 2018, 2019, and 2020 reports of the International Renewable Energy

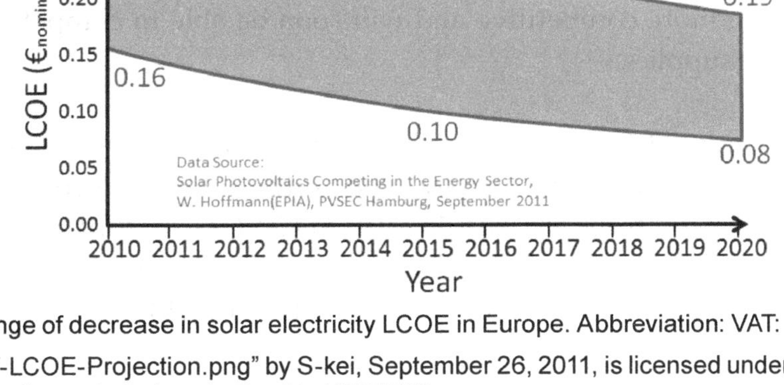

Figure 3-5. Range of decrease in solar electricity LCOE in Europe. Abbreviation: VAT: value-added tax. *Source:* "EU-PV-LCOE-Projection.png" by S-kei, September 26, 2011, is licensed under CC0 1.0, https://commons.wikimedia.org/w/index.php?curid=16702599.

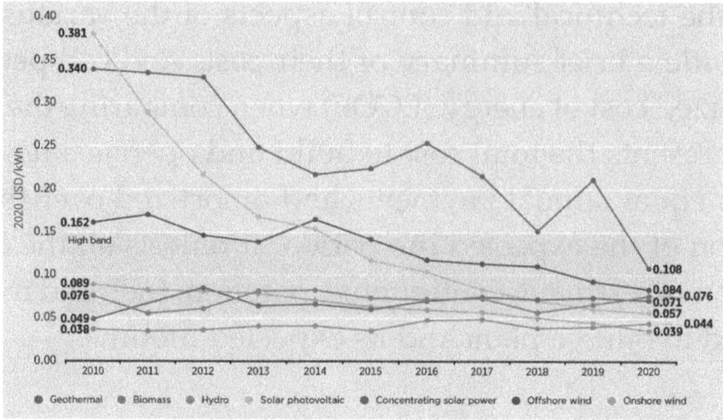

Figure 3-6. IRENA estimates of the costs of green electricity by source. (See color images after page 306.)

Source: Renewable Power Generation Costs in 2020, Figure 1.6, International Renewable Energy Agency (IRENA), Abu Dhabi, https://web.archive.org/web/20210622210701/https://www.irena.org/publications/2021/Jun/Renewable-Power-Costs-in-2020.

Agency (IRENA).[19] In those reports, natural gas, geothermal, and coal-based energies were the least expensive, but today wind and solar energies are cheaper.

The changing cost differential between green and fossil energy sources is caused by the surge in wind and solar installations. For example, the cost of onshore wind energy has fallen by approximately 23% since 2010, and the cost of solar PV electricity has fallen by 73% in the past decade (Figure 3-6 shows data published in 2021 by IRENA). Although future energy costs are highly unpredictable, solar- and wind-generated electricity costs are already less than $0.10/kWh today in the United States. In contrast, we in Connecticut pay $0.22/kWh for regular electricity from the grid.

Whereas solar and wind power are the main drivers of the shift to green energy, other green energy sources such as biomass-based, geothermal, and hydropower energy are also becoming more competitive and will soon be able to compete directly with fossil and nuclear supplies.

As shown in Figure 3-6, electricity generated by onshore wind installations is even cheaper than that from solar. The total global electricity generating capacity has increased from 564 GW in 2018 to 624 GW in 2019 and to 750 GW in 2020. Naturally, the capital costs of offshore installations are higher ($6,500/kW in 2019) than those of onshore installations ($1,600/kW in 2019).

19. "IRENA Report Predicts All Forms of Renewable Energy Will Be Cost Competitive by 2020," SDG Knowledge Hub, January 16, 2018, http://sdg.iisd.org/news/irena-report-predicts-all-forms-of-renewable-energy-will-be-cost-competitive-by-2020/.

In 2020, the cost of 1 kWh of electricity generated was 12¢ for concentrating solar, 8¢ for offshore wind, 6¢ for solar PV, and 4¢ for onshore wind sources.[20] The *2017 Annual Technology Baseline* produced by the US National Renewable Energy Laboratory (NREL) predicted that onshore wind and utility-scale solar energy would continue their rapid cost decreases. NREL estimated that the midrange LCOE of onshore wind would be $39/MWh in 2020, $33/MWh in 2030, $31/MWh in 2040, and $28/MWh in 2050. Similarly, NREL forecasted that the most likely midrange LCOE of utility-scale solar would be $51/MWh in 2020, $45/MWh in 2030, $41/MWh in 2040, and $37/MWh in 2050.

Building new wind and solar plants costs less than continuing to operate existing coal plants. The mean subsidized LCOE for utility-scale solar plants during the past decade fell 72% from $178/MWh in 2009 to $50/MWh in 2017, while the mean LCOE for wind energy fell 47% from $85/MWh to $45/MWh over the same time span. These declines outstripped the cost trends from 2009 to 2017 for natural gas combined-cycle plants (down 27%), coal (down 8%), and nuclear (up 20%).

By comparison, NREL forecasted that the LCOE of fossil-fuel generation would hold steady or increase. The most likely midrange LCOE for natural gas combined-cycle plants was predicted to rise from $43/MWh to $51/MWh in 2050. Similarly, NREL forecasted coal's most likely midrange LCOE to stay about the same between 2017 and 2050 at $70/MWh. Similarly, NREL forecasted that the nuclear power midrange LCOE would remain steady until 2050 at about $78/MWh.

According to IRENA, more than $1 trillion has been invested in renewable energy around the world since 2013 because of the drop in renewable energy costs, and the industry now provides more than 10 million jobs. Figures 3-7 and 3-8 provide data on the dropping capital costs of utility-sized solar and wind power plants up to 2019 and the capital costs of solar plants up to 2017, respectively. Although I have no way to check the accuracy of their data, the general trends they illustrate, particularly in terms of the decreasing costs of solar PV projects, are definitely correct.

3.2 Solar Collector Designs

In this section, I discuss the amount of solar energy that is available and the systems that are being used to collect it. A discussion of solar energy storage and transportation technologies is the subject of the next chapter.

---

20. "Majority of New Renewables Undercut Cheapest Fossil Fuel on Cost," IRENA website, June 22, 2021, accessed November 21, 2022, https://www.irena.org/newsroom/pressreleases/2021/Jun/Majority-of-New-Renewables-Undercut-Cheapest-Fossil-Fuel-on-Cost.

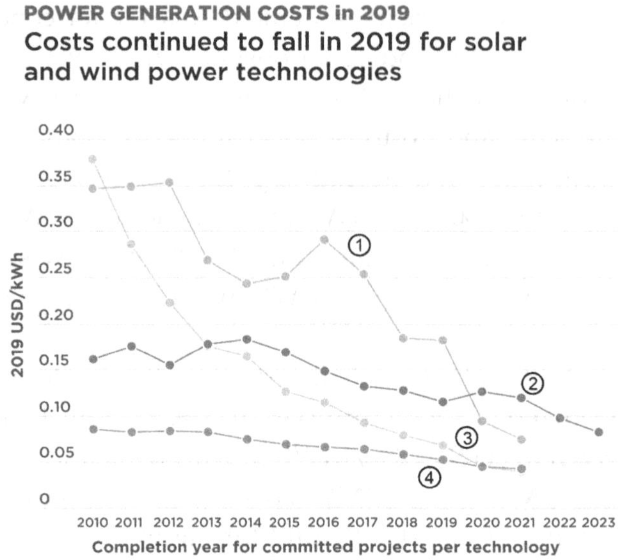

Figure 3-7. Decrease in electricity costs produced by utility-sized solar and wind plants (1, concentrating solar power; 2, offshore wind; 3, solar PV; 4, onshore wind).

Source: Reprinted with permission from "How Falling Costs Make Renewables a Cost-Effective Investment," IRENA website, June 2, 2020, https://www.irena.org/newsroom/articles/2020/Jun/How-Falling-Costs-Make-Renewables-a-Cost-effective-Investment.

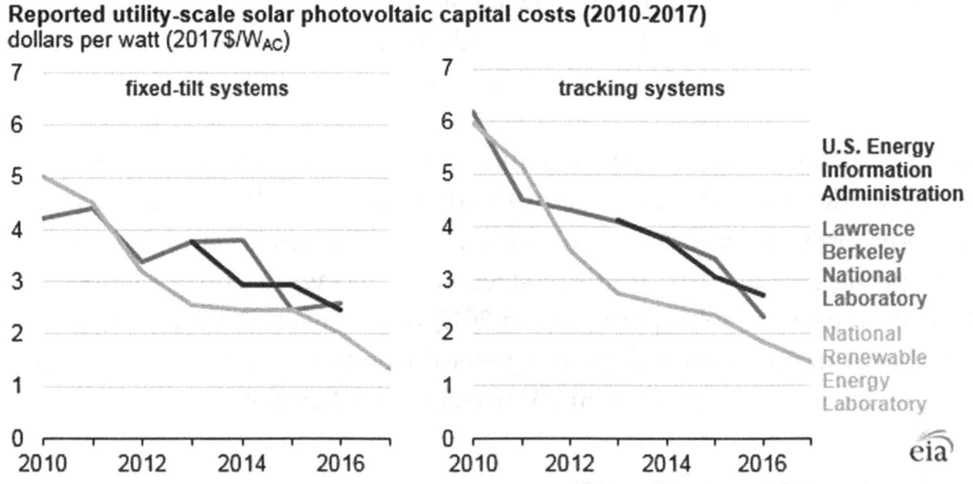

Figure 3-8. Trends in the capital costs of utility-scale, fixed-tilt, and tracking solar systems.

Source: "Solar photovoltaic costs are declining, but estimates vary across sources," *Today in Energy Blog*, March 21, 2018, https://www.eia.gov/todayinenergy/detail.php?id=35432.

The total solar energy absorbed by Earth's atmosphere, oceans, and land masses is about five million quad per year. Of that, photosynthesis in plants and biomass captures approximately 3000 to 5000 Q per year (about 0.01%). In comparison, humankind's total yearly energy use is under or about 600 Q, or about 0.001% of the total yearly amount received.

The solar energy needed to produce the total amount of electricity being used in a year (~25 PWh, equivalent to ~84 Q) is being received on Earth in less than 10 minutes. Another way to describe the immense quantity of energy received from the Sun is this: the yearly amount reaching the land and ocean surfaces of our planet is more than the energy content of all of the energy reserves on Earth (uranium, coal, oil, and natural gas combined).

Another way to discuss the availability of solar energy is by considering the insolation (intensity of solar radiation) received in various parts of the world. Figure 3-9 shows the daily energy received per square meter in various parts of the world in January and April. These amounts range from 0 to more than 8.5 kWh/($m^2 \cdot d$), which

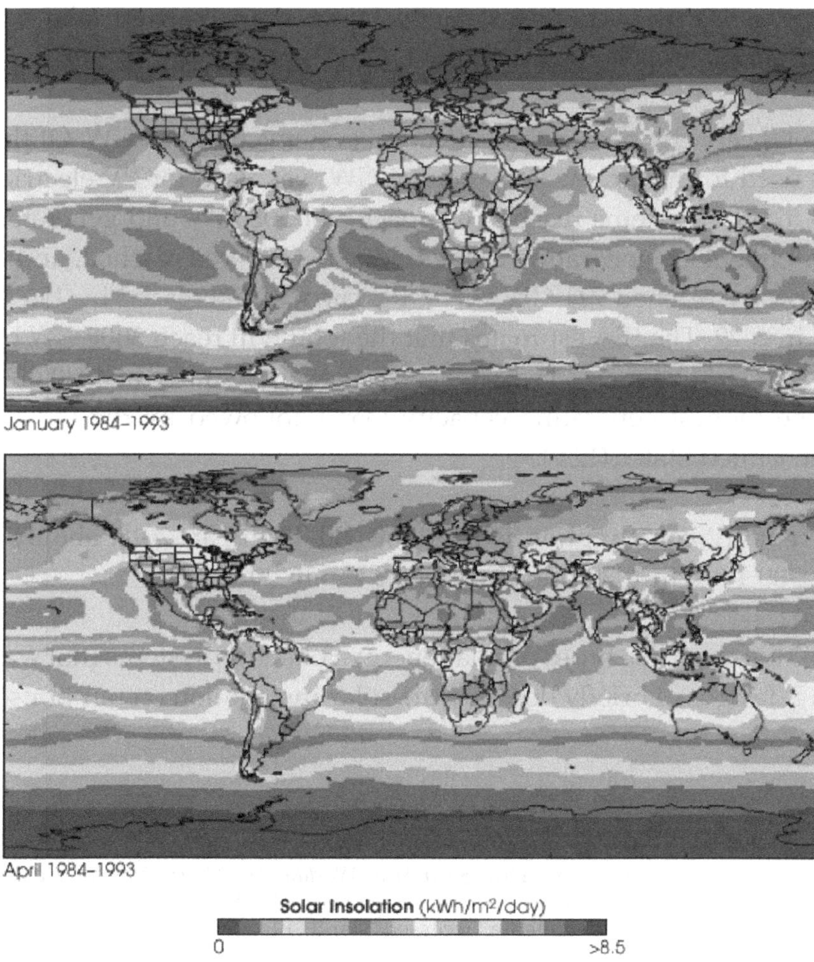

Figure 3-9. Global insolation in units of kilowatt-hour per square meter per year in January (top) and April (bottom). (See color images after page 306.)

Source: "Measuring Solar Insolation," NASA Earth Observatory, accessed March 11, 2022, https://www.earthobservatory.nasa.gov/images/1355/measuring-solar-insolation.

equals 0 to 3100 kWh/[$m^2 \cdot$ year]. Figure 3-9 also shows the effects of the seasons (due to Earth's changing axis of rotation), which causes the polar winter to move from the North Pole to the South Pole between January and April. Specifically, the seasons affect the amounts of insolation received at the various latitudes on the globe, with the high-insolation region (red) moving to the north as the seasons change from winter (January) to spring (April).

In addition to knowing how the insolation changes by the seasons, we are also interested in the yearly total amount of solar energy that can be collected. Such information is readily available for any particular location on the planet (Figure 3-10). In this figure, the color coding indicates values from less than 700 kWh/($m^2 \cdot$ year) to more than 2700 kWh/($m^2 \cdot$ year). It should be emphasized that, even in the temperate and subtropical zones, Earth receives more insolation than realized (about 1500 kWh/[$m^2 \cdot$ year] in Europe and even more in the United States).

In 2017, the total capacity of all solar collection systems was 880 GW. Of that capacity, 474 GW was used for heating water, and 404 GW was used for PV electricity generation. In 2018, the PV systems reached a capacity of 508 GW and generated 600 TWh of electricity.[21] In 2019, a generating capacity of 640 GW was reached, and in 2020, the global PV capacity reached 714 GW.[22] The PV capacity increased by 156 GW in 2021 and 200 GW in 2022. The IEA estimates that the total will reach 3000 GW by 2030, and by 2050, PV will be the world's largest electrical energy source at 4700 GW.[23] In 2020, China had the highest generating capacity (35%), followed by the European Union (22%) and the United States (12%).

Global solar energy consumption was about 600 TWh[24] in 2018, about 720 TWh in 2019,[25] about 900 TWh in 2020, and over 1000 TWh in 2022.

Before discussing the three types of solar energy collection systems (hot water, PV, and concentrating designs), I briefly describe the solar installation on our house.

---

21. Werner Weiss and Monika Spörk-Dür, *Solar Heat Worldwide 2020* (IEA Solar Heating & Cooling Technology Programme, May 2020), https://www.iea-shc.org/solar-heat-worldwide.
22. "Solar power by country," Wikipedia, accessed March 12, 2022, https://en.wikipedia.org/wiki/Solar_power_by_country.
23. "Growth of photovoltaics," Wikipedia, accessed March 12, 2022, https://en.wikipedia.org/wiki/Growth_of_photovoltaics.
24. Weiss and Spörk-Dür, *Solar Heat Worldwide 2020*.
25. Heymi Bahar, *Solar PV* (Paris: IEA, June 2020), https://www.iea.org/reports/solar-pv. License: CC BY 4.0.

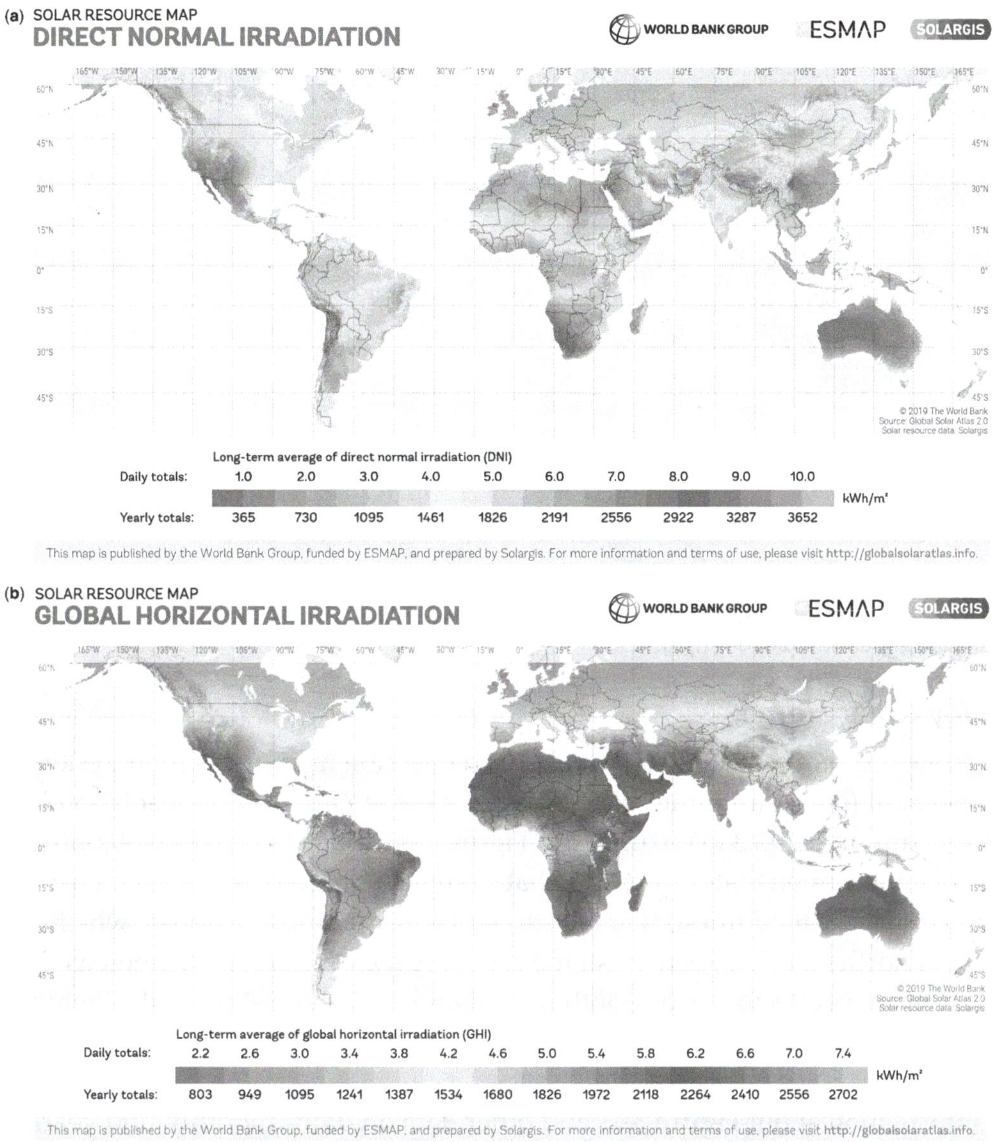

Figure 3-10. Daily and annual total insolation in April in units of kilowatt-hour per square meter per year throughout the world: (a) global horizontal irradiation, (b) direct normal irradiation. (See color images after page 306.)

Source: Maps obtained from the "Global Solar Atlas 2.0," a free, web-based application developed and operated by the company Solargis s.r.o. on behalf of the World Bank Group, utilizing Solargis data, with funding provided by the Energy Sector Management Assistance Program (ESMAP).[26]

---

26. For additional information, see: "World GHI Solar-resource-map GlobalSolarAtlas World-Bank-Esmap-Solargis.png" and "World DNI Solar-resource-map GlobalSolarAtlas World-Bank-Esmap-Solargis.png" by Solargis, ©The World Bank / CC BY 4.0, https://commons.wikimedia.org/wiki/. File:World_GHI_Solar-resource-map_GlobalSolarAtlas_World-Bank-Esmap-Solargis.png and https://commons.wikimedia.org/w/index.php?curid=73047799, respectively.

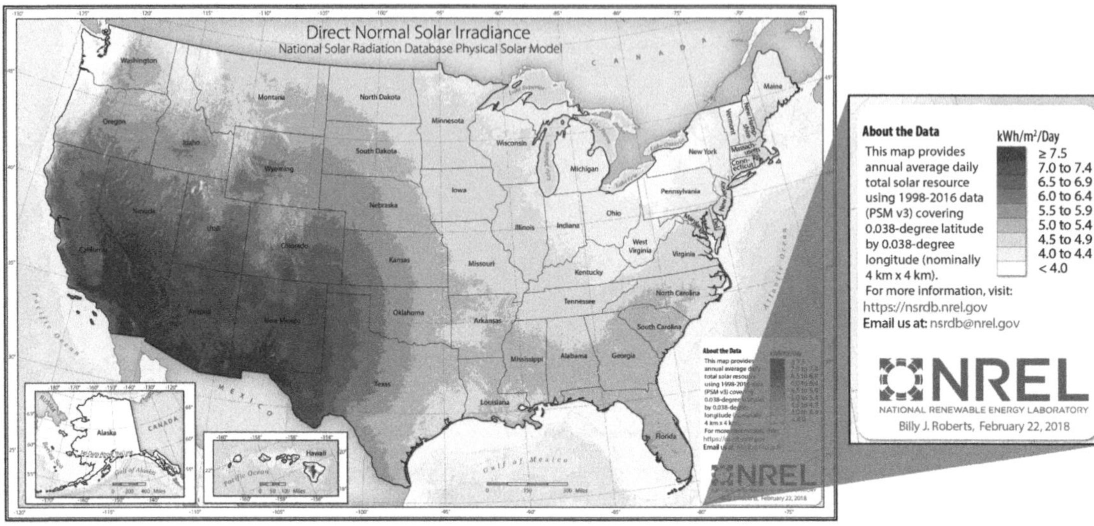

Figure 3-11. The yearly average of daily insolation in different regions of the United States.

Source: "U.S. Annual Solar DNI," NREL website, February 22, 2018, https://www.nrel.gov/gis/assets/images/solar-annual-dni-2018-01.jpg.

3.2.1 My Solar Home

To illustrate a typical home installation, I will describe my home's solar system. I live in Connecticut (northeast United States), where the average yearly insolation is about 1200 kWh/($m^2 \cdot$ year) (3.3 kWh/[$m^2 \cdot$ d]). Figure 3-11 illustrates the insolation regions throughout the United States, showing that Connecticut receives less solar energy than most other states. In addition, there are large trees shading the house, which is positioned in a north–south direction, so the roofs are facing east–west, whereas the ideal location for solar collectors is on south-facing surfaces. The system is also outdated (10 years old) and undersized.

More information about Figure 3-11 and other data on daily, monthly, and yearly average solar insolation are available from the NREL website.[27]

My family has a SolarCity (today Tesla) system that was installed on the basis that we pay nothing for the system but make the roof of the house available for the company to place it on. We also agreed to purchase the generated electricity for 20 years. The system was installed in 2014, and the purchase price we agreed to pay for the electricity was \$0.11/kWh, compared to the regular price for electricity in our location of \$0.22/

27. "Solar Resource Data, Tools, and Maps," NREL website, accessed March 8, 2022, https://www.nrel.gov/gis/solar-resource-maps.html.

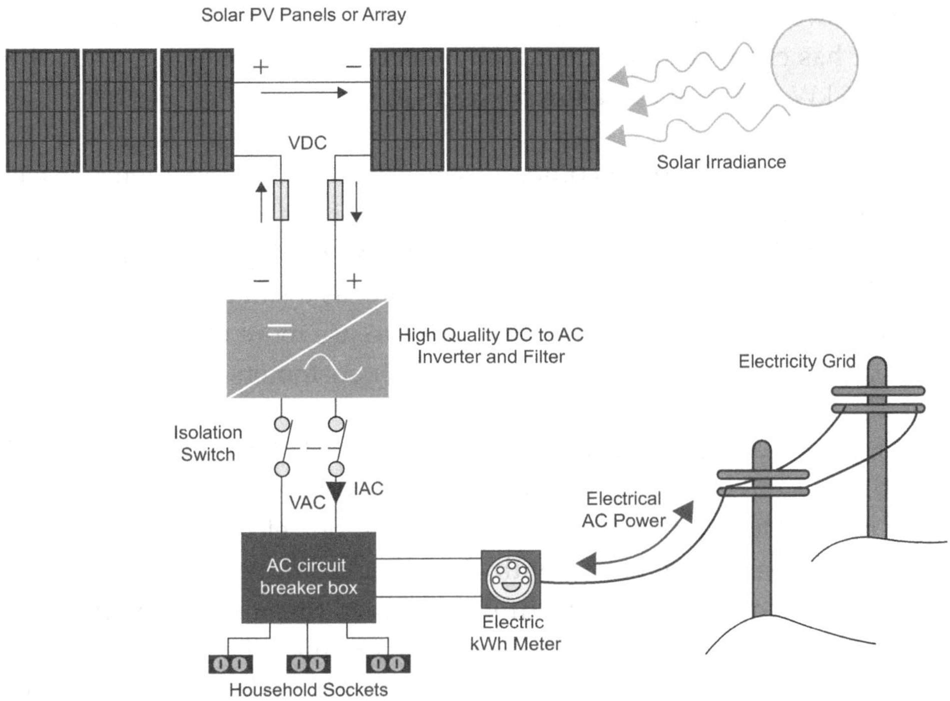

Figure 3-12. Grid-connected solar system for private homes.

kWh. SolarCity also provided the inverter, the two-directional electric meter, and the other components needed (Figure 3-12).

The installation is connected to the grid so that excess electricity can be sent to the utility. When electricity is sent to the grid, the electric meter is "running backward" in this grid-connected "buy-back" configuration. In this way, when our solar system produces more electricity than the house needs, we send the excess electricity back into the grid. When the production is below our needs (e.g., at night), we take electricity from the grid.

The reason the cost of solar electricity is so much lower than the cost of nuclear (the common electric energy source in our area) is that half of the price of regular electricity is to pay for the electricity losses in the grid cables and the construction and maintenance of the grid. As the gridless homes in California show, solar electric systems do not need the grid because their solar electricity is generated where it is used and, thus, does not need to be transported.

Last year, our need for electricity was 8320 kWh, and to generate that amount, we would have needed 14 kW solar collectors. In 2014, however, Connecticut did not allow systems larger than 10 kW on private homes (presumably because of pressure from the conventional power lobby). Therefore, we currently have a 10 kW system. I understand that this limit has changed (the green lobby has become stronger), and the limit has increased to 16 kW, but we have not expanded yet.

Part of the argument of the power companies was based on the experience in Germany, where large numbers of solar energy systems were installed in a short time. When the Sun came out, the electricity supply jumped, and the regular power plants could not reduce their production fast enough. Therefore, as a safety measure, they had to decrease the price of electricity to motivate users to use more of it. In some cases, this went so far that electric companies had to pay their customers to use more.

On our roof, we have 40 panels, which cover an area of $10 \times 6.4\ m^2$. The panels have polycrystalline silicon cells and generate about 6000 kW per year, which satisfies about 72% of our needs. During the winter and summer, our electricity needs exceed the solar capacity, and we supplement it with electricity from the grid. In the winter, this occurs when snow covers the panels. In the summer, this also occurs because of the extra loads of air-conditioning and pool pumps. On the other hand, in the fall and spring, the solar system generates more electricity than needed, so we send the excess back to the grid.

Overall, we save about $1,000 a year. We have asked for the system to be expanded to 16 kW, but as of this writing, the company has not yet installed the extra panels because it is too busy. Once the expansion has been completed, we will replace one of our old cars with an electric car, which we will recharge overnight. Also, we plan to use the batteries in the electric car to supply the electricity needed during power outages.

For the reasons I listed, our installation is not optimal. This is because the roofs are facing east and west, whereas collectors on south-facing roofs would collect more energy, and because we have big trees all around the house. In addition, our panels are only 14% efficient, whereas today's panels are better. Of course, if my home were not 71 years old and were currently being built, instead of placing collector panels on the roof, I would use a solar roof. In fact, solar roofs and energy storage are already requirements for new construction in California.

3.2.2 Thermal Solar Collector Designs

Solar collectors can be either concentrating or flat, their operation can either be stationary or track the Sun, and they can convert the solar energy into either thermal energy (heat) or electricity by the PV process. (The idea of tracking the Sun is not new, sunflowers and other plants do just that.) There are also solar collector designs that can generate both heat and electricity.

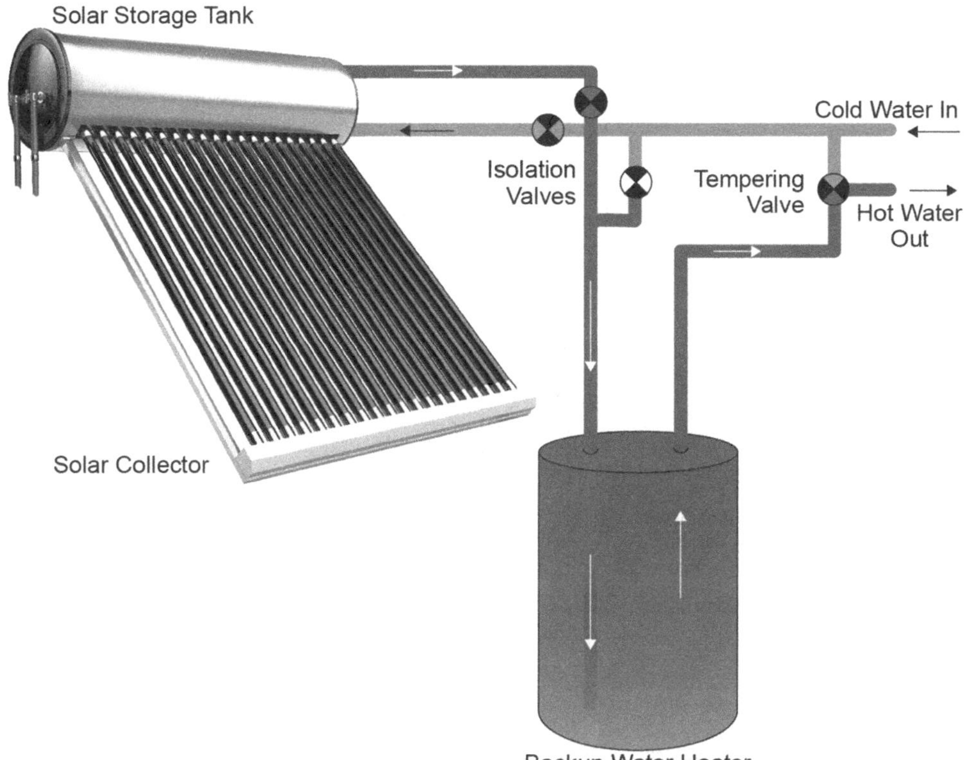

Figure 3-13. Solar hot water system without pumps.

The majority of solar collectors are still used to generate hot water. Such thermal collectors provide a residence with heat and/or hot water. An average home in the northeastern United States requires about 44,000 to 66,000 kWh (150 to 250 million Btu) for heat and hot water. This is about the same as the heat energy obtained by burning 4000 to 6000 L (about 1000 to 1500 gal) of oil.

In the United States, 46% of the energy use of the average residence is for home heating, 20% for water heating, 20% for cooling, 5% for lighting, and 11% for other needs (appliances, computers, etc.).[28] The total energy consumption of all residences in the United States is 12 Q, or 13% of the total energy consumption of 93 Q).

The simplest design of a roof-mounted collector system (Figure 3-13) consists of only a flat collector and a solar storage tank. In this configuration, cold water first passes through the collectors and the solar storage tank and then enters the regular water

28. "Use of energy explained: Energy use in homes," Today in Energy website, last updated June 23, 2021, accessed August 8, 2022, https://www.eia.gov/energyexplained/use-of-energy/homes.php.

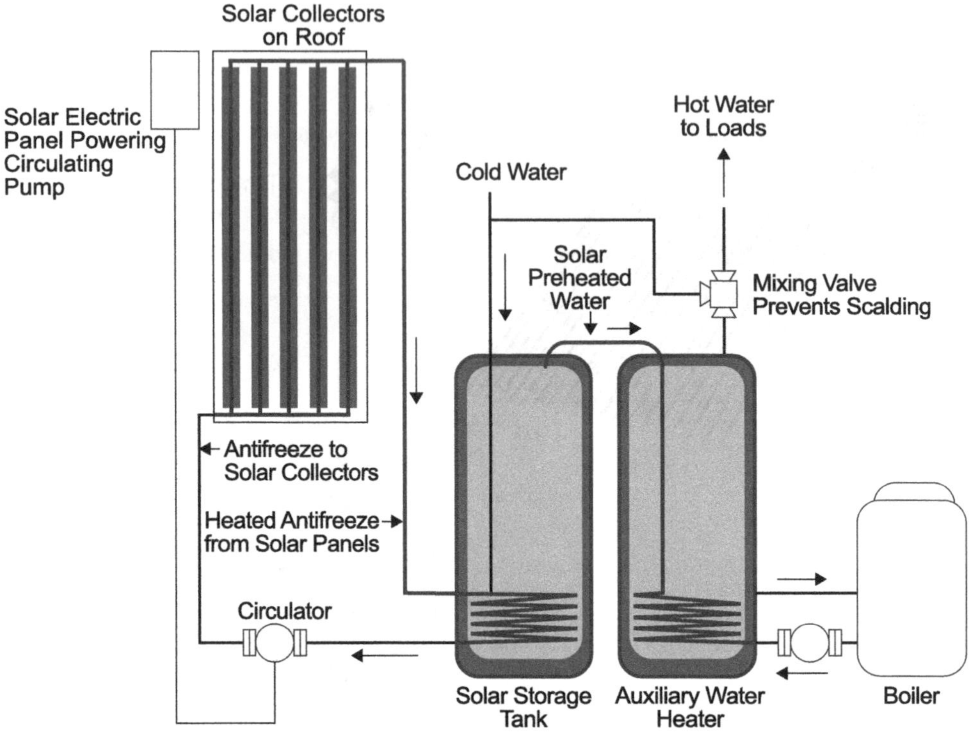

Figure 3-14. A solar hot water system that can be used year-round in all locations.

heaters of the home. In this case, when the solar energy is insufficient, the regular heater acts as a backup and provides additional heat as needed.

A dual-loop design is used in locations where the ambient temperature can drop below freezing (Figure 3-14). In that case, an antifreeze-circulating loop is added so that the fluid passes first through the solar collectors and then through a heat exchanger, where the collected solar heat is transferred to the water. This configuration is more expensive, but it can be operated year-round, as it does not need to be drained in cold weather.

The total thermal collector market in 2017 was about[29] $40 billion; today, it is about $50 billion.

29. "Global Solar Thermal Collectors Market Research Report 2018-2025," Fior Markets website, November 2018, https://www.fiormarkets.com/report/global-solar-thermal-collectors-market-research-report-2018-2025-316847.html.

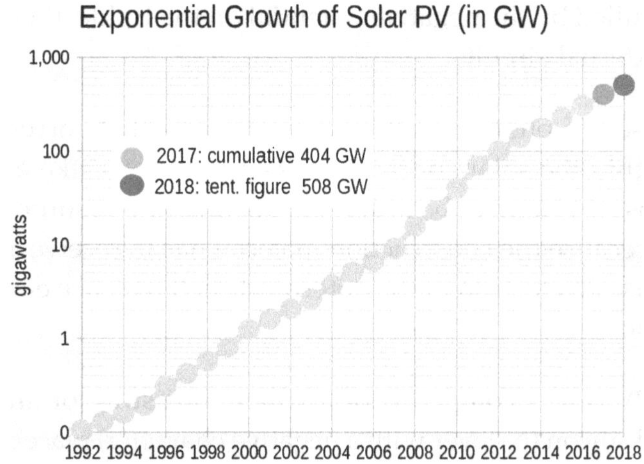

Figure 3-15. Growth of PV capacity around the world.[30]

Source: "PV cume semi log chart 2014 estimate.svg" by Rfassbind, November 8, 2014, Public Domain, https://commons.wikimedia.org/w/index.php?curid=36675622.

3.2.3 PV Collectors

The first applications of PV collectors involved powering satellites and calculators that did not need batteries. The PV cells absorb solar radiation to generate a voltage, so current is generated if connections are placed on the cell. An important feature of PV cells is that they do not need sunshine to work; they will also produce electricity using regular light.

Figure 3-15 shows the cumulative worldwide growth of PV collector capacity in gigawatts.

In 2017, of the 880 GW global solar collector capacity, 404 GW was of the PV type. In 2018, PV systems reached a capacity of 508 GW and generated 600 TWh of electricity.[31] In 2019, a generating capacity of 640 GW was reached. In 2020, the global PV capacity reached 714 GW. The IEA estimates that the total capacity will reach 3000 GW by 2030, and that PV will be the world's largest electrical energy source at 4700 GW by 2050.[32]

When a photon of light hits a piece of silicon, it either goes straight through, if its energy is lower than the band-gap energy of the silicon semiconductor, or it is absorbed by the silicon, if its energy is greater than the band-gap energy. In the latter case, an electron-hole pair is produced. The electrons are pulled by the positive potential on the N side,

---

30. For further information, see: "Photovoltaics," Wikipedia, accessed March 12, 2022, https://en.wikipedia.org/wiki/Photovoltaics#/.
31. Weiss and Spörk-Dür, *Solar Heat Worldwide 2020.*
32. "Growth of photovoltaics," Wikipedia.

and the holes are pulled by the negative potential on the P side, thereby forming a current through the external circuit.

Sunlight is composed of photons with energy contents that correspond to the range of wavelengths within the solar spectrum. When photons strike a collector cell, they can be reflected, pass through, or be absorbed, but only the absorbed photons generate electricity. This is because the construction material has to receive a certain amount of electronvolts to cause an electron in the outermost shell (valence electron) to move into the conduction zone.

A typical silicon PV cell is composed of a wafer consisting of an ultrathin layer of phosphorus-doped silicon (N-layer with a negative character) placed on top of a thicker layer of boron-doped silicon (P-layer with a positive character). These layers are connected by a P–N junction. When sunlight strikes the surface of the PV cell, an electrical field is generated, which provides momentum and direction to the light-stimulated electrons, resulting in a flow of current (not voltage) when the solar cell is connected to an electrical load (Figure 3-16).

In newer PV designs, the range of spectral wavelengths has been extended from the visible range (380 to 780 nm) toward UV wavelengths (250 to 380 nm) and the infrared range (780 to 1500 nm). Because of the stratification of different materials, each of

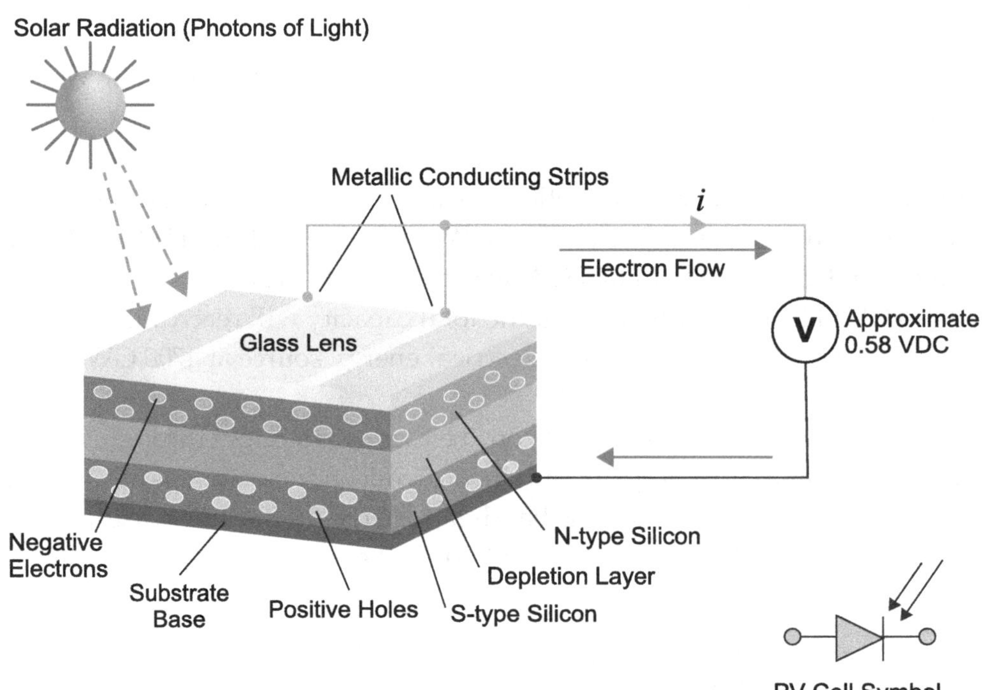

Figure 3-16. Operation of a basic PV cell.

which is adapted to a portion of the working spectral range, these novel devices present high performances over a wavelength range from 200 to 1800 nm.

3.2.3.1 Efficiency and Temperature Coefficient

Traditional solar cells with a single band gap utilize less than half the energy in the solar spectrum. Energy is lost because low-energy photons are not absorbed, and high-energy photons are absorbed inefficiently. Tandem solar cells convert a greater percentage of incidental sunlight into electricity.

The efficiency of installed solar panels is also affected by their operating temperatures. Solar panel efficiency statements refer to the efficiency when the panels are tested at a temperature of 25°C (77°F), which is their reference point. Under operating conditions, that efficiency is reduced as the operating temperature rises. Specifically, their efficiency usually falls by –0.3% to –0.5% per degree Celsius of temperature rise. This number indicates how much the efficiency will decrease when the panel operates at temperatures above 25°C (77°F). So, for example, if a 20% efficiency panel has a temperature coefficient of –0.4%/°C and operates at 35°C (95°F), its actual efficiency is 16% (20 – 0.4 × 10). The temperature coefficients of thin-film solar panels are lower than those of traditional monocrystalline or polycrystalline panels and are closer to –0.2%/°C. The efficiencies of most PV solar collectors today are in the range of 20% to 22%. The efficiencies of all three types of solar collectors (crystalline silicon, thin-film, and concentrator) have been increasing and today are about 25%.

3.2.3.2 New Designs

Perovskite solar cell (PSC) designs have further increased the efficiency that can be achieved. The PSC is a type of solar cell with a perovskite-structured compound, most commonly a hybrid organic-inorganic lead halide or a tin halide-based material, as the light-harvesting active layer.[33] Perovskite halide materials are inexpensive to produce and simple to manufacture. Solar cell efficiencies of devices using perovskites have increased from 3.8% in 2009 to 25.2% in 2019 in single-junction cells and to 28% in silicon-based tandem cells, which exceeds the maximum efficiency achieved in single-junction silicon solar cells.

PSCs have become commercially attractive. They are thus the fastest-advancing solar technology to date and have the potential to achieve even higher efficiencies and low production costs.

---

33. "Perovskite solar cell," Wikipedia, accessed March 12, 2022, https://en.wikipedia.org/wiki/Perovskite_solar_cell.

The tandem units can also be tuned to respond to different colors in the solar spectrum by changing the material composition. This band-gap flexibility opens up the potential to reach efficiencies exceeding 30%.

Bifacial solar panels are among the newest designs. Their efficiencies are better than those of monofacial units and approach 30%. The back sheet of monofacial solar panels is opaque, whereas the bifacial unit's back sheet is transparent or consists of dual-tempered glass. The collection efficiency of a bifacial unit is increased because the back side of the panel collects the solar radiation reflected from the surface on which it is mounted. Therefore, it is advantageous to mount them on light-colored (highly reflective) surfaces.

The initial costs and installation costs of these systems are higher than the corresponding costs of monofacial units, but because of their higher efficiency, they still cost less in the long run. Moreover, the amount of space required per watt requirement is reduced, and they can withstand higher operating temperatures and strong winds, have long warranties, do not need grounding, and operate well in diffused light. Whereas a typical warranty for a monofacial solar panel is about 25 years, bifacial solar panels are often sold with 30-year warranties. Bifacial units are also heavier and require precise installation to maximize the albedo effect (reflectivity) of the mounting surface.[34]

New designs under development include cells that capture visible light from all directions and cells that utilize the waste heat emitted by solar panels. For a more detailed description of the different concentrating solar collector designs and their efficiencies, see Section 3.2.6, "Concentrating Collectors."

3.2.3.3 Special Applications

One can collect solar energy on just about any surface, and there are many ways in which imaginative suppliers are exploiting this potential (window curtains, heated car seats, solar textiles, etc.). I mention here only one potentially powerful application: solar-powered roads. This development is just beginning, and common use is probably years or even decades away.

In 2019, the sidewalks along Route 66, the historic US interstate highway, were chosen to test solar-powered pavement technology. Such walkways can generate clean energy. They contain light-emitting diode (LED) bulbs that can light them at night and have the thermal heating capacity to melt snow during the winter. Also, in 2019, the first solar "roadway"

---

34. George Duval, "Bifacial Solar Panels: What Are They And How Do They Work," Semprius website, last updated November 9, 2021, accessed March 18, 2022, https://www.semprius.com/bifacial-solar-panels/.

Figure 3-17. A 1-kilometer-long solar highway in Normandy was linked to the local power grid to power the town's streetlights.

Source: "Route solaire de Tourouvre 2.jpg" by Kumkum, December 22, 2016, is licensed under CC0, https://commons.wikimedia.org/wiki/File:Route_solaire_de_Tourouvre_2.jpg.

opened in Amsterdam. It was only a 72 m (236 ft) stretch of bicycle path between two city suburbs. It generated solar power from rugged, textured glass-covered PV cells.

In Tourouvre-au-Perche, France's Normandy region, the world's first "solar highway" was inaugurated in December 2016 (see Figure 3-17). It was a 1 km (0.6 mi) stretch of highway covered with 2800 m^2 (~30,000 ft^2) of resin-coated PV solar cells linked to the local power grid that powered the town's streetlights.[35]

These early installations were fragile and uneconomical, but the concept has a future. In this technology, the Chinese are way ahead because most of their roads are made of concrete, whereas US roads are usually constructed with asphalt, which flexes slightly under pressure. Consequently, when solar panels are placed on asphalt roads, they will easily snap under the weight of cars and trucks. Concrete is a sturdier foundation for solar panels, so such installations have better durability.

Regarding the future potential of solar roads, the American Road & Transportation Builders Association determined that there are about 8,615,000 lane miles of roads in the lower 48 states, which equates to approximately 18,000 square miles of area (equal

35. Max Teodorescu, "France opens the world's first highway paved with solar panels," Electronic Products website, December 28, 2016, https://www.electronicproducts.com/Sustainable/Solar/France_opens_the_world_s_first_highway_paved_with_solar_panels.aspx?terms=solar.

to an area three times that of the state of Connecticut). NREL estimates[36] that roughly 1948 square feet per person (23,000 square miles total) would be needed to supply enough electricity to run the country on solar power. If the NREL calculations are correct, the roads of the lower 48 states could generate 78.2% of the nation's electric energy needs. If the area of parking lots were included, solar roads could probably generate 100% of the electric energy needs of the United States.

Another new design introduced by the Platio Solar Pavement company in Hungary uses solar energy and recycled plastics to create sustainable building materials for outdoor, walkable surfaces around buildings, homes, or public spaces. From rooftop terraces of commercial buildings and driveways of homes to pontoons in marinas and sidewalks of smart cities, this invention aims at generating solar electricity and reducing plastic waste by employing a special pressuring method to make bricks that the solar panels sit on.

3.2.4 Multijunction PV Cells

The market share of multijunction solar cells is growing rapidly. In their design, multiple p-n junctions are made of different semiconductor materials. The different p-n junction materials produce electric current by absorbing different wavelengths of radiation. As a result, multijunction cells capture a broader range of wavelengths than standard single-junction cells, thereby improving the efficiency of the cell.

Multijunction solar cells (Figure 3-18) are successful third-generation solar cells. They capture sunlight automatically and pass it through the cell, ensuring that each wavelength range is absorbed at its most efficient junction.

This design can improve solar cell efficiency past the Shockley-Queisser limit using a single p-n junction by increasing the range of photons absorbed from the solar spectrum while minimizing heat losses. It is important to note that the junction materials' band gaps generate similar currents. Naturally, the light absorbance efficiency and heating losses can be improved by increasing the number of junctions.

When electrons hit the junction between adjacent PV cells, the voltage losses can be minimized by adding a thin film of gallium arsenide which is naturally resistant to

---

36. Jacob Marsh, "Solar roadways: what you need to know," EnergySage website, March 25, 2019, https://news.energysage.com/solar-roadways-what-you-need-to-know/.

Figure 3-18. Triple-junction solar cell design.[37]
Source: Courtesy of the US Naval Research Laboratory.

damage from heat. The addition of this thin film is particularly valuable in concentrating photovoltaic (CPV) solar cells if their concentration exceeds 700 "suns."[38]

In the past few years, thin-film designs have been marketed to cover the roofs of buildings. For example, Uni-Solar laminates were installed on the roof of the General Motors facility in Fontana, California, one of the largest solar power installations for corporate use in the United States. This project was the second installation of Uni-Solar laminates for GM after the 1 MW solar installation in Rancho Cucamonga, California.

The PSC is a type of solar cell that includes a perovskite-structured compound, most commonly a hybrid organic-inorganic lead halide- or tin halide-based material, as the light-harvesting active layer.[39] As I mentioned previously, perovskite halide materials are inexpensive to produce and simple to manufacture.

---

37. For further information, see "Triple-junction solar cell design could break 50 percent conversion barrier," New Atlas website, January 17, 2013, https://newatlas.com/nrl-solar-cell-50-percent-conversion/25826/.

38. One "sun" is the intensity of the solar radiation reaching the collector surface (1 sun = 0.1 W/cm^2). A particular number of suns refers to the multiple of this radiation intensity that is reached through concentration.

39. "Perovskite solar cell," Wikipedia.

Figure 3-19. Home roof covered by long PV roof shingles.

Source: "Thin Film Flexible Solar PV Installation 2.JPG" by Fieldsken Ken Fields, May 17, 2008, is licensed under CC BY-SA 3.0, https://commons.wikimedia.org/w/index.php?curid=10861988.

3.2.5 Thin-Film PV

Flexible, thin-film solar cell collectors are made of amorphous silicon deposited on plastic and used for private homes and commercial buildings. For example, OkSolar.com PV shingles are 0.3 m (12 in) wide by 2.2 m (7.2 ft) long and are nailed on top of 30 lb felt sheeting on roof decks.[40] Electrical lead wires extend from the underside of each shingle and pass through the roof deck; wire connections are made in the roof space. The solar shingles are installed to match the course of the conventional shingles (Figure 3-19). They are wired together in the space below the roof, and then the array is connected to the combiner box. The wires from the combiner box are then run to the power conditioning equipment (charge controller, meters, system disconnect switches, fuses, inverter, and batteries), which converts the power from the solar array into a form suitable for subsequent use.

Solar cells are also produced by depositing ink on a flexible substrate (called *paper*), where the nanocomponents in the ink align themselves through molecular self-assembly. This approach improves the speed of production and eases the deposition of a uniform layer that provides the correct ratio of elements everywhere on the substrate. The cost of these collectors is expected to be about $1/W because the substrate material on which the ink is printed is much less expensive than the stainless steel substrates that are often used in thin-film solar panels.

---

40. SHR-17 PV Shingles, OkSolar.com website, accessed March 12, 2022, https://www.oksolar.com/lion/Item/15685/shr-17-pv-shingles.

Tandem fabrication techniques have been used to improve the performance of existing designs and to lower the cost of thin-film solar cells by using amorphous silicon, as opposed to conventional crystalline silicon, to produce a cell with about 10% efficiency that is lightweight and flexible.

In contrast to traditional crystalline silicon PV that is cut from solid crystalline material, thin-film PVs are composed of thin layers of semiconductor materials grown on a glass or flexible substrate and sandwiched between front and back contacts. The laminate can also be made from a triple-junction amorphous silicon solar cell mounted on a stainless steel substrate. The blue, green, and red light segments of the Sun's radiation band are absorbed in the corresponding layers of the cell. This technology results in better performance under low and diffused light conditions.

Another new development overcomes one major limitation of thin-film collectors, namely that they are rather unsightly. Sistine Solar, a Boston-based design firm, has developed aesthetic enhancements that provide solar panels with a customized look. This MIT start-up has created a product called SolarSkin that makes it possible for solar panels to match the appearance of a roof without interfering with panel efficiency or electricity production.

In 2019, MiaSolé, a subsidiary of one of the world's largest thin-film power solution companies, Hanergy Thin Film Power Group, announced that the US Department of Energy's NREL had independently confirmed a 20.56% cell efficiency for its thin-film device, which is deposited on a stainless steel foil.

3.2.6 Concentrating Collectors

Concentrating designs include parabolic troughs, parabolic dishes, central receivers, and Fresnel lenses, which are light and have large apertures and short focal lengths. The Fresnel lens design uses refraction instead of reflection. Heat-resistant oil or other fluids are circulated in concentrating collectors to transport the collected heat.

Of the approximately 1000 GW global capacity of PV solar collectors, the share of concentrating collectors is only 6.8 GW, but that number is growing fast. Concentrating units are claimed to provide conversion efficiencies of about 30% if the irradiation level is high enough. High irradiation is defined as 200 to 300 suns (1 sun = 0.1 W/cm^2), which corresponds to 20 or 30 W/cm^2 irradiation. To obtain such levels of solar irradiation, the ambient insolation has to be focused (concentrated) by a factor of 200 to 300.

Figure 3-20. Part of the 354 MW SEGS solar complex in northern San Bernardino County, California. *Source:* "Solar Plant kl.jpg" by US Bureau of Land Management, Public Domain, accessed March 12, 2022, https://commons.wikimedia.org/w/index.php?curid=15957890.

Concentrating solar power (CSP) technology is employed at Solar Energy Generating Systems (SEGS), a group of nine solar power plants in California's Mojave Desert.[41] The facility employs 400,000 parabolic mirror reflectors (troughs) distributed over 1000 acres (4 km^2). In this design, the troughs are used to track the trajectory of the Sun and concentrate the sunlight onto absorber tubes located at the focal line of the parabolic mirrors. Heat-resistant oil is circulated inside the absorber tubes. This heat-transfer fluid serves to transport the collected heat to boilers, which generate the steam used to drive turbine generators (Figure 3-20). The system is still operating today and has grown into one of the world's largest solar power plants, having a capacity of 354 MW.

In the past, the hot oil circulated at 400°C (752°F), but even higher operating temperatures can be achieved using molten salt. One advantage of the SEGS design is that solar energy can be conveniently stored (as thermal energy) and later used as needed to compensate for daily, seasonal, and weather-related variations in the availability of solar energy.

3.2.6.1 Central Concentrators

The thermal version of CSP systems generate electric power using mirrors or lenses to concentrate sunlight onto thermal units. The collected heat is used to generate steam, which drives a steam turbine connected to an electrical power generator. While the

41. "Solar Energy Generating Systems," Wikipedia, accessed March 12, 2022, https://en.wikipedia.org/wiki/Solar_Energy_Generating_Systems.

Figure 3-21. Utility-scale thermal CSP plant.

Source: "Crescent Dunes Solar December 2014.JPG" by Amble, December 31, 2014, is licensed under CC BY-SA 4.0, https://commons.wikimedia.org/w/index.php?curid=37624786.[42]

non-concentrating solar PV systems experienced huge growth in recent years, CSP growth has been slow due to technical difficulties and higher costs. CSP represented less than 2% of the worldwide installed capacity of solar power plants.

Thermal solar collectors can be "central concentrator" or "solar concentrator" designs. In these configurations, a large number of independently movable flat mirrors (heliostats) reflect the solar radiation onto a central receiver on top of a tall tower. Each heliostat moves about on two axes. The receiver has a vertical bundle of tubes in which a heat-transfer fluid (water, oil, or molten salt) is heated by the concentrated insolation.

One advantage of this design is that it provides energy storage so that intermittent solar energy can be used continuously. The Crescent Dunes Solar Energy Project (Figure 3-21) is a 110 MW net solar thermal power plant with a large (1.1 GWh) energy storage capacity. It was the first utility-scale CSP plant with a central receiver tower using molten salt energy storage.

---

42. For additional information, see: "Crescent Dunes Solar Energy Project," Wikipedia, accessed March 12, 2022, https://en.wikipedia.org/wiki/Crescent_Dunes_Solar_Energy_Project.

3.2.6.2 High-Concentration PV Thermal Plant

The high-concentration PV thermal (HCPVT) system developed by IBM combines the capabilities of converting 25% of the insolation into electricity and collecting 50% of the Sun's energy in the form of heat. The concentration achieved is 2000 suns. The collected energy can not only be electricity but it can also provide distilled water, the source of which can also be salt water. Therefore, this facility can supply isolated villages with electricity and safe drinking water.

The inside surface of the HCPVT system's parabolic dish is covered with 36 elliptical mirrors made of 0.2 mm recyclable plastic foil with a silver coating slightly thicker than the wrapper on chocolate bars. These are curved using a slight vacuum. The mirrored surface area concentrates the Sun's radiation by reflecting it onto several microchannel liquid-cooled receivers, each of which is populated with a dense array of tiny multijunction solar cells called PV chips. Each 1×1 cm^2 chip produces electrical power of up to 57 W on a typical sunny day. In terms of electricity production, each square meter of receiver area can generate 2 kWh in 8 hours on a sunny day.

The disks are cooled with water (can also be salt water or brine). The heat is absorbed into the PV chips, which gather the energy and are then cooled by microchanneled water, which is why they can safely concentrate such large amounts of solar energy. In the process of cooling the chips, the water is heated to 90°C (194°F); this liquid is filtered, vaporized, condensed, desalinated, and vacuum distilled, producing 30 to 40 L of drinkable water per square meter of receiver area per day.

Therefore, if a large multi-dish installation (Figure 3-22) is used, the daily production of 25 kWh of electricity and 30 to 40 L of drinking water per square meter of collector

Figure 3-22. HCPVT system developed by IBM.

Source: Reprinted with permission from Andrea Pedretti.

area could provide enough drinking water and electricity for a small town. In addition, through the use of thermally driven absorption chillers, cool air can also be produced.

3.2.6.3 Stirling Solar Power Generators

Stirling solar power generators focus the Sun's energy on hydrogen (or other substances) in a sealed Stirling engine. As the hydrogen is heated (to 732°C [1350°F]), it expands and drives the positive displacement piston at the hot end (in the expansion space of the engine) to compress it. The other piston does the work when the heat is rejected (gas is cooled at the cold end) in the compression space. As the gas is cooled, the displacement piston moves the compressed gas back into the power piston. As the gas is returned to the power piston, it is heated at this hot end by concentrated solar energy, and the cycle repeats. The work output of the Stirling cycle is then used to drive a generator to generate electric power. In this design, the working fluid in the piston, located at one end of the Stirling engine, is heated by solar radiation, and the fluid in the other piston at the other end is cooled by air or water. As the working fluid expands and cools, the pistons drive an electric generator. (Stirling engines are used on submarines because they are quiet—no combustion takes place.)

Stirling applications have existed for many decades. One of their first applications was irrigation in underdeveloped areas of the world. One solar pump design combines the Stirling engine with a large dish that concentrates solar energy to a focal point and uses the heat flow from that hot source to make electric power. Another application uses the cold water in a spring as the cold sink, thereby increasing the temperature difference across the engine, which increases its efficiency and lowers the pumping cost.[43]

Japan is planning a solar satellite to operate with a Stirling engine by 2040. In solar applications, the main advantage of Stirling engines is their high efficiency (30%). These closed systems have low maintenance requirements because they do not need to be refilled; their mirrors just need to be washed every few weeks. Their operation can be fully automated including starting up in the morning, shutting down in the evening, tracking the Sun, and monitoring activity remotely over the Internet.

Figure 3-23 is Southern California Edison's 25 kW per dish Stirling system located in Mojave Desert, California. The concentrator has an area of 944 ft^2 and consists of 82 3-foot by 4-foot spherically curved glass mirrors. The Stirling engine uses hydrogen

43. "Solar-powered Stirling engine," Wikipedia, accessed March 19, 2022, https://en.wikipedia.org/wiki/Solar-powered_Stirling_engine.

Figure 3-23. Solar Stirling engine in the Mojave Desert, San Bernardino County, California.
Source: "SolarStirlingEngine.jpg" by US Department of Energy (DOE), September 6, 2006, accessed March 19, 2022, Public Domain, https://commons.wikimedia.org/w/index.php?curid=4143670.

to provide the working pressure, with the maximum gas pressure being 2900 psig (197 atm). It delivers 25 kW output at 1000 W/m^2 insolation.[44]

3.2.7 Steam Generation

3.2.7.1 Direct Steam Generation

An advancement over the SEGSs design (Figure 3-20) is direct steam generation (DSG), which improves the system's efficiency. In this system, the hot oil collection and water/steam Rankine power circuit combination of the SEGS design is replaced with a single circuit in which the collector circuit is directly coupled to the power circuit. This eliminates the oil-to-water heat-transfer equipment and the associated loss in efficiency. With DSG systems, superheated steam at 100 bar pressure and 400°C (752°F) temperature can be generated (Figure 3-24).

---

44. Robert Sier, "Solar Dish/Stirling Power Systems," Stirling and Hot Air Engine website, accessed March 12, 2022, https://www.stirlingengines.org.uk/sun/sola3.html.

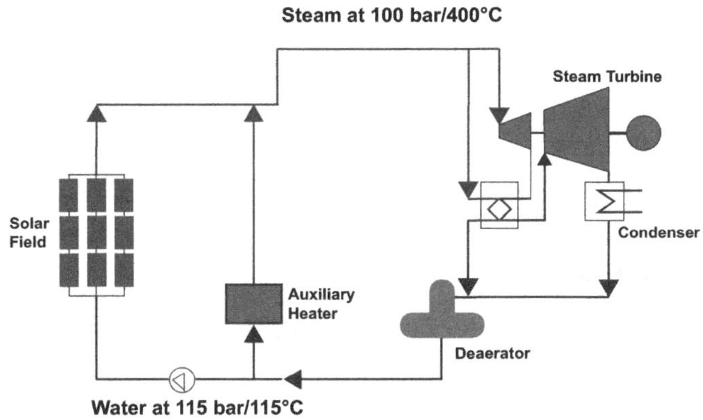

Figure 3-24. DSG solar power plant.

3.2.7.2 Combined-Cycle Plant

A combined-cycle plant utilizing a solar parabolic-trough collector system is referred to as an integrated solar and combined-cycle system (ISCCS). SEGS steam cycle plants have been constructed with capacities ranging from 15 to 80 MW, whereas ISCCS

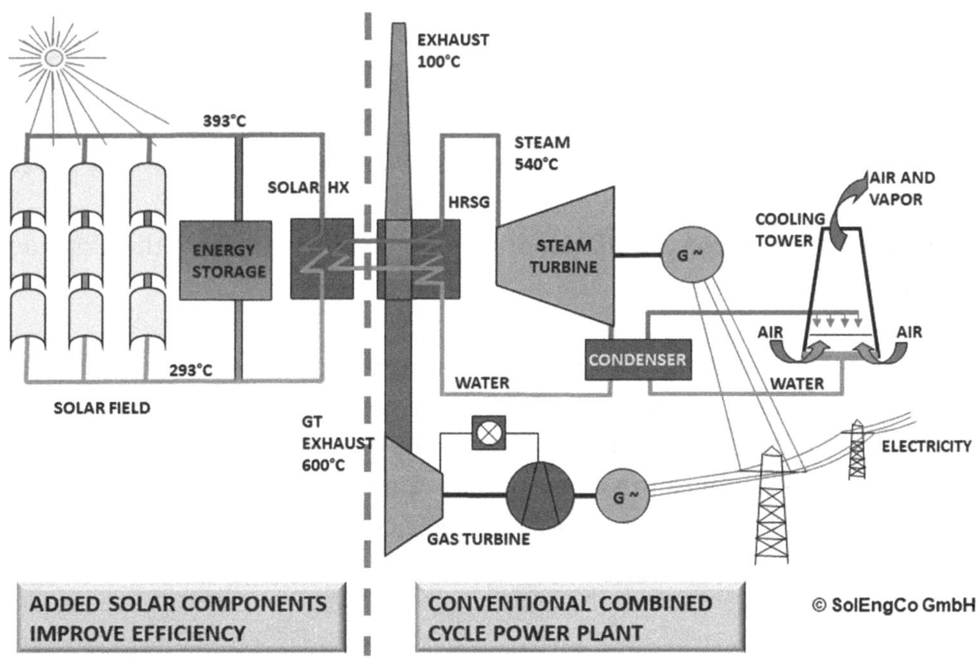

Figure 3-25. The cost of an ISCCS plant is less than that of separate solar and conventional power plants.[45] In an ISCC (integrated solar combined cycle), solar-generated heat is integrated into a conventional gas and steam power plant. The ISCC increases efficiency and reduces costs, compared to having separate solar and combined- cycle power plants.

Source: Reprinted with permission from Georg Brakmann, SolEngCo GmbH.

45. For additional information, see: "Concentrated Solar Thermal Power – Now!" Greenpeace, the European Solar Thermal Power Industry Association (ESTIA), and the IEA's SolarPACES Programme, September 2005, https://upload.wikimedia.org/wikiversity/en/c/c3/Concentrated-Solar-Thermal-Power.pdf.

plants can be larger. Correspondingly, the electricity generated by ISCCS plants is usually less expensive because, as the size of the plant increases, the unit cost of electricity decreases. When solar energy collection is coupled with a combined-cycle power plant (ISCCS configuration), the overall generating capacity can exceed 200 MW. Figure 3-25 depicts an ISCCS plant incorporating a solar field and both gas and steam turbines.

3.2.8 Solar Updraft Towers

The chimney effect of tall towers generates an updraft because the atmospheric pressure is higher at the bottom of the chimney and lower at the top. Therefore, as the Sun heats the air under the collector, the air warms, and its pressure is reduced (to less than that of the outside air); the lighter air rises upward inside the chimney. Static home cooling systems utilize the reverse of this effect (reverse draft) to pull cold air into homes from underground ducts.[46]

According to Wikipedia:[47]

> Model calculations estimate that a 100 MW plant would require a 1000 m tower and a greenhouse of 20 square kilometers (7.7 sq mi). A 200 MW tower of the same height would require a collector 7 kilometers in diameter (total area of about 38 km^2). One 200 MW power station will provide enough electricity for around 200,000 typical households and will abate over 900,000 tons of greenhouse-producing gases from entering the environment annually. The glazed collector area is expected to extract about 0.5 percent, or 5 W/m^2 of 1 kW/m^2, of the solar energy that falls upon it. If a transpired solar collector is used in place of the glazed collector, the efficiency is doubled.

Without further tests, these calculations are only theoretical, and their accuracy is therefore uncertain.

A solar updraft tower (Figure 3-26) is a solar energy converter that converts thermal energy into aerodynamic energy (wind).[48] In this system, air is heated under a circular greenhouse-like canopy. The roof of the canopy slopes upward from the perimeter toward the center, where the tower stands. Under the canopy, the Sun heats the air, which rises up the tower and generates electricity by driving an array of turbine generators.

---

46. Traveling in Iran, I was amazed to see that the ancient Persian palaces were cooled by the chimney effect 2000 years ago. They pulled in air that was cooled by water brought down from the mountains through underground ducts. They also used this system to water their beautiful gardens and to run turbines under the kitchen, which served to grind grains and nuts in the kitchens.

47. "Solar updraft tower," Wikipedia, updated March 14, 2022, accessed March 19, 2022, https://en.wikipedia.org/wiki/Solar_updraft_tower.

48. For more information, see: Mike Chino, "EnviroMission Plans Massive Solar Updraft Towers for Arizona," Inhabitat website, January 6, 2010, https://inhabitat.com/enviromission-plans-massive-solar-updraft-towers-for-arizona/solarupdraft/.

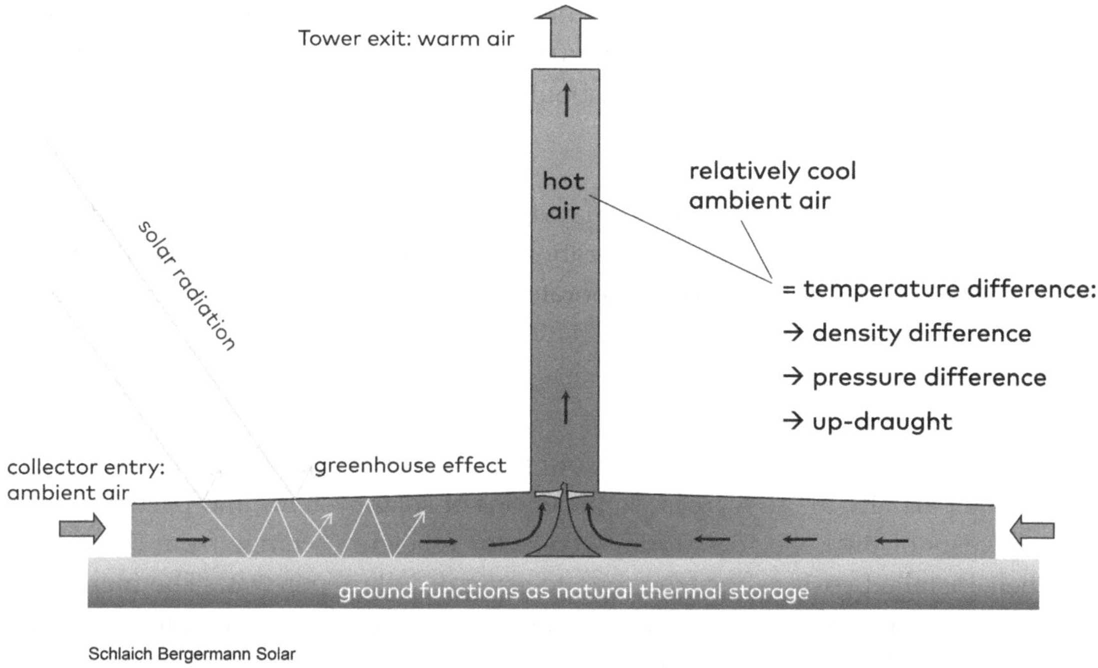

Figure 3-26. Solar thermal updraft wind turbine.
Source: Reprinted with permission from Schlaich Bergermann Solar.

This "low-tech" solar energy collector concept is more than 100 years old. Spanish Colonel Isidoro Cabanyes first proposed it in 1903, and Robert E. Lucier submitted the first patent application for this design in 1975. The first 50 kW working model was built in 1982 in Ciudad Real, Spain. The chimney of that model was 10 m (33 ft) in diameter and 195 m (640 ft) tall, and the canopy diameter was 244 m (800 ft) (about 46,000 m^2 or 11 acres in area). This prototype operated for 9 years and reached a maximum production of 50 kW.

Today, 50 to 200 MW installations are planned in Australia, China, Africa, and the southwestern United States (specifically, Arizona). As of 2021, although several prototype models have been built, no full-scale practical units are in operation. The design of an updraft tower planned for Arizona included a tower twice as tall as the Empire State Building that was expected to generate 200 MW. The planned units may allow the development of applications for agriculture or horticulture by providing water extraction and distillation or for health by reducing urban air pollution.

Commercial investment may have been discouraged by the high initial cost of building a very large novel structure and the large land area required. A comprehensive review of the theoretical and experimental aspects of *solar updraft tower power plant* (SUTPP) development recommended commercial development.[49] A plant in China is planned to consist of a 38 km^2 canopy and a 1 km (0.62 mi) tall tower to provide 200,000 homes with electricity. It is expected to abate about 1 million tons of greenhouse gases and cost some $800 million.

Design variations include adding thermal storage by covering the ground with heat-absorbing surfaces so that power generation continues at night or when the Sun is not shining. Also, the towers can incorporate PV modules on transpired collectors for additional daytime output.

3.2.9 Costs and Efficiencies

As is the case with most green energy sources, the use of solar energy is also rising. This is due partially to the increased efficiencies of solar systems and partially to the reduced costs of the collectors. Today's collector efficiencies range from 5% to 40%, and their installed costs range from one to several dollars per watt. The increase in efficiencies is particularly due to improvements in the multijunction solar cells and in concentrating collector designs (Figure 3-27).

Currently, collectors can comprise 25% or more of the total system capital costs for CSP plants. The US Department of Energy Solar Energy Technologies Office (SETO)

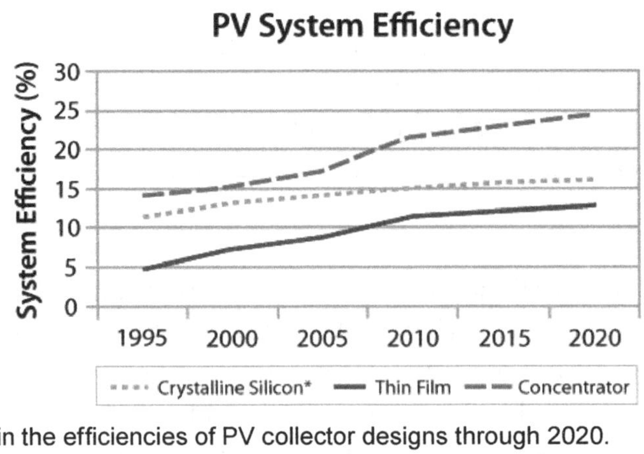

Figure 3-27. Trends in the efficiencies of PV collector designs through 2020.

Source: Reprinted with permission from Amos Han, "System efficiency vs. time for different technologies," derived from "Efficiency of Solar PV, Then, Now and Future," April 2014, https://sites.lafayette.edu/egrs352-sp14-pv/technology/history-of-pv-technology/.

49. Xinping Zhou and Yangyang Xu, "Solar Updraft Tower Power Generation," Solar Energy 128 (2016): 95–125, doi:10.1016/j.solener.2014.06.029.

is working to lower collector costs, with a target of $50 per square meter for highly autonomous heliostats,[50] to reach its goal of $0.05 per kilowatt-hour for baseload CSP plants with at least 12 hours of thermal energy storage.

The approximate efficiencies of some collector designs (including plant and tree leaves) are as follows:

- Trees, plants = 6%
- Thin-film PV = 5% to 15%
- Single-junction (standard) PV = 15% to 20%
- Multijunction PV = 20% to 22% (maximum theoretical[51] = 33%)
- Bifacial solar panel[52] = 30% to 35%
- Multijunction concentrating PV = 47% (maximum achieved in December 2019)

The efficiency of PV cell installations depends not only on the quality and characteristics of the selected PV design used but also on the installation features. In terms of the amount of solar radiation received, the orientation of the collector is an important factor. For the best results, solar cells should be installed at a height that minimizes shading and at an orientation facing south so that the cells will receive the maximum amount of solar energy. Even a small amount of shade can reduce the efficiency of the whole system.

Mounting the collectors at an angle of about 45° is also important (Figure 3-28), particularly in the case of flat roofs. If one desires to collect the maximum amount of energy, it is recommended that the inclination be decreased in about November and increased in about May.

The ideal tilt angle depends on the latitude and the time of year. When the tilt is adjustable, it can be changed with the season and tilted at an angle equal to the latitude of the city in which the collectors are installed. Regarding operating temperature, 25°C

---

50. Heliostats are computer-controlled mirrors which keep the Sun reflected on a target as the sun moves across the sky.

51. "Solar power in US.pdf," ELECTRICAL ME205 Power, Pakistan Institute of Engineering & Applied Sciences, Islamabad, Course Hero website, https://www.coursehero.com/file/p17n8ct/Efficiency-The-maximum-theoretical-efficiency-of-a-single-solar-cell-is-3316/.

52. "Bifacial solar panels: Innovative, eye-catching, and best of all, more efficient," SolarReviews website, updated January 29, 2021, https://www.solarreviews.com/blog/bifacial-solar-panels.

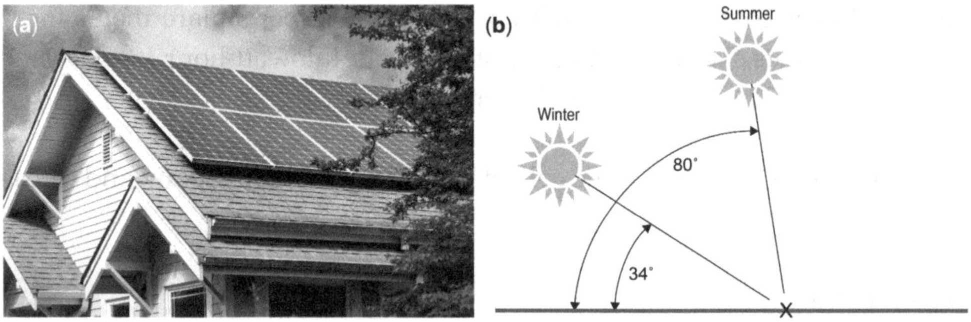

Figure 3-28. Mounting of PV collectors. 3-28a: Installation should provide free air circulation under the backside of the collectors. 3-28b: For fixed installations, the angle of inclination should preferably be between 40° and 50°, but definitely between 25° and 70°.[53]

(77°F) is best because the collector efficiency drops with rising temperature. Therefore, sufficient space should be provided between the roof and the solar modules for the free circulation of the cooling ambient air.

PV cells are arranged on a panel to form a solar module. Modules are then linked in series to form what is known as a solar array. Costs are usually provided in units of dollars per watt or dollars per watt peak because the size of a solar module or array is given in terms of its peak power[54] production. The capacity of a module is the amount of power that a PV module can supply when it receives full solar radiation (under set standard test conditions), which is 1000 W/m^2 at 20°C (68°F).

As shown in Figure 3-29 collector system costs are falling because of the development of mass markets and corresponding mass production, plus design changes such as those of the thin-film collectors. Approximate data on the costs of installed systems and their relationship to other energy sources are as follows:

- Globally, solar energy is already the least expensive energy source.[55]
- The household installation cost of a 10 kW system (without subsidy) is $30,000.
- A solar panel (18% efficient, 300 W capacity, 3 ft × 5 ft) costs $300.
- The cost of uninstalled PV panels is about $2/W$_p$.

---

53. For further information, see: "How does the tilt angle and/or orientation of the PV panel affect system performance?" *NLPIP Lighting Answers* 9, no. 3 (July 2006), https://www.lrc.rpi.edu/programs/nlpip/lightinganswers/photovoltaic/14-photovoltaic-tilt-angle.asp.

54. Solar panel peak power is the maximum electrical power that a solar panel system is capable of generating under the following standard conditions: temperature, 20°C; received irradiance, 1000 W/m^2.

55. George McAtee, "Yes, America, There Is (Some) Hope for the Environment," *New York Times*, March 15, 2021, https://www.nytimes.com/2021/03/15/opinion/environment-climate-technology.html.

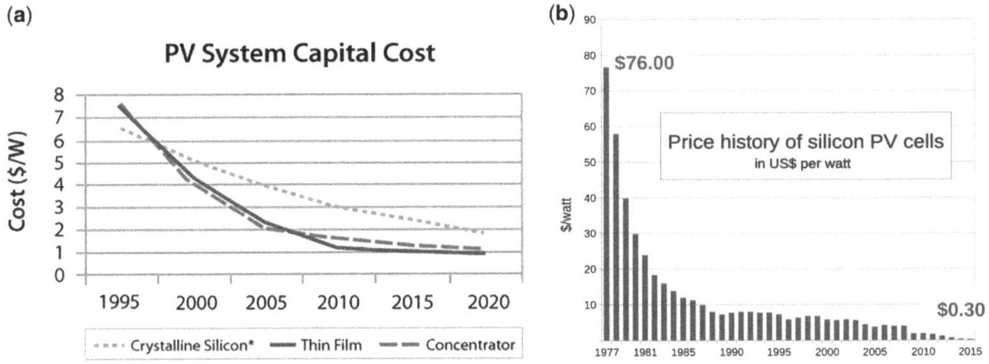

Figure 3-29. Trends in collector capital costs (not including installation) as a function of design. (**a**) PV system capital cost trends during the past 25 years. (**b**) Silicon PV cell costs dropped until 2015 but increased recently.[56]

Source of Figure 3-29a: Reprinted with permission from Amos Han, "PV System Capital Cost vs. time," derived from "Efficiency of Solar PV, Then, Now and Future," April 2014, https://sites.lafayette.edu/egrs352-sp14-pv/technology/history-of-pv-technology/.

Source of Figure 3-29b: "Price history of silicon PV cells since 1977.svg" by Rfassbind, based on Hanjin's 2013 version (in Spanish), amended with average sales prices for 2014 and 2015, May 2, 2015, Public Domain, https://commons.wikimedia.org/w/index.php?curid=33592736.

- The cost of a small unsubsidized residential system is about $3,000/kW.
- The cost of a utility-size fixed PV system is $1,100/kW.
- The cost of a utility-size tracking PV system is $1,200/kW.

During the past decade, the costs of residential, commercial, and utility-scale PV systems all declined by some 70% to 80%, driven largely by increased module efficiency and decreased hardware and inverter costs (Figure 3-30). Together with these reductions, the costs of solar energy storage systems also declined, although the soft costs of labor, taxes, and profits did not. The general trend of solar energy becoming more affordable is likely to continue as the cost of fossil fuels continues to increase.

Between 1992 and 2018, PV evolved from a small-scale niche market (compared to the mainstream electricity supply sources) to being recognized as a promising green energy technology. As a result, a number of governments implemented subsidy programs, such as feed-in tariffs, to provide economic incentives for investments. For several years, growth was mainly driven by Japan and pioneering European countries. Today, China remains the leader in this field.

56. According to some reports, after a decade of price declines, polysilicon is now at its highest price since 2011, because the material's cost to purchasers has risen. In January 2021, one kilogram of Chinese polysilicon cost about $13 and in 2022 the price is about $43/kg.

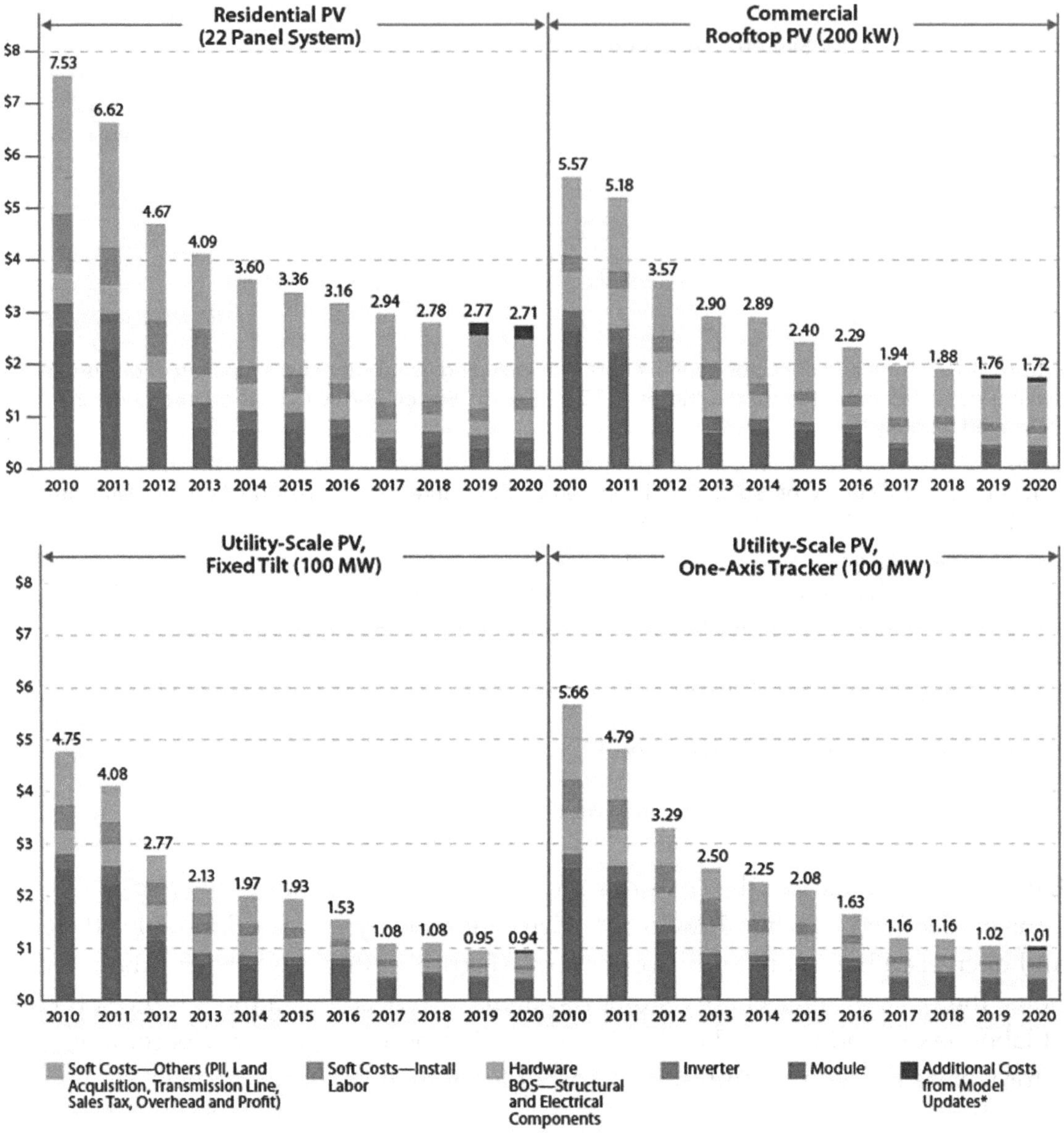

Figure 3-30. During the past decade, all cost components of solar energy installations have declined, except for their soft costs. (See color images after page 306.)

Source: "Documenting a Decade of Cost Declines for PV Systems," NREL website, February 10, 2021, https://www.nrel.gov/news/program/2021/documenting-a-decade-of-cost-declines-for-pv-systems.html.

From Q1 2020 to Q1 2021, installations in China, the United States, and Germany increased by 35% to 45%, and installations in India increased by 89%.[57] As of 2020, the

57. David Feldman, Kevin Wu, and Robert Margolis, "H1 2021 Solar Industry Update," Report NREL/PR-7A40-80427, NREL website, June 22, 2021, accessed November 21, 2022, https://www.nrel.gov/docs/fy21osti/80427.pdf.

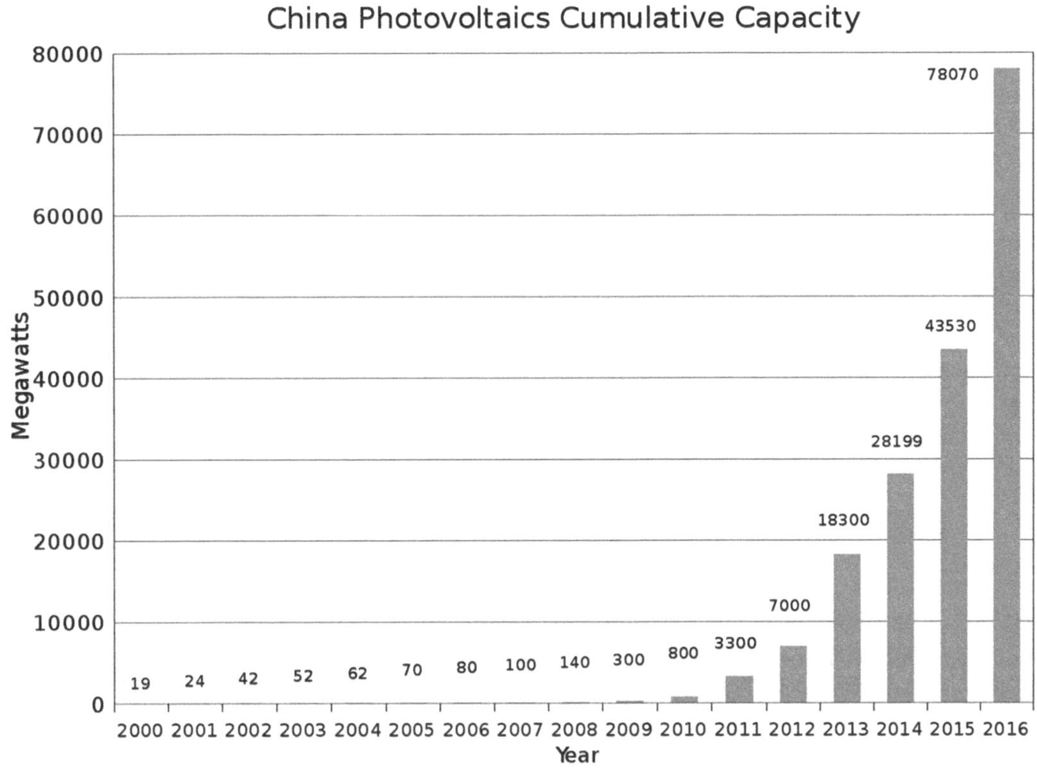

Figure 3-31. Cumulative solar capacity in China grew from near zero in 2000 to 78 GW by 2016 and over 300 GW by 2021.

Source: "China Photovoltaics Installed Capacity.svg" by Delphi234, May 11, 2015, is licensed under CC0, https://commons.wikimedia.org/w/index.php?curid=37225265.

total generating capacity of China was 254 GW, that of the European Union was 153 GW, and that of the United States was 76 GW. (Figure 3-31 shows that, in 2016, China's cumulative capacity was only 78 GW.) Compared to the previous year, the increases were 50 GW in China, 19 GW in the European Union, and 15 GW in the United States. In 2020, approximately 100 MW of solar generating capacity was added globally.

Figure 3-32 shows the projected trends in the market shares of the various solar cell designs. The share of thin-film collectors is projected to remain constant, whereas the share of single-junction designs is projected to rise and the share of multijunction designs is projected to drop. In terms of capacity, the total sales in 2019 amounted to 120 GW and increased to 168 GW in 2021.

3.3 Wind Energy

Wind energy propelled boats on the Nile River as early as 5000 BC, and it helped the Persians to pump water and grind grain between 900 and 500 BC. Later, the use of

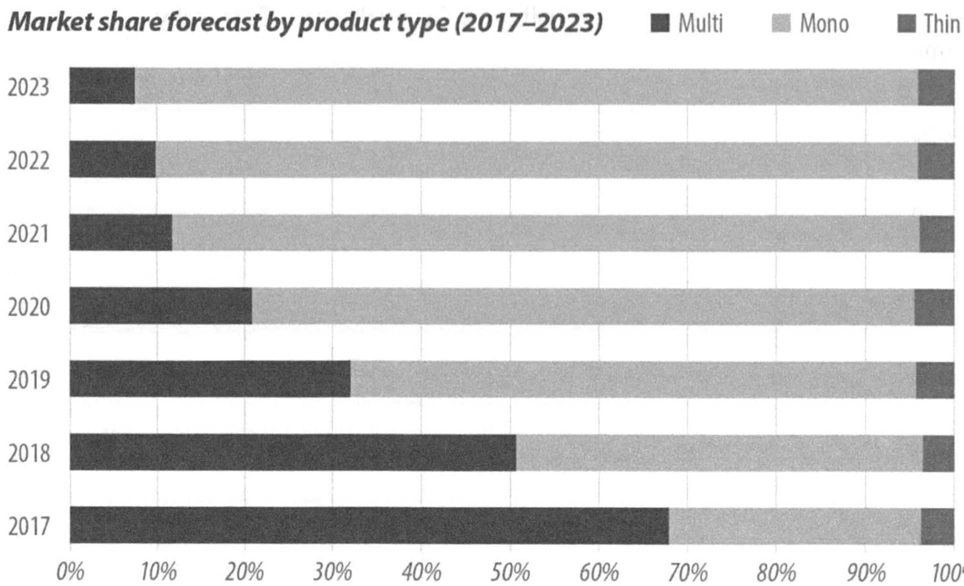

Figure 3-32. Projected market share trends of the various solar cell designs.

Source: Reprinted with permission from *PV InfoLink/pv* magazine, accessed March 12, 2022, https://16iwyl195vvfgoqu3136p2ly-wpengine.netdna-ssl.com/wp-content/uploads/2019/11/00014.jpg.

windmills spread from Persia to the surrounding areas in the Middle East, where they were used extensively in food production. Eventually, in about AD 1000, wind power technology spread north to European countries such as the Netherlands, which adapted windmills and used them to drain lakes and marshes in the Rhine River Delta (Figure 3-33).

Figure 3-33. Traditional windmill.

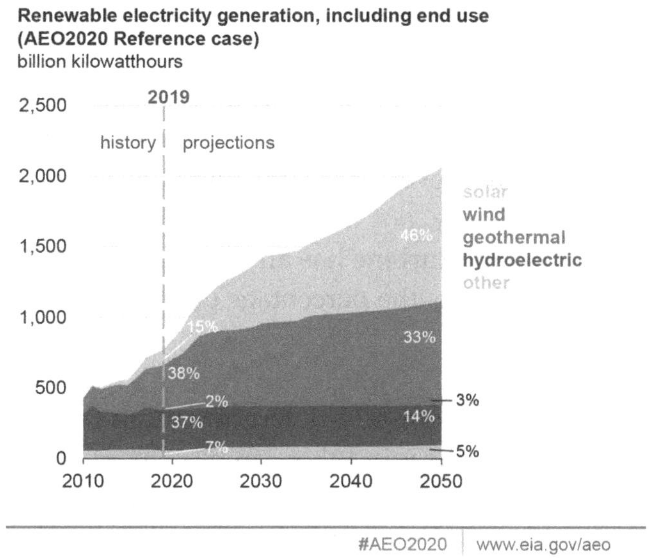

Figure 3-34. The IEA projections for renewable electricity consumption in the United States through 2050.

Source: Annual Energy Outlook 2020 (Washington, DC: US Energy Information Administration, January 29, 2020), 62, https://www.eia.gov/outlooks/archive/aeo20/.

During the past century and particularly the past several decades, wind process control and optimization technology became a major field of interest, and automation increased its efficiency by optimizing both the direction and angle of the turbine blades. Recently, wind power has made remarkable advances by evolving into a major source of electricity in some nations, such as the United States (Figure 3-34), and by becoming one of the least expensive sources. For a more detailed breakdown of all energy source trends, refer to the latest EIA report.[58]

While the consumption of solar energy rises the fastest today wind and solar generated 10% of the global electricity in 2019, and wind energy sources accounted for nearly 6.59% of that. At that time, wind was the leading non-hydro green energy source, which generated 870 TWh by 2021. The global electricity generation capacity of onshore wind turbines was 7.5 GW in 1997 and 564 GW in 2018; it reached 750 GW in 2020.

3.3.1 Design and Sizing

The amount of electricity generated by a wind turbine is a function of the air density, the sweep area of the blades, and the wind speed. The electricity generated is a linear function of the first two variables, whereas it increases with the power of the wind speed:

---

58. *Annual Energy Outlook 2022* (AEO2022), (Washington, DC: US Energy Information Administration, March 3, 2022), https://www.eia.gov/outlooks/aeo/.

power (W) = 0.5 × [air density (kg/m^3)] × (sweep area [m^2]) × [wind speed (m/s)]^3

Cold and dry air is denser, which is one of the factors considered in selecting the locations for installation sites. In terms of sweep area, the larger it is, the more wind power can potentially be captured. Moreover, each time wind speed doubles, the power generation increases 8-fold. Naturally, not all of the wind's kinetic energy can be converted into electricity. Each wind turbine has an efficiency (expressed as a *coefficient of performance*—% CoP), which equals the percentage of wind power that it can generate in the form of electricity.

In 1919, the German scientist Albert Betz determined that the absolute limit for the fraction of the kinetic energy of the wind that can be converted to mechanical energy is 59.3% (59% CoP). The CoP of a wind turbine is the combined efficiency of both the blades and the generator. For today's advanced turbine designs, the blade pitch is controlled automatically. Their efficiency at wind speeds in the range of 10 to 35 mph (4.5 to 15.6 m/s, 16 to 56 km/h) ranges from 40% to 45% CoP. Under the same conditions, fixed-blade turbine efficiencies range from 1% to 35% CoP. The three curves in Figure 3-35 (starting from the top) show the kinetic energy content of the wind, the Betz limit of recoverable maximum energy (59.3%), and the actual power (in watts) produced by each square meter of a turbine's sweep area.

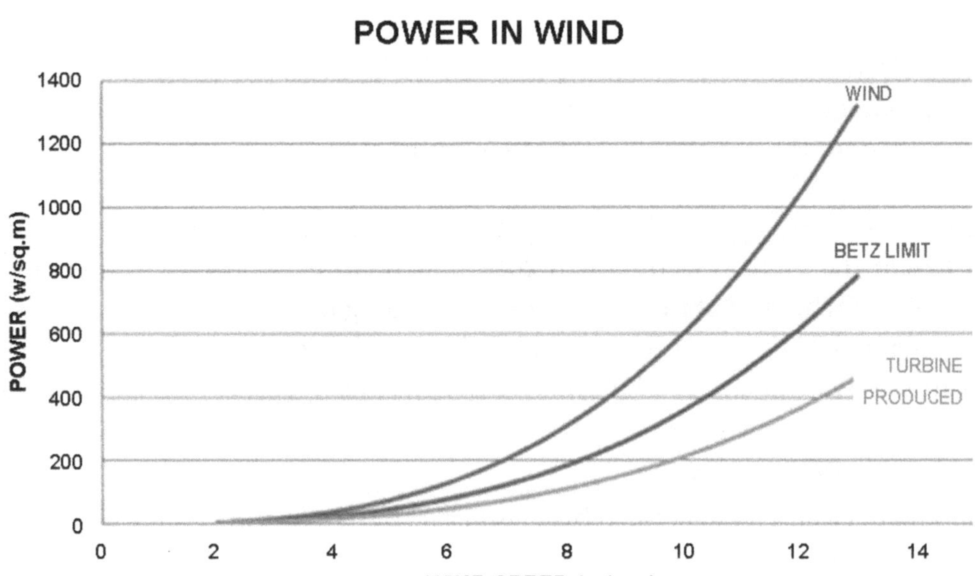

Figure 3-35. Power generation of a wind turbine as a function of wind speed.

Source: Reprinted with permission from "How Much Power Will a Wind Turbine Produce," WindyNation website, accessed March 12, 2022, https://www.windynation.com/jzv/inf/how-much-power-will-wind-turbine-produce.

3.3.2 Capacity and Cost

In May 2007, the US National Academy of Sciences determined that wind turbines could reduce the nation's CO_2 generation by 4.5% by 2025. Although the electricity generated by wind farms quadrupled between 2000 and 2006, the amount generated by them was still less than 1% of the total national electricity consumption. In 2020, the wind electricity capacity of the United States reached 8% of the total electricity generating capacity of the nation. According to a European estimate, if fully exploited, Europe could meet 30% of its electricity needs from wind energy. According to the Global Wind Energy Council (GWEC), the global generating capacity increased between 2006 and 2016 from 72 GW to 460 GW. In 2020, it reached 743 GW (700 GW of which was onshore), and it is expected to reach 4042 GW by 2050.

The generating capacity of the average wind turbine in the United States ranges from 2 to 10 kW. The electricity cost at the source, without considering the cost of transportation, for wind-turbine-generated electricity is competitive ($0.04/kWh to $0.06/kWh) and dropping. In many parts of the world, government support is available to finance wind turbine installations. In addition, in many locations, the excess electricity that wind turbines generate is accepted by power companies on a basis called *net metering*, in which case the meters are said to *run backward* when the excess electricity is sent to the grid. Regarding land area requirements for wind farms, on most large wind turbine farms, a spacing of about 15 times the rotor diameter is economical, taking into account typical wind turbine and land costs.

Wind turbine costs range from $2.6 million to $4 million for an average-sized commercial unit. The typical cost is $1.3 million per megawatt of electricity-producing capacity. Most commercial wind turbines have a capacity of 2 to 3 MW, but offshore turbines can be as large as 12 MW.[59]

Having few moving parts, wind turbines do not require much maintenance. Contingent on geographical location, well-located wind turbines can pay for themselves within 15 years or in about half of their useful life.

There are also a few upfront fees with wind power because wind turbines are tall structures that require building-related permits. Fees for permits and plot plans can range anywhere from $400 to $1,600, and public notifications, hearings, and environmental impact studies may necessitate another few hundred to several thousand dollars.

---

59. "Wind Turbine Cost: How Much? Are They Worth It In 2022?," Weather Guard Lightning Tech website, December 20, 2021, https://weatherguardwind.com/how-much-does-wind-turbine-cost-worth-it/.

3.3.3 Size and Performance

Commercial utility-connected turbines usually deliver about 80% of the Betz limit[60] of power extractable from the wind when operating at the rated operating speed. They are usually three-bladed, low-torque ripple designs[61] (Figure 3-36) that provide good reliability. The blades are usually colored white for daytime visibility by aircraft and range in length from 20 to 80 m (66 to 260 ft). The sizes and heights of turbines increase year by year. Offshore wind turbines built today have capacities of up to 8 MW and blade lengths of up to 80 m (260 ft). Tubular steel towers of multimegawatt capacity are 70 to 120 m tall and, in extreme cases, are up to 160 m.

Most horizontal-axis turbines have rotors upwind of the supporting tower, but downwind machines have also been built because they do not need additional controls to keep them in line with the wind. If the United States is to meet its goal of having 20% of its electricity supplied by wind power by 2030, the US Geological Survey estimated that it would require 50,000 km^2 (19,300 mi^2) of onshore land area or 11,000 km^2 (4250 mi^2) of offshore area.

Figure 3-36. Three-bladed, low-torque designs provide good reliability.

---

60. The Betz limit is the theoretical maximum efficiency for a wind turbine. Its value was determined by the German physicist Albert Betz in 1919. Betz concluded that at most, only 59.3% of the kinetic energy from wind can be used to spin the turbine and generate electricity.

61. For a detailed diagram of a wind turbine, see: Department of Energy, "How a Wind Turbine Works," June 20, 2014, https://www.energy.gov/articles/how-wind-turbine-works.

3.3.4 Applications

Currently, the world's largest wind turbine is the GE-designed offshore Haliade-X, which has a power generating capacity of up to 14 MW. The prototype was part of a $400 million investment aimed at reducing the cost of offshore wind energy. It was installed at the port of Maasvlakte-Rotterdam in the Netherlands in 2019. The three 107-meter LM Wind Power blades were manufactured in France, and the tower sections were manufactured in Spain.

Many other wind power projects are in progress, including some in the United States. In 2010, only 2% of US electricity was generated by wind energy. In 2021, wind farms generated 9.2% of electricity in the US, according to the US Energy Information Administration (EIA). The state of New York has committed to eliminating all greenhouse gas emissions by 2050, and offshore wind turbines (some floating) will play a major role in achieving that goal. In the United Kingdom, in 2010, only 1% of the electricity was generated by wind power; today, it is about 10%.

3.3.5 Advantages and Disadvantages

Wind turbines produce electricity at between $0.04/kWh and $0.06/kWh, which makes them one of the lowest-cost sources. As technology improves, the prices will decrease further. Wind turbines provide a clean energy source, emitting no greenhouse gases and generating no waste products. For each megawatt of electricity generated, more than 1500 metric tons of CO_2 emissions per year can be eliminated when wind turbines are used for generation instead of fossil fuels.

The disadvantages of wind energy include the fact that, because wind turbines can be more than 140 m (460 ft) tall and have blades that are 55 m (60 yd) long, they can interfere with air traffic, damage birds, and have an undesirable visual impact. Thousands of birds, including rare species, have been killed by the blades of wind turbines. Another limitation is that their energy generation is intermittent; therefore, they require energy storage to serve as a continuous source of electricity, and the storage and transportation of the generated electricity are still under development. (Chapter 4 discusses this topic in detail.)

3.4 Geothermal Energy

Earth has immense geothermal energy potential. So far, utilization of this energy has been limited to areas where geological conditions permit a carrier (water or steam) to transfer the heat from deep hot zones to the surface. There are two main types of geothermal resources: convective and hydrothermal. Hydrothermal is where natural hot water or steam is present and can be brought to the surface. Geothermal is where

Figure 3-37. Larderello geothermal power station.

Source: "Larderello 001.JPG" by Janericloebe, August 2009, Public Domain, https://commons.wikimedia.org/w/index.php?curid=7823040.

hot rock is present. Geothermal areas can be low- or high-temperature fields. High-temperature fields are found around tectonic plate boundaries where volcanic activity is high. Low-temperature fields can be found in most locations, and heat can be obtained from the hot rocks or water below the ground.

Lord Kelvin invented the heat pump in 1852, and Heinrich Zoelly patented the idea of using it to draw heat from the ground in 1912. The first commercial geothermal electric power plant opened in Larderello, Italy, in 1911. It was the world's only industrial producer of geothermal electricity until New Zealand built a plant in 1958. The Larderello power plant is still operating today (Figure 3-37).

In 1960, Pacific Gas and Electric began operating the first successful geothermal electric power plant in the United States at The Geysers geothermal field in California. The original turbine lasted more than 30 years and produced 11 MW of net power. The technology became popular in Sweden as a result of the 1973 oil crisis and has been growing in acceptance worldwide since then. Among other regions, Greenland and Iceland have good geothermal potential. Geothermal heat pumps are also gaining in use,[62] particularly since the 1979 development of polybutylene pipe that greatly aug-

62. For more information, see: "Geothermal Heat Pumps," Office of Energy Saver, Office of Energy Efficiency & Renewable Energy, US DOE, accessed March 12, 2022, https://www.energy.gov/energysaver/heat-and-cool/heat-pump-systems/geothermal-heat-pumps.

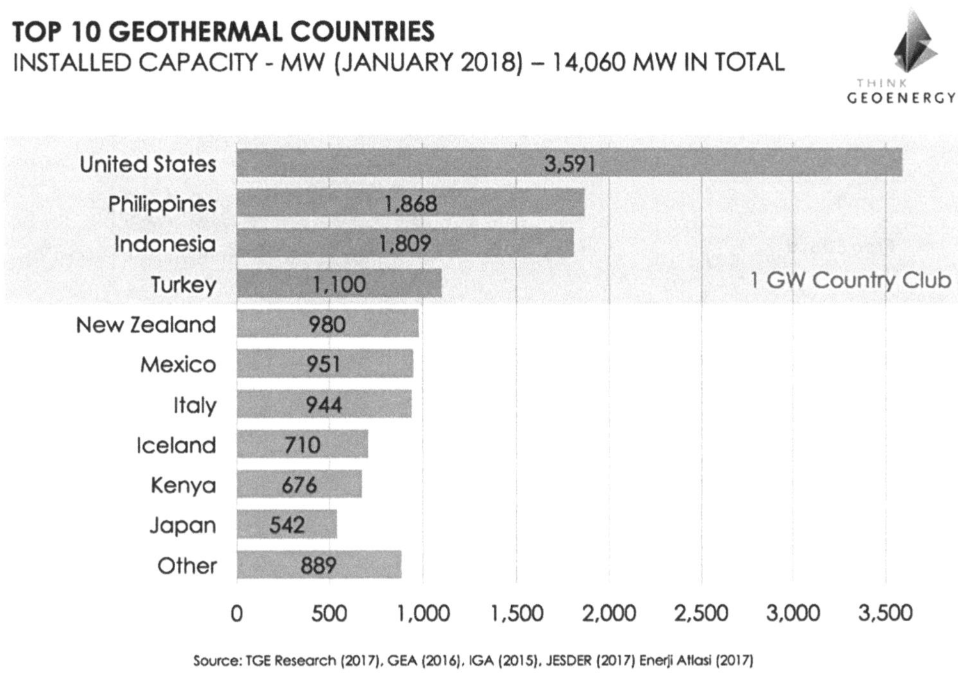

Figure 3-38. Installed geothermal power generation capacities in 10 countries.
Source: Reprinted with permission from Think GeoEnergy.[63]

mented the heat pump's economic viability. As of 2019, the global geothermal capacity was about 15.4 GW (Figure 3-38), and by the end of 2020, it reached 15.6 GW. Wikipedia provides a complete list of geothermal power plants at https://en.wikipedia.org/wiki/List_of_geothermal_power_stations.

Earth's interior continuously supplies an enormous quantity of heat to the crust. This geothermal energy comes from the natural heat of the Earth, primarily due to the decay of the naturally radioactive isotopes of uranium, thorium, and potassium. The temperature difference between the planet's core and its surface drives the continuous conduction of heat from the core to the surface. Temperatures at the core-mantle boundary can reach 4000°C (7200°F). Rock and water are heated in the crust (Figure 3-39) up to 370°C (700°F). Bringing that heat to the surface requires a "carrier" (usually hot water or steam).

The technological limit for drilling depth today is based on the experience of the oil industry at approximately 5000 m. This limit could be exceeded if the research efforts in progress are successful. According to researchers at the Norwegian University of

63. For additional information, see: Alexander Richter, "Top 10 Geothermal Countries based on installed capacity – Year End 2017," ThinkGeoEnergy website, January 9, 2018, accessed November 21, 2022, https://www.thinkgeoenergy.com/top-10-geothermal-countries-based-on-installed-capacity-year-end-2017/.

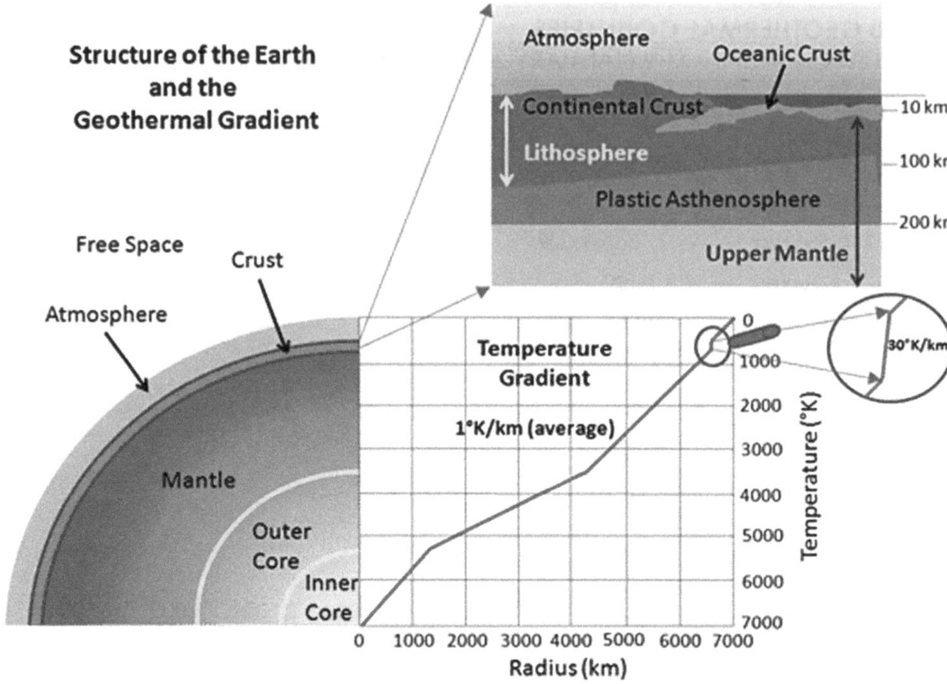

Figure 3-39. Temperature gradient as a function of the depth in the Earth.

Source: "Geothermal Energy," Electropaedia, accessed March 12, 2022, https://www.mpoweruk.com/geothermal_energy.htm.

Science and Technology (NTNU), the University of Bergen, the Geological Survey of Norway (NGU), and SINTEF, it may be possible to collect geothermal energy from depths of 10,000 m, where temperatures can reach at least 374°C (707°F) and the water is under a pressure of at least 220 bar. The main challenge is to find materials that are suited for drilling at such high temperatures because steel, for example, becomes brittle and most other materials will weaken or melt.

Geothermal energy has been used for bathing and space heating since ancient Roman times. Today, geothermal heat is also used to generate electricity through steam-driven turbines. In the past quarter century, the geothermal electric capacity of the world has increased by about 10-fold (Figure 3-40).

The total global installed geothermal power generation capacity at the end of 2020[64] was 15.6 GW, which represented an increase of 0.2 GW compared to 2019. In 2020, the six countries with capacities exceeding 1.0 GW were the United States (3.7 GW), Indonesia (2.1 GW), the Philippines (1.9 GW), Turkey (1.6 GW), and New Zealand (1.0 GW).

64. Richter, "ThinkGeoEnergy's Top 10 Geothermal Countries 2020."

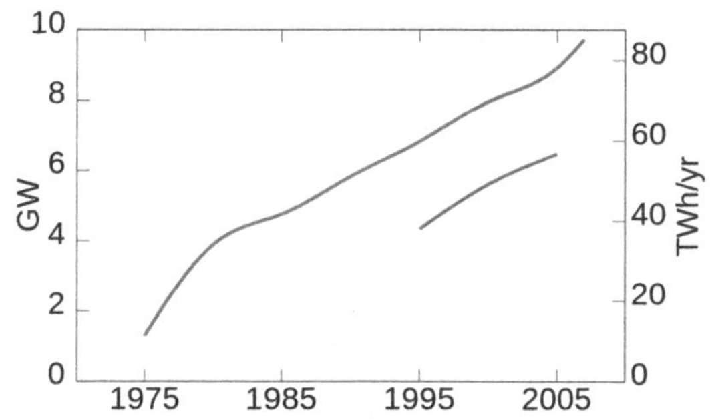

Figure 3-40. Global geothermal electric capacity and consumption. The red line shows the installed capacity, the green line represents the realized production through 2005. (By the end of 2020, the installed capacity exceeded 15.6 GW.[65])

Source: "Geothermal capacity.svg" by Ytrottier, April 11, 2009, is licensed under CC BY-SA 3.0, https://commons.wikimedia.org/w/index.php?curid=6498985.

The Geysers, located in California, is the largest complex of geothermal power plants in the world. However, Indonesia's 30 GW geothermal reserve capacity is the largest in the world. Its production is predicted to overtake that of the United States in the next decade.

The global use of geothermal heat (direct and storage) is much larger than that of geothermal electricity. In 2016, the installed global capacity of direct heat use was estimated to be 20.6 GWe (equivalent electric power).[66] Roughly 40% represents direct use; the rest is energy from heat pumps. China dominates heat usage, with more than half of the world's consumption. Europe is the second largest user, with 30% of the world's consumption. Direct heat use is geographically concentrated in regions above latitude 35° N in areas where heating is required in the winter.

3.4.1 Geothermal Heat Energy

The temperature of thermal springs can reach 350°C (662°F). At that temperature, direct heating is possible. If the groundwater temperature exceeds 150°C (302°F), flash-steam power plants can be built, and if the water temperature is between 100°C and 150°C (212°F to 302°F), binary-cycle power plants can be operated.

---

65. Alexander Richter, "ThinkGeoEnergy's Top 10 Geothermal Countries 2020 – installed power generation capacity (MWe)," ThinkGeoEnergy website, January 7, 2021, https://www.thinkgeoenergy.com/thinkgeoenergys-top-10-geothermal-countries-2020-installed-power generation-capacity-mwe/.

66. Jon Limberger, Thijs Boxem, Maarten Pluymaekers, David Bruhn, Adele Manzella, Philippe Calcagno, Fred Beekman, Sierd Cloetingh, and Jan-Diederik van Wees, "Geothermal energy in deep aquifers: A global assessment of the resource base for direct heat utilization," *Renewable and Sustainable Energy Reviews* 82, Pt. 1 (Feb 2018): 961–975, https://www.sciencedirect.com/science/article/pii/S1364032117313345.

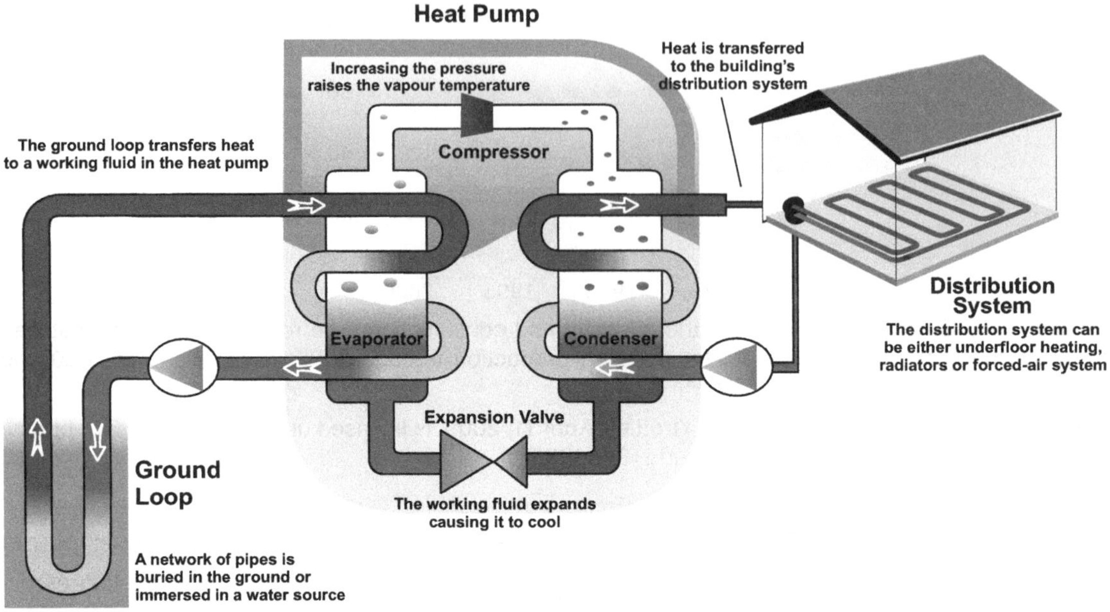

Figure 3-41. Schematic of GHP operation in winter mode. In summer mode, the flow direction of the compressor is reversed so that heat is removed from the building and rejected into the groundwater in the well.[68]

Source: Reprinted with permission from the Geothermal Heat Pump Association of New Zealand (GHANZ) of the New Zealand Geothermal Association (NZGA).

When the groundwater temperature is less than 100°C (212°F), *geothermal heat pumps* (GHPs) can be used to heat buildings in the winter and cool them in the summer. Air conditioners (hot to cool) and refrigerators (cool to warm) are both heat pumps.[67] Two-directional air pumps that cool the home in the summer and heat it in the winter are also available. In the winter, they take the heat from a relatively cooler area and move it into a relatively warmer one. In the winter, GHPs can take heat from the groundwater in a well outside the building and use it to heat the air inside the building, whereas in the summer, they do the opposite by taking the heat from inside the building and reinjecting it into the water in the well outside (Figure 3-41).

---

67. Heat pumps use a refrigerant (usually Freon), that is circulated by a compressor. The refrigerant removes heat from the cold end of its circulatory loop, where it is evaporated under the low pressure generated by the suction of the compressor. Next, the compressor compresses the refrigerant and moves it to its high-pressure discharge side. At the higher pressure, it is condensed, and the heat it carries is passed on to the receiving medium (air or water). After condensing, the refrigerant returns to the suction side of the condenser, where it evaporates again and the cycle repeats.

68. For additional information, see: "How a Ground Source Heat Pump Works," Aris Vourvoulias, *GreenMatch Blog*, last updated March 25, 2021, accessed November 21, 2022, https://www.greenmatch.co.uk/blog/2016/01/how-a-ground-source-heat-pump-works.

In heating mode, the GHP removes the heat from groundwater and transfers it to the hot water or hot air that is used to heat the building. I examined the operation of such a system at Bard College, Annandale-on-Hudson, New York, where the buildings are heated by heat pumps taking the heat from groundwater that is 10°C to 16°C (50°F to 60°F) year-round and transporting it into the buildings to supply warm air and hot water at approximately 38°C (100°F). GHPs are also applicable for use in large systems, such as the central heating of Budapest, Hungary.[69] After the heat content of the geothermal water is removed, the spent fluid is returned to the underground water reservoir to maintain the water level in the well.

A GHP can be operated as an open loop or a closed loop. The loop shown in Figure 3-41 is closed because the ground loop is closed. If a lake or other clean water source is

Figure 3-42. Closed loops can be installed (a, b) underground (horizontally or vertically) or (c) placed on water, such as floating in a pond.

Source of Figure 3-42b: "3-ton Slinky Loop.jpg" by Mark Johnson (Marktj), November 15, 2006, Public Domain, https://commons.wikimedia.org/w/index.php?curid=7926670.

Source of Figure 3-42c: "Pond Loop Being Sunk.jpg" by Mark Johnson (Marktj), November 15, 2006, Public Domain, https://commons.wikimedia.org/w/index.php?curid=7952259.

---

69. "Geothermal District Heating Project Starting in Budapest," Hungary Today website, March 31, 2021, accessed November 21, 2022, https://hungarytoday.hu/geothermal-district-heating-project-starting-co2-emission-budapest/.

available, the ground loop could be open, and the system could circulate the lake water directly through the building. Closed loops do not need to be covered by groundwater. They can also be installed in the ground, or they can float on a pond (Figure 3-42).

Solar and geothermal systems can be combined when the size of the system is small, such as in the case of "energy-free" private homes, where the goal of combining the solar and geothermal systems is to provide electricity for the home continuously. In such a system, the changes in the amount of solar radiation received result in changes in the temperature of the water already present in the solar hot water tank. Therefore, a simple control strategy is to depend on the solar heat only as long as the hot water in the tank is at the desired temperature and to start the geothermal heat pump only when the solar heat is insufficient and the temperature in the tank starts to fall.

3.4.2 Geothermal Power Plants

Iceland and the Philippines lead the world in geothermal energy use. They get about 30% of their power from geothermal sources, compared with approximately 1% in the United States.

Geothermal power plants require high-temperature (~150°C to ~370°C, or 300°F to 700°F) hydrothermal resources that come from either dry steam wells or hot water wells. In the past few decades, we saw the introduction of binary-cycle power plants that significantly reduced the required temperature. Geothermal power is generated by drilling wells into the ground and bringing up steam to operate steam turbines that generate electricity. Some geothermal wells are as much as 2 miles deep.

There are three basic types of geothermal power plants. The type selected depends on the state and temperature of the water in the well, which can be dry steam, water that is hot enough to flash into steam, or hot water. The configuration differences among dry steam, flash steam, and binary-cycle power plants are shown in Figure 3-43.

3.4.2.1 Dry Steam Plants

Dry steam plants use the steam from geothermal reservoirs. The first geothermal power plant of this type was built in 1904 in Tuscany, Italy, where natural steam erupted from the Earth.

Other dry steam plants are located at The Geysers, a geothermal field in California's Mayacamas Mountains. It is the world's largest single geothermal field, comprising 22 geothermal power plants. The nameplate capacity of the field is about 1600 MW (about the size of a nuclear power plant), and its annual generation is about 6500 GWh. The

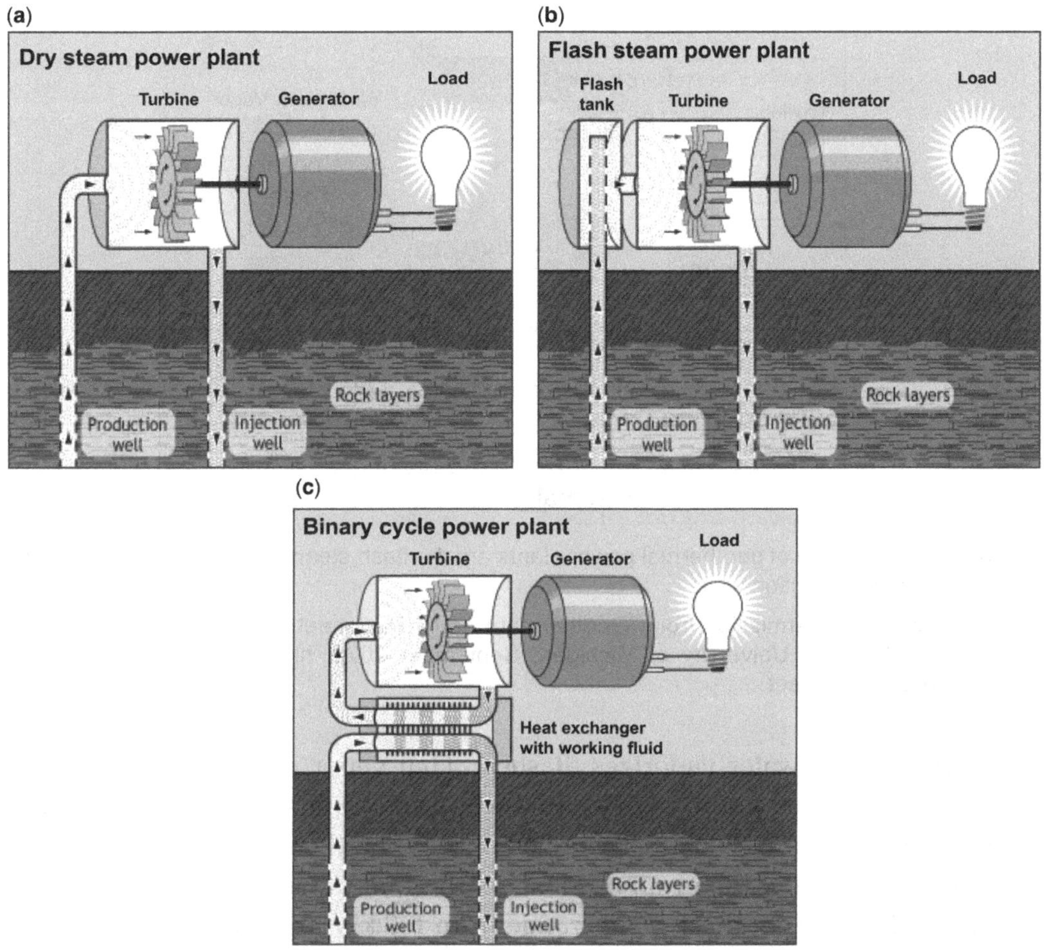

Figure 3-43. The three basic configurations of geothermal power plants.

Source: "Geothermal explained," US Energy Information Administration, last reviewed November 19, 2020, Public Domain, https://www.eia.gov/energyexplained/geothermal/geothermal-power-plants.php.

Geysers is where the first geothermal power plant in the United States was put into commercial operation. The number of geothermal power plants currently operating in the region is 18, and the present combined generation capacity of these plants is 900 MW, enough to power 900,000 homes. A new geothermal power plant is currently being constructed in The Geysers region. In these plants, the only emissions are excess steam.

3.4.2.2 Flash Steam Plants

The wells in flash steam plants contain high-pressure hot water at temperatures greater than ~150°C (300°F). The wells pump the water from deep in the hot rock layer below into a low-pressure flash tank on the surface. In this tank, which is at a much lower

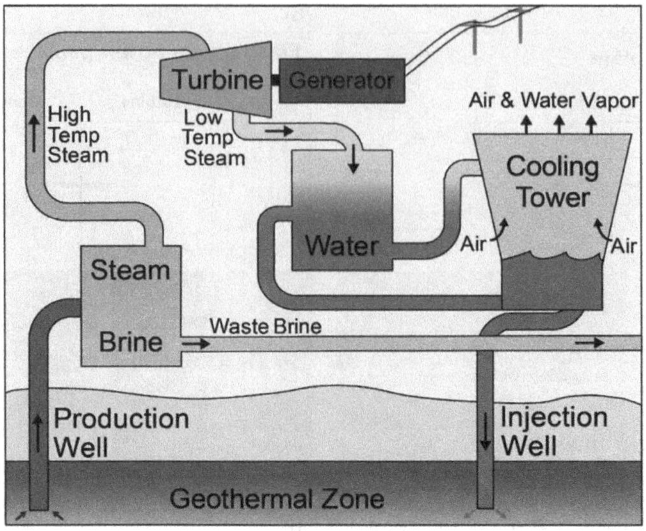

Figure 3-44. The majority of geothermal power plants are the flash steam type that uses high-pressure hot water from their production wells.

Source: Reprinted with permission from "Geothermal Energy Factsheet," Pub. No. CSS10-10, Center for Sustainable Systems, University of Michigan, September 2020, http://css.umich.edu/factsheets/geothermal-energy-factsheet.

pressure, part of the water vaporizes (flashes). That vapor is then used to drive the turbine generator. If any liquid remains in the tank, it can be flashed again in a second even lower-pressure tank to extract more energy. When the steam cools, it condenses and is injected back into the ground. Most geothermal power plants are of this type (Figure 3-44). The sizes of these plants range from 100 kW to 150 MW.

The control system that can be used for a flash steam power plant is shown in Figure 3-45. Here, a variable-speed pump transfers the hot water from the production well into the steam separator. The speed of the pump is set by the tank level (LC). The level control signal is corrected for steam pressure (PT) variations. This control configuration is called a *two-element* feed water system.

The high-pressure steam (HP) from the separator is sent to the steam turbine, the speed of which is controlled (SC) by throttling the flow of the high-pressure steam. The energy of this high-pressure steam rotates the turbine generator to generate electricity, which is sent to the grid or other users. The pressure of the low-pressure (LP) steam that is exhausted from the turbine is controlled by the pressure controller (PC), which modulates the cooling-water flow through the condenser.

The system can be optimized by realizing that the lower the low pressure is, the more energy the turbine will generate. Therefore, if there are no other considerations (such

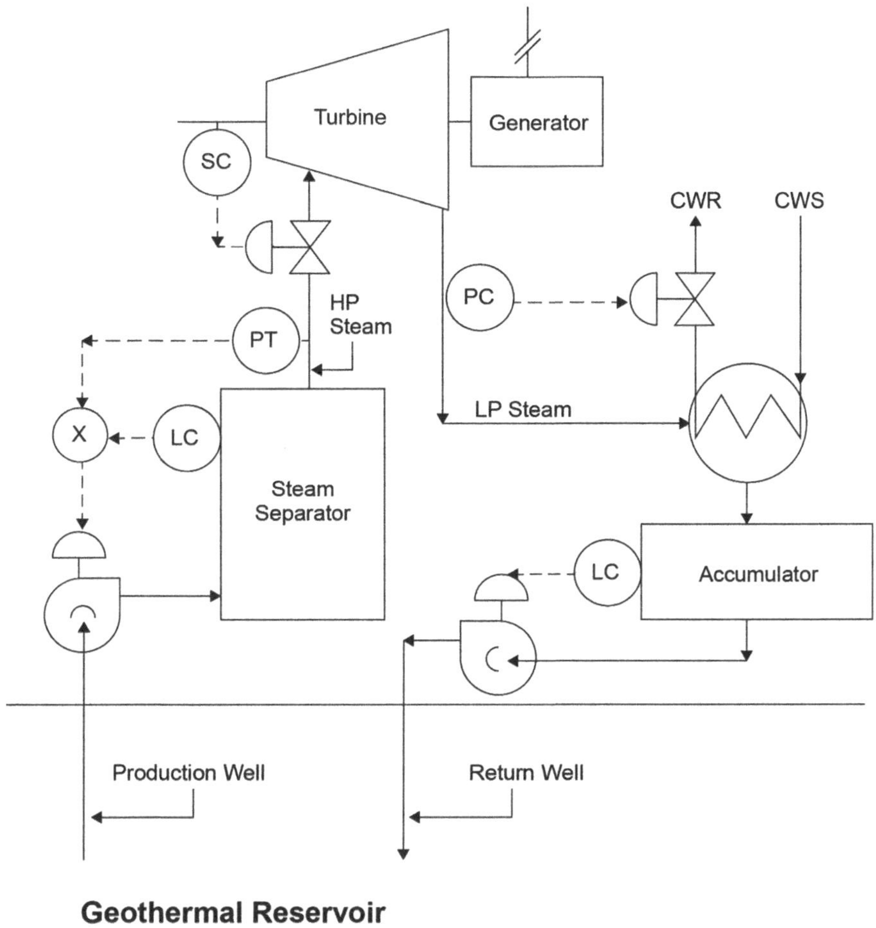

Figure 3-45. Optimized controls of a flash steam geothermal power plant. Abbreviations: CWR, cooling-water return; CWS, cooling-water supply; LC, level controller; PC, pressure controller; PT, pressure transmitter; SC, speed controller; X, multiplier.

Source: Courtesy of *Control Global.* Béla Lipták, "3 technologies to capture geothermal energy," October 1, 2020, Figure 2, https://www.controlglobal.com/articles/2020/3-technologies-to-to-capture-geothermal-energy/.

as the limited availability of cooling water), electricity production can be maximized by fully opening the valve on the cooling-water return (CWR). Yet, the system will be optimized only when the cost of operation is at a minimum. That minimum is found by making sure that the incremental energy gain obtained by lowering the turbine exhaust pressure exceeds the increase in the pumping cost of cooling water that is required to achieve that reduction. The condensate is collected in an accumulator tank and is returned under level control (LC) to the return well.

3.4.2.3 Binary-Cycle Power Plants

Binary-cycle power plants (Figure 3-46) transfer the heat from geothermal hot water to another liquid, which is turned into the steam that drives the turbine generator.

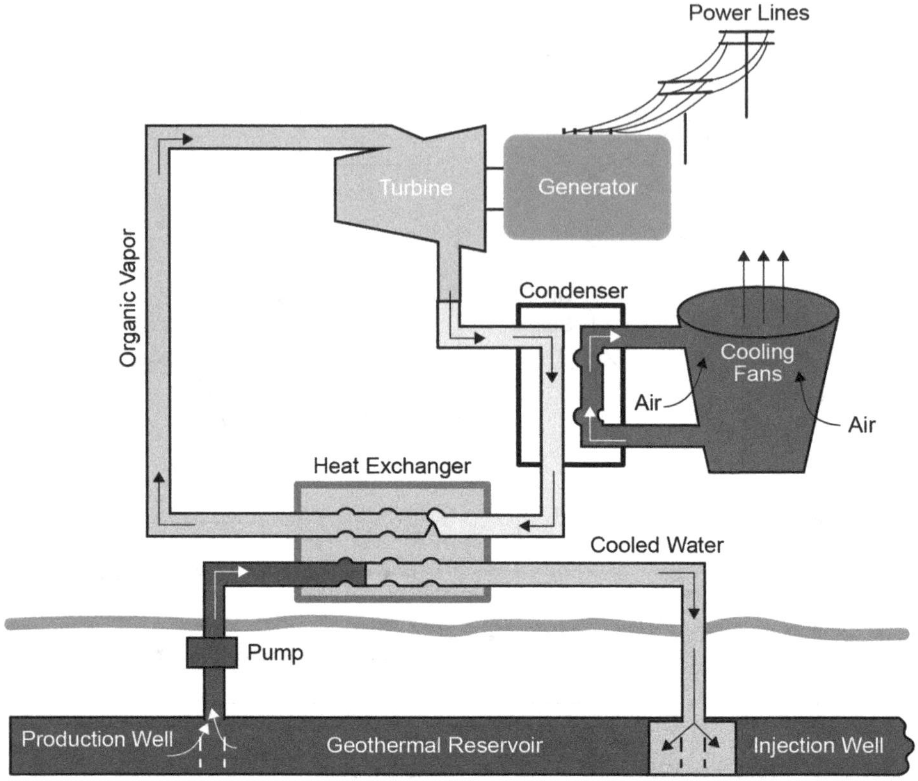

Figure 3-46. Binary-cycle power plant.

The binary-cycle power plant was first demonstrated in 1967 in the Soviet Union and introduced to the United States in 1981. This technology enables the generation of electricity from much lower-temperature resources than was previously possible. In 2006, a binary-cycle plant in Chena Hot Springs, Alaska, came online, producing electricity from a record-low fluid temperature of 57°C (135°F).

Binary-cycle geothermal systems differ from dry steam and flash steam systems in that the water from the geothermal reservoir never comes into contact with the turbine generator units. The two fluids used in binary-cycle systems are a low to moderately warm (temperature below ~150°C [300°F]) geothermal fluid and a secondary fluid with a boiling point much lower than that of the water that passes through a heat exchanger. Heat from the geothermal fluid causes the secondary fluid to flash to vapor, which drives the turbines and, subsequently, the generators.

Binary-cycle power plants are closed-loop systems; they release essentially no emissions (except water vapor) into the atmosphere. Because most geothermal heat sources are at temperatures below ~220°C (400°F), a significant proportion of geothermal electricity in the future could come from binary-cycle plants.

3.4.3 Geothermal System Costs

Geothermal energy is just as inexhaustible and zero-carbon renewable as solar energy, plus it has the added advantage of being a continuously available energy source. Its production costs are lower than those of fossil energy, but its front-end expenses are often high, especially if the production wells are deep and the ground is rocky. Still, the cost of generating geothermal power has decreased by 25% over the past two decades.

The thermal efficiency of geothermal plants is a function of the temperature of geothermal fluids, which tends to be low. On the other hand, their capacity factor is usually high because they provide a reliable and stable source, unlike the intermittently available wind or solar plants. As an example, The Geysers geothermal complex in California is the largest geothermal field in the world. It is located near San Francisco and spreads over an area of about 116.5 km^2 (45 mi^2) several kilometers above a large magma chamber.[70] The complex, comprising 18 geothermal power plants drawing steam from more than 350 wells, has an installed capacity of 1590 MW and produces an average of 955 MW.

The initial cost for a geothermal field and power plant is approximately $2,500 per installed kilowatt in the United States, probably $3,000 to $5,000/kW for a small (<1 MW) power plant. Most geothermal power plants can run at greater than 90% availability (i.e., producing more than 90% of the time). Higher-priced electricity justifies running the plant 98% of the time because the resulting higher maintenance costs are recovered.

On average, the LCOE of geothermal electric power is about $0.06/kWh, which is about half the LCOE of fossil-fuel- or nuclear-based electricity. The only renewable-energy-based electricity that is less expensive is wind energy, which has an LCOE of $0.04/kWh to $0.06/kWh. Although solar and wind power are the main drivers of the shift to green energy, other green energy sources such as biomass-based, geothermal, and hydropower are also becoming more competitive and can compete directly with fossil and nuclear supplies. (See Figure 3-6.)

Geothermal power is inexhaustible, cost-effective, reliable, sustainable, and environmentally friendly, but its use has historically been limited to areas near tectonic plate boundaries. Recent technological advances have dramatically expanded the range and size of geothermal resources, especially for applications in areas such as home heating.

---

70. "The Geysers Geothermal Field," US Geological Survey website, accessed November 29, 2022, https://www.usgs.gov/volcanoes/clear-lake-volcanic-field/geysers-geothermal-field.

As a result, the US geothermal industry alone already employs an estimated 25,000 workers.

3.5 Nuclear Energy

Although they have zero carbon emissions, nuclear power plants are undesirable because they are unsafe and expensive. I discuss them in this chapter dealing with green energy sources because they do not emit GHGs; therefore, some believe they are alternatives to fossil energy sources.

Fission breaks atomic nuclei into smaller pieces; fusion joins atomic nuclei together. The fission type in operation today obtains its energy by splitting a radioactive isotope of uranium. The main concerns about these nuclear plants are that they are potential nuclear bombs, and we have not found a way to dispose of their waste safely. The other type of nuclear energy is obtained by the fusion of two light nuclei into a heavier nucleus at a temperature that is approximately six times that of the Sun's core, which has safety implications. I oppose developing and using fusion power plants not only because their operating temperatures are very high but also because they are only in the research and development stage. We need clean energy sources now, not in the next century.

My discussion here focuses on the safety aspects of fission plants, with emphasis on cyberterrorism and nuclear waste storage. I describe some past accidents, as well as some potential safety improvements. I also detail my reasons for opposing building new nuclear power plants. I do not oppose temporarily operating existing plants because they can mitigate GW while the infrastructure is developed to convert to a green energy economy.

3.5.1 History and Status

In 1933, Hungarian-American physicist Leó Szilárd discovered the nuclear chain reaction.[71] However, it was not until 1951 that electricity was first generated by a nuclear reactor at an experimental station near Arco, Idaho. In 1954,[72] the Obninsk nuclear power plant in the Soviet Union became the world's first nuclear power station to generate electricity for a power grid, and in 1956, the Calder Hall nuclear power station[73]

---

71. Rosemond Ho, "Leo Szilard and the Nuclear Power Patent," Stanford University website, June 24, 2017, accessed December 1, 2022, http://large.stanford.edu/courses/2017/ph241/ho2/.

72. "Obninsk Nuclear Power Plant," Wikipedia, accessed October 27, 2022, https://en.wikipedia.org/wiki/Obninsk_Nuclear_Power_Plant.

73. "Calder Hall Nuclear Power Station," Wikipedia, accessed October 27, 2022, https://en.wikipedia.org/wiki/Calder_Hall_nuclear_power_station.

in the United Kingdom became the world's first full-scale electricity-producing nuclear reactor.

In addition to being used for power generation, globally, 240 reactors are being used in research and 150 in ocean transportation. As to the power generating units, there were 439 nuclear reactors in operation around the globe (93 in the United States) in 2022. Globally, 63 are under construction; 200 are permanently shut down, 12 of these are closed due to damage in accidents or serious incidents; *and over 200 commercial reactors are projected to be decommissioned by 2040. Globally,* of the nuclear power plants that have already been shut down (because of aging), only about 17 have been decommissioned and decontaminated to the extent that they no longer need radiation protection. In the United States, 28 plants have been shut down, and an additional 13 are in the process of being decommissioned. *Of the 439 operating reactors, 70% are nearing 40 years, and 25% of them are older.* As of April 2021, the United States had permanently shut down 39 nuclear power plants.

In 2020, 6 GW of additional nuclear-generating capacity was added to the grid (the comparable value for 2018 was 11.2 GW), and 5.4 GW of nuclear-generating capacity was removed, bringing the total global capacity to about 415 GW.

As of 2021, the installed capacity of all operating nuclear power plants worldwide was more than 393 GW[74] (about 120 GW in the United States). In 2020, nuclear reactors supplied 2553 TWh of electricity globally. Global energy consumption is about 176 PWh (600 Q), and global electricity consumption is about 27 PWh. The consumption of nuclear-generated electricity is about 2.6 PWh or 9.6% of the total electricity consumed. Looking at these numbers in another way, the total global nuclear electricity generating capacity of 415 GW is about 5.5% of the total global electricity generating capacity of 7.65 TW. In the United States, nuclear energy is the source of about 8% of the total energy consumed and about 20% of the electricity consumed.

The nations that are most dependent on nuclear electricity are France (75%), followed by Belgium and Slovakia (52%), whereas one of the least nuclear-dependent nations is China (approximately 2%). Yet, China and Russia continue to lead the world in new grid connections and construction launches. In fact, 22% of the nuclear reactors under construction are being built in China.

---

74. "World Nuclear Generation and Capacity," Nuclear Energy Institute website, updated June 2021, https://www.nei.org/resources/statistics/world-nuclear-generation-and-capacity.

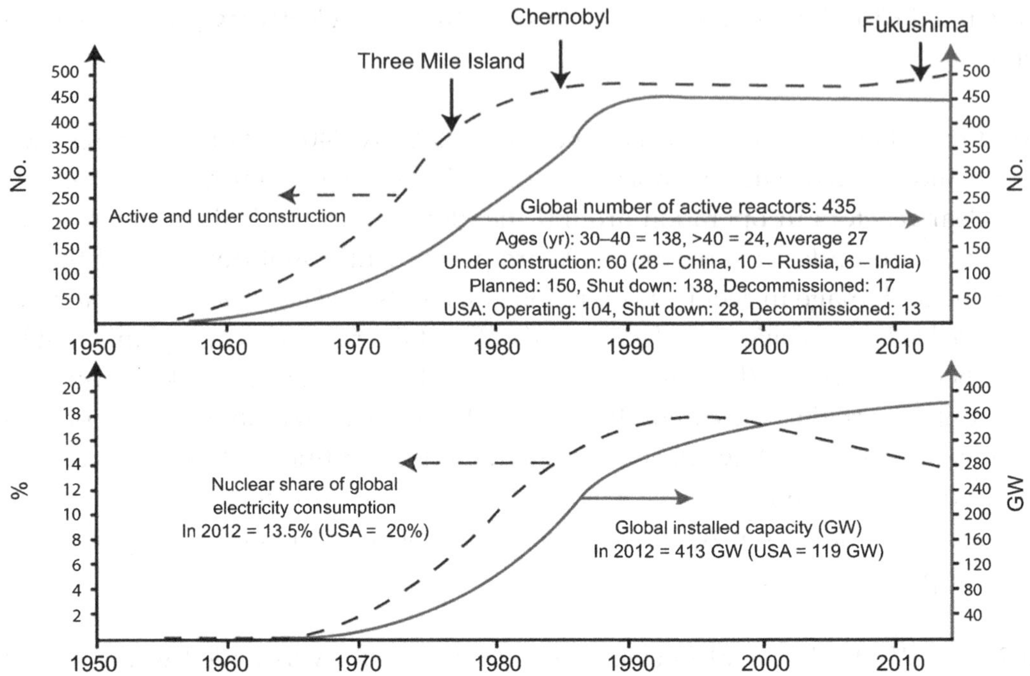

Figure 3-47. Evolution of nuclear reactor fleet and its contribution to global electricity generation until about 2012. (Top) The red line shows the approximate number of active reactors, whereas the blue dotted line shows the number of reactors that are active or under construction. (Bottom) The red line shows the current installed nuclear capacity, and the blue dotted line shows the approximate percentage share of global electricity consumption satisfied by nuclear energy.

Source: Béla Lipták, *Automation Can Prevent the Next Fukushima,* Figure 1-1 (Research Triangle Park, NC: ISA, 2014).

Figure 3-47 shows that new nuclear power plant construction slowed after the accidents at Three Mile Island and Chernobyl. The percentage of global electricity consumption met by nuclear power dropped from 18% in the early 1990s to 13.5% by 2012 and to approximately 8% as of 2021,[75] while American electricity consumption is about 20%. A plot of global energy consumption by energy source (1990 to 2040) according to the US EIA is shown in Figure 3-1.

Earlier nuclear plants were often located near populated areas (Figure 3-48) and used boiling water reactors (BWRs) that were designed for life spans of 30 to 40 years. Newer plants usually employ pressurized-water reactor (PWR) designs and have operating lives of 40 years or more. Today, the average age of the operating plants is over 30 years. The ages of 138 operating plants are between 30 and 40 years, whereas 24 plants are more than 40 years old. (The world's oldest operating nuclear power plant is the

75. "Energy consumption by source, World," Our World in Data website, accessed March 12, 2022, https://ourworldindata.org/grapher/energy-consumption-by-source-and-region?country=~OWID_WRL.

Figure 3-48. Nuclear power plants are often located near populated areas.

54-year-old plant in Beznau, Switzerland.) According to the US EIA, the average age of the US commercial nuclear power reactors that were operational as of October 31, 2020, was about 39 years. The oldest operating reactor is Nine Mile Point 1 in New York, which entered commercial service in December 1969. Today, due to the rise in energy costs, some nuclear power plants are allowed to operate until the age of 60 years.

Globally, to date, the cost of building nuclear power plants has totaled about $3 trillion (approximately $0.75 trillion in the United States). The expected cost of their eventual decommissioning will exceed $1 trillion. Therefore, the total cost of the nuclear power industry, including uranium mining, construction, decommissioning of plants, and temporary and eventually permanent nuclear waste storage, can be estimated at $5 to $10 trillion. For comparison purposes, the global gross domestic product (GDP) in 2019 was about $88 trillion (US gross domestic product [GDP] was approximately $22 trillion). In 2018, nuclear plants supplied 2563 TWh of electricity, up from 2503 TWh in 2017 (Figure 3-49). In 2020, nuclear reactors supplied 2553 TWh of electricity globally, down from 2657 TWh in 2019.

3.5.2 Reactor Designs

The purpose of this section is to show how automation can substantially improve the safety of nuclear power plants. After familiarizing the reader with the process, I describe the most likely potential causes of accidents and how automation can decrease these risks.

3.5.2.1 The Process

All traditional power plants are similar in that they use their energy sources to boil water and then use the energy content of the steam to generate electricity. In fossil-fuel plants, the fuel is continuously charged from the outside, whereas in nuclear

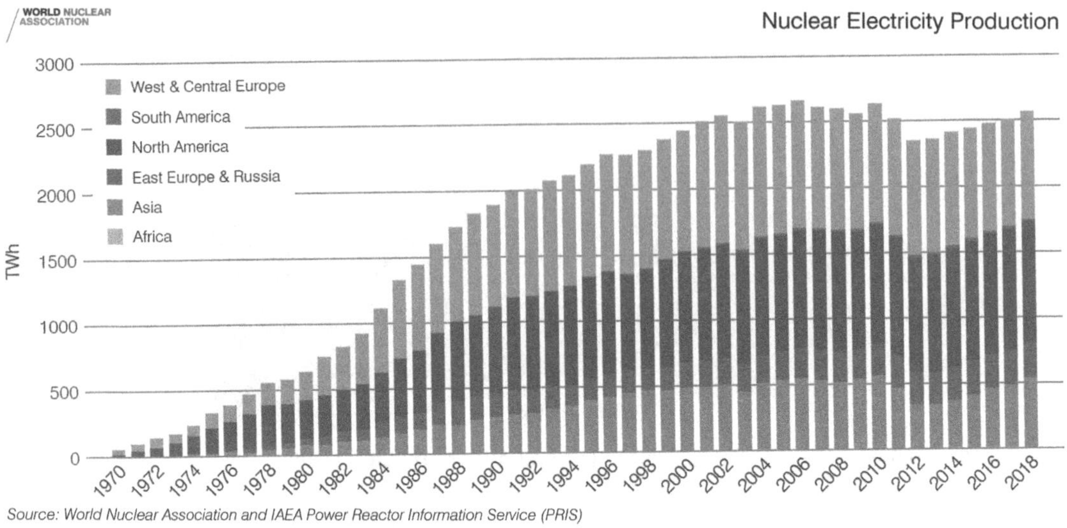

Figure 3-49. Nuclear power production worldwide.

Source: "Nuclear Power in the World Today," World Nuclear Association, updated March 2021, http://world-nuclear.org/information-library/current-and-future-generation/nuclear-power-in-the-world-today.aspx.

power plants, the fuel is intermittently charged and is stored inside the boiler (reactor). Because of this difference, the fuel flows to fossil-fuel boilers are easy to shut down, whereas turning off the nuclear process takes longer. In both cases, the controlled variable is the amount of electricity to be generated, and the manipulated variable is the rate at which the fuel is burned or the atoms are split. The means of throttling the manipulated variable in the case of fossil-fuel power plants is the positioning of the fuel supply opening. In nuclear plants, the means of changing this rate is repositioning the control rods in the core, where the fuel is stored.

Control rods are made of neutron-absorbing materials; they are placed in the core and are raised and lowered to modulate the reaction. If more heat is needed, the control rods are lifted from the core to allow more neutrons to split more atoms. If the heat release is to be reduced, the rods are lowered into the core to absorb more of the neutrons before they have a chance to come into contact with the uranium fuel (usually uranium-235, denoted ^{235}U). In case of emergency, the control rods are automatically shoved down into the core (using gravity, hydraulics, or mechanical spring force to do so), causing the chain reaction to nearly stop. This is called *scramming* the reactor.

In some reactor designs, moderators are provided to increase the chain reaction efficiency by slowing the neutrons so that the uranium fuel captures more neutrons and the reactor operates at higher efficiency. Graphite and heavy water are good moderators, and because they do not absorb too many neutrons, they can be used even with natural (non-enriched) uranium. Normal water is not as good a moderator, but it still can be used if the uranium fuel is enriched to about 4% ^{235}U.

3.5.2.2 Boiling Water Reactor

The nuclear power generation process is a relatively simple one. It consists of a number of energy transfer steps in series. In the earlier BWR designs (Figure 3-50), the hot fuel rods served to heat the water circulating through the core and turn this water into steam. The steam was sent to the turbine building, where the turbine generators converted the steam energy into electricity. The condensate from the turbines was returned to the reactor in the form of cooling water. In this cycle, about two-thirds of the energy was lost, and these reactors required about 3 MW_t (megawatt thermal) to generate 1.0 MW_e (megawatt electric).

In the BWR design, the generated steam can be radioactive. Therefore, when this steam is sent from the "contained" reactor building into the "uncontained" turbine building,

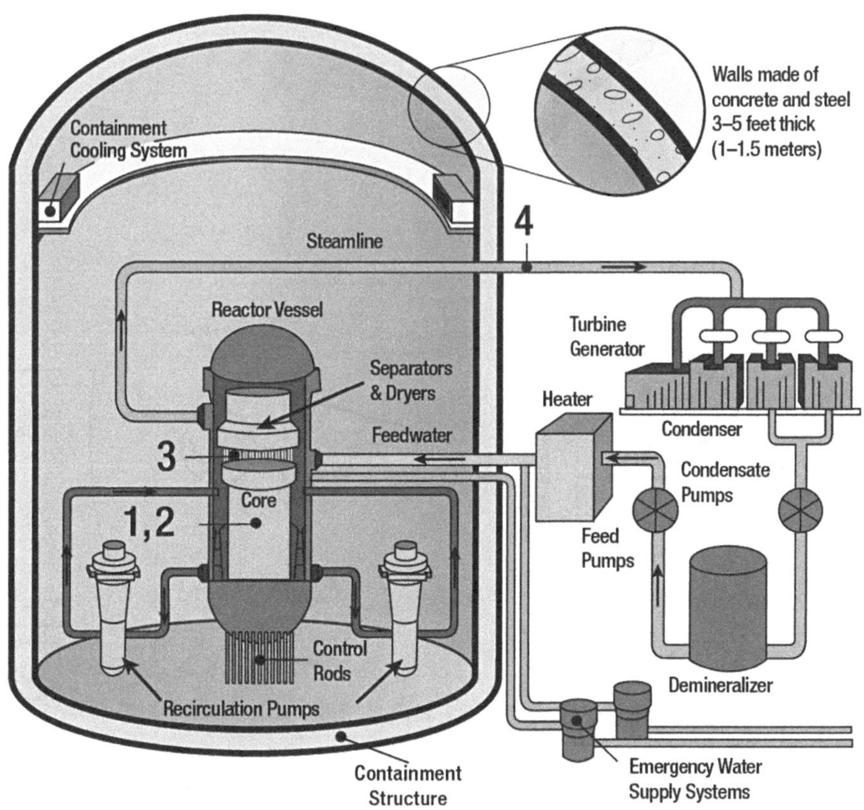

Figure 3-50. Schematic of a BWR.

Source: "Boiling Water Reactors," US Nuclear Regulatory Commission, last updated January 15, 2015, https://www.nrc.gov/reactors/bwrs.html.

radioactivity can enter with it. The reactor pressure vessel (RPV) of a BWR is usually inside the primary containment, which often has the shape of an inverted light bulb. It is a strong structure (in most cases, missile- and earthquake-proof) and is often partially or fully filled with inert gas or with low-oxygen-content inert gas. This primary containment is also called the *dry well* because its secondary purpose is to collect and vaporize any water that might leak from the reactor equipment and send the resulting leakage steam into the wet well, where it condenses.

3.5.2.3 Pressurized-Water Reactor

Pressurized-water reactors (PWRs) improve the safety and efficiency of the previously described BWR designs by inserting another heat-transfer loop (Figure 3-51) between the reactors and the turbines. In this design, the heat from the fuel rods is transferred to a circulating high-pressure water loop, which does not leave the reactor building but carries this heat into a boiler inside it. This design can provide better efficiency if the pressure of the generated steam is increased. It improves safety because both the

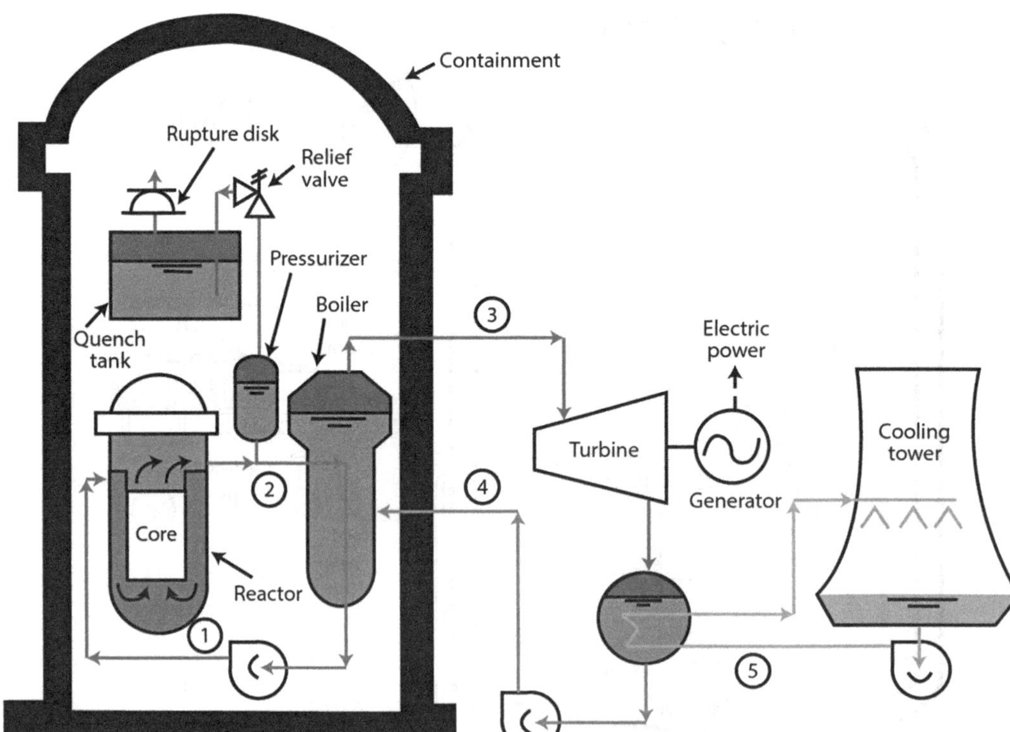

Figure 3-51. The PWR design is an improvement over the BWR design because the potentially radioactive high-pressure water loop is kept inside the reactor building; therefore, the steam sent to the turbine building does not contain radioactive isotopes.

Source: Béla Lipták, *Automation Can Prevent the Next Fukushima*, Figure 1-6 (Research Triangle Park, NC: ISA, 2014).

circulating high-pressure water loop and the boiler are kept within the primary containment, and the generated steam does not carry radioactive isotopes into the turbine building.

In this design, the hydrogen and gaseous radioactive fission products can directly enter the reactor building once the RPV is breached. If the seals deteriorate as a result of overheating, these gases can also enter the turbine buildings. The deterioration of the seals is probable if radioactivity is detected in the turbine building, and it is an indirect indication that a meltdown is in progress.

3.5.2.4 Newer, Safer Design

Although there is no such thing as a safe nuclear reactor, the safety of the present designs could be improved. The main cause of all nuclear accidents is the loss of coolant. In all present designs, pumping is required to get the cooling water into the reactor, and some form of energy is required (electric, steam, diesel, etc.) to operate the valves and pumps. Accidents occur when these energy sources fail. Therefore, a safer design would result if cooling could be provided for the nuclear reactor without an external energy source.

Before any nuclear accident (no matter the cause), the reactor first starts to heat up, and the rise in temperature causes thermal expansion. Without any external energy source, this thermal expansion (like mercury in a thermometer) can be used to open the coolant valve, and if the reactor is located below the cooling-water source[76] (lake, river, or artificial water storage), gravity would send the water into the reactor. Therefore, a safer nuclear reactor design would result if cooling started automatically whenever the reactor warmed to a certain temperature. In this design (Figure 3-52), if the temperature rose, the thermal expansion fluid would expand and the safety valve would open. As a result, gravity would send the cooling water down into the reactor. In this way, meltdown could not occur under any condition because nothing and no one could turn off either gravity or thermal expansion. Naturally, this is only a design concept of mine; no such nuclear reactor control system exists today.

3.5.3 Major Accidents

Since Chernobyl, 57 nuclear accidents have occurred, including 22 meltdowns.[77] Here, I focus on only three major accidents (Three Mile Island, 1979; Chernobyl, 1986; and

---

76. Edward Teller recommended that every reactor be 20 to 25 m below its cooling water source.

77. For additional information, see: "Nuclear and radiation accidents and incidents: Nuclear meltdown," accessed March 12, 2022, https://en.wikipedia.org/wiki/Nuclear_and_radiation_accidents_and_incidents#Nuclear_meltdown.

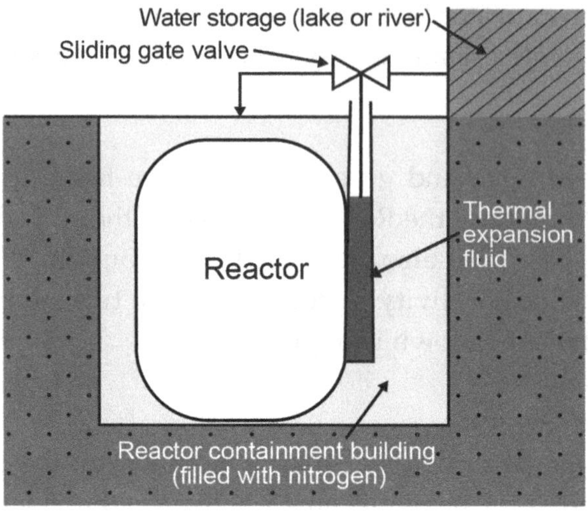

Figure 3-52. A safer reactor design is one in which reactor cooling is triggered by such energy sources (gravity and thermal expansion) that cannot be turned off.

Source: Courtesy of *Control Global*. Béla Lipták, "Cybersecurity of nuclear reactors," *Ask the Experts Blog*, February 27, 2017, https://www.controlglobal.com/articles/2017/cybersecurity-of-nuclear-reactors/.

Fukushima, 2011). The first two (Three Mile Island and Chernobyl) are discussed only briefly because they occurred further in the past. As Fukushima is a more recent accident, I will discuss it in more detail. In all of these discussions, I emphasize that the whole industry is aging (the average age of operating reactors is more than 30 years), state-of-the-art automation is often lacking, and, essentially, no reactors are designed for safe shutdown in the case of long-term and simultaneous loss of external (grid) and internal (backup) electric power supplies.

I should also note that human error is the main cause of nuclear accidents, such as those at Three Mile Island and Chernobyl. In contrast, the Fukushima accident could have been prevented if the operators had shut down the reactors as soon as they learned that an earthquake occurred and if they noticed that the water from the "wet leg" of their cooling-water level detector has evaporated.

I emphasize the critical role of the operators in guaranteeing nuclear plant safety because today Ukraine's Zaporizhzhya nuclear power plant is in a war zone; it operates on its last remaining external power source, and its operators—harassed by the Russian occupiers—have either left or are working under unsafe conditions.

3.5.3.1 Three Mile Island

At 4 a.m. on March 28, 1979, Unit 2 of the 900 MW reactor at Three Mile Island (TMI-2) in Pennsylvania experienced a partial core meltdown. Between 13 and 43 million

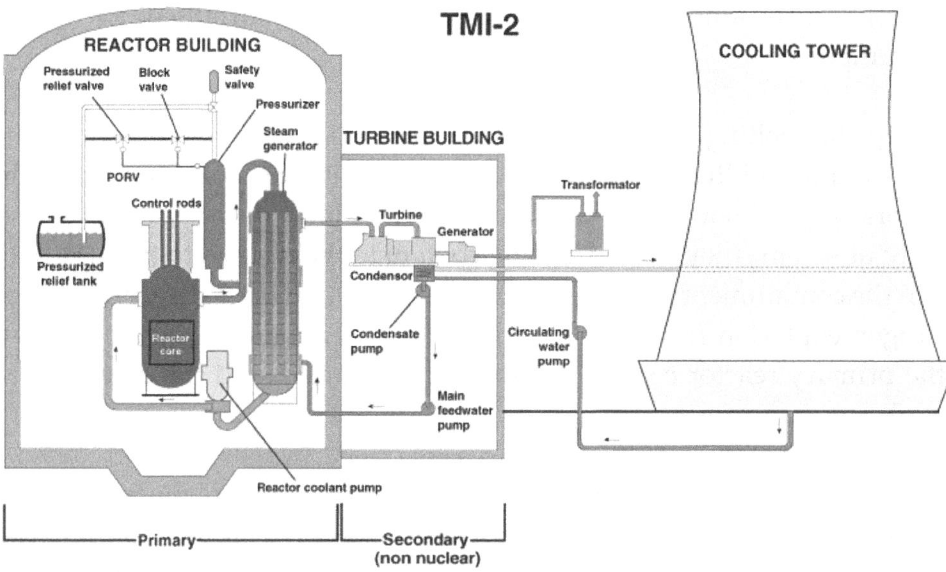

Figure 3-53. A simplified view of the equipment and piping configuration of the Three Mile Island nuclear power plant.

Source: "Three Mile Island Accident," World Nuclear Association, accessed April 23, 2020, http://www.world-nuclear.org/information-library/safety-and-security/safety-of-plants/three-mile-island-accident.aspx.

curies of radioactive krypton gas was released, half the core melted, and 90% of the fuel rod cladding was destroyed. The maximum offsite radiation reached 83 millirems (mrem), but the radiation dose received by the community was small.

The process consisted of three heat-transfer loops (Figure 3-53). The first or primary heat-transfer system transferred the heat generated by nuclear fission into the circulated high-pressure cooling water, which carried it into the steam boiler. There, the high-pressure and high-temperature primary water transferred the heat into the secondary water loop where steam was generated. The steam turned the turbines, and the waste heat from the turbine condensers was sent to the atmosphere by the cooling towers.

The accident occurred because operators working on a demineralizer at 4 a.m. accidentally admitted water into the instrument air supply to several pneumatic instruments. As a consequence, they accidentally closed the valve on the suction side of the secondary water pump. In addition, the control system did not work because the secondary water flow stopped. Therefore, plant management was not warned of the heat, nor was there an automatic interlock to start a backup cooling system. The heat from the pressurized-water loop was no longer being removed, which caused the water to boil, and the badly trained operators interpreted the resulting rise (as water starts to boil, its level rises) as if the amount of water increased (in fact it dropped) and shut down the water pump.

This sped up the melting of the core. By 6:00 a.m., the reactor core had overheated, and the zirconium cladding of the fuel rods reacted with the steam to form hydrogen. The operators still did not realize that the low water level in the reactor exposed the core. Finally, at 11 a.m., they started adding coolant to the reactor. In the afternoon, the pressure in the containment building spiked to 29 psig (almost 2 atm), probably caused by a hydrogen explosion from the reaction between zirconium and steam/water. At 8 p.m., the primary reactor coolant pumps were restarted, and the core temperature began to fall.

The Three Mile Island accident did result in the partial meltdown of the fuel rods, but fortunately, it did not result in the release of large amounts of radioactive isotopes. The local population of 2 million people received only an insignificant dose (1 mrem) of radiation. Yet, this was an important accident because it showed the degree of operator unreliability (admitting water into air supplies, closing valves that should be open, and not understanding the swelling effect of boiling). It also showed that it is dangerous to depend on manual controls because automatic safety controls would have prevented the accident.

3.5.3.2 Chernobyl

This accident at the Reactor Bolohoj Moshosztyl Kanalnyj (RBMK) nuclear power plant at Chernobyl in Ukraine occurred on April 26, 1986, at 1:23 a.m., an hour or so after the shift change of the operators. The meltdown was caused by a runaway condition during which it is estimated that the power generation reached more than 100 times the design capacity and caused a steam explosion that blew the 2500 T top off the reactor. This was followed by hydrogen explosions, and as air entered the reactor, the graphite in the core also ignited.

As a result of the explosion and fire, 20 million curies of radioactivity were released. This is 30 times the nuclear fallout that occurred at Hiroshima and Nagasaki. The 1986 meltdown destroyed Unit 4, killed 31 operators and firefighters, and required the establishment of an exclusion zone 30 km in diameter from which more than 100,000 people were evacuated. The released radioactivity cloud spread as far away as Norway, and the atmosphere in the area is expected to remain radioactive for some 300 years (the ground even longer). The nuclear power plant site cleanup is scheduled for completion in 2065.

Figure 3-54 schematically describes the design of the Russian RBMK reactor. The reactor core consisted of hundreds of pressure tubes (in the figure, only one is shown for

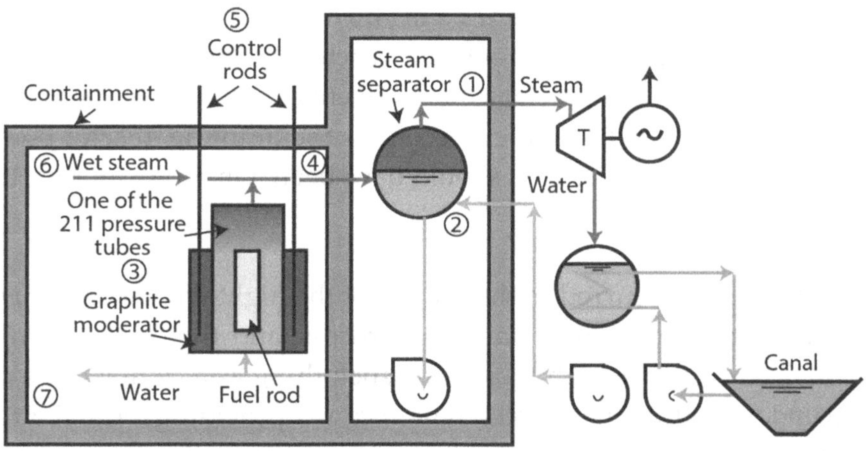

Figure 3-54. Numerous design errors existed at Chernobyl: (1) no automatic pressure relief valve, (2, 3) no reliable level controls, (4) no pressure relief on the roof, (5) operators should not have been able to disable scramming controls, and (6, 7) hydrogen detection should have triggered venting and the introduction of an inert gas.

Source: Courtesy of *Control Global*. Béla Lipták, "Why Nuclear Needs Process Automation," January 10, 2013, Figure 2, https://www.controlglobal.com/articles/2013/liptak-pauto-control-chernobyl/.

simplicity). The low-concentration (2% ^{235}U) fuel rods were located in the center of the pressure tube (which was surrounded by a solid graphite neutron moderator) and were immersed in water that was pumped from the bottom up the tube. The fission (the splitting of the ^{235}U molecule, which releases heat and neutrons) in the fuel rods heated the water until it turned to steam, which was sent to the steam separator and then to the turbine generators.

At Chernobyl, the events that caused the meltdown occurred during a safety test. This test caused the disaster because the operators disabled most of the automatic safety systems and the reactors were inherently dangerous (poorly designed). The purpose of the test was to determine whether, in the case of a power grid failure (loss of the external electric power supply to the plant), the power generated by the turbines (the residual rotational energy or inertia) would be sufficient to provide the electric power needed until the backup diesel generators could take over. The Chernobyl power plant had been in operation for 2 years, and the operators still did not know if it could ride through the first 60 to75 seconds of a total loss of electric power so that the backup diesel generator would have time to take over. The purpose of the test was to determine whether it could.

The basic design error in these reactors was that the sudden insertion of the control rods (scramming) could initially cause a dramatic and self-accelerating power surge. An even more significant flaw was that the lower part of each control rod was made of graphite, and the rods themselves were 1.3 m (4.3 ft) shorter than necessary. With this design, when the rods were inserted into the reactor, the graphite parts initially displaced some water (which absorbs neutrons), effectively causing fewer neutrons to be absorbed initially. Thus, for the first few seconds of scramming, reactor power output was *increased* rather than *reduced* as it should have been. On top of all this, the control rods also jammed.

The Chernobyl accident occurred in the middle of the night when the plant was operated by a crew that did not understand the process, and the control computer was disabled. It showed that (although there is no such thing as a completely safe nuclear power plant) understanding the process dynamics and providing redundant automatic controls would have prevented this accident. The resulting disaster also showed that the use of manual controls and overrides must be minimized to maintain safe operation, and operators must not be allowed to bypass the redundant automatic safety interlocks.

3.5.3.3 Fukushima

The accident at Fukushima released up to 10,000 TBq (terabecquerels)[78] of radioactive iodine-131 per hour during the first days, reaching a total of 630,000 TBq, about one-tenth of the 5.2 million terabecquerels released at Chernobyl. The accident was caused by an earthquake that occurred at 2:46 p.m. on March 11, 2011. The earthquake generated a tsunami, which hit the plant at 3:30 p.m. This time difference of 44 minutes provided a time window that, if the plant had been automatically controlled, could have been sufficient to not only shut down the reactors but also activate all the cooling systems.

The tsunami height was 15 m (49.2 ft), 5 m (16.4 ft) above sea level (Figure 3-55). The turbine building was not strong enough to keep out the seawater, and its basement flooded. The reactor building was strong enough to withstand the seawater invasion because it was designed to prevent the escape of radioactivity even under such conditions.

The reactors themselves were automatically scrammed by the earthquake-detection-based reactor trip interlock. Because this took some time, the reactors did not shut down immediately, and the heat generation of the fuel rods was reduced only to the level of the 5% "decay heat" (the heat from natural radioactive decay that continues to

78. The becquerel is a unit used to measure the rate of radioactive decay. Radioactive decay is measured in terms of the rate at which the atoms making up a radioactive substance are transformed into different atoms. One becquerel (1 Bq) is equal to one of these atomic transformations per second.

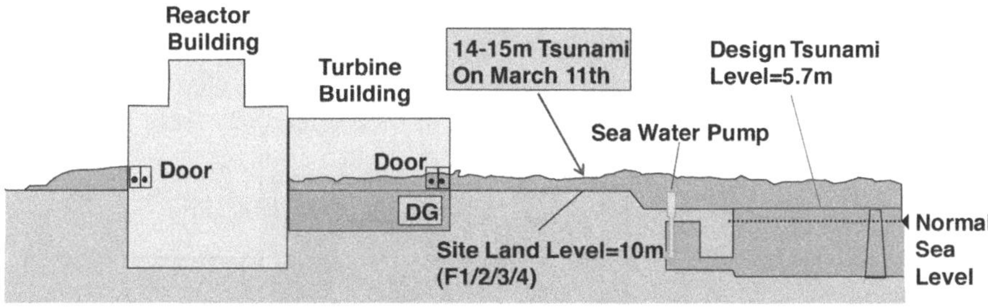

Figure 3-55. The tsunami waves reached 5 m (16.4 ft) above sea level. They flooded the turbine building, where the backup diesel generators (DG in the figure) and batteries were located, but they did not enter the reinforced reactor building.

Source: Courtesy of Ritsuo Yoshioka, President, Japan Functional Safety Laboratory.

be generated even after a reactor is shut down). Shortly after the external electricity supply from the grid failed, the internal backup systems (battery and diesel generator located in the flooded basement of the turbine building) also failed, and the cooling-water pumps stopped.

The Fukushima plant (and most other nuclear power plants worldwide) was not designed to safely shut down under conditions of long-term failure of both internal and external electricity supplies. This is unfortunate because the steam generated by the 5% decay heat would have been enough to drive the main cooling-water pumps.

After the tsunami hit, the main cooling-water pumps stopped, and all electricity was lost. As a result, the operators had to work with emergency lights on the Unit 1 side of the control room and in total darkness on the Unit 2 side unable to read vital reactor parameters or observe the results of their interventions. One hundred minutes after the earthquake, the water level recorders for Unit 1 also stopped, and the shift supervisor finally declared an emergency.

There was another 4- to 5-hour window of opportunity to prevent a meltdown. At that time, the cooling-water level in Unit 1 was still 4 m above the top of the fuel rods. By 9 p.m., the level had dropped by 8 m, fully uncovering the fuel rods. During that same time period, the core temperature increased from about 300°C (572°F) to nearly 3000°C (5432°F). By the morning of the next day (March 12), the reactor core had melted, dropped to the bottom of the reactor vessel, and probably either ruptured or leaked through it into the primary containment vessel (PCV).

The RPV at Fukushima (Figure 3-56) produced steam at a pressure of approximately 70 atm.[79] If the pressure increased to 75 atm, the steam was relieved into the wet well

79. Conversions among some common units for pressure are as follows: 1 atm (standard) = 14.7 psi = 1.014 bar = 101,325 Pa = 0.101 MPa.

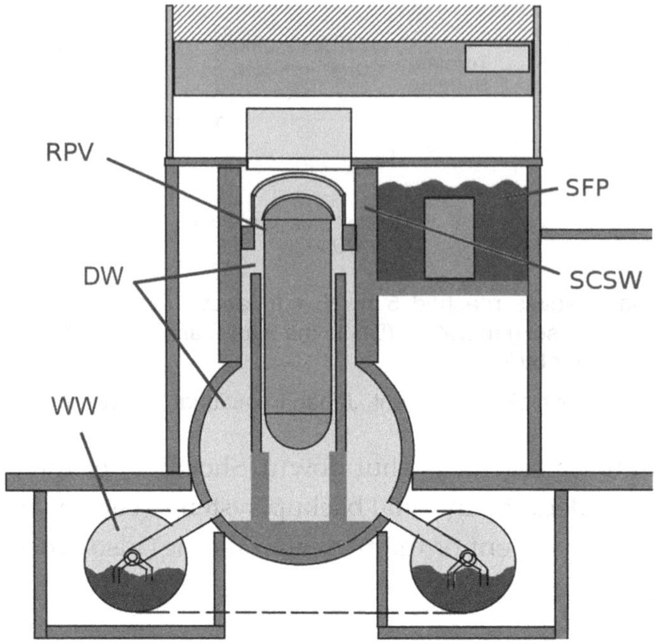

Figure 3-56. Cross section of the Fukushima containment used in Units 1 to 5.

Source: "BWR Mark I Containment sketch with downcomers.png" by User:84user, March 16, 2011, Public Domain, https://commons.wikimedia.org/w/index.php?curid=14616400.

(WW or torus), which contained 3000 T of water, serving to condense the steam. The spent fuel was stored in a water pool (spent fuel pool—SFP) above the RPV. The reactor was contained in a dry well (DW) filled with nitrogen. This PCV was protected from overpressure by a rupture disk set to relieve at 5.5 atm. The secondary containment shield wall (SCSW) was the wall of the building itself, which was filled with air under a slight vacuum. The roof of the building was weakened so that it would rupture and relieve the pressure in case of an explosion, thereby protecting the building from collapsing.

Given this somewhat outdated, manually controlled design and poorly installed, unreliable instruments, it is understandable that the operators were guessing half the time about what was happening in the plant. Now imagine that the lights went out because both the battery backup and the diesel generator were down, and imagine that these panicked people were trying to figure out what to do in the dark using flashlights. I should also emphasize that automatic shutdown systems could have continued functioning even when external and internal electric power supplies were lost.

Finally, the process to measure the cooling-water level failed because the condensate pot and the reference leg of the level gauge were located inside the dry well, and the water in the reference leg boiled off as the temperature increased. Figure 3-57 shows

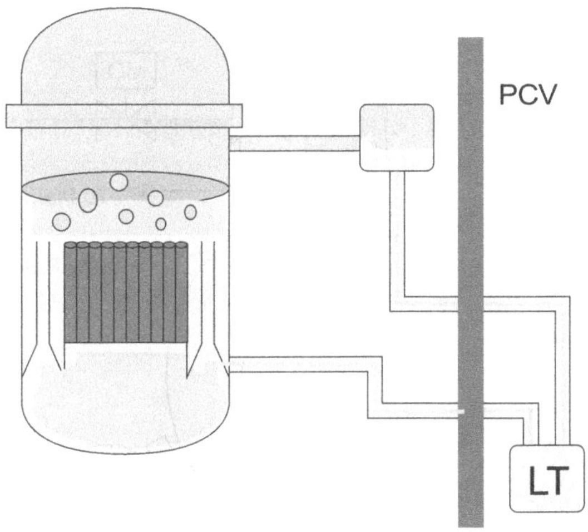

Figure 3-57. Water level detection failed because, as the temperature inside the PCV increased, the condensate in the reference leg of the level transmitter (LT) boiled off, causing the gauge to report a rising level when, in fact, the level was dropping.

Source: Courtesy of Ritsuo Yoshioka, President, Japan Functional Safety Laboratory.

that the level was detected with a differential pressure gauge. The high-pressure side of this gauge was connected to the steam region in the RPV through an uninsulated condensate pot, which (under normal conditions) would have contained condensate at ambient temperature. The level was measured by comparing the weight of this reference leg to the weight of the water level in the reactor. With this reverse configuration, as the level inside the reactor (and therefore the weight of the water column in the RPV) decreased, the detected level (pressure difference) should have increased.

Therefore, when the reactor overheated, the condensate in the reference leg started to boil off, and the differential pressure cells started to falsely report that the level was rising. As a result, the operators made the disastrous decision to disable the high-pressure coolant injection system, stopping the cooling of the reactor at the very time when it was essential to keep cooling it.

The safety system at Unit 2 operated properly and saved Reactor 2. The reactor-core-isolation cooling system (although it had to be started manually) operated correctly because its cooling-water pumps were driven by steam turbines, and steam was available (Figure 3-58).

The other cooling system that activated automatically and worked for a short time—with limited capacity—was the isolation condenser system, which utilized gravity (elevation) as its energy source. In the Fukushima design, the actuation of the system required opening motor-operated isolation valves (MO in Figure 3-58), which

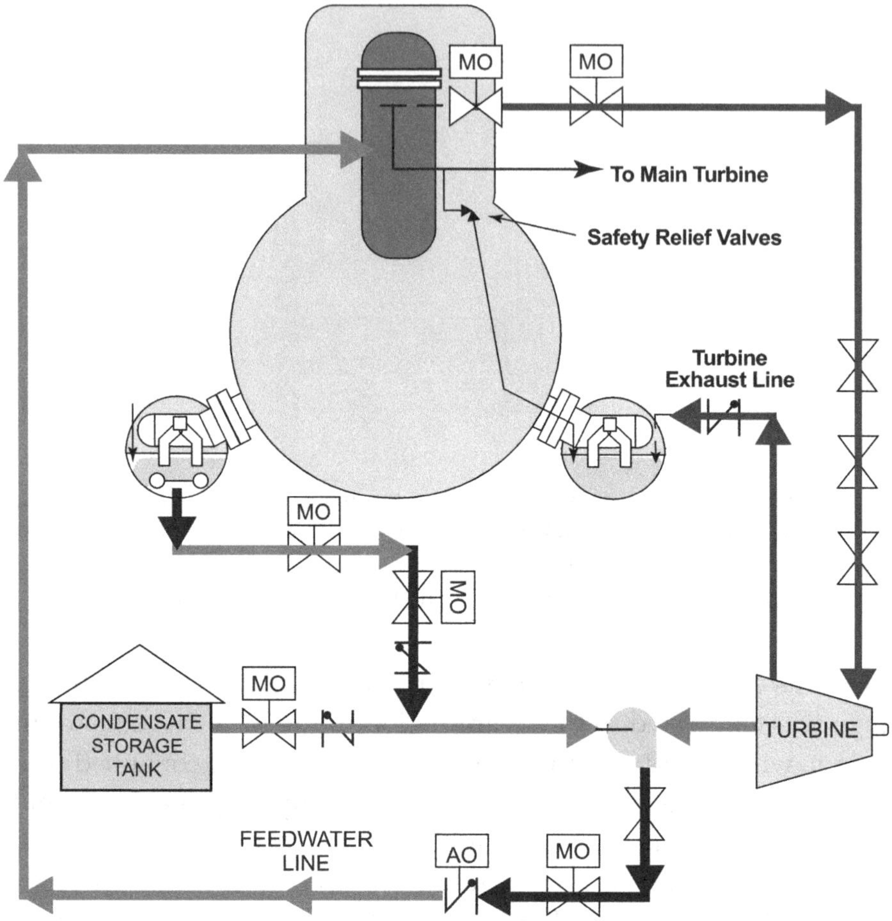

Figure 3-58. The cooling-water pumps of the reactor-core-isolation cooling system operated correctly because they were driven by steam turbines (which did not require electricity). Abbreviations: AO, air-operated isolation valve; MO, motor-operated isolation valve.

Source: Courtesy of Ritsuo Yoshioka, President, Japan Functional Safety Laboratory.

introduced safety limitations (Figure 3-59). The reason for providing these double isolation valves was to allow the operators to control the rate of pressure and temperature change because fast changes can crack the reactor walls. When the control room crew noted that the reactor pressure and temperature were dropping too fast in Reactor 1, they closed the isolation valves which stopped the cooling of the reactor.

As the reactor temperature increased from about 300°C (572°F) to nearly 3000°C (5432°F), the pressure increased as well. When that pressure reached the set point of about 75 atm, the pressure safety valves (PSV in Figure 3-60) on the reactor started to open, and the steam was released into the wet well. These valves were designed to reseat (close) automatically at about 69 atm, and they probably did.

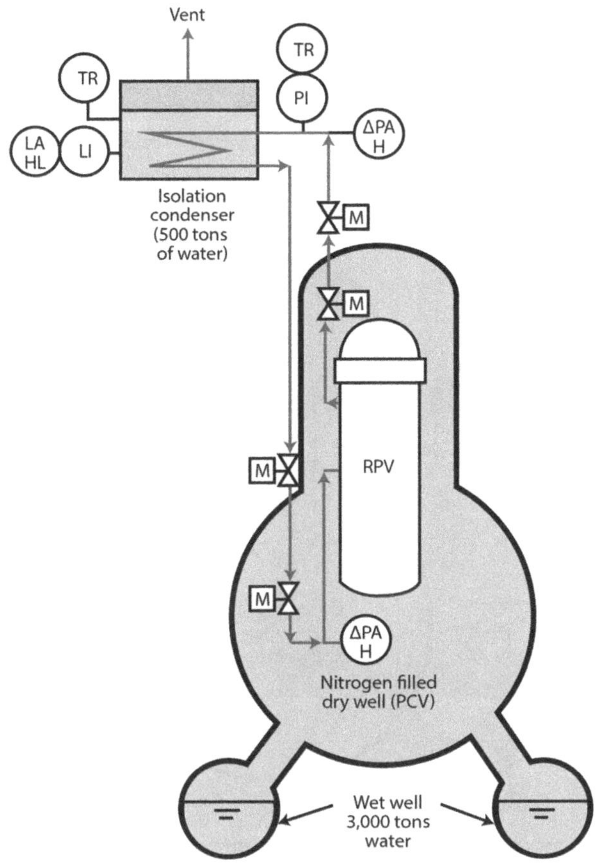

Figure 3-59. The isolation condenser system started automatically but failed because the isolation valves (M) were accidentally set to a minimum opening (provided little cooling) and were turned off later. Abbreviations: H, high; HL, high-level; LA, level alarm; LI, level indicator; PI, pressure indicator; TR, temperature recorder; ΔPA, differential pressure alarm.

Source: Courtesy of *Control* magazine, Béla Lipták, "Automation Could Have Prevented Fukushima, 2," April 30, 2013, Figure 3, https://www.controlglobal.com/articles/2013/liptak-automation-fukushima-prevention-2/.

The secondary containment (the reactor building) was filled with air and operated under a slight vacuum to prevent out-leakage. The roof and side walls of the top floor were intentionally weakened so that they would rupture (as they did) if the pressure inside increased (such as in the case of a hydrogen explosion) but the building itself would not be damaged.

3.5.3.4 Hydrogen Explosion

Hydrogen formed when the fuel rod cladding melted. For an explosion to occur, the generated hydrogen had to accumulate in an area with oxygen and an ignition source. At Fukushima, the fuel rods in Reactor 1 probably started melting about midnight

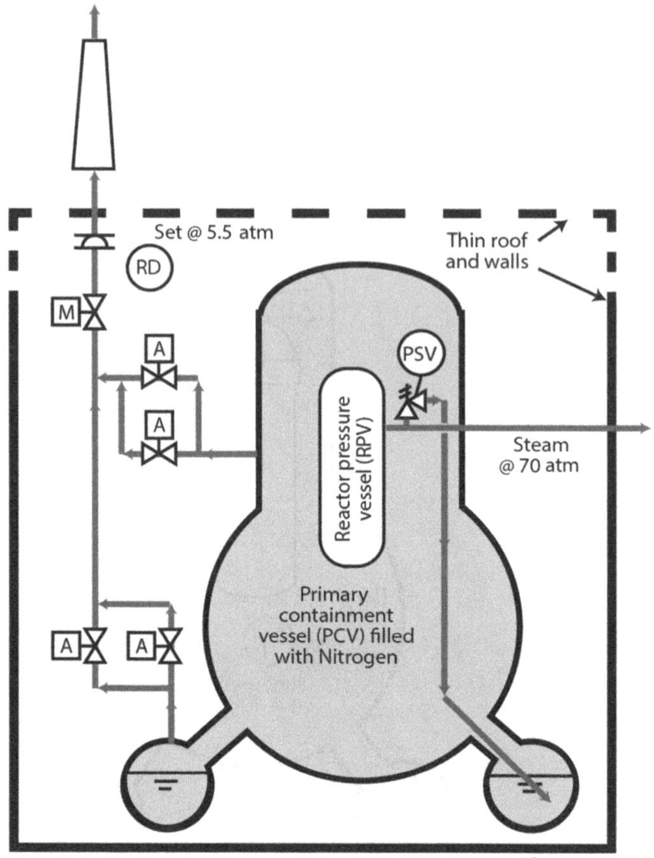

Figure 3-60. The cracking of the PCV could have been prevented if the rupture disk (RD) had ruptured as soon as the pressure in the PCV reached 5.5 atm, but it did not because block valves (A and M), which should not have existed in the first place, blocked the flow.

Source: Courtesy of *Control* magazine, Béla Lipták, "Automation Could Have Prevented Fukushima, 2," April 30, 2013, Figure 2, https://www.controlglobal.com/articles/2013/liptak-automation-fukushima-prevention-2/.

on March 11; the hydrogen explosion occurred some 13.5 hours later at 3:36 p.m. on March 12. In the other reactors, the accumulation of hydrogen took longer (days) before similar explosions occurred. So, how did the hydrogen find its way to an area where oxygen was present, and how did it accumulate there?

First, the hydrogen had to leave the RPV. This could have started by the PSV sending steam into the wet well. Next, hydrogen, being a low-molecular-weight gas (that is not soluble in water), is moved from the wet well into the PCV. Later, after the fuel rods had melted and leaked out from the bottom of the RPV, the melted material accumulated at the bottom of the PCV, and hydrogen continued to form there (Figure 3-61).

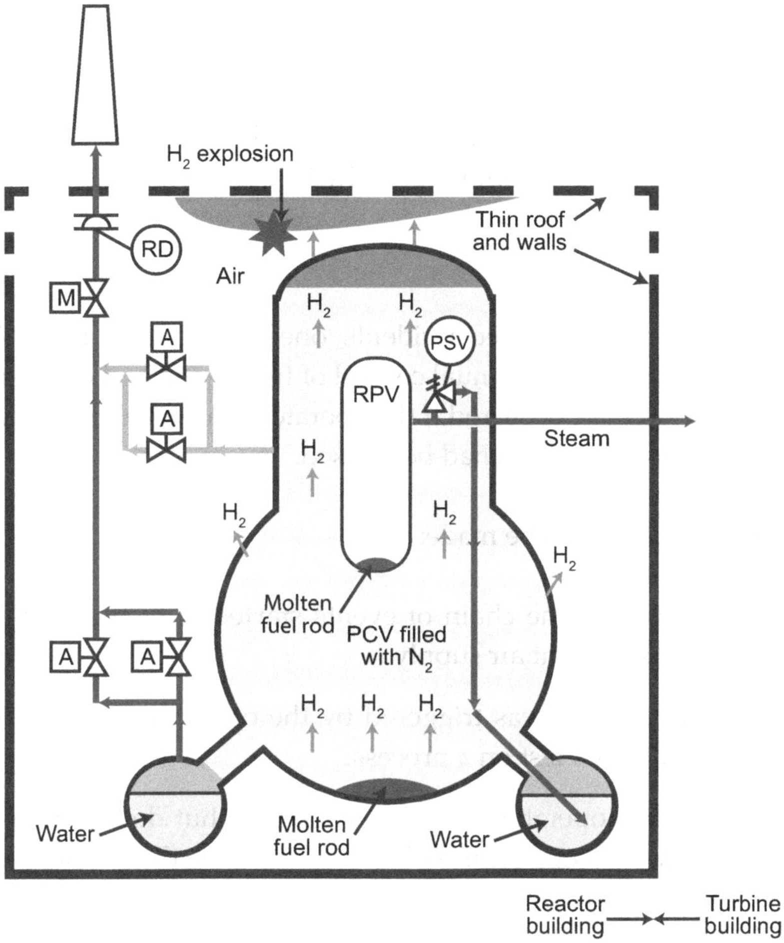

Figure 3-61. The hydrogen explosion at Fukushima could have been prevented if the hydrogen from the wet well of the PCV had been vented through the rupture disk (RD, set at 5.5 atm), but it was not because the isolation valves (M and A) blocked that flow path. Light blue represents water, red represents molten fuel rods, and dark blue represents hydrogen.

Source: Courtesy of *Control* magazine, Béla Lipták, "Automation and Fukushima – Part 3," *Control* magazine 26, no. 7 (July 2013): 14–16.

When the hydrogen was inside the PCV, it could not explode because the building was filled with nitrogen. Under these conditions, the low-molecular-weight hydrogen rose and accumulated in the upper part of the PCV (Figure 3-61). It could not be vented from there because the vent connection was below. Therefore, it could only escape into the reactor building by leaking through cracks in the damaged walls of the PCV.

The unfortunate aspect of this sequence of events is that hydrogen did not need to accumulate either in the PCV or the reactor building; it could have been vented directly into the atmosphere. The hydrogen was not removed by venting because the isolation valves between the PCV and the RD could not be opened. If the RD had been on the top of the PCV and vented directly to the outside, the explosions would have been prevented.

Naturally, a safe and automatic design would have also included hydrogen detectors that would have signaled the accumulation of hydrogen and immediately initiated the venting, even before the RD ruptured.

3.5.3.5 Conclusions

By reviewing the causes of the three accidents, one can conclude that they were partially caused by being under the manual control of frightened and confused operators. The accidents would not have occurred if the operators had acted properly and if state-of-the-art automatic safety controls had been used.

The following observations can be made:

- At Three Mile Island, the chain of events started with the operators pouring water into the instrument air supply.
- The Chernobyl disaster was triggered by the operators turning off the safety interlocks to conduct a test on a process.
- At Fukushima, the operators failed to cool and shut down the reactors in the 45-minute window of opportunity between the time the earthquake hit and the time the tsunami arrived.

I also believe that the use of nuclear power will eventually be phased out because

- more and more accidents will occur due to cyberterrorism and aging,
- the permanent storage of nuclear waste will remain unresolved,
- the cost of decommissioning will become prohibitive,
- renewable energy will become competitive, and
- the health effects of living in an environment where atmospheric radiation has doubled in the past century will not be tolerated.

Yet, I also believe that it will take decades before nuclear and fossil energy use will be completely phased out. My conclusion is that the nuclear industry today is unsafe, that each nuclear reactor is a potential bomb, and that it is not a question of if but only of when and where the next accidents will occur. On the other hand, there is no question that, although we cannot eliminate these risks, using state-of-the-art automation can make this industry safer.

3.5.4 Cybersecurity

In this section, I discuss the cybersecurity of nuclear power plants. For a more detailed treatment of all aspects of cybersecurity, see Section 1.3.3, "Cyberattacks by Nation-States," in Chapter 1.

In the past, bombs could not think for themselves; they were just dropped. In contrast, today's malicious computer worms can think, learn, and adapt. Today, Watergate-type break-ins can be accomplished without humans climbing through windows. Nuclear attacks can be accomplished without using missiles because the potential bombs (nuclear reactors) are already in place.[80]

Today, when artificial intelligence (AI) triggers a device to act, it does not consider the impacts that its action may cause. In terms of nuclear industry safety, autonomous weapons hold the greatest risk. AI-operated software or drones can be trained not only to kill or assassinate but can also to blow up nuclear power plants.

Malicious computer worms have two levels of sophistication. Lower-level worms travel through the Internet. Protection against this type is relatively simple. The entry of such a worm into our control computers can be blocked by eliminating all direct and indirect connections to the Internet and other networks. This is easier said than done, not only because it denies plant management's convenient access to information but also because all components in the control network must also be isolated, and all digital tools used by the operators, such as USB flash drives, CDs, DVDs, and mobile phones, must be kept out of the control room. Yet, this isolation can and should be done.

We know that both military and industrial teams are developing cyber weapons and cyber safety systems. We also know that the bad guys are just as smart as the good guys. Therefore, when it comes to nuclear reactors, no design can be considered completely safe.

Cyberattacks can also be carried out by more sophisticated software that does not use the Internet. Therefore, we should no longer assume that a computer that is isolated from the Internet or Ethernet and that has no Wi-Fi, CD, or USB connections is safe.

---

80. In my book *Automation Can Prevent the Next Fukushima* (Research Triangle Park, NC: ISA, 2014), I discussed how malicious computer worms can reach the PLC controls or SCADA interfaces of nuclear reactors and trigger their meltdown by turning off their cooling systems.

This is because the more sophisticated, higher-level computer worms (Flame, Duqu, Agent.btz, etc.) are more malicious, as they need no path to travel on. They are called air-gap worms because they travel through the air.

The average age of the nuclear reactors currently in operation is over 30 years old, and some plants still operate with analog or even pneumatic controls and/or safety systems. When they were built, there were no data highways; therefore, the data transfer between the plants and corporate offices was secure from cyberattacks. Today, in the newer plants, digital systems monitor critical operating conditions (valve openings, pump status, temperatures, pressures, levels, flows, radiation, loading, etc.).

The old, noncomputerized plants are actually safer and less vulnerable to cyberattacks. Increasing digitization and reliance on commercial software have increased the risks that the nuclear industry faces.

3.5.4.1 Attacks That Have Already Occurred

As to cyberattacks on nuclear power plants, I am convinced of their danger, but I have no firm conclusion about why they have not been used more often or if they were, why do we not know about them? The ones I know about are these:

- **2003** – An example of vulnerability is the Davis–Besse nuclear power plant in Ohio, which is patrolled by armed guards and surrounded by a double row of tall fences that are electronically monitored, just as are the grounds of all other nuclear power plants. Such steps reduce the probability of somebody driving a truck full of explosives into the plant. Yet, fences do not protect against computer worms or viruses, and armed guards cannot prevent computer crashes.

  Slammer (also known as Sapphire, Helkern, *SQL Hell,* or SQLExp) was a worm on Microsoft Windows that appeared in January 2003 and was the fastest-spreading worm of its time. On January 25, 2003, a Slammer worm penetrated the private computer network of Ohio's Davis–Besse nuclear power plant. The worm entered by first penetrating the unsecured network of a contractor and then squirmed its way into the Davis–Besse corporate business network. Because that network was connected to the plant's network, it bypassed the firewall and spread to the plant network. *The Slammer worm infected more than 90% of the vulnerable hosts within 10 minutes and reached its full scanning rate (of more than 55 million scans per second) in 3 minutes.*

  After the penetration, the following sequence occurred: At 4:00 p.m., the operators noticed the slowing of the plant network, and at 4:50 p.m., the safety parameter display system (SPDS) crashed. The SPDS monitors the operation of the coolant system, core temperature, radiation levels, and other critical conditions. At 5:13 p.m., the plant process computer (PPC) also crashed. Therefore, although the plant's network was protected by a firewall, both the plant's SPDS and PPC were disabled for about 5 hours. Fortunately, the plant was not operating because a hole in the reactor head was being repaired. Another reason no harm was done is that the worm could not attack the SPDS and the PPC analog backups. It should be noted that the design and practices of the operator of the Davis–Besse plant (FirstEnergy) were not in violation of the Nuclear Regulatory Commission's (NRC's) cybersecurity regulations of the time.

- **2007** – In November 2007, the US government and Israel used a malicious computer worm (Stuxnet) to stop Iran's nuclear program by infiltrating the controls of a large number of its ultracentrifuges. Stuxnet entered the targets through the computers' USB flash drives and infected their programmable logic controllers (PLCs). At that time, Iran did not have such sophisticated worms, so it could not retaliate.

  Air-gap-type worms can enter not only through USB flash drives but also through any of the digital control components in a plant's network. Such components can be wireless sensors (where the worm enters by faking the sensor signals), speakers and microphones, or iOS and Android devices. Malware can be injected into the operators' smartphones or tablet PCs (including iPhone, iPad, iPod Touch, and Android counterparts), which will then "hit" the control computers.

Here I list a few more accidents, but in less detail:

- **2008** – A Georgia nuclear power plant went into emergency shutdown for 48 hours due to a cyber incident.
- **2010** – A Stuxnet worm is suspected of having attacked a Russian facility.
- **2014** – Yukiya Amano, a nuclear power plant in Germany, suffered a "disruptive" cyberattack.
- **2014** – On January 2, one of the eight computers in the control room at Monju Nuclear Power Plant in Tsuruga, Japan, was compromised by a cyberattack.
- **2014** – In December, the South Koran government revealed that a nuclear plant in the country was hacked.
- **2016** – In April, the German BR24 news agency reported that a computer virus was discovered at the Gundremmingen nuclear power plant.
- **2019** – On October 30, a cyberattack occurred at India's Kudankulam nuclear power plant.

- **2022** – For years, Ukraine has been a proving ground for Russian cyber weapons, but to my knowledge, they have not been used against nuclear power plants, probably because of the risk that the wind could blow the radiation onto Russian territory.

3.5.4.2 Backups and Redundancy

The protection of nuclear power plants can be improved by both redundancy and automation. Redundancy is different from a simple backup. For example, a triple-redundancy voting system (implemented for both the hardware and software systems) means that in all nuclear power plants, all critical measurements and status indicators are served by three accurate sensors (preferably utilizing different principles of operation). If one sensor disagrees with the other two, the control system always acts on the majority view, automatically disregarding the disagreeing sensor and scheduling it for maintenance and recalibration. The same should apply to all software packages, including supervisory control and data acquisition (SCADA), SPDS, and PPC networks in the plant.

In terms of protecting the plant from intentional or unintentional operator errors, one can provide hardwired interlocks on all critical safety systems and configure them in such a way that the operators cannot bypass or shut them down.

3.5.4.3 Global and National Risks

On November 12, 2018, French President Emmanuel Macron issued the Paris Call as he brought together like-minded countries, private-sector entities, and international and civil society organizations to issue a consensus statement on growing concerns about cyber threats. Signatories agreed to take steps to guarantee a peaceful cyberspace, to develop international law, and to act against the threats posed by malicious cyber activities.

A 2016 study by the Nuclear Threat Initiative (NTI)[81] revealed that the situation was grim, as 20 countries had extensive nuclear power systems that were vulnerable to cyberattacks. In addition, 24 states had weapons-usable nuclear materials at hand. It was also reported that a new piece of dangerous malware,[82] TRISIS, was found in the Middle East that specifically targets the industrial controllers used to protect the safety of critical control systems, including nuclear power plant controls.

---

81. Catalin Cimpanu, "Nuclear Power Plants from All Over the World Are Vulnerable to Cyberattacks," Softpedia News website, January 15, 2016, https://news.softpedia.com/news/nuclear-power-plants-from-all-over-the-world-are-vulnerable-to-cyberattacks-498955.shtml.

82. Malware is software that is designed specifically to damage a computer, server, client, or computer network. (Software that causes unintentional harm is called a software bug.) A wide variety of types of malware exists. The term malware is a blending of the terms "malicious" and "software."

Therefore, commissions such as the NRC should forbid linking plant operations networks with corporate local area networks (LANs). The convenience and cost savings associated with these corporate links cannot justify the risk they present. Similarly, the corporate networks of utilities and the SCADA networks of plants should also be separated because SCADA networks control the remote terminal units (RTUs) sprinkled throughout the plants that directly monitor and/or control the operation of power plant equipment.

In 2018, the US Department of Homeland Security reported that Russia's military intelligence agency had infiltrated the control rooms of power plants across the country.[83] In theory, that could enable that military agency to take control of parts of the grid by remote control (Figure 3-62). Russian military hackers appear to have directed their efforts at implanting malware into the electrical grid and the systems of intelligence officials and executives of the companies that oversee the world's computer networks. The *New York Times* also reported that Russia was conducting a series of cyberattacks aimed at US and European nuclear power plants, water supplies, and electric grid systems between 2015 and 2017. (There must be later reports, but I have not seen them.)

This trend can have disastrous consequences not only in newly built nuclear power plants but also throughout industry if unsecured SCADA systems and unprotected connections between plant and company business networks are permitted.

Figure 3-62. The grid is also vulnerable to cyberattacks.[84]

83. See: David E. Sanger, "Russian Hackers Appear to Shift Focus to US Power Grid," *New York Times*, July 27, 2018, accessed November 21, 2022, https://www.nytimes.com/2018/07/27/us/politics/russian-hackers-electric-grid-elections-.html.

84. For more information, see: Dan Lohrmann, "How Vulnerable Is America's Power Grid?," Govtech.com website, May 15, 2014, accessed November 21, 2022, https://www.govtech.com/security/How-Vulnerable-Is-Americas-Power-Grid.html.

Also, the attacks are not necessarily limited to interfering with the operation of a plant. For example, in 2019, India's Kudankulam nuclear power plant (KKNPP) suffered a malware attack. An investigation by the national agency for responding to cybersecurity incidents found that a malware-infected personal computer caused the attack, which was aimed at the plant's administrative network.

Work is being done to assess the risk of various nuclear power plant attacks and protect against them. For example, the Integrated Cyber-Physical Impact Analysis framework developed by Sandia National Laboratories can be used for risk assessment in the cyber-energy arena. Multiple attack scenarios were used to demonstrate the extended consequences of cyberattacks on plant operations and at the regional and national levels. This framework evaluates the causes, consequences, and potential mitigation strategies in the areas of threats, cyberattacks, component effects, physical system response, extended consequences, and recovery.

3.5.5 Nuclear Waste Storage

In addition to a loss of cooling, leaks, earthquakes, aging, and terrorist attacks, errors in transporting and storing nuclear wastes can also cause accidents. When the useful life of a nuclear power plant is reached, the owner is faced with the costs of plant decommissioning and permanent waste storage.

Today, each nuclear reactor produces 20 tons of high-level nuclear waste per year. In the United States alone, there are approximately 30,000 tons of spent fuel rods in storage and some 380,000 m^3 of high-level radioactive wastes in temporary storage. In addition, weapons-grade plutonium from dismantling nuclear weapons is stored in various locations, such as Amarillo, Texas. Only about 20 pounds of separated plutonium is needed to build a dirty nuclear bomb, and there is already some 300 tons of plutonium in storage, enough to build more than 30,000 bombs.

Radioactive waste is classified as low-level waste (LLW), intermediate-level waste (ILW), high-level waste (HLW), and transuranic waste[85] (TRUW). Here, I concentrate on the long-lived HLW fission products technetium-99 (half-life of 220,000 years) and iodine-129 (half-life of 17 million years), the TRUW neptunium-237 (half-life of 2 million years), and plutonium (half-life of 20 to 24,000 years).

Today, the Treaty on the Non-Proliferation of Nuclear Weapons allows all nations to concentrate uranium-235 to about 3%, which is required to fuel nuclear power plants,

85. Transuranic waste (TRUW) is any waste that has been contaminated with α-emitting substances, has a half-life longer than 20 years and is present in concentrations greater than 100 nCi/g (3.7 MBq/kg).

but allows only members of the UN Security Council to concentrate it to 90%, as needed to build nuclear weapons. (Yet, nonmembers, such as India, Pakistan, and Israel, also have nuclear weapons, and others are developing the technology to reach this concentration.) As with the risks associated with the waste from nuclear power plants, waste from nuclear weapons also poses a risk.

Although the planet will not be safe until all nuclear weapons are dismantled and all energy sources are converted to free, inexhaustible, and safe ones, this cannot occur overnight. As a result, we will be living with various concentrations of radioactive materials for about another century. During this period, process control can play a major role in making decommissioning nuclear power plants and reprocessing, transporting, and storing nuclear waste safer than it is today. First I describe some waste handling processes, then the safety tools that can be used to make them safer.

3.5.6 Reprocessing and Temporary Storage

3.5.6.1 Reprocessing

Instead of storage, spent fuel from reactors or dismantled nuclear weapons can be taken to reprocessing plants to recover their residual fuel value. Initially, reprocessing served to recover plutonium only for use in nuclear weapons (e.g., at the Hanford Site in Washington state[86]), and the process used bismuth phosphate. This was followed by the solvent extraction process, which can also recover uranium (e.g., at the Savannah River Site in South Carolina).

During reprocessing, the civilian fuel can be dissolved in nitric acid to separate the fissionable uranium and plutonium. Reprocessing increases the volume of radioactive waste more than 20-fold and requires long-distance transportation from the source to the reprocessing plant. For these reasons, and because of the risks associated with the potential for plutonium to be stolen and used to build dirty nuclear bombs, reprocessing in the United States stopped three decades ago. However, France, India, Japan, Russia, and the United Kingdom continue to operate such plants. Today, an international partnership is evolving to reprocess spent nuclear fuel, but in such a way as to make the product usable only as fuel for power plants and not for use in building nuclear weapons.

---

86. The financial liability for waste treatment at the Hanford Site amounts to some $500 billion to manage the 57 million gallons of nuclear waste from its past production of plutonium for nuclear weapons during World War II and the Cold War era. This waste is stored in aging, leaking underground tanks that present a growing safety risk to the Columbia River and residents living nearby.

3.5.6.2 Temporary Waste Storage

Ninety-nine percent of all radiation in nuclear waste comes from spent fuel rods. After the fuel pellets are loaded into fuel rods, the core assembly is installed in the nuclear reactor and is used until it is exhausted in a year or two. After that, the used fuel rods are replaced with fresh ones, and the extremely hot spent fuel rods are usually stored in storage water ponds (Figure 3-63) filled with boric acid, which absorbs radiation.

This is dangerous because, as the ponds fill with waste, the rods approach and can eventually reach criticality, at which point they can start a nuclear chain reaction. Therefore, neutron-absorbing materials must be placed between the rods, and their integrity must be closely monitored.

After at least 6 months, the rods can be moved into steel-reinforced helium-cooled casks, but they often stay in the ponds for years. Dry storage in casks (Figure 3-64) or concrete entombing in bunkers usually occurs after some 5 years of cooling. In the United States, nuclear waste is stored at 122 temporary sites. Most of the casks used can be penetrated by regular weapons, in which case they can release radioactive cesium

(**a**) Spent fuel rods stored in temporary storage ponds. In the case of a loss of coolant, these spent fuel rods can cause meltdowns and hydrogen explosions.

(**b**) Spent nuclear fuel rods are stored underwater at the Hanford Site in Benton County, Washington.

Figure 3-63. The temporary storage of nuclear fuel rods.

Source of Figure 3-63a: "Fuel pool.jpg" by US DOE, Public Domain, accessed March 12, 2022, https://commons.wikimedia.org/w/index.php?curid=8548517.

Source of Figure 3-63b: "Spent nuclear fuel hanford.jpg" by US DOE, Public Domain, accessed March 12, 2022, https://commons.wikimedia.org/w/index.php?curid=5650481.

Figure 3-64. Dry storage of nuclear waste.

Source: "Backgrounder on Dry Cask Storage of Spent Nuclear Fuel," US Nuclear Regulatory Commission, last updated August 29, 2020, https://www.nrc.gov/reading-rm/doc-collections/fact-sheets/dry-cask-storage.html.

gas. One might mention that in the 2022 war in Ukraine, missiles landed within a few hundred yards of Zaporizhzhia nuclear plant's nuclear waste storage casks.[87]

The long-range plan was to place the nuclear waste in deep geologic repositories, but no such underground facilities have been opened to date. Therefore, the described methods of interim (up to 50 years) storage of high-level nuclear waste are continuing, and these storage facilities are filling up. As of 2019, no country in the world had a final disposal site for high-level nuclear waste in operation. Most countries have yet to develop and implement a functioning waste management strategy for all kinds of nuclear waste.

The latest development regarding nuclear repositories in the United States is that, in 2021, the US NRC issued a license allowing commercial nuclear waste to be stored in Texas. The NRC gave its official blessing for a "consolidated interim storage facility" to be built and operated in Andrews, Texas, to temporarily store the millions of pounds of waste currently stored at many sites around the country. Naturally, Texans fear that no permanent site will ever materialize and that they will be stuck with all of the country's radioactive waste forever.

About 36 miles west of the contested Texas site is something that resembles a solution: the nation's only permanent, deep geologic repository. The Waste Isolation Pilot

87. Patrick Smith and Josh Lederman, "Fears of nuclear catastrophe grow after attacks on power plant in Ukraine," NBC News website, updated August 8, 2022, accessed November 9, 2022, https://www.nbcnews.com/news/world/russia-ukraine-war-nuclear-disaster-fears-zaporizhzhia-plant-rcna41959.

Plant has been disposing of federal waste from nuclear weapons production and the like since 1999. Waste is stored more than 2000 ft (610 m) underground in large panels mined from salt rock beds.

3.5.7 Transportation and Permanent Depository

When a nuclear power plant reaches the limit of its useful life, it must be *decommissioned.* The technology required for the permanent and safe disposal of the components of dismantled nuclear power plants is unresolved. All materials exposed to radioactivity (LLW) are mostly stored at dedicated dry burial sites, and the decommissioned plant itself is often entombed in concrete.

3.5.7.1 Transportation

The transportation of waste to storage sites (Figure 3-65) can be short, as the sites are usually located near the plants, but if residual fuel is to be recovered, HLWs might need to be transported to remote reprocessing plants.

3.5.7.2 Permanent Depository

Hardened interim storage of spent fuel or HLW in dry casks is not a permanent solution. To date, no permanent disposal site has been found because of concerns about leakage, earthquakes, and volcanic activity. The US government planned to build a nuclear waste repository in the Yucca Mountains. It was originally intended to open in

(a) Fuel rod transportation by train **(b)** Transporting nuclear waste in casks

Figure 3-65. Transportation of nuclear waste to storage sites.

Source Figure 3-65a: "Nuclear waste flask train at Bristol Temple Meads 02.jpg" by Chris McKenna (Thryduulf), June 15, 2006, is licensed under CC BY-SA 4.0, https://commons.wikimedia.org/w/index.php?curid=1339908.

Source Figure 3-65b: "Integrated Waste Management," Office of Nuclear Energy, US DOE, Office of Nuclear Energy, Public Domain, accessed March 12, 2022, https://www.energy.gov/ne/consent-based-siting/integrated-waste-management.

Figure 3-66. Seal of the Yucca Mountain Repository.

Source: Office of Civilian Radioactive Waste Management, US Department of Energy, Public Domain, accessed March 21, 2022, https://www.atlasobscura.com/places/yucca-mountain-repository.

1998, but that target date was moved, and some believe it will never be opened because of public resistance.[88]

The Yucca Mountain Nuclear Waste Repository was planned to be a deep geological repository in Nye County, Nevada. The federal government spent more than $10 billion on what was to be the permanent disposal site for the country's vast stockpile of commercial nuclear waste, even as Nevada state officials vehemently opposed it. The building of this repository was discontinued during the Obama administration after several decades of work, debate, and immense investment. Figure 3-66 shows the seal door of the repository, just to illustrate the scale of the project.

Finland is also working on a waste repository at Olkiluoto, but as of this writing, no permanent storage facility is in operation anywhere in the world.[89]

3.5.8 Sensors and Monitoring

Monitoring requires sophisticated systems with the main goal of protecting the public from the accidental release of radioactive materials during all phases of nuclear power

88. This leaves US utilities and the US government, which currently disposes of its transuranic waste (TRUW) 660 m (2150 ft) below the surface at the Waste Isolation Pilot Plant (WIPP) in New Mexico, without any designated long-term storage site for the high-level radioactive waste stored on site at various nuclear facilities around the country.

89. The only purpose-built deep geological repository that is currently licensed for disposal operations is the WIPP in the United States. Plans for the disposal of spent fuel are well advanced in Finland, Sweden, France, and the United States, although there have been political delays in the United States. In Canada and the United Kingdom, deep disposal has been selected, and the site selection process has commenced.

plant operation, including plant operation and dismantling and nuclear waste transportation and storage. In addition, these systems must also detect any materials from which dirty bombs can be built.

3.5.8.1 Dosimeters

A *dosimeter* measures and displays the cumulative dose of radiation received. It can be as simple as a pen-like device clipped to one's clothing. Dosimeters can transmit their readings to a central control room where the exposure of all employees is continuously monitored. Similarly, central displays and alarms can be provided to monitor stationary radiation sensors installed throughout the plant. These monitors can be standard radiation environment monitors (SREMs), proton monitors (PMs), scintillation fiber detectors (SFD), or real-time radiation-sensitive field-effect transistor (RADFET)[90] units. The readings of both the dosimeters and the monitors can be transmitted (via satellites) to monitoring centers around the world. Process control engineers can design the area display and alarm systems required.

Of the three types of nuclear radiation (alpha—α, beta—β, and gamma—γ), γ radiation (X-rays) is the most harmful. Radiation exposure is measured in terms of the roentgen (R), which is equal to the quantity of ionizing radiation that will produce one electrostatic unit of electricity in 1 cm^3 of dry air at 0°C (32°F) and standard atmospheric pressure. The reading on a radiation range meter in an area provides an estimate of the likelihood of death. For example, at a reading of 10 R/h, death occurs in about 10% of cases; at a reading of 500 R, the death rate is about 50%; and at a reading of 800 R, the death rate is about 99%.

3.5.8.2 Satellite Monitoring

Satellite monitoring is performed by the International Satellite Monitoring Agency (ISMA) of the UN's International Atomic Energy Agency (IAEA),[91] which can detect at a ground resolution of about 1 m. The IAEA monitors the worldwide flow of nuclear materials using some 350 cameras producing 150,000 images that are fed to 50 radiation detection and 90 surveillance systems. Some sites (such as those in Armenia, Brazil, Chernobyl, and Hungary) are already connected to the central hub in the IAEA

---

90. A RADFET is a microelectronic chip that is specially engineered to be sensitive to high-energy (ionizing) radiation. RADFETs are small and robust, require no power to detect radiation, can be easily integrated into electronic systems, and are cost-effective.

91. According to the IAEA website, "The International Atomic Energy Agency is the world's central intergovernmental forum for scientific and technical co-operation in the nuclear field. It works for the safe, secure, and peaceful uses of nuclear science and technology, contributing to international peace and security and the United Nations' Sustainable Development Goals." International Atomic Energy Agency Overview, accessed March 14, 2022, https://www.iaea.org/about/overview.

headquarters. Process control can help integrate these systems into a worldwide satellite communication network.

Special sensors (developed by Science Applications International Corporation—SAIC) can detect high-velocity spin-off particles and pinpoint the locations of enriched uranium throughout the world. These sensors can detect the leakage of radioactive materials and the transport of all nuclear devices.

Until total nuclear disarmament, decommissioning of all nuclear power plants, and the safe and permanent storage of all nuclear wastes, the process control approach discussed in this section will play a critical role in improving nuclear safety.

3.5.9 Conclusions

The currently operating nuclear power plants are of the fission type, obtaining their energy by splitting a radioactive isotope of uranium. I disapprove of their continued use for safety reasons, including the fact that they are potential nuclear bombs because they cannot be fully protected from accidents, attacks, or cyberattacks. Also, I consider all currently operating designs to be unsafe because they will melt down if the energy sources (including their backups) that operate the cooling-water supplies to the reactors fail. Finally, I oppose their use because there are no permanent sites anywhere in the world for disposing of their high-level waste or used fuel rods.

The other method of generating nuclear energy is fusion, in which hydrogen atoms are fused into helium atoms. This process has been investigated for a century, and in August 2021, scientists at Lawrence Livermore National Laboratory (LLNL) managed to trigger this process for 100 trillionths of a second using giant lasers. The energy obtained was only 70% of the energy needed to produce the reaction.

In addition to laser fusion units, doughnut-shaped tokamaks have also been tested, using magnetic fields to compress the hydrogen fuel. I oppose the development or use of all types of fusion reactors because all materials on Earth evaporate at over 100 million degrees K; therefore, this process can be contained only by means such as magnetic fields. Also, this process is not currently operational, whereas the transition from fossil fuels is an immediate task. Moreover, fusion reactors are prohibitively expensive.

3.6 Hydropower

Hydropower is one of the oldest and best-understood energy sources. In 2017, the total generating capacity of hydropower plants was 1.1 TW, and by 2019, it had increased to

1.3 TW.[92] Its global generating capacity in 2021 was 1.34 TW, which was about 17.5% of the total global electricity generating capacity of about 7.65 TW.

Global hydroelectricity consumption in 2019 was 4.22 PWh, which was about 15.6% of the total global electricity consumption of about 27 PWh and about 2.4% of the total energy consumption of about 176 PWh (600 Q). Many opportunities for hydropower development remain throughout the world. Although there is no clear consensus, it is estimated that the annual hydroelectric electricity production could still be doubled because nearly 14 PWh of hydroelectric energy is considered to be technically and economically obtainable worldwide (Asia, 6.8 PWh; South America, 2.6 PWh; Africa, 1.7 PWh; North and Central America, 1.5 PWh; Europe, 1.2 PWh).

Hydropower is a flexible and consistent renewable energy resource, capable of meeting both base load electricity requirements and, with pumped-storage technology, unexpected power demands. There are still many opportunities for hydropower development throughout the world, but there are also limitations, such as the blocking of rivers which stops the free movement of ships, alluvial deposits, and fish. Another undesirable consequence of damming occurs when the reservoir created upstream floods villages and large areas of agricultural land, which occurs when the vertical drop of the river is low.

There are three types of hydropower stations: conventional, where power is generated through the release of stored water from an elevated reservoir; run-of-river, where there is no reservoir and the electricity is generated just by the flow of the river; and pumped storage, where some of the water is pumped up to a storage reservoir, which is at a higher elevation, to be released later.

3.6.1 Conventional Hydropower Plants

Conventional hydropower plants range in size from less than 100 kW to greater than 22 GW, with individual turbines reaching 1000 MW in capacity. As conventional plants use the potential energy of dammed water, the amount of energy extracted depends on the volume and the difference between the height of the water level in the reservoir and the water outflow level, which is called the *water head* (Figure 3-67).

The dam is constructed to ensure sufficient water storage at a given height, which determines how much potential energy is stored there. The control gate sets the water

92. "Installed electricity capacity worldwide in 2019, by source," Statista website, released 2021, https://www.statista.com/statistics/267358/world-installed-power-capacity/.

flow from the reservoir through the penstock to the turbine. When the control gate is fully opened, maximum generating capacity is reached. The penstock is a huge steel pipe that carries water from the reservoir to the turbine, where the potential energy of the water is converted into kinetic energy. The turbine is mechanically coupled to an electric generator. The kinetic energy of the water drives the turbine and, consequently, the generator. There are two main types of water turbines: (1) impulse and (2) reaction. Impulse turbines are used for large water heads, and reaction turbines are used for low and medium water heads. In both cases, the nameplate (rated) power, that is, the production capacity that is reached only at maximum flow, is greater than the actual power. Because the flow varies throughout the year, only about half of the nameplate production is reached on a yearly average basis.

The generator is mounted in the powerhouse and is mechanically coupled to the turbine shaft. The rotation of the turbine blades drives the generator.

Surge tanks (not shown in Figure 3-67) are usually provided in high- and medium-head power plants when considerably long penstocks are required. A surge tank is a small reservoir or a tank that is open at the top. It is fitted at the top of the penstock between the reservoir and the powerhouse. The water level in the surge tank rises or falls to reduce the pressure swings in the penstock. When there is a sudden reduction in load on the turbine, a governor closes the gates to reduce the water flow, and the pressure in the penstock rises. This pressure rise is moderated by the surge tank, whose water level rises to reduce the pressure. On the other hand, the surge

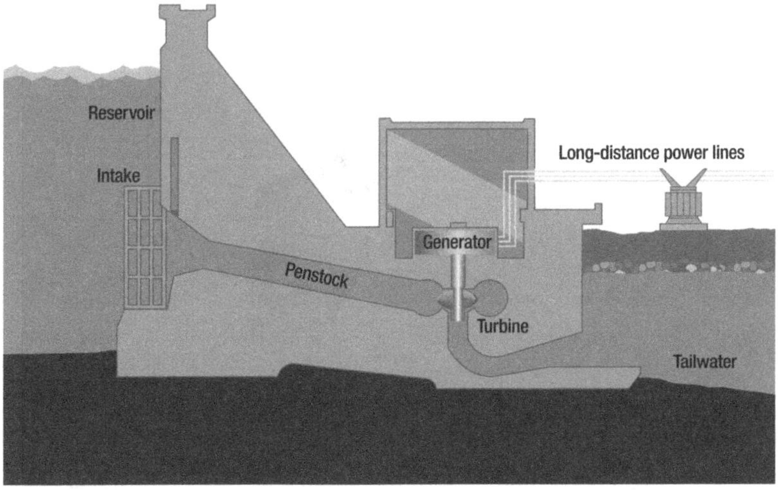

Figure 3-67. Typical layout of a hydroelectric power plant and its basic components.

Source: Reprinted with permission from the Tennessee Valley Authority. "How Hydroelectric Power Works," accessed March 12, 2022, https://www.tva.com/Energy/Our-Power-System/Hydroelectric/How-Hydroelectric-Power-Works.

tank provides the excess water needed when the gates are suddenly opened to meet increased load demand and the pressure in the penstock drops. Proper tuning of the pressure control algorithms is a key requirement for stable and efficient operation.

3.6.2 Pumped-Storage Plants

In pumped-storage plants, a second reservoir is constructed near the water outflow from the turbine (Figure 3-68). When the demand for electricity is low, the water from the lower reservoir is pumped into the upper (main) reservoir. This is done by using power from a power grid to run electric generators that spin the turbines backward to act as pumps. This causes the turbines to pump the water from the lower reservoir to the main (upper) reservoir. The system's response time can be as short as a few minutes. This system verifies for the grid dispatcher that a sufficient amount of water will be available in the main reservoir when the load rises to peak levels. This is valuable in stabilizing and controlling the grid and providing a combination of low latency, high power, and high response.

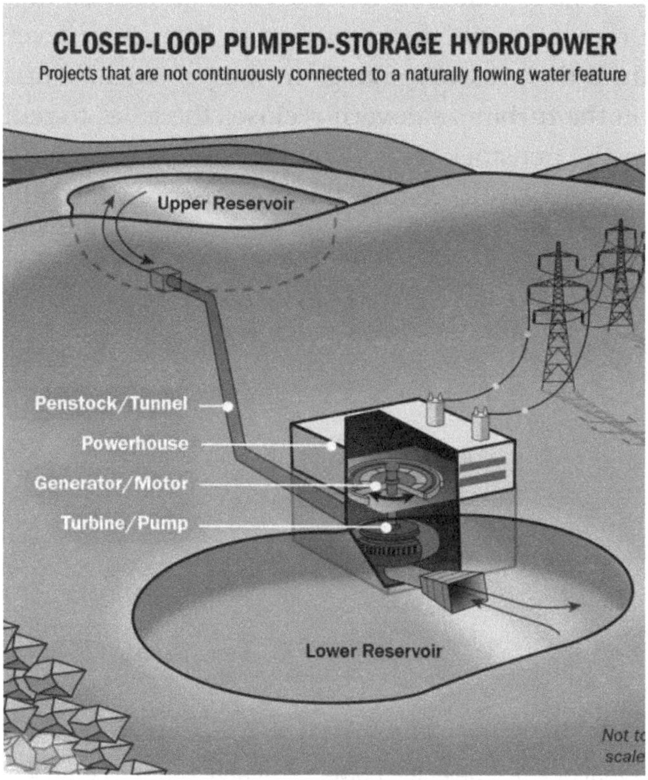

Figure 3-68. Pumped-storage hydropower plants operate with two reservoirs.

Source: "Pumped-Storage Hydropower," Office of Energy Efficiency and Renewable Energy, US DOE, accessed March 12, 2022, https://www.energy.gov/eere/water/pumped-storage-hydropower.

3.6.3 Run-of-River Plant

This type of plant requires no dam or reservoir (Figure 3-69). It operates by diverting a portion of a river into a penstock or canal leading to the turbines. Not having a reservoir limits the availability of the water to the flow that happens to exist at the time, so this design is better suited to rivers with steady flows. Because part of the river flow is diverted, the water level in the original bed downstream of that point is reduced, which often has undesirable environmental impacts.

Similarly, when the flow in the river is high, the oversupply of water passes unused. Still, because run-of-river hydroelectric plants operate without dams or other obstructions to the river's flow, they usually are less disruptive to the natural state of the river.

So-called microhydro systems of the run-of-river type are also available for private home use if a spring or river is nearby. To use such microhydro systems, a water flow of at least two gallons per minute is required, which necessitates that the home is at an

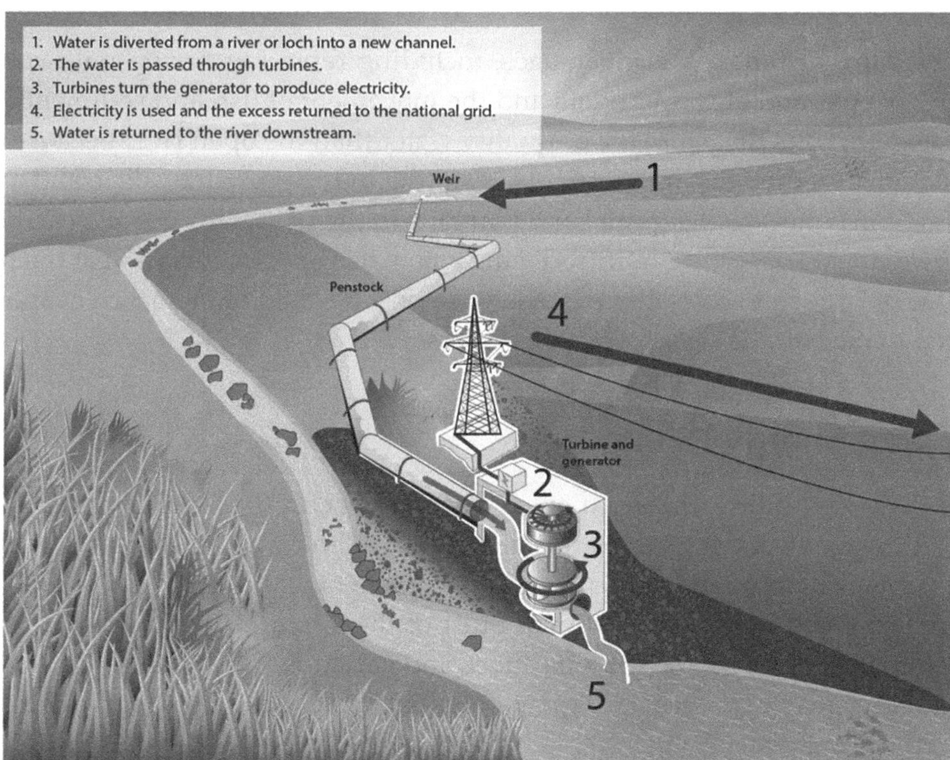

Figure 3-69. A run-of-river hydropower plant takes only part of the water from a river; hence, the traditional environment can survive, and the riverbed is not blocked, so that fish and debris can move freely. *Source:* Reprinted with permission. Helyette Geman, "Electricity in Eastern Africa: The Case for Mini Hydro," Policy Brief PB-17/13, OCP Policy Center, April 12, 2017, https://www.policycenter.ma/publications/electricity-eastern-africa-case-mini-hydro.

elevation that is a few meters below the point from which the water is taken. In addition, a small shed is needed to protect the turbine from inclement weather.

According to Scott Davis, author of *Microhydro: Clean Power from Water*,[93] microhydro systems can generate 75 to 350 kWh per month. One advantage of these systems is that they operate 24/7 and, therefore, require little battery storage. They can also supply 10 to 100 times more power than PV panels or wind turbines for the same amount of capital investment (according to Energy Alternatives Ltd.)

3.6.4 Advantages and Disadvantages

For all hydroelectric plants, the fuel is free, inexhaustible, and nonpolluting. The plant lifetimes are long (about 100 years), and the maintenance needs are relatively low. For conventional or pumped-storage plants, the capital costs are high, mainly because of the cost of dam construction. In addition, if the plant is located at a distance from the users and in a hilly area, the construction costs of building the required transmission systems can be high.

Naturally, dams also have disadvantages, including replacing nature with concrete, relocating people, and blocking rivers and the movement of fish and sediments. They also create reservoirs that store lower-quality water than the open river because of the accumulation of sediments and nutrients, resulting in the growth of algae and aquatic weeds that can suffocate the animal and plant life in the river. This process can be controlled by manual harvesting or introducing a new fish species that can eat the plants. In the case of large reservoirs, there is also water loss because of increased evaporation compared to the natural levels of flowing rivers.

3.7 Ocean Energy Plants

The ocean is used indirectly as an energy source when offshore wind turbines are installed or when floating solar islands are built. The energy of the ocean can also be used directly by tapping the immense energy content of its currents, waves, and tides. Tides are generated by the gravitational pulls of the Sun and Moon. Ocean currents are generated by the wind, the Coriolis effect, and variations in water temperature and salinity. These energy sources have great potential, but the technology to utilize them is still in the development stage.

---

93. Scott Davis, *Microhydro: Clean Power from Water* (Gabriola Island, BC, Canada: New Society Publishers, 2003),

3.7.1 Wave-Power Buoys

Wave-power buoys capture the energy in the up-and-down movement of the waves and transmit the generated electricity to grids on shore by underwater cables. Waves are generated by wind passing over the surface of the sea. As long as the waves propagate more slowly than the wind speed just above the waves, energy transfer occurs from the wind to the waves. Air-pressure differences between the upwind and lee sides of a wave crest and friction on the water surface cause the growth of the waves. Wave height is determined by the wind speed, the duration of time the wind has been blowing (i.e., the distance over which the wind excites the waves), and the depth and topography of the seafloor. In general, larger waves are more powerful, but wave power is also determined by wave speed, wavelength, and water density. The oscillatory motion is highest at the surface and diminishes exponentially with depth.

The highest-energy waves are concentrated off the western coasts in the latitude range between 40° to 60° both north and south of the Equator. The power in the wave fronts varies in these areas between 30 and 70 kW/m, with peaks of up to 100 kW/m in the Atlantic southwest of Ireland, in the Southern Ocean, and off Cape Horn (Chile). If this energy resource were harnessed, it could meet 10% of the global electricity demand.

The world's first wave farm, Aguçadoura Wave Farm, was located 5 km (3 mi) offshore near Póvoa de Varzim, Portugal. Three Pelamis wave energy converters (nicknamed *sea snakes*) were to convert the motion of the ocean's waves to electricity, producing up to 2.25 MW of power. The farm opened on September 23, 2008, and was shut down 2 months after the official opening because of issues with the bearings.[94]

In this setup, each Pelamis sea snake consisted of five tube sections, linked by universal joints, that flexed in two directions and faced the onrushing waves. As the waves passed down the length of the tubes, the sections bent, and that movement was converted to electricity. Another design is Ocean Power Technologies' "smart" oceangoing buoy that captures and converts wave energy into electricity. A power take-off converts the mechanical power of the up-and-down motion of the passing waves to drive an electrical generator. The resulting electricity is transferred back to shore through underwater power cables.

---

94. "Aguçadoura Wave Farm," Wikipedia, accessed March 12, 2022, https://en.wikipedia.org/wiki/Agu%C3%A7adoura_Wave_Farm.

The European Marine Energy Centre (EMEC) provides a variety of test sites under real sea conditions to develop these systems. Its grid-connected wave test site is located at Billia Croo, on the western edge of the Orkney mainland off Scotland.[95]

3.7.2 Underwater Turbines

Underwater turbine systems use water currents to turn their blades underwater and generate electricity (Figure 3-70). Because water is so much denser than air, the amount of energy produced by water turbines is much greater than that produced by wind turbines. Although they are based on similar technology, underwater turbines are more efficient than wind turbines. At a velocity of 16 km/h (10 mph, or about 8.6 knots in nautical terms), an ocean current would generate the same amount of electricity as a comparable wind turbine at a wind velocity of 145 km/h (90 mph).

Unlike conventional hydroelectric turbines that rely on dams or diversions that direct water flow, underwater turbines can also rely on the unconstrained currents found in rivers, tidal areas, and the open ocean.

Land constrictions such as straits and inlets can also create high water velocities at specific sites that can be captured using turbines. These turbines can be horizontal, vertical, open, or ducted.

3.7.3 Tidal Energy

Tidal energy uses the rise and fall of tides to convert the kinetic energy of incoming and outgoing tides into electrical energy. Some tidal generators can be built into the structures of existing bridges or entirely submersed, thus avoiding concerns about their impact on the natural landscape. Their huge costs and the limited availability of sites are two of the drawbacks of the use of tidal energy. Tidal power is harnessed by using double-effect (two-directional) turbine generators; the water flow through them is caused by the difference between the levels of the ocean and a storage basin. The water flows into the basin at high tide, and at low tide, it flows from the basin back into the ocean. There are more than 40 sites in the world where the difference between high- and low-tide levels is sufficient for these power plants to be profitable.

The world's most successful tidal energy installation, as well as the first and still one of the largest, is La Rance, a 240 MW installation on the Rance Estuary at Ille-et-Vilaine

95. For more information, see: *Environment Description for the EMEC Wave Test Site Billia Croo, Orkney*, Report REP096-04-03 (Orkney, UK: EMEC, June 25, 2009), https://tethys.pnnl.gov/sites/default/files/publications/Billia%20Croo%20Environmental%20Description.pdf.

Figure 3-70. Underwater turbine generators.

in the Brittany region of France. Operated by Électricité de France (EdF),[96] construction began in 1943, and the plant was commissioned in November 1966.

The project consists of a 330 m long dam that creates a 22 km^2 basin and a large reservoir of 184 million cubic meters. With an average of more than 8 m of tidal range, the system generates approximately 540 GWh annually from its 24 bulb turbines.[97] Each turbine is rated at 10 MW, weighs 470 T, and has a diameter of 5.4 m. Bulb turbines generate power on both the flood and ebb cycles of the tide, which is when the lagoon is filling and emptying, respectively, and across a wide range of flows. At La Rance, the flow range is between 4000 m^3/s and 18,000 m^3/s. (This flow range is about the same as that of the river Danube.)

After more than 50 years, each turbine had run about a quarter million hours on average, with a cumulative gross output of about 21.6 TWh produced during those 50 years, and none of the 24 bulb turbines had been replaced. A program for renovating and modernizing the plant was undertaken in 2013 and will run through 2022, during which time the control system (among other things) will be upgraded.

La Rance was the oldest tidal barrage scheme, but since the April 2012 commissioning of the 254 MW Sihwa Tidal Power Plant in South Korea, it is no longer the largest.

---

96. This description is largely based on the following article: David Appleyard, "Turning the Tide on Barrage Technology," April 15, 2014, Renewable Energy World website, https://www.renewableenergyworld.com/storage/turning-the-tide-on-barrage-technology.

97. The term *bulb* describes the shape of the upstream watertight casing, which contains a generator located on a horizontal axis.

Sihwa is a single-effect flood-type arrangement that cost $355 million to develop. The project is located in the 12.7 km long Sihwa embankment, a seawall constructed in early 1994 for flood mitigation and agricultural purposes that created a 56.5 km^2 lake. It generates approximately 552 GWh annually and saves an estimated 315,000 T of CO_2 emissions annually.

Plans for other tidal power plants in the world included two more in South Korea: a 520 MW plant at Garolim Bay and a plant near Incheon with a potential capacity of 700 to 1000 MW. Other countries with large tidal ranges and suitable geography include China, Russia, the United Kingdom, the United States, India, and Mexico. In the United Kingdom, the Severn Estuary's 14 m tidal range has a potential generating capacity of more than 8 GW (about 5% of the current power consumption in the Severn Estuary). It has been the site of proposed tidal barrage schemes for more than 100 years.

Dynamic tidal power (DTP) is an untried but promising technology that would exploit the difference between the potential and kinetic energies of tidal flows. In this case, very long dams (e.g., 30 to 50 km in length) would be built from the coast straight out into the sea or ocean without enclosing an area. The tidal phase differences would then be introduced across the dam, leading to a significant water level differential in shallow coastal seas featuring strong coast-parallel oscillating tidal currents, such as are found in the United Kingdom, China, and Korea.

3.8 Bioenergy

The oldest source of heat energy on Earth is burning wood and other organic materials. Biomass energy is produced from organic material and is commonly used worldwide. Chlorophyll in plants captures the Sun's energy by converting CO_2 from the air and water from the ground into carbohydrates through photosynthesis. When the plants are burned, the water and CO_2 are again released back into the atmosphere. Biomass generally includes crops, plants, trees, yard clippings, wood chips, and animal wastes. Biomass energy is used for heating and cooking in homes and as a fuel in industry, including power generation.

Burning vegetation-based biomass fuels emit as much CO_2 as the plant absorbs during its growth. In addition, this burning also emits considerable amounts of pollutants, such as particulates and polycyclic aromatic hydrocarbons. Even modern pellet boilers generate many more pollutants than oil or natural gas boilers. Pellets made from agricultural residues are usually worse than wood pellets because they produce much larger emissions of dioxins and chlorophenols.

Biomass can be any plant or animal material, including wood, sawdust, grass trimmings, domestic refuse, charcoal, agricultural waste, nonfood energy crops, and dried manure. It can be used for energy and heat production and in various industrial processes as raw material. It can be purposely grown as energy crops such as switchgrass, or it can be wood or forest residues; wastes from food crops (wheat straw, bagasse), horticulture (yard waste), food processing (corn cobs), or animal farming (manure); or even human waste from sewage plants.

The energy obtained from burning biomass is often classified as renewable (even by such respected sources as the European Union and United Nations) because photosynthesis eventually cycles the CO_2 back into new crops. In some cases, this recycling of CO_2 from plants to the atmosphere and back into plants can even be CO_2-negative, as a relatively large portion of the CO_2 is moved to the soil during each cycle. I disagree with classifying biofuels as renewable energy sources without indicating that their renewal is delayed because it takes a long time (sometimes decades) to regrow the plants that are burned.

There are multiple biofuel sources. Most fossil fuels can also be viewed as ancient biofuels that formed millions of years ago. Here, I discuss only the biofuels that have formed recently. These biofuels, such as agricultural, municipal, and industrial wastes, release their energy content when they are burned. Burning them is desirable because such waste (such as those of paper and pulp plants) would eventually decompose anyway and release the same amount of CO_2 as if they were burned.

Burning waste is desirable and can be called carbon-neutral because it increases our energy supplies (recovers energy content) without increasing the CO_2 emissions over the amount that would be emitted if they were left to decompose naturally. Using agriculture-based fuels (ethanol, biodiesel) for transportation or similar purposes is even better because, even though their formation is also a delayed carbon-neutral process, the delay is short, only 1 year instead of decades.

Cutting down forests to obtain biofuels is different from burning waste because, by burning trees, we are not only removing a carbon sink and adding CO_2 to the atmosphere, but we are also destroying the plants and animals that live in the forests. Seventy percent of Earth's animals and plants live in forests and cannot survive deforestation. The practice of clear-cutting converts forests into farmland or pastures for cattle grazing and involves building roads to accommodate the bulldozers and heavy equipment needed to remove the old-growth trees. We all know about the events in the Amazon rainforest and the fires in California, so I will not belabor this topic.

3.8.1 Biomass Power Plants

The preindustrial source of energy was wood. Wood is called a renewable energy source, but it is not a zero-carbon one. In regard to renewable energy, it is worth repeating that it takes a few minutes to fell a tree but a long time to regrow a forest. In process control language, we would call this type of process a long-dead-time, low-gain one. When we cut down a tree, we remove a carbon sink from the environment, and when we burn it, we emit a large quantity of carbon into the atmosphere. Later, when we plant a tree (if we do), it takes years before it becomes a carbon sink similar in size to the old tree. In fact, cutting and burning forests around the planet has the immediate consequence of removing operating carbon sinks while simultaneously emitting millions of tons of carbon into the atmosphere. Even if the forests were immediately replanted, fertilized, and watered, we would not see a carbon benefit for many years.

In terms of the present situation, it is not only that the trees are burned, but that branches of the trees are often left on the forest floor. The energy content of these potential fuels is not recovered at all, and as the branches decompose, they release their stored carbon into the atmosphere. In addition, in case of wildfires, they support and spread them. Moreover, burning wood is very inefficient because of its moisture content. Actually, the carbon release per megawatt-hour of electricity generated by burning wood is even higher than that of coal. Therefore, in terms of an immediate effect, cutting down trees reduces the carbon-sink capacity on the planet, and burning them increases carbon emissions.

Cofiring with biomass has increased in coal power plants because it does not require building new infrastructure. Cofiring is not without issues, however, because an upgrade for burning biomass is often necessary. The use of biomass fuel in power plants started earlier in the United States than elsewhere, and today, it is increasing worldwide. Most of these plants produce both electricity and heat (cogeneration), and the generated steam or hot water is often used for district heating.

Burning biofuels is desirable because biomass releases the same amount of CO_2 when it rots as when it is burned for energy. Therefore, if forests are managed in a sustainable manner, the cycle of harvesting and growing can be continued indefinitely in a *delayed carbon-neutral* manner. From a process control perspective, this is a delayed (lag-dominant) but controllable process.

If organic waste decomposes in the open, it releases methane; if the waste is burned, these methane emissions are eliminated. This is one advantage of burning organic waste. Other advantages of the biomass power industry are that it improves forest health (forest fires are reduced if the millions of tons of undergrowth and forest debris are removed), and it diverts material from open burns and landfills into becoming power supplies. Biomass is also a dependable energy source because its storage and continuous availability are guaranteed. In the United States, about 100 power plants burn biomass, and they generate more than half of the renewable electricity production in the country.

In 2017, several initiatives promoted sustainable biofuel development. These included establishing the BioFuture Platform and the Sustainable Biofuels Innovation Challenge organizations with the participation of 20 countries. In 2017, the global bioelectricity capacity was 122 GW, and it reached 136 GW by 2020. The amount of bioelectricity generated in 2017 was 555 TWh; a year later, in 2018, it reached 637 TWh; and it is still growing.

Denmark is probably the world's leader in the use of biomass-based energy, but woodchip fuels are widely used throughout the world. This trend is likely to continue because several large-scale combined heat and power plants are still in the process of converting from fossil fuels to the use of woody biomass. In the 1980s and 1990s, the types of solid biomass used largely consisted of straw, biodegradable waste, and fuel wood, but after 2000, the use of wood pellets and wood chips increased significantly. In Denmark, more than half of the woody biomass is imported. This mainly consists of wood pellets originating from the Baltic countries (Estonia and Latvia), the United States, Russia, and other European countries. Several types of solid biomass are used in Denmark, and most of the biomass is produced domestically. The share of imports amounts to 20% of the total use of solid biomass.

Biofuels already provide a significant portion of global energy consumption (twice that of nuclear), but the use of coal is still much greater. Power plants that previously burned coal can and should be converted to using biomass, as conversion is much less expensive than building new plants.

Fluidized-bed combustors (FBCs) are good for burning biomass for two main reasons. The first is that they can handle a wide variety of solid fuels, including ones that are difficult to burn in regular boilers. The second is that the FBCs can achieve low emissions of both nitric oxides and sulfur dioxide (SO_2).

The controls of circulating fluidized-bed (CFB) boilers for burning biomass (or combinations of biomass and other fuels) are similar to those of coal-burning boilers, but because they are newer, manual controls are slowly being replaced by automatic ones.

The key pollution controls are aimed at reducing the emissions of SO_2 and nitrogen oxides (NO_x), as well as CO_2, mercury, volatile organic compounds (VOCs), particulates, stack opacity, and other greenhouse gases and pollutants. Electrostatic precipitators are commonly used to remove particulates, and scrubbers are commonly used to remove both particulates and SO_2.

Reducing NO_x emissions is a major goal because of their role in forming ground-level ozone and acid rain. NO_x analyzer designs employ IR, UV, chemiluminescence, colorimetric, electrochemical, and chromatographic approaches. NO_x formation depends on the fuel's nitrogen content and the temperature of the flame because, at high flame temperatures, more dissociated nitrogen combines with the oxygen present in the excess air. Therefore, the best way to minimize NO_x formation is to reduce the flame temperature, reduce the excess oxygen, and/or burn low-nitrogen-containing fuels. Once NO_x is formed, selective catalytic reduction (SCR) can be applied by injecting mixtures of ammonia and air or ammonia and steam into the exhaust gas, which then passes through a catalyst and triggers the NO_x reduction reaction according to the equation:

$$4NO + 4NH_3 + O_2 \rightarrow 4N_2 + 6H_2O \tag{3-1}$$

Like NO_x, SO_2 is a major concern because of the relatively high sulfur content of most biomass fuels. Wet SO_2 scrubbers are the most widely used tools for flue gas desulfurization (FGD). The most popular sorbent is limestone, which is favored because of its wide availability and relatively low cost. The overall chemical reaction that occurs with limestone or lime sorbents is as follows:

$$SO_2 + CaCO_3 = CaSO_3 + CO_2 \tag{3-2}$$

3.8.2 Combined-Cycle Cogeneration

Biomass-based power plants often serve the dual functions of generating electricity and producing heat (usually steam) for industrial plants and district heating. For these plants, it is important that the biomass fuel be available year-round, which requires either large storage facilities or fuels such as wood chips (or garbage) that are available year-round. This means that seasonal energy customers, such as sugar mills or district heating, are not as desirable as users that operate year-round like paper mills and tire production plants.

To reduce steam transportation and heat losses, the steam users should be near the power plant (within 300 to 1500 m [1000 to 5000 ft]). Smaller combined-cycle power

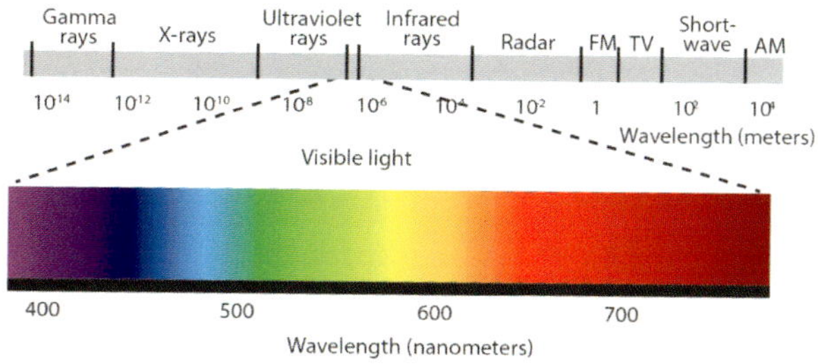

Figure 1-22. Potential wavelength ranges of sensors. (Béla Lipták, courtesy of *Control* magazine.)

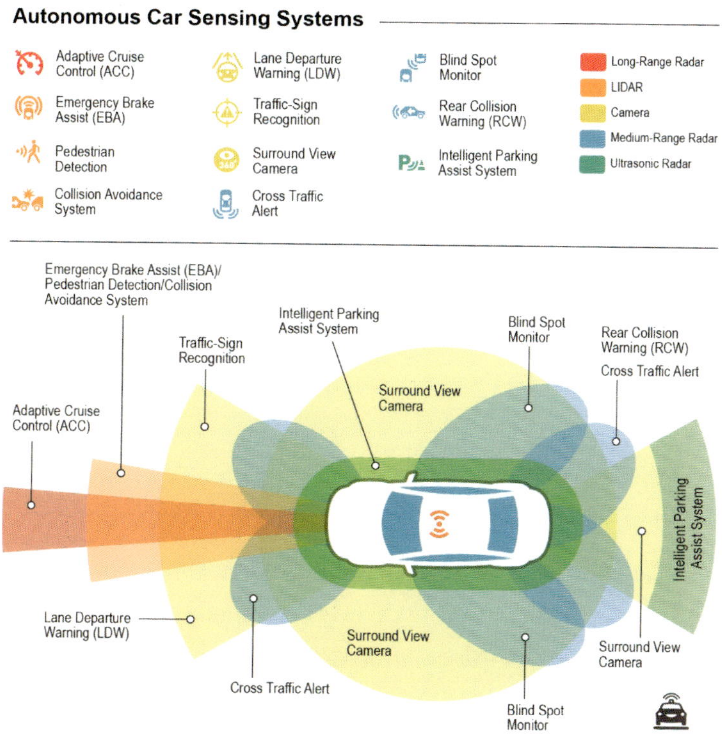

Figure 1-23. In autonomous driverless cars, different wavelength ranges are used for different measurement functions, including camera, radar, and LiDAR.

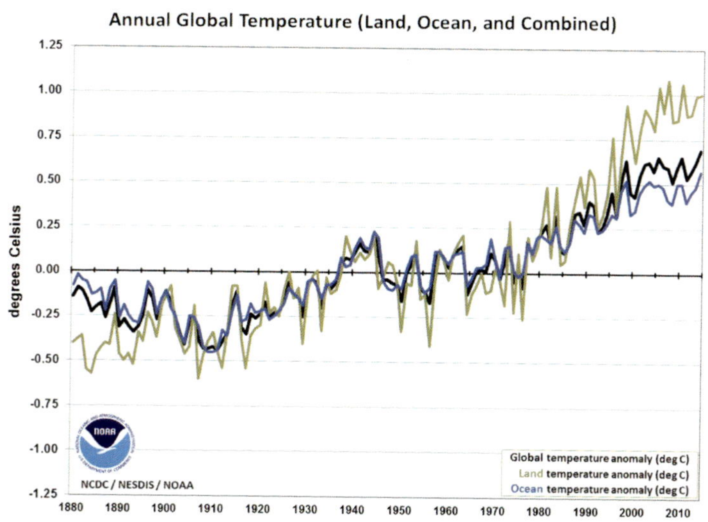

Figure 2-8. Increase in the average temperature anomaly of Earth between 1880 and 2010: (black) entire globe, (tan) land area, and (blue) oceans. During the past decade, GW increased even faster. ("Global Climate Report – Annual 2014: Various Global

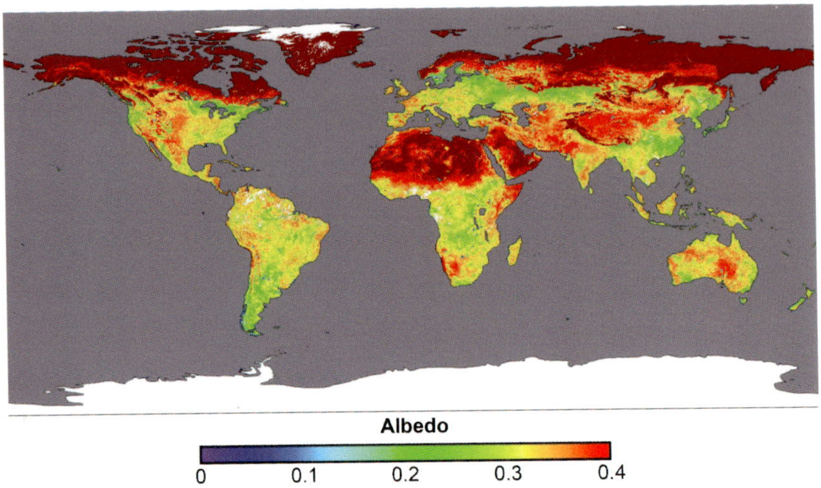

Figure 2-10. Values of the albedo (reflectivity) of the various surfaces on the planet according to the NASA Earth Observatory. ("Global Albedo," NASA Global Observatory.)

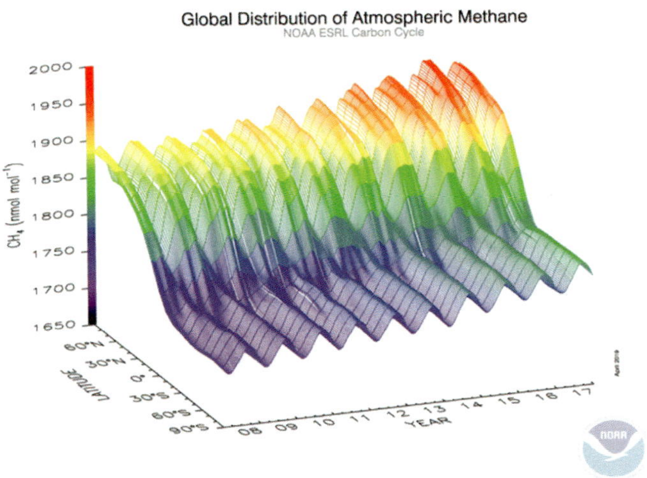

Figure 2-17. CO_2 concentrations in the marine boundary layer at different latitudes between the North and South Poles measured between 2008 and 2017. ("Global Distribution of Atmospheric Carbon Dioxide," NOAA.)

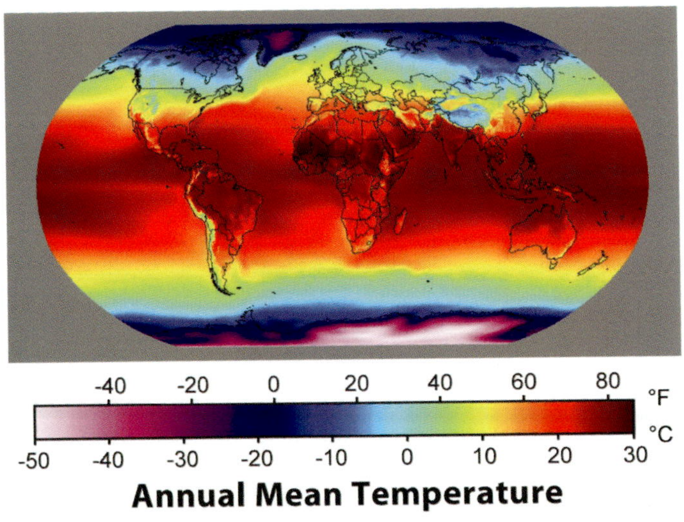

Figure 2-18. Annual mean temperatures around the planet. (Robert A. Rohde, "Annual Average

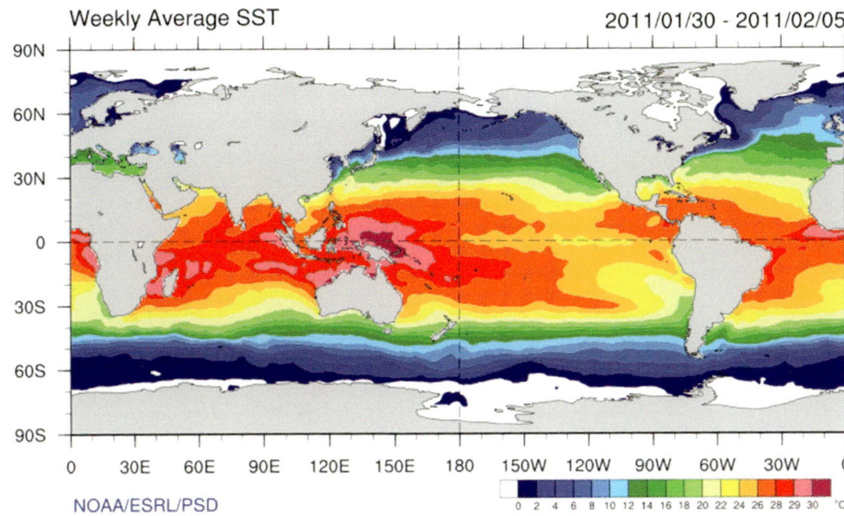

Figure 2-19. SSTs during the first 5 days of February 2011. ("Weeklysst.gif," Earth System Research Laboratory Physical Sciences Division.)

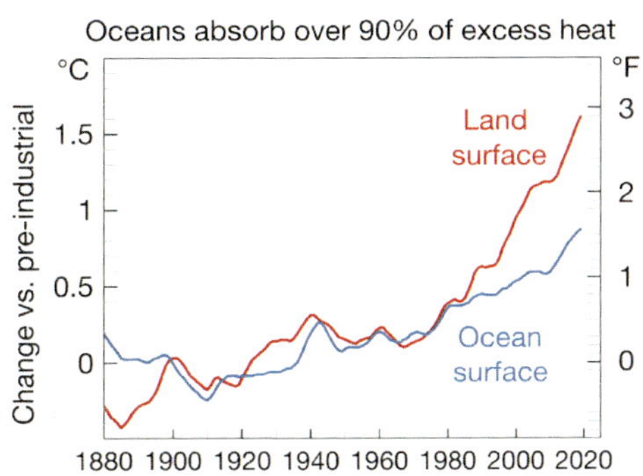

Figure 2-21. NASA data shows that land temperature increased faster than sea surface temperatures. The baseline for these curves is the preindustrial average temperature. (Efbrazil, "Land vs. Ocean Temperature.")

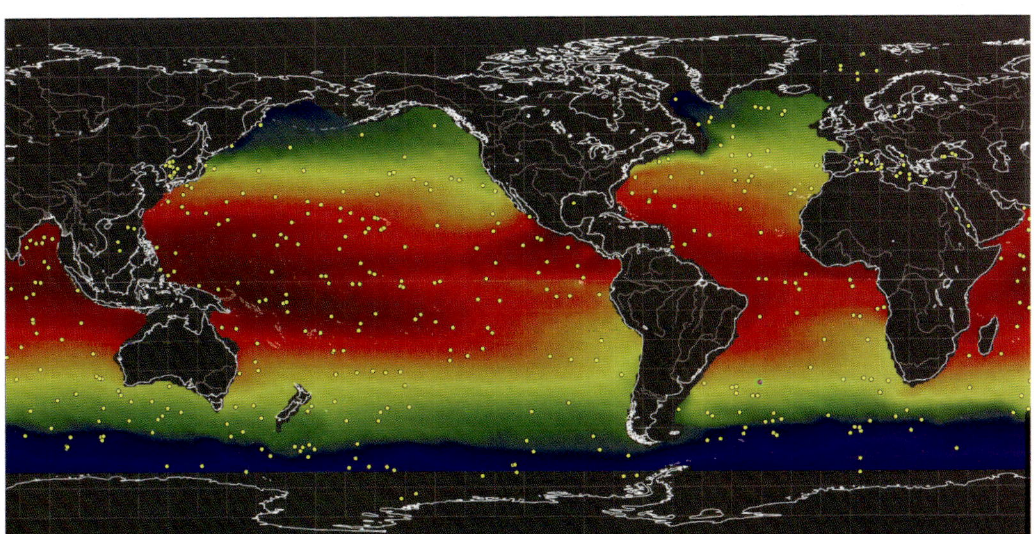

Figure 2-23. As of October 2020, nearly 4000 Argo floats (yellow dots) had been installed around the world. The background colors indicate ocean temperatures (red, warmest; blue, coldest). ("Argo's Status: Implementation Status," Argo Program website.)

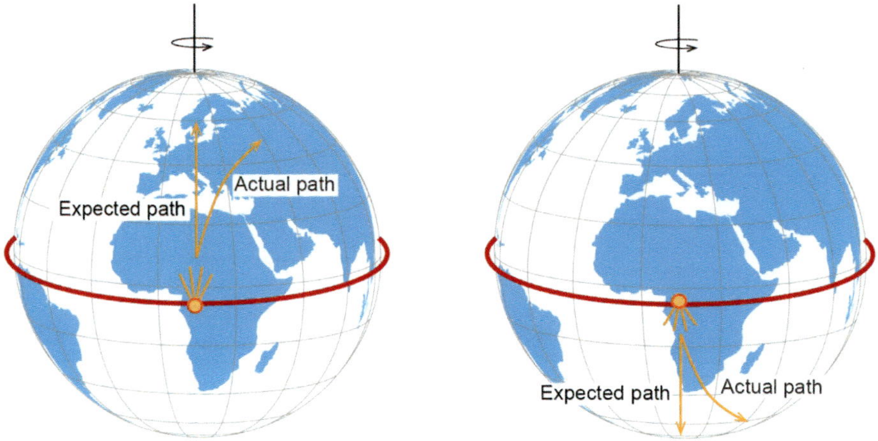

A Projectile fired northward **B** Projectile fired southward

Figure 2-27. The Coriolis force causes air and water currents to bend toward the east.

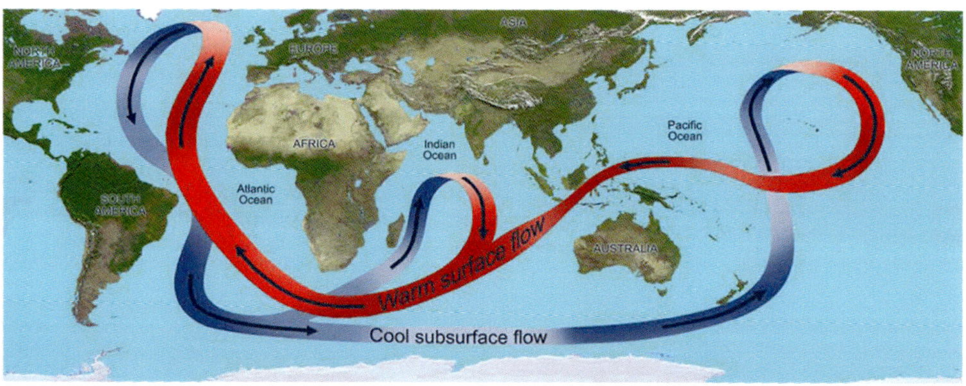

Figure 2-28. The dominant current, called the great ocean conveyor belt or the great heat conveyor belt (GHCB), moves ocean water around the globe. Warm water on the surface (red) moves northwest (from the Pacific to the Atlantic Ocean). ("Why does the ocean get colder at depth?," National Ocean Service website.)

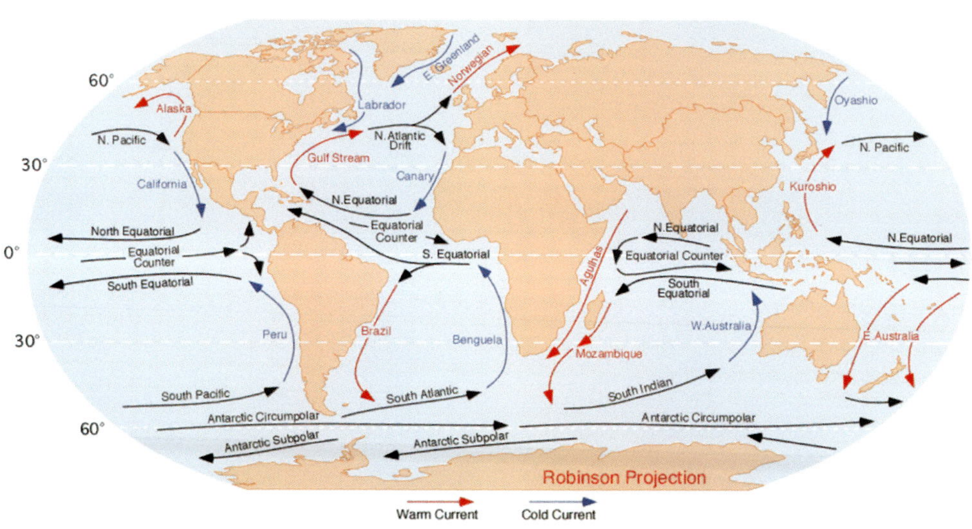

Figure 2-29. Schematic of oceanic currents, where red is warm and black is cold. Warming currents (red) in both hemispheres move heat toward the poles. (Michael Pidwirny, "Corrientes-oceanicas.png.")

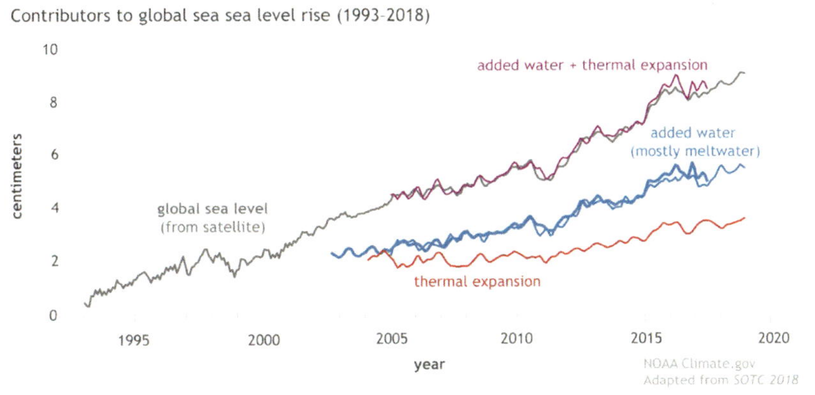

Figure 2-30. Contributions to ocean level rise during the past quarter century. (Rebecca Lindsey, "Climate Change: Global Sea Level," Climate.gov website.)

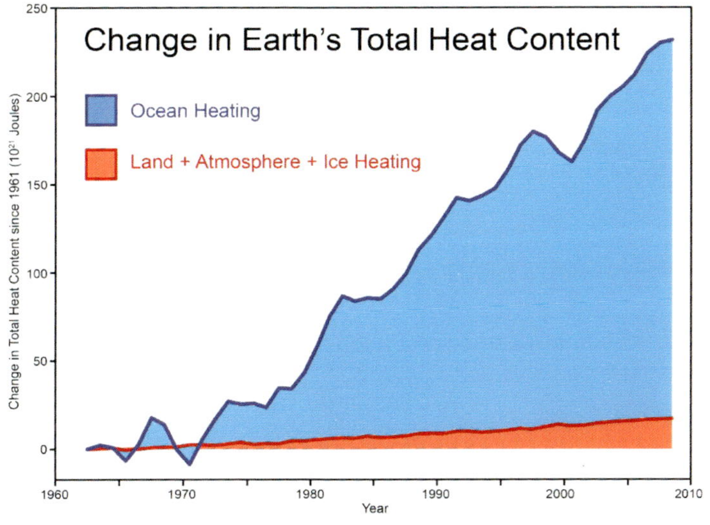

Figure 2-32. Global heat accumulation: Some 90% of the GW energy accumulates in the oceans. By 2010, GW caused the accumulation of more than 2.0 × 1023 J (2.0 × 105 Q); by the time of this writing, this accumulation had increased by another 25%. ("Total Heat Content [2011 update]," Skeptical Science.)

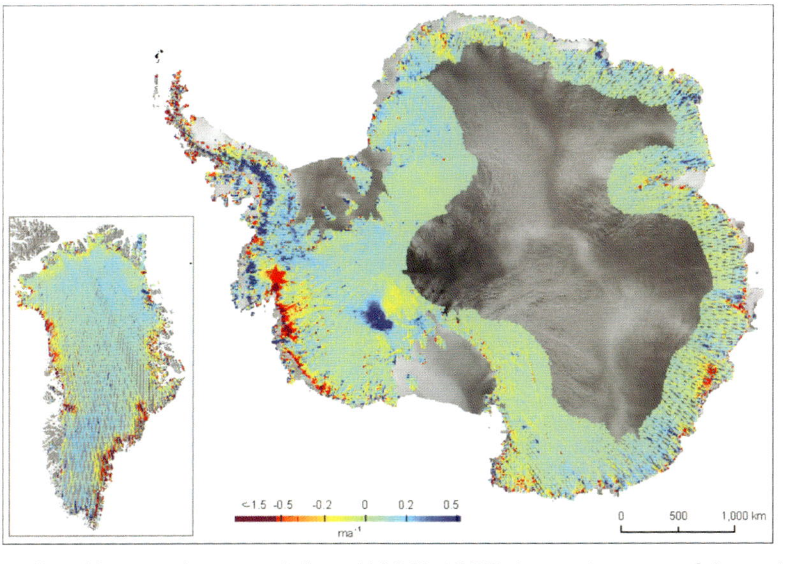

Figure 2-37. Satellite altimetry photograph from NASA's ICESat reveals areas of dynamic thinning (red) in Antarctica (main image) and Greenland (bottom left). The colors indicate ice loss (red) and gain (blue) in meters. ("Fast-Flowing Glaciers Thin Greenland and Antarctic Ice Sheets," NASA.)

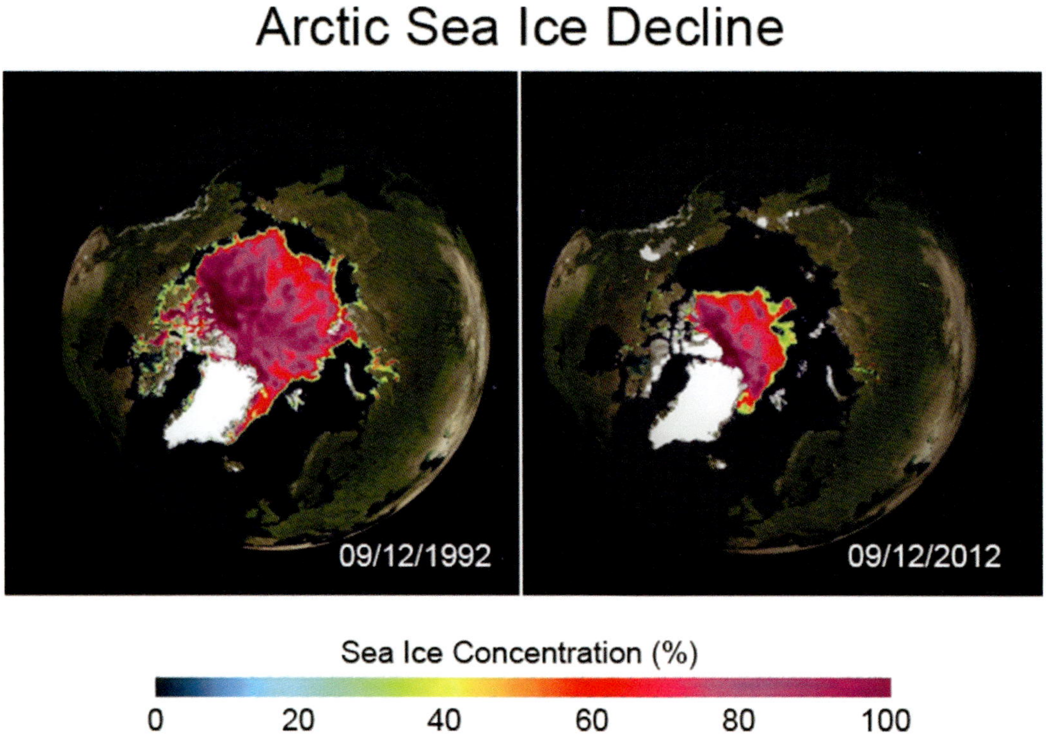

Figure 2-40. The area of high sea ice concentration (the amount of ice in the seawater in September) in the Arctic decreased by 50% between 1992 and 2012. (Courtesy of the Department of Atmospheric Sciences, the University of Illinois at Urbana-Champaign; from "Arctic Sea Ice Decline.")

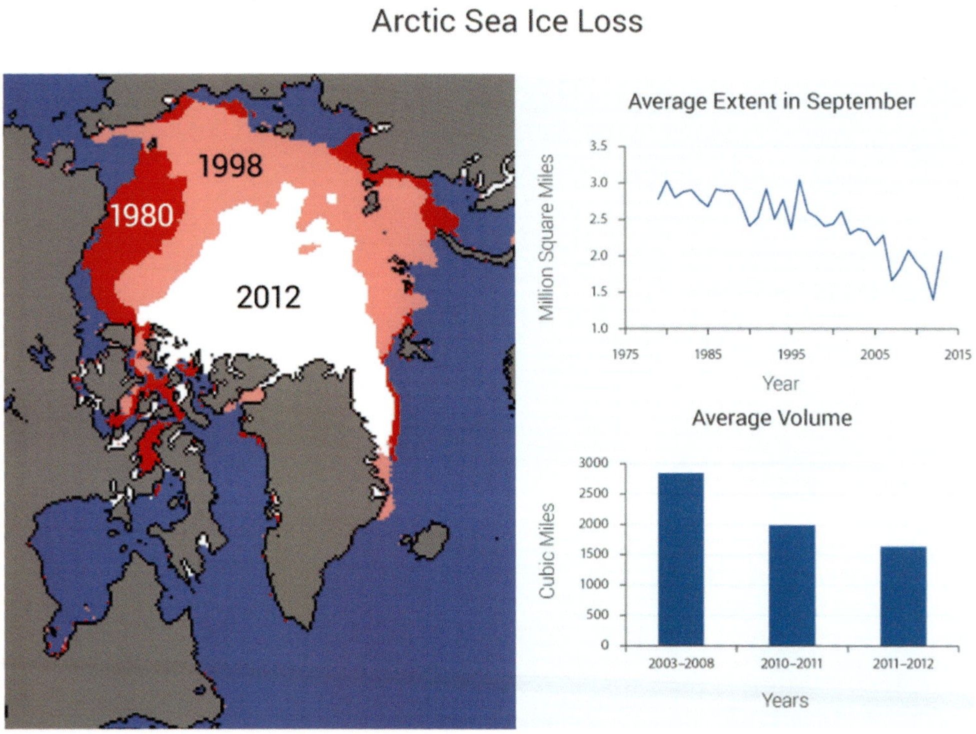

Figure 2-41. The area of the ice cover of Greenland is shrinking. ("Decline in Arctic Sea Ice Extent," National Snow and Ice Data Center.)

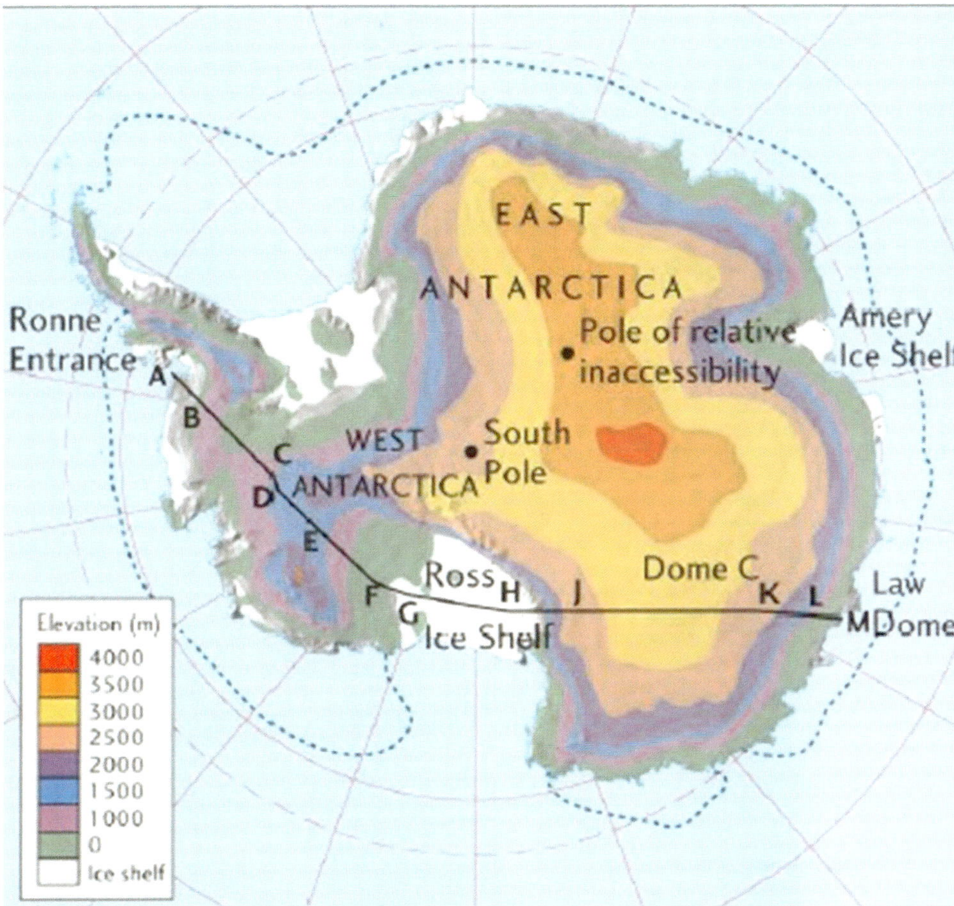

Figure 2-45. Antarctic ice sheet. East Antarctica is much higher in elevation than West Antarctica. (Erik Conway, "Is Antarctica Melting?," NASA/Jet Propulsion Laboratory.)

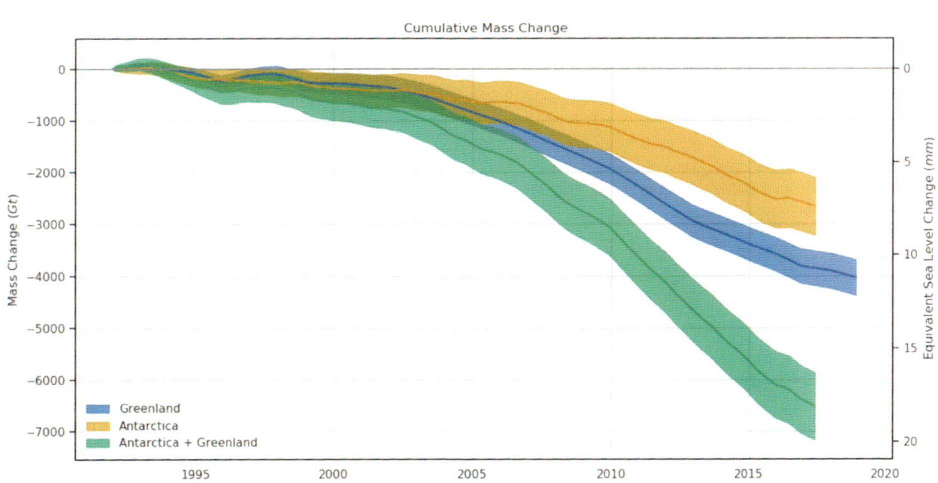

Figure 2-48. The amount of ice loss and resulting sea level change caused by the melting across Antarctica (yellow), Greenland (blue), and if these losses are combined (green). (Courtesy of Andrew Shepherd.)

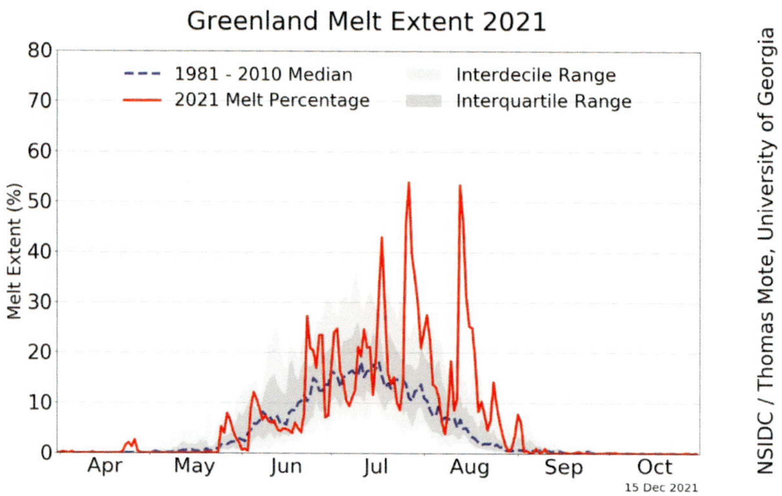

Figure 2-49. Percentages of melting by month for the years 1981 (blue) and 2018 (red) showing that, in Greenland, almost all glacial melting occurs between May and September. ("Greenland Melt Extent 2018," National Snow & Ice Data Center.)

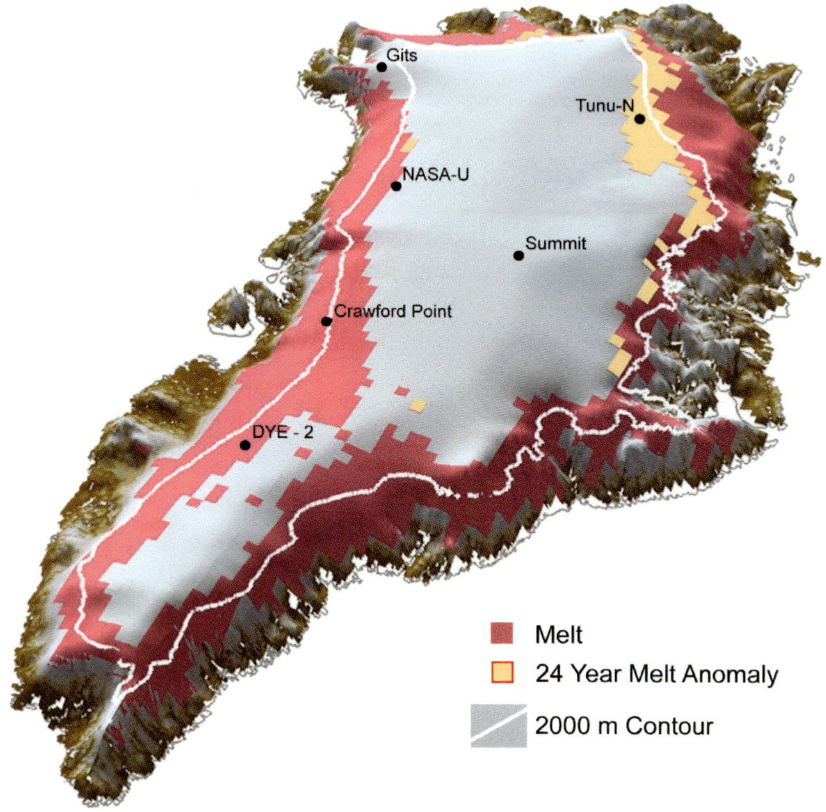

Figure 2-50. Computer graphic showing the 2002 extent of melting of Greenland's ice sheet in pink. (Courtesy of Konrad Steffen and Russell Huff, CIRES, the University of Colorado at Boulder.)

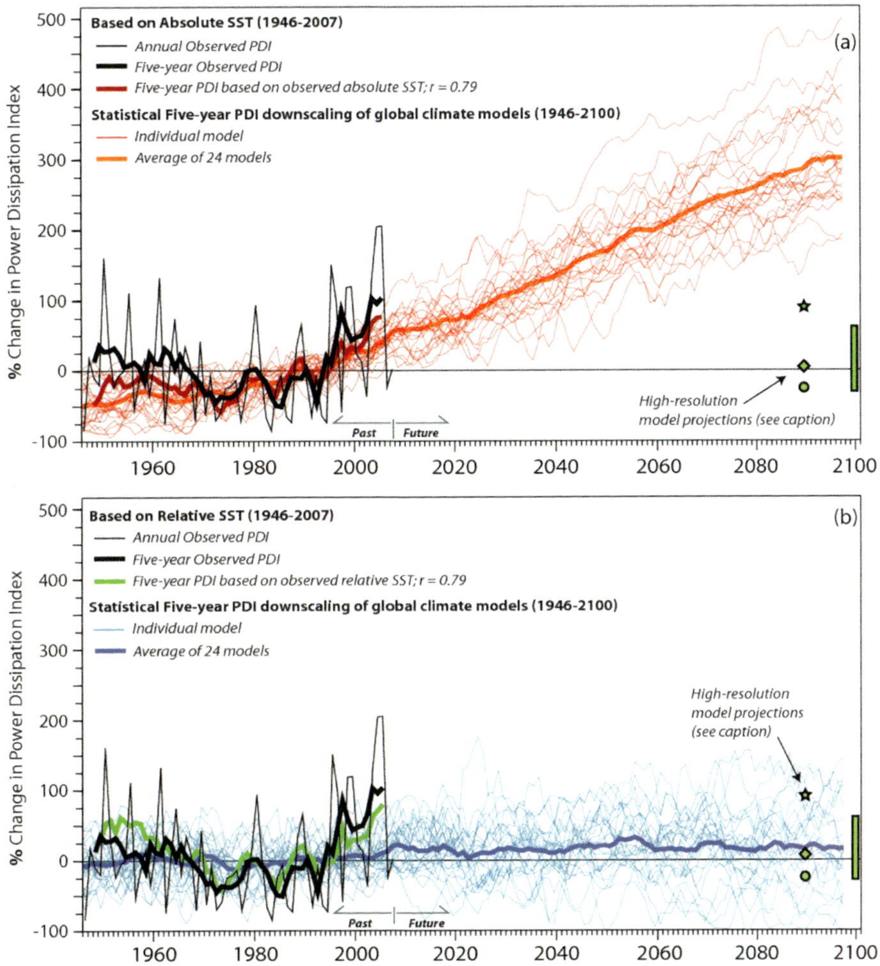

Figure 2-53. Past and extrapolated variations in the PDI of the Atlantic Ocean basin as calculated from the absolute (a.) and relative (b.) SSTs. (Gabriel A. Vecchi, Kyle L. Swanson, and Brian J. Soden, "Whither Hurricane Activity?," NOAA.)

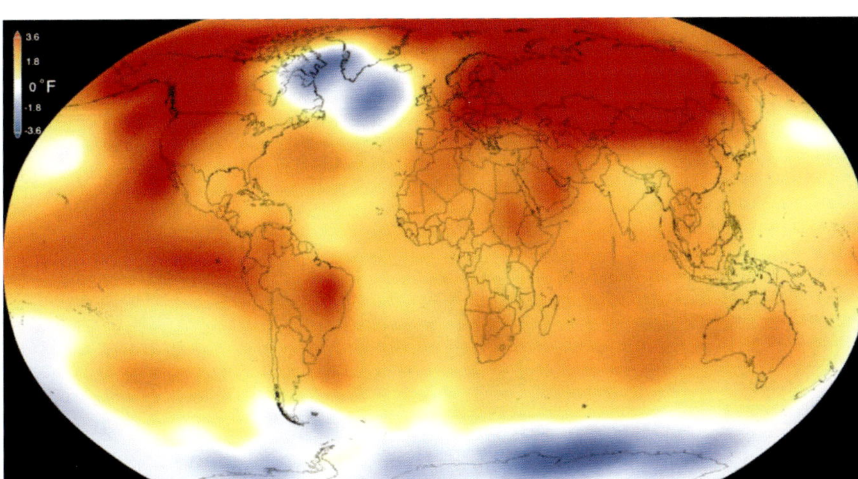

Figure 2-57. The warming of the globe is not uniform. ("NASA, NOAA Analyses Reveal Record-Shattering Global Warm Temperatures in 2015," NASA Scientific Visualization Studio/Goddard Space Flight Center.)

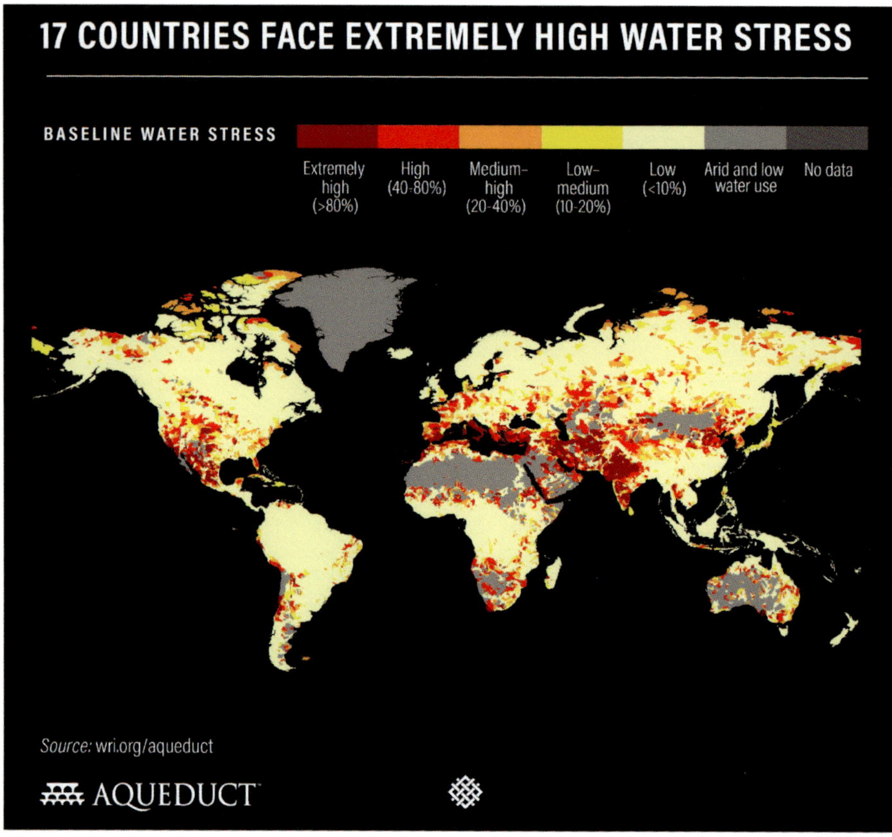

Figure 2-59. According to the World Resources Institute, about one-quarter of the world's population in the indicated regions is experiencing extreme water shortages. (Courtesy of Rutgers Willem Hofste, Paul Reig, and Leah Schleifer; Aqueduct, World Resources Institute; from "17 Countries, Home to One-Quarter of the World's Population, Face Extremely High Water Stress.")

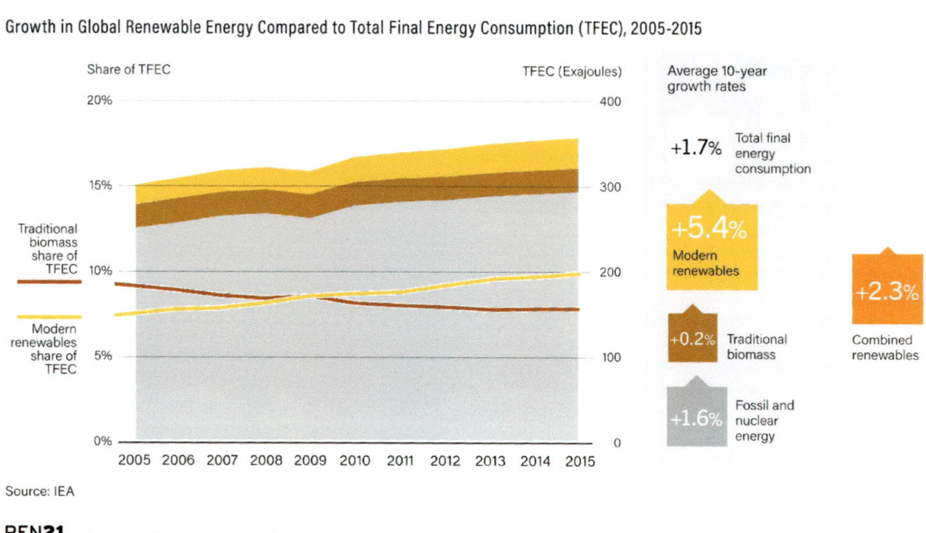

Figure 3-2. From 2005 to 2015, the total final energy consumption (TFEC) increased at about 2% per year, whereas the rate of increase of modern energy sources (solar, wind, etc.) was more than 5% per year. (Courtesy of REN21; from *Advancing the Global Renewable Energy Transition: Highlights of the REN21 Renewables 2018 Global Status Report in Perspective.)*

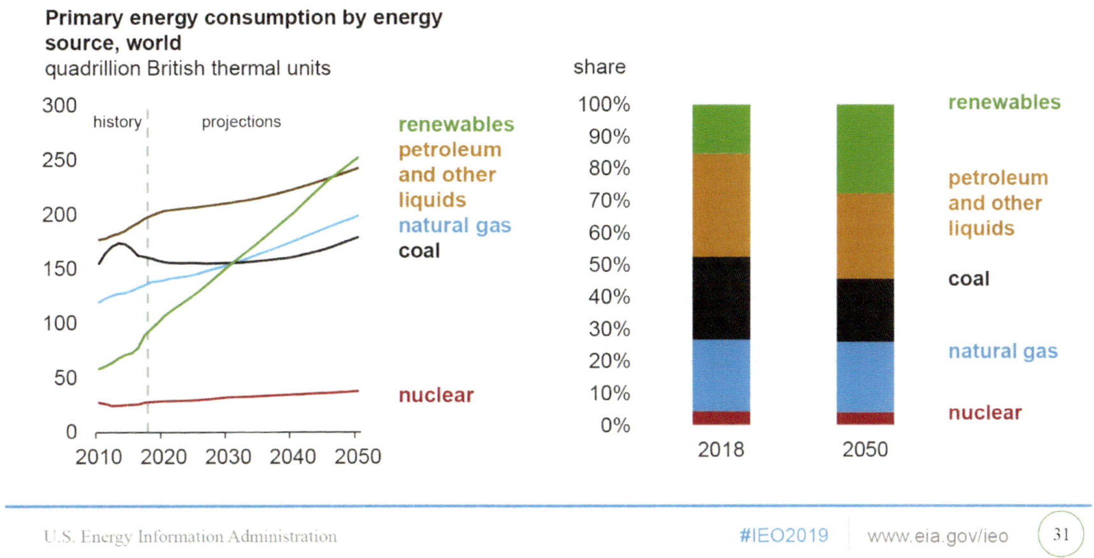

Figure 3-3. Projection of global primary energy consumption between 2010 and 2050 in (quads) (1Q = 1015 Btu). ("EIA projects nearly 50% increase in world energy usage by 2050, led by growth in Asia," US EIA.)

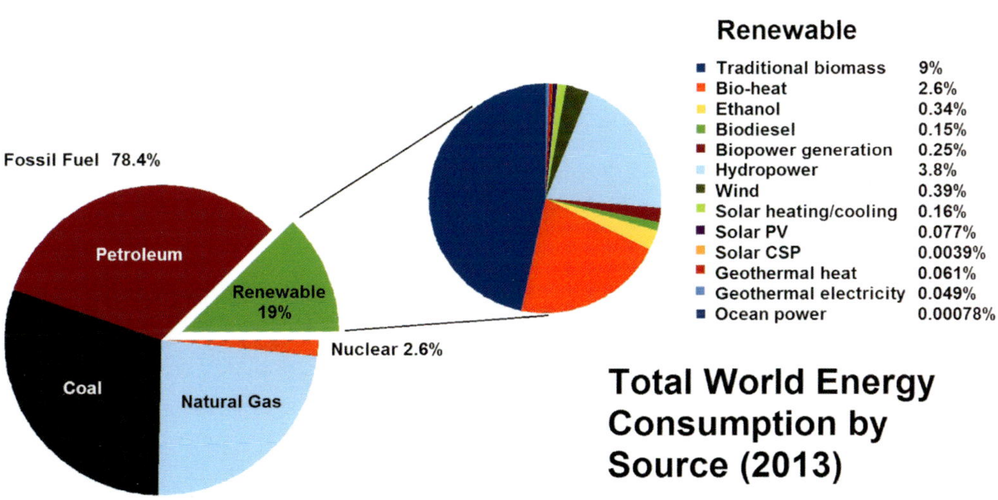

Figure 3-4. Illustration of the confusing terminology used to describe global renewable energy consumption in 2013. (Delphi234, "Total World Energy Consumption by Source 2013.png.")

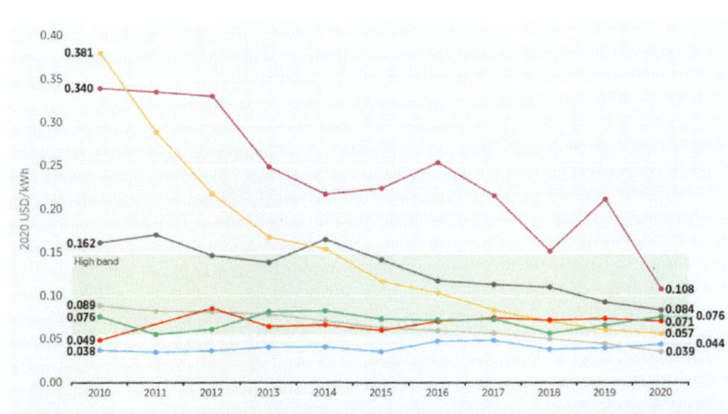

Figure 3-6. International Renewable Energy Agency (IRENA) estimates of the costs of green electricity by source. (*Renewable Power Generation Costs in 2020*, IRENA.)

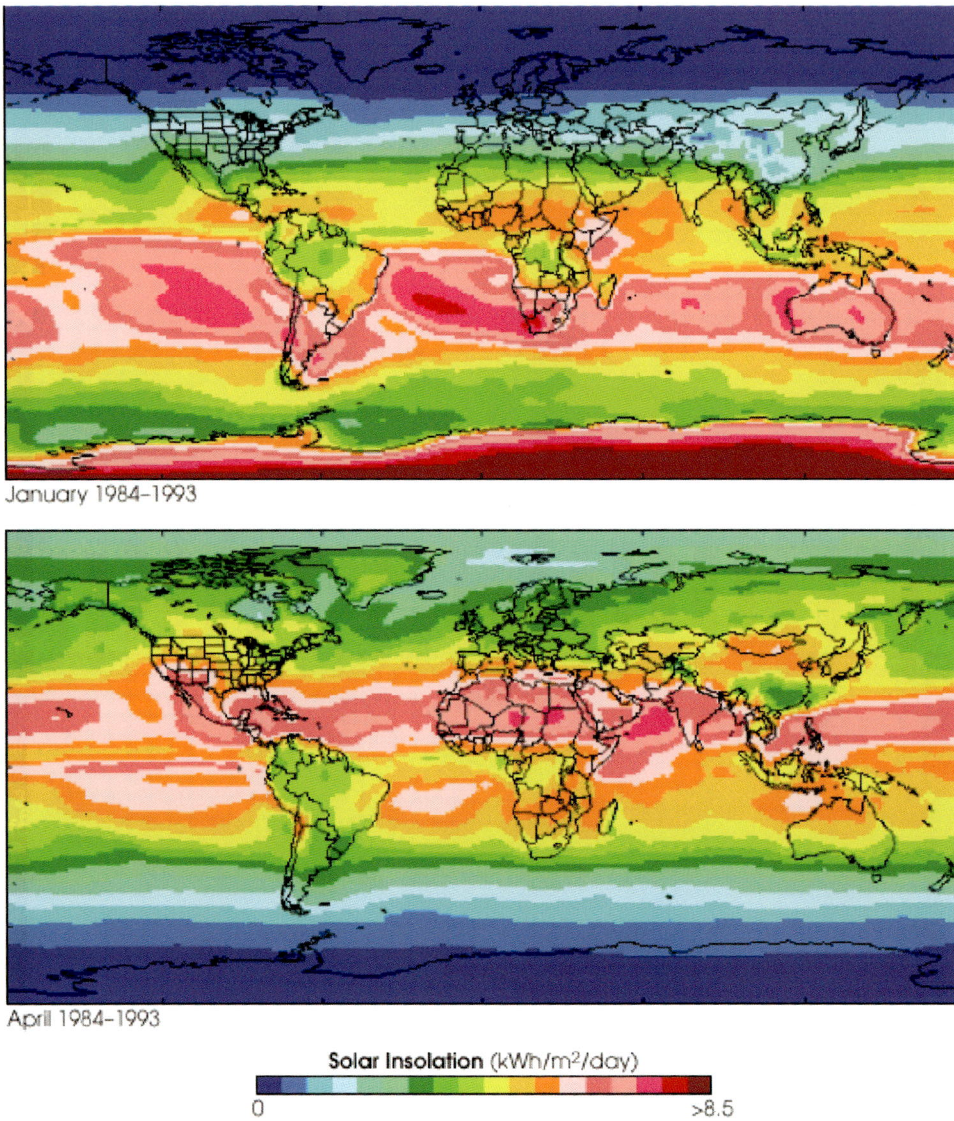

Figure 3-9. Global insolation in units of kilowatt-hour per square meter per year in January (top) and April (bottom). ("Measuring Solar Insolation," NASA Earth Observatory.)

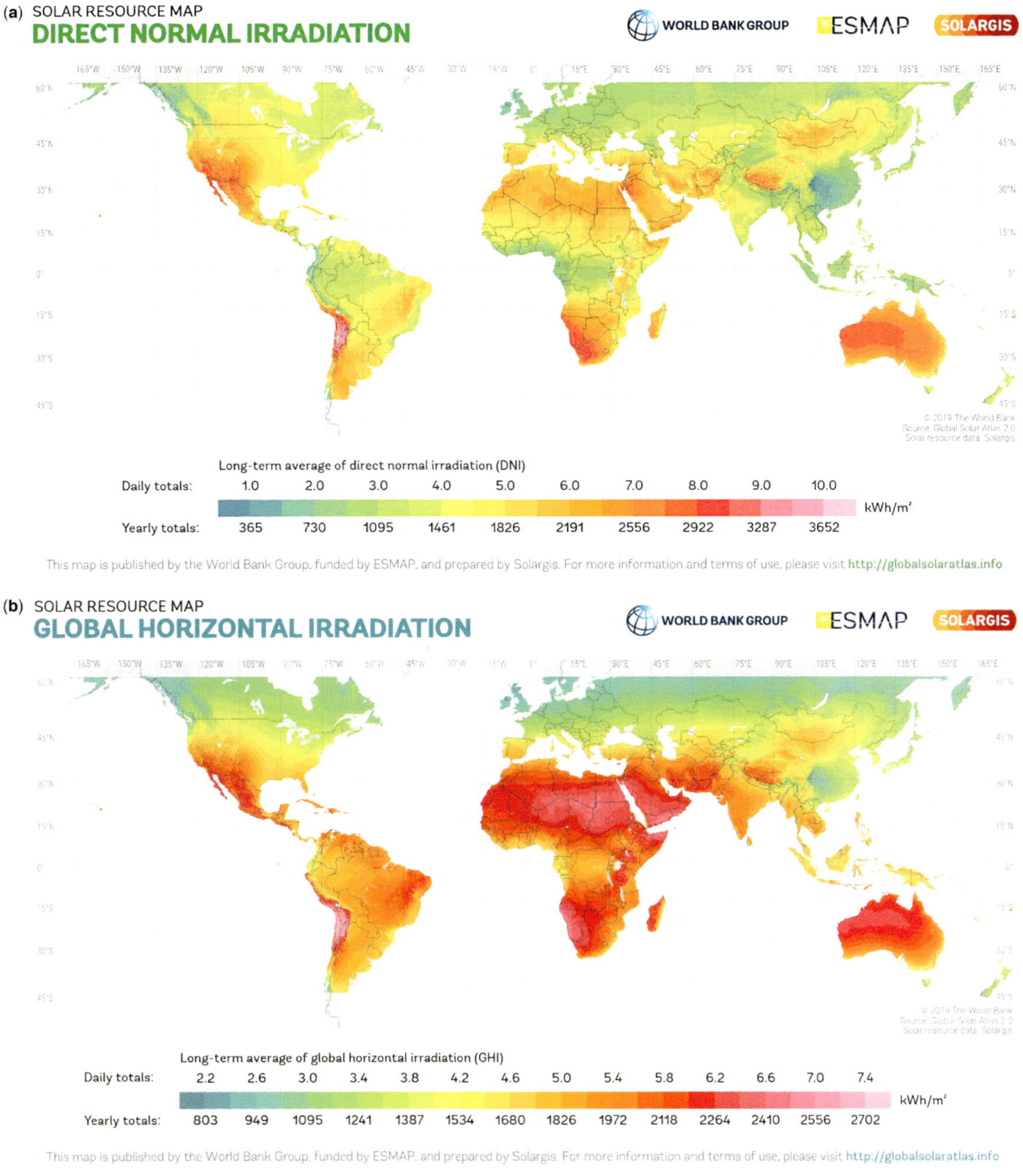

Figure 3-10. Daily and annual total insolation in April in units of kilowatt-hour per square meter per year throughout the world: (a) global horizontal irradiation, (b) direct normal irradiation. (Global Solar Atlas 2.0, Solargis s.r.o. on behalf of the World Bank Group.)

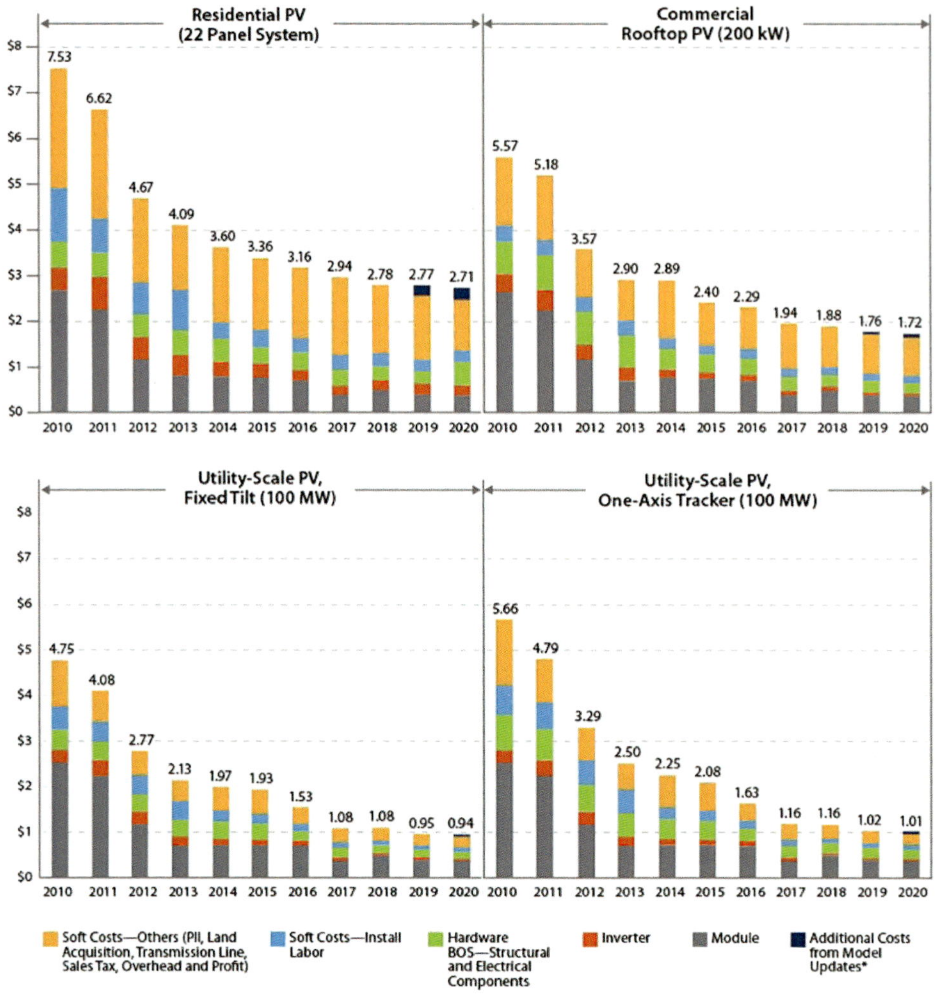

Figure 3-30. During the past decade, all cost components of solar energy installations have declined, except for their soft costs. ("Documenting a Decade of Cost Declines for PV Systems," NREL.)

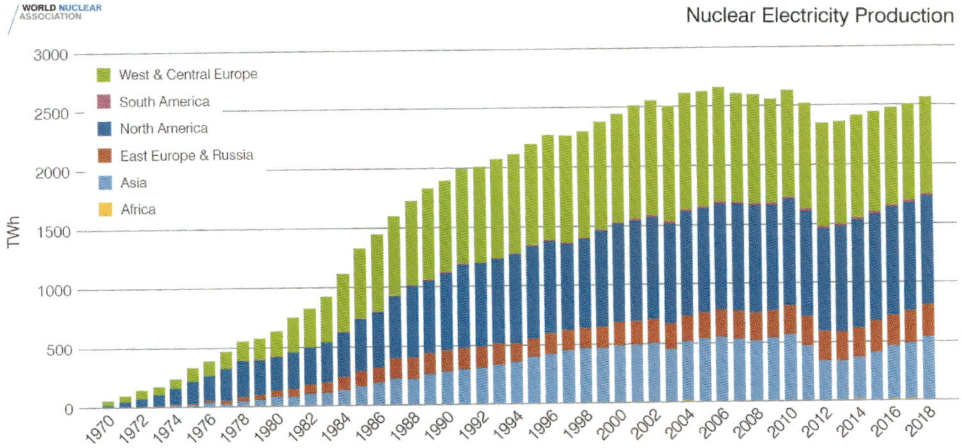

Source: World Nuclear Association and IAEA Power Reactor Information Service (PRIS)

Figure 3-49. Nuclear power production worldwide. ("Nuclear Power in the World Today," World Nuclear Association.)

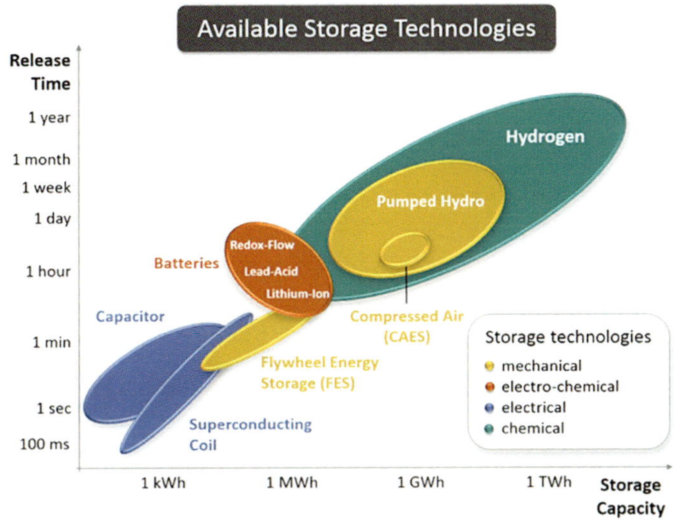

Figure 4-2. Energy storage capacities and release-time ranges of various energy storage systems. (Courtesy of Paul O'shea; from "Solar Storage Moves Out of the Lab into Mainstream: Special Report," *Power Electronics News*.)

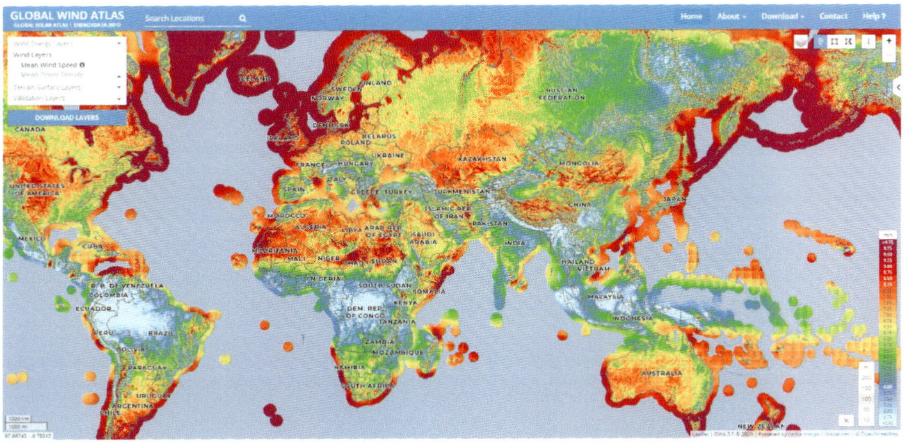

Figure 4-7. Map from the Global Wind Atlas. (Note that yellow and darker areas have good potential for wind energy.) (Global Wind Atlas 3.0, Technical University of Denmark.)

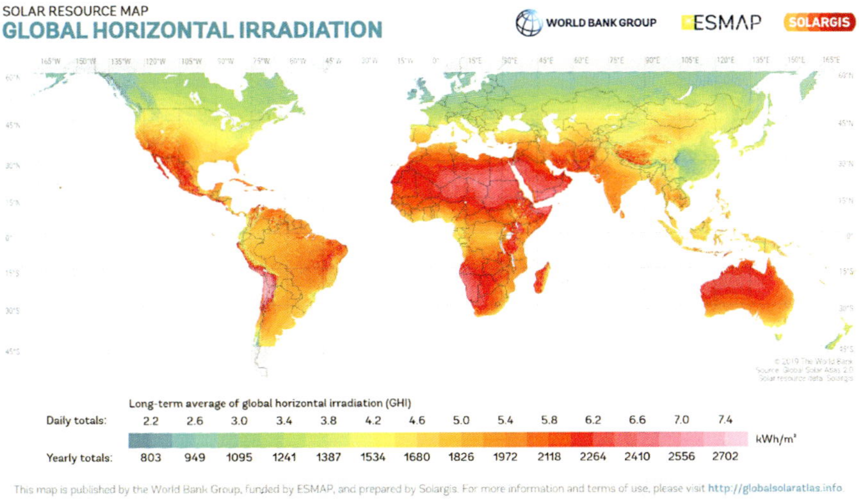

Figure 4-18. The yearly average global horizontal irradiation (GHI) received per unit area of Earth's surface ranges between 500 and 2500 kWh/m^2. (Global Solar Atlas 2.0, Solargis s.r.o. on behalf of the World Bank Group.)

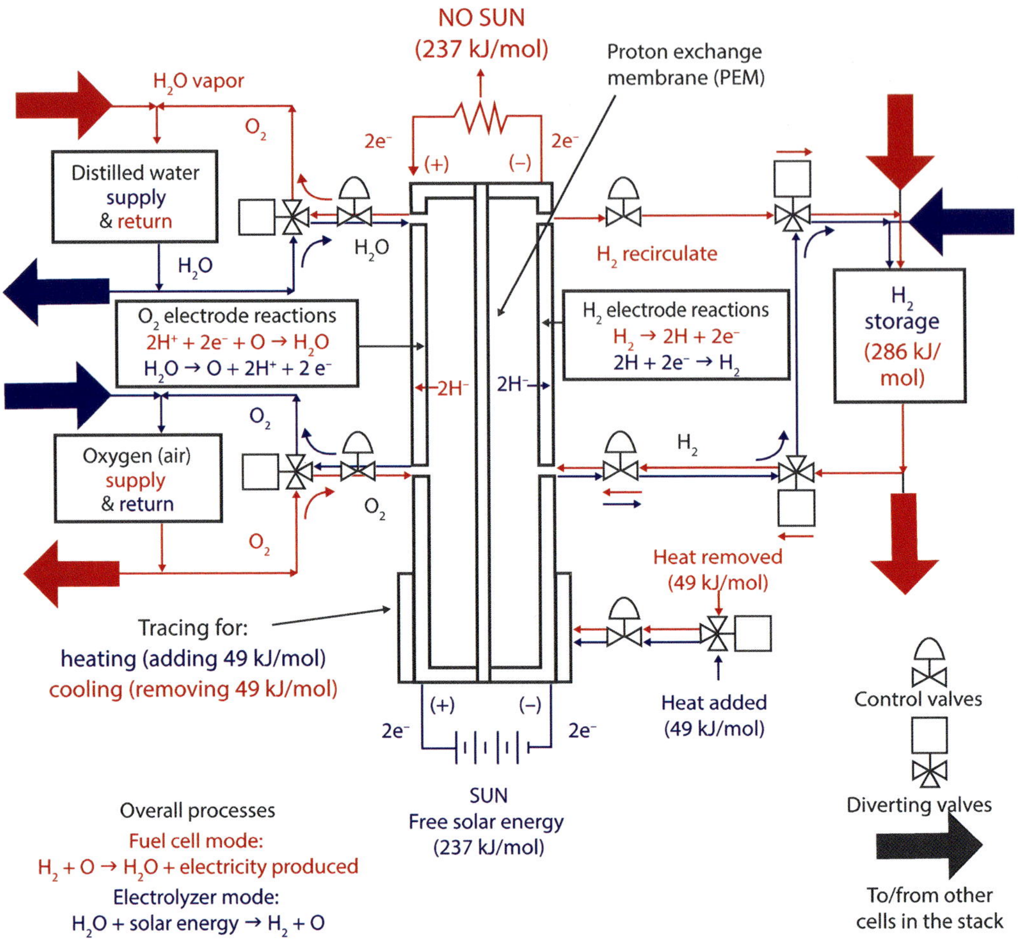

Figure 4-40. Developing a fully automated, optimized, and mass-produced RFC package will take a decade or more to finalize, and once finalized, it might look like this. The blue lines show flow directions for daytime mode (hydrogen generation), and the red lines show the flow directions for nighttime mode (hydrogen consumption). (Béla Lipták, courtesy of *Control* magazine.)

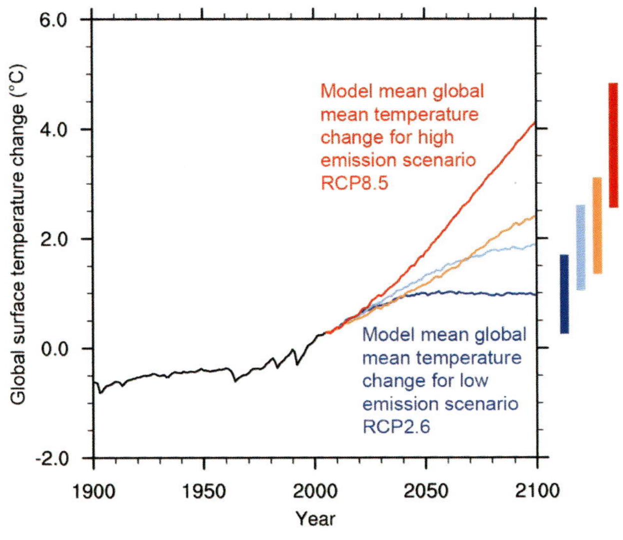

Figure 6-2. Scenarios for future GW trends under four emissions trajectories, called RCPs, where the vertical colored bars show the ranges of uncertainty. (Courtesy of IPCC; from *Climate Change 2013: The Physical Science Basis.*)

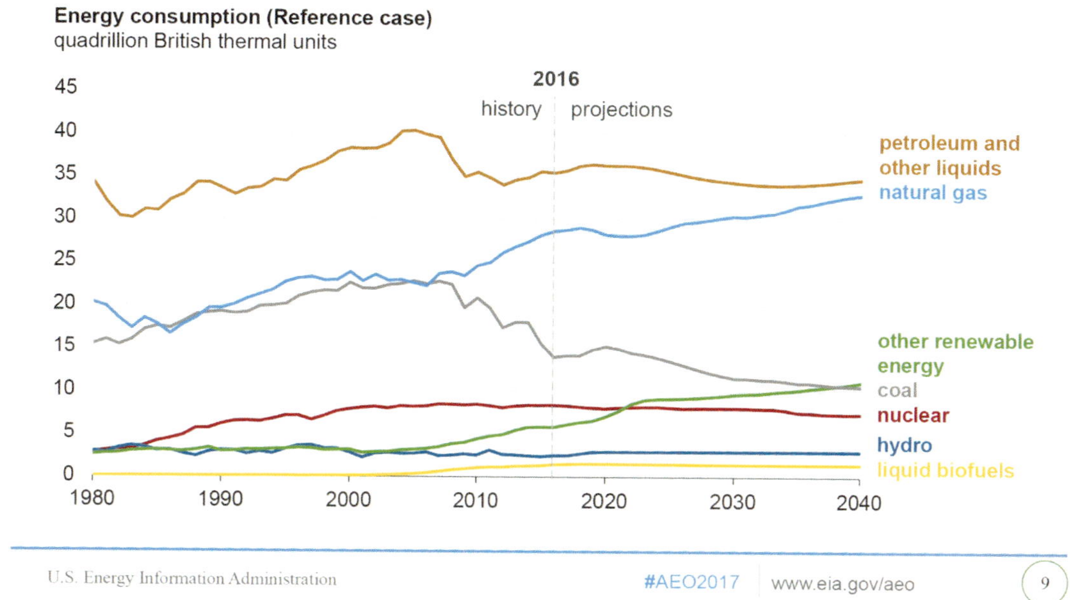

Figure 6-3. In *International Energy Outlook 2017*, the US Energy Information Administration (EIA) projects that world energy consumption will grow by 28% between 2015 and 2040. (*International Energy Outlook 2017*, EIA.)

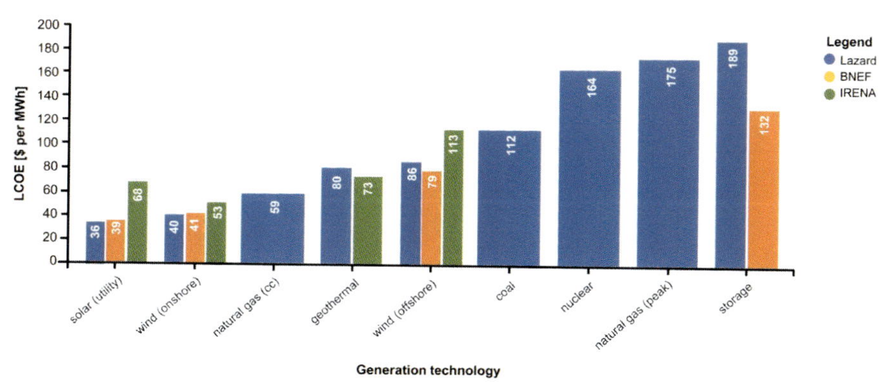

Global levelized cost of generation (US$ per MWh)

Source	Solar (utility)	Wind onshore	Gas CC	Geothermal	Wind offshore	Coal	Nuclear	Gas peaker	Storage (1:4)
BNEF	39	41			79				132
Lazard	36	40	59	80	86	112	164	175	189
IRENA	68	53		73	113				
Lazard (ranges)	29-42	26-54	44-73	59-101	86	65-159	129-198	151-198	132-245

Figure 6-4. Global LCOE cost ($/MWh) of electricity. Sources of information: Bloomberg New Energy Finance (BNEF), 2021; Lazard and the International Renewable Energy Agency (IRENA), 2020.

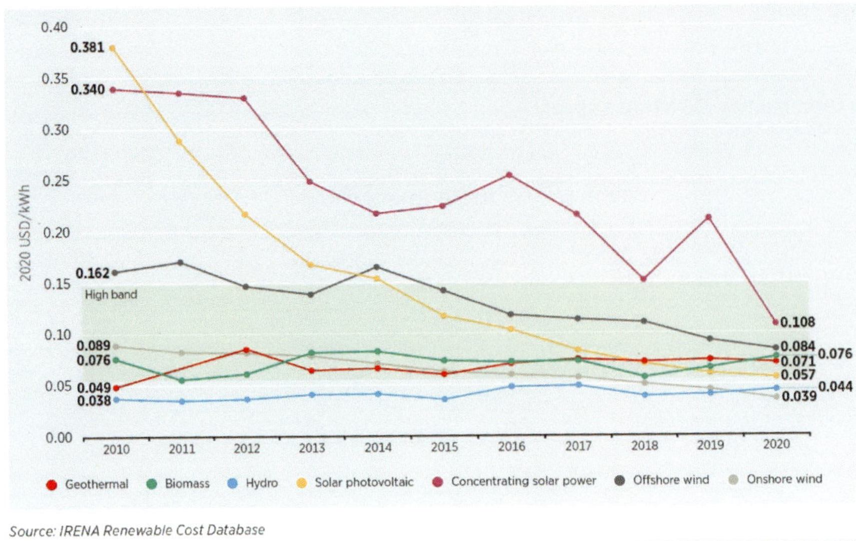

Figure 6-5. IRENA estimate of green electricity costs. (*Renewable Power Generation Costs in 2020*, IRENA.)

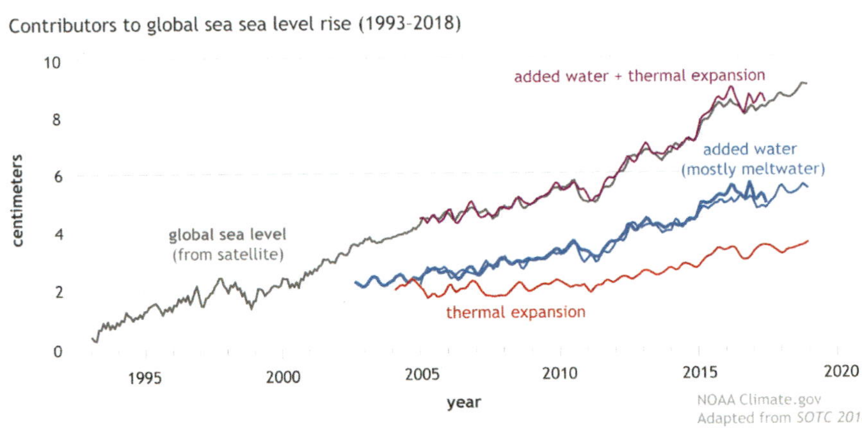

Figure 6-6. Contributions to ocean level rise during the past quarter century. (Rebecca Lindsey, “Climate Change: Global Sea Level,” Climate.gov news.)

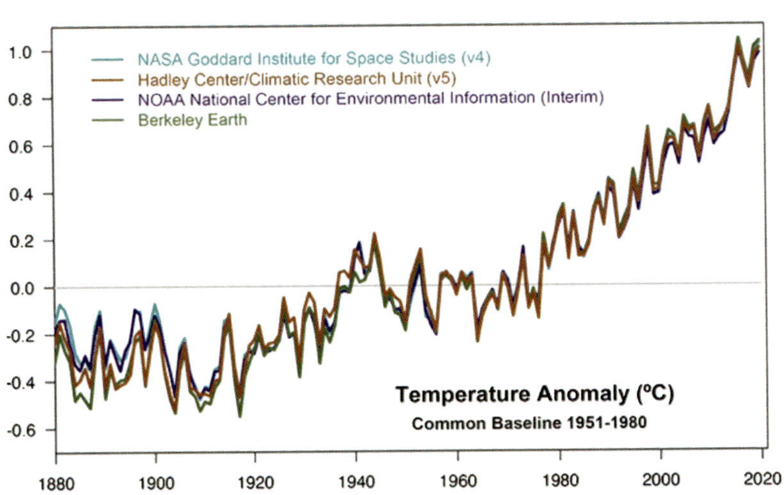

Figure 6-10. Comparison of four independent methods for estimating global temperature anomalies. (NASA.)

plants (up to 60 to 100 MW) are particularly suitable for cogeneration because they can be located very close to the thermal user or even within the fence of an industrial plant.

A load-following, cogeneration control system is shown in Figure 3-71. The control system follows the demand for steam in such a way that when the demand for steam heating drops, it sends the excess steam to the turbine to make electricity. When the demand exceeds the available supply, the control system supplements the steam from the turbine.

In this control configuration, the low-pressure steam supply pipe has two pressure-indicating controllers (PICs) in series and a fail-closed (FC) control valve (PV-2) between them. The set point of PIC-1 is higher than that of PIC-2, which provides the driving force for the steam to pass through the valve. The direct-acting (DA) PIC-1 controls the excess steam to the turbine by throttling the control valve (PV-1), whereas the reverse-acting (RA) PIC-2 controls the steam supplement from the turbine by throttling the fail-closed control valve (PV-3). PV-2 between the two controllers is controlled by the lower of the two controller output signals, the selection of which is made by the low-signal selector (PY-2).

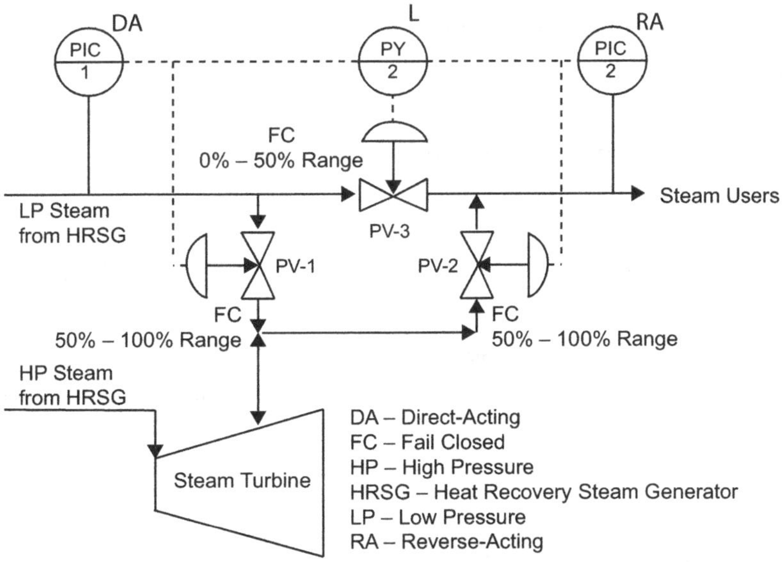

Figure 3-71. Control system for the cogeneration of steam and electricity. When the demand for LP steam is low, the excess steam is sent to the turbine (PV-1) to generate electricity. When the demand for LP steam exceeds the supply, the supply is supplemented by steam from the turbine (PV-2). The low-signal selector (PY-2) selects the controller, which throttles the fail-closed control valve (PV-3). Additional abbreviations: L, low-signal selector; PIC, pressure indicator controller; PV, pressure valve.

Source: Béla Lipták, *Process Control and Optimization, Vol. 2 of Instrument Engineers' Handbook*, 4th ed. (Boca Raton, FL: CRC Press, 2006), Chapter 8, Section 8.33.

All three control valves are fail-closed, but their throttling ranges differ: PV-1 and PV-3 valves throttle between 50% and 100%, so they are closed when their control signals drop below 50%, whereas the PV-2 valve throttles between 0% and 50%, and is fully open when its control signal rises to 50%.

When there is no demand for low-pressure steam, the pressure in the pipe rises, and the output signal of the DA PIC-1 rises to its maximum (100%), fully opening PV-1, which sends all the steam to the turbine. At this time, the low-signal selector (PY-2) blocks the PIC-1 output and selects the output of PIC-2 because that output signal is lower (0%), which closes PV-3.

When the demand for steam begins, the same configuration remains in operation, but as the pipeline pressure drops, PIC-1 starts to close the previously fully open PV-1 while also reducing its signal to PY-2 (down from 100% to 50%). At the same time, as the pipeline pressure starts dropping, the PIC-2 output signal starts rising. As long as the signal is below 50%, PV-2 remains closed, and the PIC-2 output continues to control PV-3. At this point, the PIC-1 output is dropping but is still above 50%, and the PIC-2 output is rising but still below 50%. This condition is maintained until the demand for low-pressure steam starts to exceed the supply.

At that point, the steam pressure in the supply pipe starts to drop, which signals that the steam supply to the users must be supplemented. As the pressure drops, the PIC-1 output signal also drops, and when it reaches 50%, it fully closes PV-1. At the same time, as the pipeline pressure drops, the output signal of PIC-2 rises, and when it reaches 50%, it starts to open PV-2.

At this point, the low-signal selector (PY-2) switches the control signal to PV-3, and PIC-1 starts to control it. This state of operation remains in effect as long as the demand for low-pressure steam exceeds the supply, and it returns to the previous configuration when the demand drops below the supply. This automatic load following increases operation efficiency by 10% or more.

3.8.3 Biofuels

In addition to directly obtaining its heat energy content, we can also use wood or other types of biomass to produce biofuels. In 2011, the IEA set the goal of meeting more than a quarter of the world's demand for transportation fuels from biofuel sources by 2050. Biofuels can be produced from plants or agricultural, commercial, domestic, and industrial wastes. Renewable biofuels are formed by contemporary carbon fixation that occurs in plants and microalgae through photosynthesis. Biomass can be converted to a better and more practical fuel in three ways: thermal, chemical, and biochemical conversion. The conversion can produce fuel in solid, liquid, or gaseous form.

Biofuel is produced directly by the anaerobic digestion of energy crops or indirectly from agricultural, commercial, domestic, and/or industrial wastes. In theory, biofuels are carbon-neutral because the quantity of CO_2 absorbed by the plants is equal to the amount released when the fuel is burned. However, in practice, whether or not a biofuel is carbon-neutral also depends greatly on whether the biofuel plant is regrown. Even if regrown, it becomes a carbon sink only sometime after it has been burned.

Whereas first-generation biofuels were made from sugars and vegetable oils found in arable crops and second-generation fuels were made from lignocellulose biomass, woody crops, agricultural residues, and waste plant materials, it is likely that the source of third-generation fuels will be algae from algae ponds in wastewater treatment plants.

3.8.3.1 Ethanol

Biologically produced alcohols, most commonly ethanol and less commonly propanol and butanol, are produced by the action of microorganisms and enzymes through the fermentation of sugars and starches (easiest) or cellulose (which is more difficult). Biobutanol (also called biogasoline) is often claimed to be a direct replacement for gasoline because it can be used directly in a gasoline engine.

Biofuels, such as ethanol, can be made from soy oil, animal fat, and agricultural wastes, but the largest source is corn. Cellulosic ethanol promises higher yields than corn-based ethanol, but processes to produce it in any substantial quantity are still being developed. Cellulosic biomass, derived from nonfood sources such as trees and grasses, is also being developed as a feedstock for ethanol production.

Ethanol is widely used as a fuel additive and infrequently used as fuel for electric power generation. The first ethanol-fired power plant (Minas Gerais) was started up on January 19, 2010, in Brazil by the Petrobras company. The plant generates electricity on a commercial scale using sugar cane-derived ethanol. It reuses the equipment used in natural gas power plants with some revisions. Its capacity is 87 MW, enough to serve a city of 150,000 with electricity.[98]

---

98. "Ethanol Power Plant, Minas Gerais," Power Technology website, March 21, 2013, accessed October 10, 2022, https://www.power-technology.com/projects/ethanol-power-plant/.

Ethanol is a grain alcohol that can be blended with gasoline and used in motor vehicles. Ethanol advocates are pushing for high (20% to 30%) ethanol blends, but of the 280 million motor vehicles registered in the United States, only about 6 million were certified to run on blends higher than 10% in 2020. Also, some opponents claim that an increase in ethanol concentration decreases mileage and increases pollutant emissions. Ethanol can be used as a fuel for vehicles in its pure form (E100), but it is usually used as a gasoline additive to increase octane and improve vehicle emissions.

In 2018, the United States produced the greatest amount of fuel ethanol in the world, generating some 16.1 billion gallons. With nearly eight billion gallons, Brazil was ranked second. US flexible-fuel vehicles can run on gasoline or flex fuel (also called E85 gas), which is a high-level combination of anhydrous ethanol and gasoline that consists of between 51% and 83% ethanol blended with gasoline. This limit on the ethanol content is set to reduce ethanol emissions at low temperatures and to avoid starting problems during cold weather temperatures lower than 11°C (52°F). The alcohol content is reduced during the winter in regions where temperatures fall below 0°C (32°F) to a winter blend of E70.

The European Union wants to have 10% ethanol in its transportation fuel, and China is aiming for 15%. The US Senate is proposing a biofuel production target of 36 billion gallons by 2022, of which 21 billion would come from corn-based ethanol. At the same time, the Organisation for Economic Co-operation and Development (OECD) estimated that replacing 10% of the motor fuels used in the United States with biofuels would require about one-third of all of the cropland that is now devoted to the production of cereals, vegetable oil seeds, and sugar crops.

Brazil is the current leader in biofuel technology, with 70% of the new cars in Brazil being flex fuel cars that can also run on biofuels. Large biofuel plants have been built in other regions, such as Hawaii. Today, of the approximately 170,000 gas stations in the United States, only about 4,000 are equipped with pumps for E85. Ten times that number would be needed for most US motorists to have access to a pump within 5 miles of their homes.

In many parts of the world, wood, sugarcane, and similar materials are still burned, and potential sources of biofuels, such as switchgrass, corn husks, prairie grass, soybean, and corn, are mostly unused. Various waste materials can also be used to make ethanol, butanol, biodiesel (Figure 3-72), and other substitutes for gasoline.

Figure 3-72. Biodiesel-fueled bus in England.[99]

Source: "Cribbs Causeway - First 39401" by Geof Sheppard, May 3, 2019, is licensed under CC BY-SA 4.0, accessed March 12, 2022, https://commons.wikimedia.org/w/index.php?curid=78751753.

As already mentioned, the largest users of biofuels are the United States and Brazil. The use of ethanol reduces the dependence on oil, but it does not reduce CO_2 emissions. It also increases the price of food and the health risks due to ozone emissions. Ethanol from corn also increases the use of nitrogen-based fertilizers, which cause water pollution. Another disadvantage is that ethanol production from corn is inefficient because nearly 80% of the energy content of the ethanol is used in producing it (its energy balance[100] is 1.3:1), and often, the energy source for making ethanol is imported oil. In addition, ethanol cannot be transported by pipeline because it absorbs water and other impurities and is corrosive. Even gasoline pipelines cannot be used to distribute ethanol; it requires corrosion-resistant and leak-proof distribution pipelines, trucks, railcars, and barges.

In contrast to the 1.3:1 energy balance of corn-based ethanol, sugarcane-based ethanol has an energy balance of 8:1, cellulosic ethanol made from prairie and switchgrass has an energy balance of 6:1, and even gasoline has an energy balance of 5:1. Therefore, the overall environmental impact of corn-based ethanol can even exceed that of gasoline. Cellulosic ethanol is advantageous because it can come from perennial plants, and the whole plant, not just the seeds, can be used.

---

99. For further information, see: "Biofuel," Wikipedia, accessed March 12, 2022, https://en.wikipedia.org/wiki/Biofuel.

100. Energy balance is the ratio of the usable energy produced by a fuel to the energy invested in making it. It is also called the energy returned on energy invested (ERoEI).

Figure 3-73. Ethanol pump station in São Paulo, Brazil.

Source: "Sao Paulo ethanol pump 04 2008 74 zoom.jpg" by Mariordo Mario Roberto Duran Ortiz, April 28, 2008, is licensed under CC BY 3.0, https://commons.wikimedia.org/w/index.php?curid=3988176.

The farm industry naturally supports the production of corn-based ethanol (which could not replace more than 10% of US consumption), whereas the food industry opposes it because the increased demand for corn is raising the cost of animal feed and, therefore, the cost of dairy and poultry products.

The production of sugarcane- or cellulose-based ethanol has less of an effect on food prices, but their expanded production can end up destroying wildlife habitats and forests, threatening the survival of the rain forests, and polluting water supplies. The United States imposes a tariff on each gallon of Brazilian ethanol (Figure 3-73), whereas US ethanol producers receive tax breaks. The US Department of Energy has established four bioenergy research centers[101] to evaluate the various processes of turning cellulose into fuel.

---

101. The four centers are as follows:
    1. The Great Lakes Bioenergy Research Center, led by the University of Wisconsin-Madison in partnership with Michigan State University, whose goal is to develop the science and technological advances to ensure sustainability at each step in the process of creating biofuels and bioproducts from lignocellulose
    2. The Center for Bioenergy Innovation, led by the Oak Ridge National Laboratory, whose goal is to accelerate the domestication of bioenergy crops and target consolidated bioprocessing innovations to improve cost efficiencies within the bioenergy supply chain
    3. The Joint BioEnergy Institute, led by the DOE's Lawrence Berkeley National Laboratory, whose goal is to broaden and maximize the production of economically viable fuels and chemicals from plant biomass to enable biorefinery development
    4. The Center for Advanced Bioenergy and Bioproducts Innovation, led by the University of Illinois at Urbana-Champaign, whose goal is to enable the direct production of fuels and chemicals in plants as sustainable biofactories for a range of bioproducts.

    "Bioenergy Resource Centers," Genomic Science Program website, accessed March 12, 2022, https://genomicscience.energy.gov/bioenergy-research-centers/.

3.8.3.2 Biodiesel, Biogas, and More

Biodiesel fuels are made from vegetable and seed oils, wood chips, soybeans, animal fats, and recycled restaurant grease. Biodiesel can be used as a fuel for vehicles in its pure form (B100), but it is usually used as a diesel additive to reduce levels of particulates, carbon monoxide, and hydrocarbons in the exhaust from diesel-powered vehicles. Biodiesel is the most common biofuel in Europe. Biodiesel is also an oxygenated fuel, meaning that it contains a reduced amount of carbon and higher hydrogen and oxygen contents than fossil diesel. This improves the combustion of biodiesel and reduces the particulate emissions from unburned carbon. However, the use of pure biodiesel may increase NO_x emissions.

In New York State, a tax credit supports the use of biodiesel. Federal policy is still tilting toward ethanol even though turning biomass into gasoline is simpler because it does not require changes in the pipeline and car designs.

Byproducts of biofuel production are glycerin and lignin. The production of each gallon of biodiesel also produces a pound of glycerin. These materials can be used to replace oil-based products with bio-based ones. In the decades to come, it is possible that the developing biofuel industry will result in the building of multiproduct biorefineries.

In the past, plastics have been made from hydrocarbons. Now, with the development of new catalysts, plastics can also be made from agricultural products. The organic alternatives to petrochemical polymers can contribute to lowering the dependence on oil and replacing nondegradable plastic products with degradable ones. As demonstrated in the discussion of the conditions of the oceans, this is a critical consideration because it is projected that there will be more plastic than fish in the oceans in a decade or two.

Gasification is another way to produce biofuels. One such gas is methane, produced by the anaerobic digestion of organic material. It can be produced from biodegradable waste materials or energy crops fed into anaerobic digesters to supplement gas yields. Farmers can produce biogas from manure from their cattle by using anaerobic digesters. Another biogas is syngas, a mixture of carbon monoxide, hydrogen, and other hydrocarbons. It is produced by the partial combustion of biomass when the amount of oxygen provided in the combustion is not sufficient to convert the biomass completely to CO_2 and water.

3.9 Conclusions

In the previous chapters, we saw that the GW process is a lag-dominant one, having multiple gains (low gain in the ocean-related subprocesses but higher gain in the atmospheric ones). We have also seen that GW involves mostly self-accelerating and high-inertia processes, with some positive feedback subsystems. Further, the dead times of these processes are at least years, and their time constants can be decades or longer.

The important positive feedback subprocesses include the melting of permafrost, which releases immense amounts of methane and CO_2, further increasing the greenhouse effect and accelerating melting. Because more carbon is stored in permafrost than the total quantity that has accumulated in the air—and the amount of all fossil-fuel reserves on the planet—these feedback effects can greatly impact the rate at which global temperatures rise. In process control, the time at which a self-accelerating process reaches its maximum impact is called a *tipping point*. After this point, a process can accelerate until it becomes hard or impossible to control.

In addition to permafrost melting, ocean degassing can create a similar tipping point because, once the temperature rise reduces the solubility of the ocean waters, the CO_2 in the water is released in a self-accelerating manner. The effects of changes in the albedo and ocean evaporation also form positive feedback cycles because, as temperature rises, they self-intensify, resulting in a further increase in temperature.

In this chapter, I described the methods for collecting energy from the various green energy sources and the equipment available to do that, including newer designs that are in the development or testing stages. In addition, I described cooling methods, such as increasing the reflectivity of global surfaces, that directly cool the planet. I also discussed the newer AI-based controls and other control algorithms, newer sensors and equipment designs, and a number of safe installation practices.

I have also shown that the various green strategies and systems can serve more functions than just solving our energy problems. For example, I described a community (Babcock Ranch in Southwest Florida) that, on September 2022, was untouched by a Category 4 hurricane that devastated Southwest Florida. The example shows that the task ahead is not just to mitigate GW by carbon emission reduction but also to convert to a green lifestyle, and we do not need to wait for the twenty-second century to do that. Babcock Ranch already calls itself "America's first totally solar-powered town," but they achieved more because their homes did not flood and did not lose electricity, water, or the Internet. The point is that the carbon-free energy economy of the future will impact not only our energy supply but also make our lifestyle safe and sustainable.

This analysis has clearly shown that the available green energy sources are free and plentiful and that the conversion to a carbon-free energy economy is feasible. It has also shown that the GW process is currently out of control, and if left that way, it will bring disaster. On the other hand, this analysis has also shown that green energy sources are plentiful and that, after an initial investment, they will need only maintenance; otherwise, the energy sources are free.

The coming chapters of this book will focus on the technologies that are already available to store, transport, and continuously distribute the intermittently available green energy. These chapters will also show the timing and investments required to reach this goal.

4

Storing Green Energy

Table of Contents

Energy has played a critical role in the evolution of humankind. In the first stage, the discovery of fire gave us the means to obtain warmth and populate some of the colder regions of the planet. The next revolutionary step occurred at the beginning of the industrial age when electrical energy became available to many of us through the burning of fossil fuels. Today, we are approaching the end of this second phase because the climate change caused by the burning of fossil fuels can no longer be tolerated.

In this book, I described not only the ultimate goal of conversion but also the steps to be taken during periods like the present one, when Russia is extorting Europe by cutting off its natural gas supplies. In this situation, Europe is justified to temporarily switch back to burning coal and suspend her efforts to cut CO_2 emission, but she must return to these longer-range efforts as soon as the energy crisis ends.

The previous chapters of this book used well-established scientific methods of process control analysis to determine how much corrective action is needed to reestablish the global heat balance, which can be achieved by either reduced heating or increased cooling. We have seen that global warming (GW) can be reduced by burning less fossil fuel and by cooling the planet by making the Earth's surface more reflective.

The main goal of this chapter is to describe the methods that make intermittent green energy sources (solar, wind, etc.) continuous. I also show that most regular power plants are too slow to respond to sudden demand changes, while systems depending on stored green energy will be able to reestablish energy stability using inexhaustible,

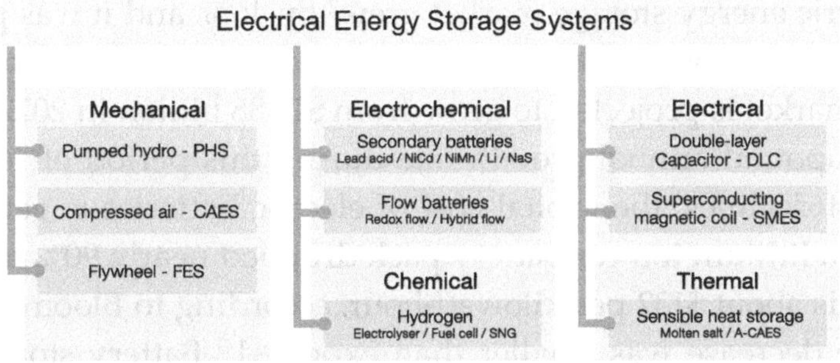

Figure 4-1. Methods for storing electrical energy. Abbreviations: (A-) CAES, (advanced) compressed-air energy storage; FES, flywheel energy storage; NaS, sodium-sulfur; NiCd, nickel-cadmium; NiMH, nickel-metal hydride; PHS, pumped hydro storage; SMES, superconducting magnetic energy storage.

Source: Reprinted with permission from "Electrical Energy Storage," IEC White Paper, December 1, 2011, Figure 2-1.[1]

high-energy-density fuel (hydrogen) from a free source (the Sun) that can be burned without emitting any carbon.

4.1 Introduction

Whereas some green energy sources (such as geothermal and hydraulic) are continuous, the primary ones (i.e., solar and wind) are intermittent. Therefore, converting to using green energy sources will necessitate the building of energy storage and distribution facilities. Today, electricity is almost exclusively stored in batteries, but batteries are of limited value if large quantities of electricity are to be stored and transported. The purpose of this chapter is to describe the capabilities and limitations of several energy storage systems, as well as their use in a future green energy system.

To stop GW, we must end our dependence on fossil fuels and nuclear energy and convert to technologies that can store, transport, and continuously distribute the intermittently available green energies among users. Storage and transportation are the main hurdles that must be overcome to convert from using dirty and exhaustible energy sources to clean and inexhaustible ones. The description of the nature, dynamics, and control of these green storage processes is the subject of this chapter.

Of the large variety of energy storage methods (Figure 4-1), I discuss only those that are either already in use or being intensively developed. Therefore, I include storage methods that take advantage of gravitational, thermal, electrical, pressure, electrochemical, and mechanical methods, with particular emphasis on the coming solar-hydrogen energy economy and the role of the reversible fuel cell (RFC).

1. The author thanks the International Electrotechnical Commission (IEC) for permission to reproduce Information from its white paper, Electrical Energy Storage. All such extracts are copyright of IEC, Geneva, Switzerland. All rights reserved. Further information on the IEC is available from www.iec.ch. IEC has no responsibility for the placement and context in which the extracts and contents are reproduced by the author, nor is IEC in any way responsible for the other content or accuracy therein.

In 2017, the electric energy storage market grew[2] by 46%, and it was projected to continue growing several-fold during the next decade. Specifically, the global battery energy storage market is projected to grow from $10.88 billion in 2022 to $31.20 billion by 2029 at a compound annual growth rate during this period of 16.3%. After more than a decade of declines, the capital cost of electric batteries is getting pricier. The average price of a lithium-ion car battery pack dropped nearly 90% from 2010 to 2020. Today, the price is about $132 per kilowatt-hour, according to Bloomberg New Energy Finance, but the decrease was smaller than expected.[3] Battery storage in stationary applications was about 175 GW in 2017 and is projected to reach 235 GW by 2030.[4] It is possible that lower installation costs, longer lifetimes, and improved performance will further drive down the cost of electricity storage.

Here, I summarize the capabilities of the various energy storage systems (Figure 4-2) and discuss their process-control- and automation-related challenges.

Aside from the capacities of energy storage systems, other important considerations include their costs, safety characteristics, and energy densities. The energy densities of the various energy sources are important because they determine the weights and volumes of the corresponding storage systems, which, in turn, determine costs in applications in which the energy storage cannot be stationary, such as transportation. Table 4-1 lists the gravimetric and volumetric energy densities of several energy sources, including fuels and batteries. From Table 4-1, we can see, for example, that a unit weight of hydrogen can store three times as much energy as the same weight of gasoline and up to 150 times as much as a battery of the same weight. The energy densities of batteries vary with their design, but the ranges of both their mass-based and volume-based energy densities fall below those of all fuels.

Projections for the future costs of electrical energy storage systems are unreliable, but they all show downward trajectories as the demand for storage systems rises. One

---

2. Brian Eckhouse, "US Energy Storage Surges 46% Led by Big Project in Windy Texas," Bloomberg, December 7, 2017, accessed November 21, 2022, https://www.bloomberg.com/news/articles/2017-12-07/u-s-energy-storage-surges-46-led-by-big-project-in-windy-texas.

3. For more information, see: "Battery Energy Storage Market Size, Share & COVID-19 Impact Analysis, By Type (Lithium-Ion Battery, Lead Acid Battery, Flow Battery, Others), By Connectivity (Off-Grid, On-Grid), By Application (Residential, Non-Residential, Utility, Others), By Ownership (Customer-Owned, Third-Party-Owned, Utility-Owned), By Capacity (Small Scale {Less Than 1 MW}, Large Scale {Greater Than 1 MW}), and Regional Forecast, 2022–2029," Fortune Business Insights web page, March 2022, https://www.fortunebusinessinsights.com/industry-reports/battery-energy-storage-market-100489.

4. *Electricity storage and renewables: Costs and markets to 2030* (Abu Dhabi: International Renewable Energy Agency [IRENA], October 2017), https://www.irena.org/-/media/Files/IRENA/Agency/Publication/2017/Oct/IRENA_Electricity_Storage_Costs_2017.pdf.

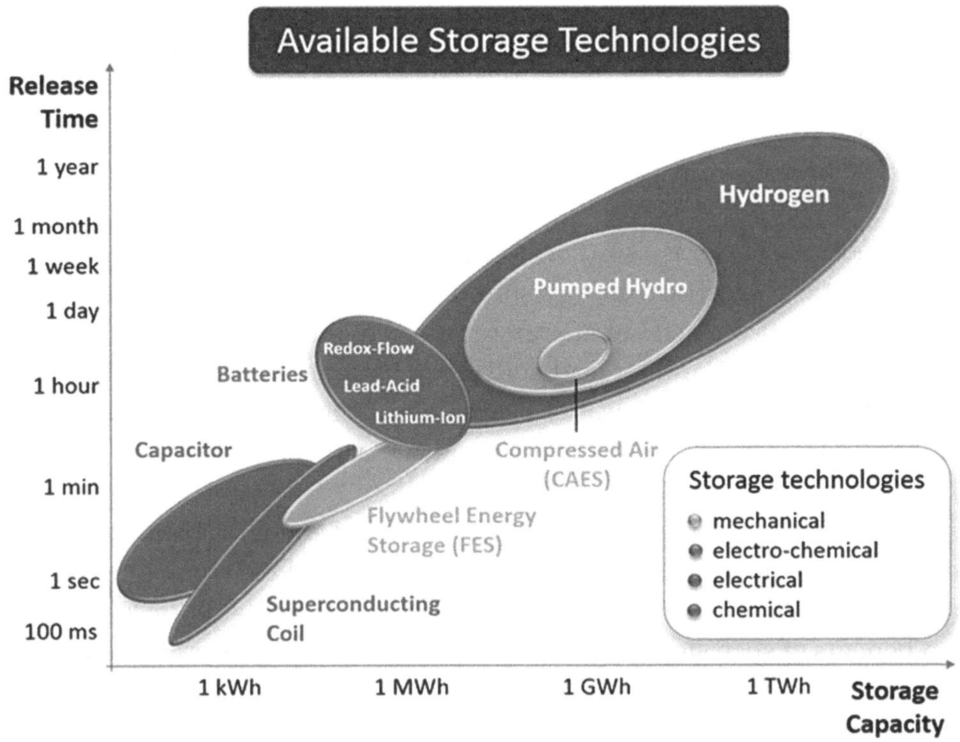

Figure 4-2. Energy storage capacities and release-time ranges of various energy storage systems. (See color images after page 306.)

Source: Reprinted with permission from Paul O'shea, "Solar Storage Moves Out of the Lab into Mainstream: Special Report," *Power Electronics News*, March 21, 2017, https://www.powerelectronicsnews.com/problems-solutions/solar-storage-moves-out-of-the-lab-into-mainstream.

Table 4-1. Energy densities of fuels and batteries.

Energy Source	Energy Density	
	Mass-Based (kWh/kg)	**Volume-Based (kWh/L)**
Liquid hydrogen	39.6	2.8
High-pressure hydrogen gas at 700 bar (10,000 psia)	39.6	1.55
Liquefied natural gas (LNG)	12.7 to 14.7	6.1
High-pressure natural gas at 250 bar (3500 psia)	12.7 to 14.7	3.05
Gasoline or diesel fuel	12.7 to 13	9.7 to 11.6
Batteries	0.26 to 0.27[a]	0.29[b] to 0.45[c]

[a]"Why Does Energy Density Matter in Batteries?," Dragonfly Energy website, December 22, 2021, accessed October 13, 2022, https://dragonflyenergy.com/why-does-energy-density-matter-in-batteries/.
[b]"Data Reveals Tremendous Growth in Volumetric Energy Density of EV Batteries," InsideEVs.com, accessed October 13, 2022, https://insideevs.com/news/581729/volumetric-energy-density-ev-batteries-growth/.
[c]US Department of Energy Office of Energy Efficiency & Renewable Energy, "FOTW #1234, April 18, 2022: Volumetric Energy Density of Lithium-ion Batteries Increased by More than Eight Times Between 2008 and 2020," April 18, 2022, accessed October 13, 2022, https://www.energy.gov/eere/vehicles/articles/fotw-1234-april-18-2022-volumetric-energy-density-lithium-ion-batteries.

estimate[5] projected that annual additions to the global utility-scale, electrical energy storage capacity would grow from 1159 MW in 2017 to 30,473 MW by 2026. It also projected that the annual additions to the global capacity for distributed energy storage systems (DESSs), also called the "behind-the-meter sector," would grow from 684 MW in 2017 to 19,700 MW by 2026.

The energy efficiency of storage systems (energy stored per energy invested—ESoEI) is defined as the ratio of electrical energy stored over the lifetime of a storage system to the amount of embodied electrical energy required to build it. For Li-ion batteries, the ESoEI is approximately 10, and for lead-acid (PbA) batteries, it is about 2. The values for other forms of storage vary widely. For example, the ESoEI values of pumped-storage hydroelectric (PSH) systems are generally greater than 200.

Currently, the most common means of storing electrical energy is batteries. In terms of costs, research by Bain & Company estimates that the cost of large-scale battery storage[6] could be competitive with that of conventional peaking plants by 2025.[7] Naturally, the added benefits of such storage include lower carbon emissions[8] and reduced maintenance. The goal is to achieve lower costs for electricity generated from renewable sources and stored in batteries than for electricity generated from coal.

As of this writing, Li-ion batteries are in the best position to provide energy storage during the next decade. In the long run, however, other storage technologies will be needed for large-scale and long-duration applications because transporting energy using such batteries is neither practical nor economical.

Because of the reduced costs, the use of storage systems based on Li-ion batteries is growing rapidly.[9] This is due to the better economy of large systems. Li-ion storage systems have reached 20 to 100 MW in size and are successfully stabilizing large power

5. Joshua S. Hill, "Global Annual Utility-Scale & Distributed Energy Storage Capacity Additions to Exceed 50 GW by 2026," CleanTechnica website, July 24, 2017, accessed November 21, 2022, https://cleantechnica.com/2017/07/24/global-annual-utility-scale-distributed-energy-storage-capacity-additions-exceed-50-gw-2026/.

6. Note that conventional peaking plants are fossil-fuel-burning power plants that are normally on standby and are started up only during periods of peak energy demand.

7. Julian Critchlow and Aaron Denman, *Embracing the Next Energy Revolution: Electricity Storage* (London: Bain & Company, 2017), https://www.bain.com/contentassets/3b1a3a1b900a486eb6e6ca7188839d9a/bain_brief_embracing_the_next_energy_revolution_electricity_storage.pdf.

8. If the batteries are charged by green energy, their carbon emissions are zero, whereas if non-green power is used, the associated carbon emissions are those of the power plant.

9. "Levelized Cost of Energy and Levelized Cost of Storage 2018," Lazard website, November 8, 2018, https://www.lazard.com/perspective/levelized-cost-of-energy-and-levelized-cost-of-storage-2018/.

grids.[10] Grid stabilization is an important consideration in power distribution systems because if the intermittent green electricity supply (so-called renewable penetration) exceeds 20%, the conventional power plants are not fast enough to keep up with the speeds at which such penetrations can occur.

In the following sections of this chapter, I discuss the need to develop large-scale, long-duration, and transportable energy storage technologies as our overall energy economy is gradually converted to a green one. In the next few decades, I expect that the primary means of energy storage will shift from batteries to solar-hydrogen systems. This transformation will take decades, but it will probably occur because hydrogen can store 150 times the energy per unit weight than batteries; therefore, storing, transporting, and distributing zero-carbon green energy is much more efficient.

4.2 Compressed-Air Energy Storage

Storing energy in high-pressure air is a stationary storage method. It is useful for helping intermittent green energy sources meet continuous loads or stabilizing electric grids by being available when peak loads evolve.

As shown in Figure 4-2, the capacity of this type of energy storage is in the gigawatt-hour range, and depending on the size of the equipment used, the time required for it to be loaded or unloaded can range from less than an hour to up to a day. Compressed-air energy storage (CAES) plants are similar to PSH power plants in terms of their energy storage capacities and rates of energy output. In a CAES plant, ambient air is compressed and stored under pressure in underground caverns. When electricity is required, the pressurized air is heated and sent to expansion turbines to drive power-production generators.

Because of the low-energy density of compressed air (Table 4-1), large-volume storage spaces are required. Locations such as artificially constructed salt caverns in deep salt formations are preferred because they have high flexibility, no pressure losses, and no reaction with the oxygen in the air or the salt in the host rock. If no such formations are available, natural aquifers can also be used; however, tests must be performed to determine whether the oxygen reacts with the rocks or with any microorganisms in the rock formation because that could lead to oxygen depletion or the blockage of the pore spaces in the reservoir. The use of depleted natural gas fields for CAES has

10. Tesla's 100 MW energy storage system in South Australia serves to stabilize a large electric grid that became unstable because of the intermittent and sudden changes in the supply of large quantities of green electricity. See: Rob Verger, "Tesla actually built the world's biggest battery. Here's how it works," *Popular Science*, December 2, 2017, accessed November 1, 2022, https://www.popsci.com/tesla-building-worlds-biggest-battery-how-it-will-work/.

been suggested, but in addition to the depletion and blockage problems, the mixing of residual hydrocarbons with compressed air can cause problems.

During the cavern-charging phase of operation, the air is compressed and it heats up. At full charge, the cavern pressure is about 1100 psig (about 75 barg), and at full discharge, the pressure is about 650 psig (about 45 barg). The typical pressure is about 70 bar (1015 psia), which results in a discharge temperature of about 149°C to 177°C (300°F to 350°F). Before air is injected into the storage cavern, the compression heat must be removed and the air cooled to 43°C to 49°C (110°F to 120°F).

During the energy generating phase, the heat energy extracted during the charging phase must be restored to the high-pressure air coming from the cavern to achieve the required expansion. The two basic CAES designs are distinguished by the method of heat transfer used. Isothermal systems use natural gas burners to return this heat; adiabatic systems use thermal storage. Both systems are described in the next section.

The electricity used during the compression phase can come from green energy sources or from the grid during off-peak hours when the electricity is cheaper. During peak hours, when the cost of electricity is high, the electricity generated from the stored air is sent to users to stabilize the grid.

Compared to other energy storage methods, CAES systems can store energy for more than a year. The typical storage efficiency ranges from 60% to 80%. Capital costs for CAES facilities vary depending on the type of underground storage but typically range from $400 to $1,000 per kilowatt-hour of power generation capacity. The lifetimes of CAES systems can range from 20 to 40 years.

The following subsections explain the two main designs of CAES systems, compare their costs and efficiencies, and describe the operation of the current operating and the proposed systems.

4.2.1 Isothermal Design

Isothermal[11] *compression* occurs when air remains at a constant temperature throughout the compression and expansion phases. In an isothermal cycle, the internal energy added by mechanical work during compression is removed in the form of heat in the cavern-charging phase. The energy is returned to the stored air when it is taken from

---

11. An isothermal process is a constant-temperature process. In such a process, when the air is compressed, the heat of compression is removed by cooling, and when the air is allowed to expand, the heat is returned by heating the air.

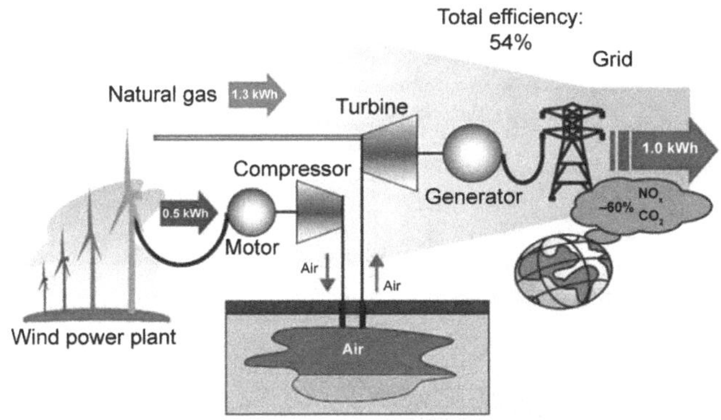

Figure 4-3. Illustration of a large-scale, wind-compressed-air hybrid system that converts the intermittently available wind energy into a continuously available energy source. Compared to fossil power plants, this configuration reduces GHG emissions by approximately 60%.

Source: Hussein Ibrahim and Adrian Ilinca, "Contribution of the Compressed Air Energy Storage in the Reduction of GHG—Case Study: Application on the Remote Area Power Supply System," in *Air Pollution—A Comprehensive Perspective*, ed. Budi Haryanto (London, IntechOpen, August 22, 2012), Figure 7, https://www.intechopen.com/books/air-pollution-a-comprehensive-perspective/contribution-of-the-compressed-air-energy-storage-in-the-reduction-of-ghg-case-study-application-on-. Used under the Creative Commons Attribution 3.0 License.

the cavern during the power generation (expansion) phase. Isothermal compression and expansion occur in compressors that have large heat-exchanging surfaces and compress a small amount of air. Otherwise, compressors employ cooling between the compression stages to achieve isothermal operation.

In the case of isothermal CAES, the heat of compression is removed before the high-pressure air enters the cavern, and this heat energy is either wasted or used, for example, to heat water (Figure 4-3). When the cooled high-pressure air is taken from the cavern in the power generation phase, the previously removed heat is returned, usually by natural gas burners, before the air is expanded in the turbines.

In other applications, compression heat is not used for heating, so it is wasted. When that is the case, the efficiency of the storage-recovery cycle is degraded. Although this approach is relatively simple, burning fuel adds to the cost of the generated electrical energy and compromises the ecological benefits associated with using renewable energy sources. Nevertheless, thus far, this is the most often used system implemented commercially (Figure 4-4).

CAES technology is several decades old. It was 1991 when a 110 MW CAES plant was built in McIntosh, Alabama. This plant requires an electric energy input of 2.5 MJ and a thermal energy input of 1.2 MJ for each megajoule of energy output, corresponding to an energy recovery efficiency of about 27%.

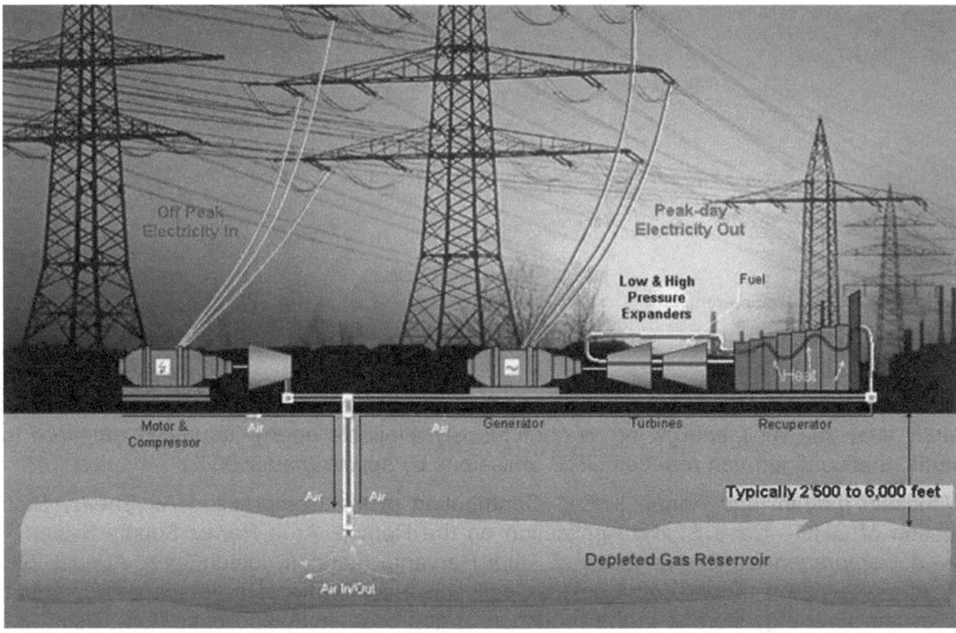

Figure 4-4. Isothermal CAES systems in which burning fuel adds to the cost of the generated electricity are still widely used.[12]

Source: Reprinted with permission from Edvard Csanyi, "Reliability of Compressed Air Energy Storage Plant," Electrical Engineering Portal, August 4, 2012, https://electrical-engineering-portal.com/reliability-of-compressed-air-energy-storage-plant.

4.2.2 Adiabatic Design

In adiabatic[13] compression, no heat energy is transferred to or from the gas during the compression process because all mechanical work is added to the internal energy of the gas. Adiabatic compression or expansion occurs when a compressor is well insulated and the gas flow is large. In practice, there will always be a certain amount of heat loss because a perfectly adiabatic compressor would require perfect heat insulation for all parts of the machine, which is impossible.

Adiabatic CAES systems keep the heat produced by compression and return it to the air when the air is expanded to generate power. This technology is mainly in the demonstration and pilot stages. Thermal oil and molten salt heat storage (Figure 4-5) are being investigated in the United States. In Germany, the ADELE demonstration plant[14] (with

---

12. For additional information, see: Nanoneophyte, "Nine Disruptive Technologies Changing the World … and Why You Should be Paying Attention," *Genesis Nanotechnology Blog*, October 7, 2015, accessed November 21, 2022, https://genesisnanotech.wordpress.com/2015/10/07/nine-disruptive-technologies-changing-the-world-and-why-you-should-be-paying-attention/.

13. During adiabatic operations, no heat leaves or enters the air from the outside. Therefore, the air heats up during compression, and that heat needs to be returned to the air during the expansion phase.

14. "ADELE – Adiabatic Compressed-Air Energy Storage (CAES) for Electricity Supply" video, RWE Power AG, accessed March 12, 2022, www.youtube.com/watch?v=K4yJx5yTzO4.

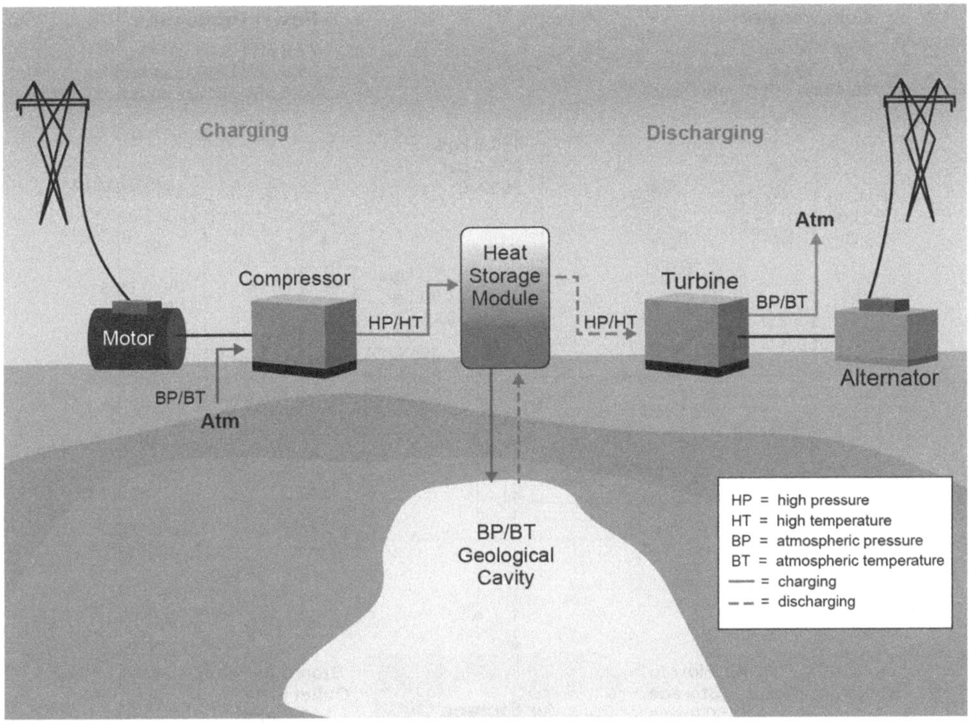

Figure 4-5. The heat storage module provides good efficiency in adiabatic CAES systems by storing the compression heat and returning it to the air during the expansion phase.

a storage capacity of 360 MWh) aimed to approach the theoretical maximum efficiency of approximately 70% (as mentioned, 100% is impossible as it would require perfect insulation). In this system, the compressed air is stored in a salt cavern. A recuperator is operated to reuse the exhaust heat energy. This reduces fuel consumption by 22% to 25% and improves the cycle efficiency from about 42% to 54%, compared with the Huntorf plant (see Section 4.2.4).

The key to CAES system operation is storing the compression heat between the time the cavern is charged and the time the stored energy is used. Research is being conducted to evaluate the methods of storing the heat in solids such as concrete and stone and fluids such as hot oil (up to 300°C [572°F]) and molten salt solutions (600°C [1112°F]). A numerical simulation of a packed-bed CAES system found that its efficiency under continuous operation could be about 70.5%.[15] Therefore, some sources expect future adiabatic CAES systems to operate at efficiencies greater than 70% with the flexibility to provide various capacities from 2 MW to more than 300 MW with specific costs significantly lower than those of batteries. They also have more than 30 years of life expectancy compared to 10 to 15 years for batteries.

15. "Compressed-air energy storage: Adiabatic," Wikipedia, accessed May 25, 2022, https://en.wikipedia.org/wiki/Compressed-air_energy_storage#Adiabatic.

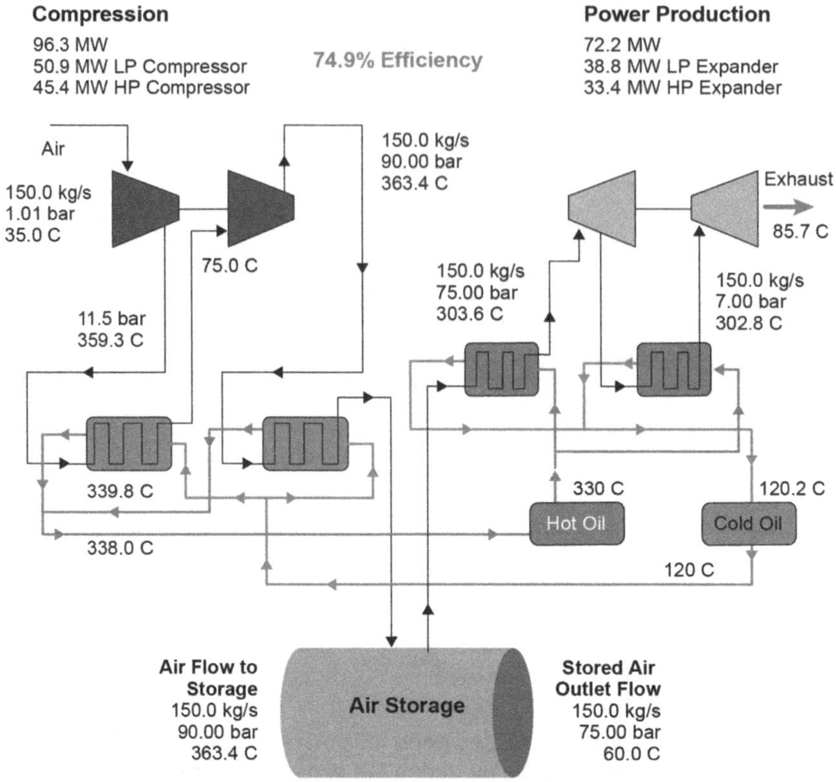

Figure 4-6. Adiabatic CAES system with hot-oil heat storage. Abbreviations: HP, high-pressure; LP, low-pressure.

Figure 4-6 shows a 72 MW adiabatic CAES system with the corresponding operating data (flow rates, pressures, and temperatures). This system is claimed to be able to operate at 74.5% efficiency.

4.2.3 Costs and Efficiencies

The efficiency of isothermal CAES systems (about 80%) is better than that of adiabatic ones (about 70%), but the adiabatic systems are superior because they eliminate both the cost of burning natural gas and the carbon emissions associated with combustion.

CAES is the lowest-cost storage technology,[16] but its applicability is highly dependent on the availability of local naturally occurring caverns. If such caverns are available, a

16. Kendall Mongird, Vilayanur Viswanathan, Jan Alam, Charlie Vartanian, Vincent Sprenkle, and Richard Baxter, "2020 Grid Energy Storage Technology Cost and Performance Assessment," Technical Report DOE/PA-0204, US Department of Energy, December 2020, https://www.pnnl.gov/sites/default/files/media/file/Final%20-%20ESGC%20Cost%20Performance%20Report%2012-11-2020.pdf.

100 MW installed system providing storage for 10 hours could cost under $400/kWh, which is significantly below the costs of PSH installations or Li-ion battery systems.

It is the view of the Electric Power Research Institute (EPRI) that the geology is suitable for CAES installations in up to 80% of the United States.[17] EPRI has also estimated that a 300 MW CAES plant having a storage capacity of about 0.62 million cubic meters (22 million cubic feet) can generate 300 MWh of electricity for about 8 hours.

For CAES installations using wind energy to compress the air, EPRI prepared a map showing favorable locations. EPRI considered all locations where the wind energy is higher than 4 on the Beaufort scale[18] to be favorable. According to EPRI, these relatively inexpensive installations can use intermittent renewable wind energy to provide reliable and continuous loads for stable power distribution.

Also, the Wind Energy Department of the Technical University of Denmark (DTU) prepared a Global Wind Atlas (Figure 4-7). This free web-based tool is now maintained in partnership with the World Bank. Policymakers and investors use it to identify promising areas for wind power generation virtually anywhere in the world.

4.2.4 Operating and Proposed Projects

The oldest isothermal CAES plant was commissioned in 1978 in Huntorf, Germany (Figure 4-8). It is a 290 MW plant operating at 90% availability and high reliability. It uses about one-third of the fossil fuel that a conventional fossil-fuel power plant of the same size would use. It stores 580 MWh of energy and operates at 42% efficiency. The fixed-price turnkey cost for this first-of-a-kind plant was about $400/kW in constant 1988 dollars ($864/kW in 2019 dollars).

A limitation of this design is that the compressor and the expander must operate simultaneously and have the same air mass flow rate. The plant requires 4 hours of compression per hour of electricity generation. At a power output of 110 MW, the power ratio is

---

17. S. Eckroad, "Handbook of Energy Storage for Transmission or Distribution Applications," EPRI, Palo Alto, CA, December 18, 2002, https://www.epri.com/research/products/000000000001007189.

18. Admiral Sir Francis Beaufort, a British naval officer, created one of the first scales used to estimate wind speeds and their effects. The scale, developed in 1805, was designed to help sailors estimate the winds based on visual observations. The Beaufort scale starts with 0 and goes to a force of 12, and it is still used today to estimate wind strengths. In the United States, winds with a Beaufort number of 6 or 7 result in the issuance of a small craft advisory; 8 or 9 results in a gale warning; 10 or 11 results in a storm warning (with a "tropical storm warning" being issued instead if the winds relate to a tropical cyclone), and 12 results in a hurricane-force wind warning (or hurricane warning if related to a tropical cyclone). Red warning flags (daylight) and red warning lights (nighttime) coinciding with the various levels of warning are displayed at shore establishments.

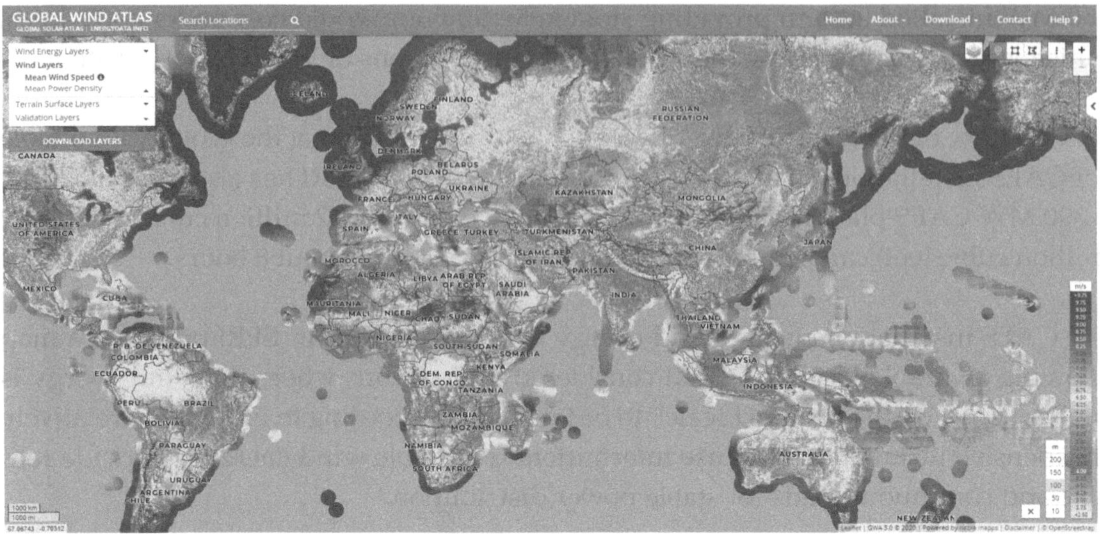

Figure 4-7. Map from the Global Wind Atlas. (Note that yellow and darker areas have good potential for wind energy.) (See color images after page 306.)

Source: Illustration obtained from the Global Wind Atlas 3.0 (accessed March 12, 2022), a free, web-based application developed, owned, and operated by the Technical University of Denmark (DTU). The Global Wind Atlas 3.0 was released in partnership with the World Bank Group, using data provided by Vortex, with funding provided by the Energy Sector Management Assistance Program (ESMAP). For additional information, see https://globalwindatlas.info.

Figure 4-8. The 290 MW adiabatic CAES plant in Huntorf, Germany.[19]

Source: T. M. I. Mahlia, T. J. Saktisahdan, A. Jannifar, M. H. Hasan, and H. S. C. Matseelar, "A review of available methods and development on energy storage; technology update," *Renewable and Sustainable Energy Reviews* 33, May 2014, pages 532–545, https://doi.org/10.1016/j.rser.2014.01.068.

0.818 kW, meaning that the output for each kilowatt of input is 0.818 kW. In addition, an input of heat energy is also needed. This is 4122 Btu/kWh if the fuel is natural gas and 4089 Btu/kWh if it is oil. The partial-load operation of this plant is also very flexible.

19. "Compressed Air Energy Storage (CAES)," *Energy-Power Blog*, August 25, 2012, accessed November 21, 2022, http://interimpulse.blogspot.com/2012/08/compressed-air-energy-storage-caes.html.

Another existing CAES plant is the Alabama Electric Cooperative McIntosh plant, which stores compressed air in a 19 million cubic foot cavern mined from a salt dome. This 110 MW isothermal plant has a storage capacity of 26 hours and requires 1.7 hours of compression for each hour of generation. The plant can store 2860 MWh of energy and operates at 54% efficiency.

More recent and some proposed installations include the following:[20]

- In November 2009, the US Department of Energy (DOE) awarded $24.9 million in matching funds for phase one of a 300 MW, $356 million Pacific Gas and Electric Company (PG&E) installation using a saline porous rock formation being developed near Bakersfield, California. On August 2, 2013, the California Energy Commission (CEC) supported the project with conditional approval of a $1 million grant.
- In December 2010, the US DOE provided $29.4 million in funding to conduct preliminary work on a 150 MW, salt-based CAES project by Iberdrola USA in Watkins Glen, New York. The project incorporated smart grid technology to balance renewable intermittent energy sources.
- In 2013, the first adiabatic CAES project, a 200 MW facility called ADELE, was planned for construction in Germany, but the project has been delayed.
- In 2017, Storelectric Ltd. started planning the construction of a 40 MW, 100% renewable energy pilot plant in Cheshire, UK, with a storage capacity of 800 MWh.
- In 2016, Apex planned for a CAES plant in Anderson County, Texas, to go online. However, this project has been delayed and will not be operational until Winter 2025.
- In Larne, Northern Ireland, a 330 MW CAES plant project received €90 million in European Union support for building a system using two caverns in a salt deposit.
- The European Union is also funding the RICAS 2020 (adiabatic) project in Austria, which uses crushed rock to store heat from the compression process to achieve 70% to 80% efficiency.

---

20. "Compressed-air energy storage: Projects," Wikipedia, accessed March 12, 2022, https://en.wikipedia.org/wiki/Compressed_air_energy_storage#Projects.

- Canadian company Hydrostor planned to build Advance plants in Goderich, Angas, and Rosamond (2020). Some included partial heat storage in water, improving efficiency to 65%.

- A 300 MWh compressed-air energy storage system utilizing salt caverns was connected to the grid in Jiangsu, China, on May 26, 2022. It can provide 60 MW of peak energy with an efficiency of 60%.[21]

- Energy Dome's 2.5 MW/4 MWh CO2 energy storage solution in Sardinia, Italy, became operational in 2022.[22]

4.3 Pumped-Storage Hydropower

These systems were already discussed in relation to hydroelectric energy in Chapter 3, Section 3.6.2, "Pumped-Storage Plants," but in that case, the focus was not on storage. Here it is. The first known PSH installations were built in Italy and Switzerland in the 1890s and in the United States in about 1930. According to the 2021 edition of the *US Hydropower Market Report*,[23] the global installed PSH generating capacity at the end of 2019 was 158 GW, with 50 projects amounting to an additional 53 GW capacity under construction.

In the United States, the PSH capacity grew between 2010 and 2019 by almost as much as the installed capacity of all other energy storage systems combined. In this case, "other" systems consisted almost entirely of battery systems. PSH continues to grow in the United States, doubling approximately every 5 years.

PSH is one of the oldest methods of energy storage. It involves pumping water into artificial lakes at high elevations and then releasing the water when electricity is needed. This method is used with fresh water in places such as Austria, the United States, and Switzerland, whereas in Germany, seawater is being used to form artificial lakes (Figure 4-9). The global capacity of pumped water storage installations worldwide in 2020 was 181 GW, of which about 29 GW was in the United States, and the total installed storage capacity was over 1.6 TWh, of which about 250 GWh was in the

---

21. Andy Colthorpe, "China's compressed air energy storage industry makes progress," June 1, 2022, accessed October 29, 2022, https://www.energy-storage.news/chinas-compressed-air-energy-storage-industry-makes-progress/.

22. Cameron Murray, "Energy Dome launches world's first multimegawatt 'CO_2 Battery' project in Sardinia, Italy," June 8, 2022, accessed October 29, 2022, https://www.energy-storage.news/energy-dome-launches-4mwh-demonstrator-project-in-italy-plans-series-b/.

23. Rocío Uría-Martínez, Megan M. Johnson, and Rui Shan, *US Hydropower Market Report*, Oak Ridge National Laboratory, Oak Ridge, TN, January 2021, accessed November 21, 2022, https://www.energy.gov/eere/water/downloads/us-hydropower-market-report.

Figure 4-9. The upper reservoir at the Helms Pumped Storage Facility in Fresno County, California.[24]
Source: "Courtright Reservoir 2.jpg" by the US Forest Service, January 17, 2012, Public Domain, https://commons.wikimedia.org/w/index.php?curid=18083112.

United States. By 2030, the global capacity is projected to reach 235 GW. The potential for pumped water storage is limited to mountainous regions, and the electricity generated is usually sent to the grid. The capacity and release-time range of pumped-storage facilities are compared with those of other storage systems in Figure 4-2.

Hydroelectricity generated by pumped storage is most often used for load balancing. The method stores energy in the form of water pumped from a reservoir at a lower elevation to one at a higher elevation. Usually, the pumps are run when the cost of electricity is low (off-peak). During periods of high electrical demand, the stored water is released through turbines (Figure 4-10). Because the energy consumption of the pumps is higher than the energy recovered by the turbines, the plant is a net energy consumer. Nevertheless, its use is beneficial because the plant generates its electricity during periods when the cost of electricity is higher (i.e., during periods of peak demand). Pumped storage is the largest-capacity form of grid energy storage now available.

Compared to other storage systems, pumped-storage systems have very low-energy densities, and therefore, they require either very large bodies of water or high storage elevations. For example, 1.0 m^3 of water at an elevation of 100 m can generate only

24. The Helms Pumped Storage Facility is located 80 km (50 mi) east of Fresno, California, in the Sierra Nevada Mountain Range's Sierra National Forest. Planned in the early 1970s, construction started in June 1977 and commercial operations began on June 30, 1984. The plant uses Helms Creek canyon on the North Fork of the Kings River for off-river water storage, has an installed capacity of 1212 MW, and is owned by PG&E.

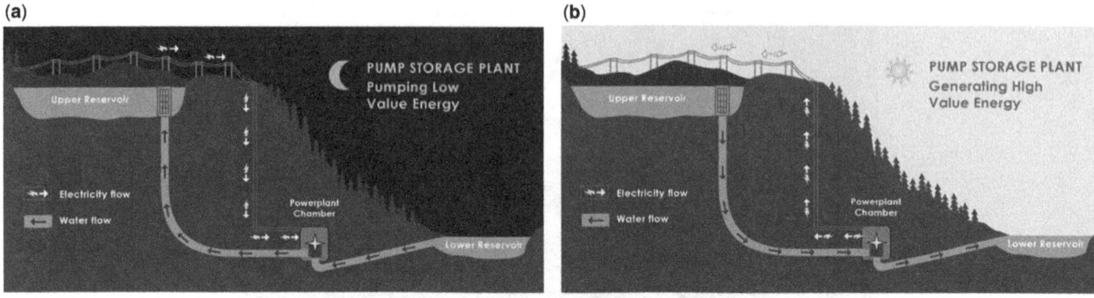

Filling the upper reservoir using off-peak electricity

Generating electricity during periods of peak demand

Figure 4-10. Operation of a pumped-storage system.

Source: Reprinted with permission from "Pumped Hydroelectric Storage," Clean Balance Power Inc. website, 2012, http://www.cleanbalancepower.com/about-pumped-storage.html.

0.272 kWh[25] (928 Btu). Therefore, the upper and lower water reservoirs must be very large and have a significant elevation difference to provide a high electricity production capacity.

The reversible pump–turbine/generator assemblies used in hydroelectric plants can act as both pumps and turbines. Pumped-storage systems are unlike hydroelectric power plants in that they are net consumers of electricity. They incur hydraulic and electrical losses when they pump water from the lower reservoir to the upper reservoir. However, these plants are still about 80% efficient (round-trip) and can be very beneficial in balancing the load on the grid while also being economical because of the price differential between peak and off-peak energy.

Pumped-storage systems have operated in the United States since the 1930s. In 2017, the more than 40 pumped-storage systems in the United States provided 22 GW, or 2%, of the total electricity supply. Table 4-2 lists the countries that have the largest pumped-storage power capacities, and Table 4-3 lists some of the world's largest storage systems.

In 2017, the European Union's storage capacity was 38.3 GW, whereas its total hydropower capacity was 140 GW, representing 5% of its total net electrical capacity. Japan's storage capacity was 10%. In June 2018, the Australian federal government announced that 14 sites had been identified in Tasmania for pumped storage, with the potential of adding 4.8 GW to the national grid if a second interconnector was constructed beneath the Bass Strait.

---

25. "Pumped-Storage Methods Hydroelectricity," *Hydro Electric Blog*, October 4, 2009, accessed November 21, 2022, http://hydroelectric-energy.blogspot.com/2010/12/pumped-storage-methods-hydroelectricity.html.

Table 4-2. Nations that have the largest pumped-storage generating capacities as of 2017.

Country	Pumped-Storage Generating Capacity (GW)
China	32
Japan	28.3
United States	22.6
Spain	8
Italy	7.1
India	6.8
Switzerland	6.4
France	5.8

Source: "Pumped-storage hydroelectricity," Wikipedia, accessed August 12, 2021, https://en.wikipedia.org/wiki/Pumped-storage_hydroelectricity.

In the United States, California's goal is to obtain 100% of its electricity from carbon-free sources by 2045. This will require an unprecedented amount of energy storage. California is already one of the leaders in this technology. For example, the capacity of the 35-year-old plant in the Sierras, PG&E's Helms Pumped Storage Facility, is more than 1200 MW. Just in California, there are seven such plants. The Federal Energy Regulatory Commission has issued 34 preliminary permits for companies and government agencies exploring projects in New York, Pennsylvania, Wyoming, and elsewhere. Other potential projects would incorporate existing reservoirs. For example, the Los Angeles Department of Water and Power has proposed converting the Hoover Dam into a giant pumped-storage facility.

Germany is also planning to expand its pumped-storage hydroelectric capacity, which is substantial already. One of the studies projects that, by 2030, the pumped-storage capacity can reach 13 GW. The study also suggests that, by 2050, the share of renewable energy can reach 80% of the total energy consumption of Germany, and the capacity of pumped-storage systems can reach 16 GW.

Although there are many advantages to PSH, building new plants is not easy. Such plants are cheap once they are up and running, but the upfront price is huge: They can cost more than $1 billion to build. In addition, the construction time is long. Securing permits, building dams, and boring tunnels can take 10 years. In addition, most projects face resistance because the construction requires flooding valleys. Environmentalists are often right in opposing these construction projects because of their impact on

Table 4-3. Pumped-storage hydroelectric power plants with power capacities of at least 1500 MW.

Station	Country	Power Capacity (MW)	Storage Capacity (MWh)
Fengning Pumped Storage Power Station	China	3600	40,000
Bath County Pumped Storage Station	United States	3003	24,000
Huizhou Pumped Storage Power Station	China	2448	
Guangdong Pumped Storage Power Station	China	2400	
Okutataragi Pumped Storage Power Station	Japan	1932	15,546
Ludington Pumped Storage Power Plant	United States	1872	19,548
Tianhuangping Pumped Storage Power Station	China	1836	
Tumut-3	Australia	1800	
Grand'Maison Dam	France	1800	34,800
Dinorwig Power Station	Wales	1728	9100
Raccoon Mountain Pumped Storage Plant	United States	1652	36,344
Mingtan Pumped Storage Hydro Power Plant	Taiwan	1602	
Okukiyotsu Pumped Storage Power Station (Plant 1 and Plant 2)	Japan	1600	
Castaic Power Plant	United States	1566	12,470
Liyang Pumped Storage Power Station	China	1500	
Xianju Pumped Storage Power Station	China	1500	
Okumino Pumped Storage Power Station	Japan	1500	

Source: "List of pumped-storage hydroelectric power stations," Wikipedia, accessed March 12, 2022, https://en.wikipedia.org/wiki/List_of_pumped-storage_hydroelectric_power_stations.

wilderness and because they use a lot of water. For example, in the United States, a plant proposed for construction next to Joshua Tree National Park, using abandoned mining pits for reservoirs, is facing fierce opposition.

4.4 Battery Storage

Between 2000 and 2020, the cost of lithium-ion batteries dropped from $2200 to $180 per kWh,[26] and their global revenue—which in 2021 was $210 billion—is projected to rise to $426 billion by 2030. This explosion is largely caused by the fact that, as of this writing, batteries are the most widely used means of electricity storage. This is mainly

26. "Global Battery Market Report – Market Analysis, Size, Share, Growth, Outlook, Industry Trends, and Forecast to 2027," Blackridge Research & Consulting website, accessed October 29, 2022, https://www.blackridgeresearch.com/reports/global-battery-market.

due to the increased use of electric vehicles (EVs) and solar backup storage units. Facts like these illustrate this trend: newly built solar homes in California must have battery storage; in the second quarter of 2022, Tesla sold a quarter million cars; and in China in 2022, there were 4 million EV charging stations, and in certain cities (Chengdu) only EVs may be driven on certain days of the week.

Yet, storing electrical energy in batteries is probably only a temporary remedy because:

- Batteries are typically recharged using "gray" electricity (electricity produced by burning fossil fuels) and indirectly cause CO_2 emissions.
- The weight-based energy densities of batteries are low; therefore, they are not suited for storing and transporting large quantities of electricity.
- They are suited to serve stationary users.
- In the last 5 years, some 25,000 EVs were recalled because of the potential for catching fire,[27] and a battery fire occurred on an airplane in 2013.[28]
- The cost of lithium will likely rise in the long range, and this factor might favor fuel-cell-operated electric vehicles (FCEVs) over battery-operated electric vehicles (BEVs).

The main limitations of batteries are their low mass-based energy densities, which range from about 50 to 250 Wh/kg (Table 4-4), and their long recharging times. Unless stationary, their substantial weights limit the amount of energy they can transport. An average electric car carries more than a half-ton of weight in batteries. This, plus the long battery recharging times, results in a practical limit of about 300 miles between recharging. In comparison, fuel cell cars have nearly twice the driving range and refueling times on the order of minutes.

As shown in Figure 4-1, electricity can be stored in many ways. Batteries store electrical energy in the form of chemical energy and can convert the stored energy back into electricity when needed. Batteries are small chemical reactors in which the reaction products are energetic electrons, ready to flow through external devices. Their

27. Peter Valdes-Dapena, "Electric car batteries are catching fire and that could be a big turnoff to buyers," CNN Business website, November 10, 2020, accessed October 29, 2022, https://www.cnn.com/2020/11/10/success/electric-car-vehicle-battery-fires/index.html.

28. "Design flaws led to 2013 lithium-ion battery fire in Boeing 787: U.S. NTSB," *Reuters*, updated December 1, 2014, accessed October 29, 2022, https://www.reuters.com/article/us-boeing-787-battery-idUSKCN0JF35G20141202.

Table 4-4. Energy and power densities of some standard batteries.

Battery Type[a]	Energy Density		
	Mass-Based (Wh/kg)	**Volume-Based (Wh/L)**	**Power Density (W/L)**
PbA	30 to 50	60 to 80	230
Nickel-metal hydride (NiMH)	60 to 120	100 to 150	330
Li-ion	100 to 265	250 to 670	275

[a]Flow batteries are not included here. Data on their energy and power densities can be found in Table 4-5.

Source: Bhaskar Garg, "Introduction to Flow Batteries: Theory and Applications," paper submitted as coursework for Introduction to the Physics of Energy, Stanford University, March 22, 2012, http://large.stanford.edu/courses/2011/ph240/garg1/.

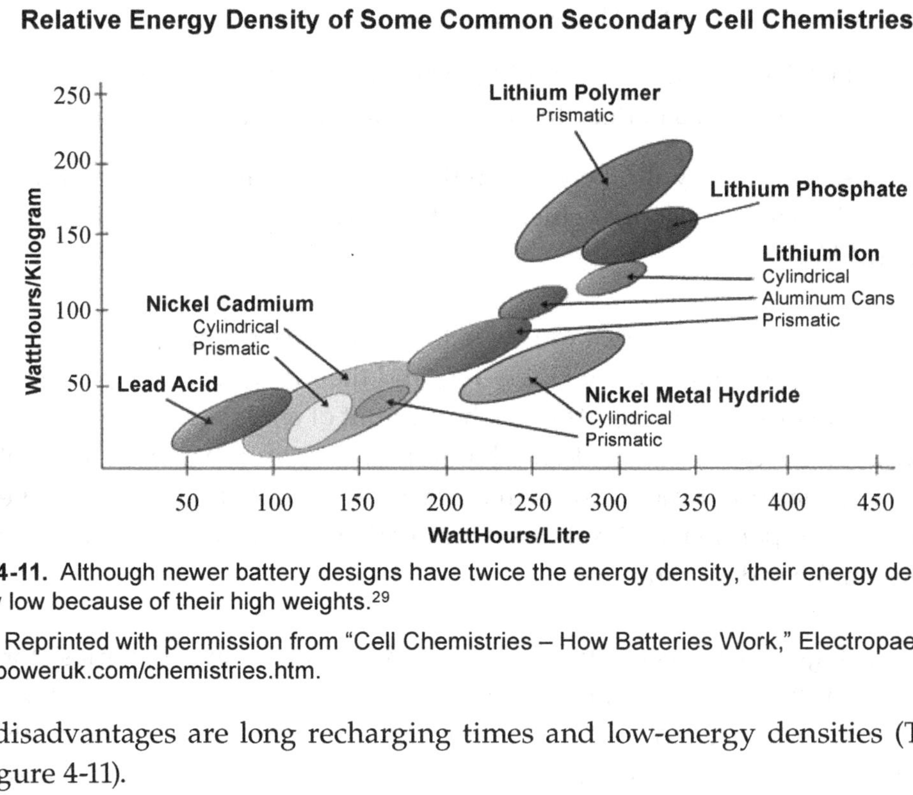

Figure 4-11. Although newer battery designs have twice the energy density, their energy densities are still very low because of their high weights.[29]

Source: Reprinted with permission from "Cell Chemistries – How Batteries Work," Electropaedia, 2005, www.mpoweruk.com/chemistries.htm.

main disadvantages are long recharging times and low-energy densities (Table 4-4 and Figure 4-11).

According to NASA, rechargeable battery types include nickel-cadmium (NiCd), nickel-hydrogen (NiH_2), Li-ion, and lithium polymer (LiPo), all of which have been

29. For further information, see: *State of the Art Small Spacecraft Technology*, NASA/TP—2018–220027, NASA, December 2018, accessed August 8, 2022, https://www.nasa.gov/sites/default/files/atoms/files/soa2018_final_doc.pdf.

used extensively on small spacecraft.[30] Lithium-based rechargeable batteries are generally connected to a primary energy source (e.g., a solar array) and can provide rechargeable power on demand. Each of the battery types listed is associated with specific applications that depend on performance parameters, including energy density, cycle life, and reliability.[31]

Of the electric vehicles on the roads today, 99% are BEVs, and most of their batteries are recharged using gray electricity. Only about 4% or even fewer of the BEVs are recharged with green electricity. In the case of FCEVs, a similarly low percentage use green hydrogen, although these percentages are slowly rising.

Ten years ago, only about 25,000 BEVs were on the roads. Today, they number 5.6 million, and by 2030, the number is projected to rise to about 200 million (10% of the total, 20% in China). In comparison, the current total number of hydrogen fuel cell cars (FCEVs) in the world is roughly 25,000, which is similar to the number of BEVs a decade ago. The rate of rise in EV use is primarily a function of infrastructure. For example, there are 125,000 regular gas stations in the United States, and the global number of EV charging stations in 2019 was only about 25,000. In 2022 the Biden administration proposed building a national network of 500,000 EV charging stations by 2030. In contrast, the global number of hydrogen fueling stations in 2019 was under 500. This number more than doubled by 2021 when 218 hydrogen refueling stations were added in China and 157 were added in Japan.

Compared to FCEVs, BEVs are superior in terms of cost and infrastructure availability. In Europe, as of 2022, driving 100 kilometers in a hydrogen car costs about €9 to €12, whereas driving a battery-operated car costs only €2 to €7 per 100 kilometers (depending on electricity prices in individual countries). The advantages of FCEVs include that their fuel availability is unlimited, they emit only water, and their driving range is high. In addition, their weights, refueling times, and noise levels are all lower than those of BEVs.

Today, lithium-ion batteries are the dominant designs. They contain lithium, cobalt, and nickel in the cathode; graphite in the anode; and aluminum and copper in the other cell and pack components. The required battery capacity per vehicle currently averages about 60 kWh for regular BEVs and 12 kWh for plug-in hybrid EVs (PHEVs). Between 2020 and 2050, it is estimated that the demands for lithium

30. *State of the Art Small Spacecraft Technology*, NASA/TP—2018-220027, NASA.
31. *State of the Art Small Spacecraft Technology*, NASA/TP—2018-220027, NASA.

and cobalt will increase 20-fold and demands for nickel will increase 30-fold. If the number of EVs on the roads is to increase to 200 million by 2030, keeping up with this growing demand will probably cause an increase in battery costs and require battery recycling. Today's demand for batteries is mainly in China, Japan, and Korea.

4.4.1 Operating Principles and Early Designs

Although the term *battery* was first used in 1749 by Benjamin Franklin, it was Alessandro Volta in 1800[32] who invented the first true battery by stacking disks of copper and zinc separated by cloth soaked in salty water. He connected wires to both ends of the stack and produced a continuous, stable current. Each cell consisting of one copper disk and one zinc disk separated by brine-soaked cloth produced a potential difference of 0.76 V.

The basic power unit inside a battery is called a *cell*, and there are three main components of a battery: the anode, the cathode, and the electrolyte that separates them. The two terminals are made of different chemicals (typically metals), and the electrolyte is a chemical medium that allows the flow of electrical charge between the cathode and anode. When a device, such as a light bulb, is connected to a battery, chemical reactions are triggered, and the electrodes generate a flow of electrical energy to the bulb or other electrical device.

The metal or compound that discharges the electrons is called the *anode*, and the metal or compound that accepts the electrons is called the *cathode*. When the battery is discharging electricity, the chemical on the anode releases electrons that pass through the outside circuit to the cathode (negative terminal) as the ions travel through the electrolyte. This is called an *oxidation reaction*. These two processes occur simultaneously: the electrons flow in the external circuit, and the ions transport the current through the electrolyte.

Batteries use various materials and chemical reactions, generating cell voltages between 1.0 and 3.6 V. Stacking the cells in series increases the battery voltage, whereas connecting them in parallel enhances the current that can be supplied. In this way, megawatt-size battery packs can be built to store solar and wind energy to provide power when those intermittent supplies are unavailable.

---

32. "History of the Battery," Wikipedia, accessed March 12, 2022, https://en.wikipedia.org/wiki/History_of_the_battery#Invention.

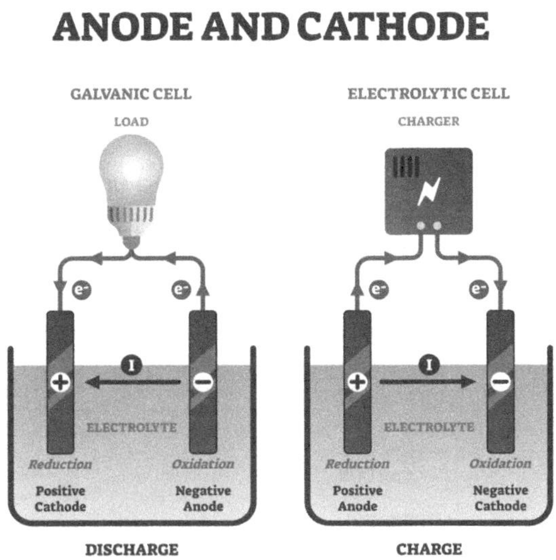

Figure 4-12. Structure and operation of a PbA battery.

4.4.2 PbA Batteries

The PbA battery (Figure 4-12) was invented in 1859 by French physicist Gaston Planté; it is the oldest of the rechargeable battery designs. It has a very low energy-to-weight ratio (mass-based or gravimetric energy density) of 30 to 50 Wh/kg and a low volume ratio. Its ability to supply high surge currents means that the cells have a relatively large power-to-weight ratio (power density) of 180 W/kg. These features, along with their low cost, make PbA batteries attractive for use in motor vehicles to provide the high current required by automobile starter motors.

4.4.3 Primary and Rechargeable Batteries

When the chemical reaction that produces the flow of electrons cannot be reversed, the battery is referred to as a *primary battery*. When one of the reactants is consumed, the battery is called *flat*. Common batteries include the zinc-carbon design, which lasts longer with alkali electrolytes. The NiCd was one of the earliest rechargeable batteries, and later, the longer-lived nickel-metal hydride (NiMH) batteries were developed. One limitation of these batteries is that they can easily be overcharged, in which case they overheat. These batteries are suitable for mobile phones and laptop computers, which require the most compact energy storage. They are not recommended for larger applications because of the risk of violent discharge.

4.4.4 Li-Ion Batteries

In 1980, the Li-ion battery (in which lithium migrates through the battery as a positive ion) was invented by Professor John Goodenough of the University of Texas at Austin. This was an important invention because lithium is light and has a large

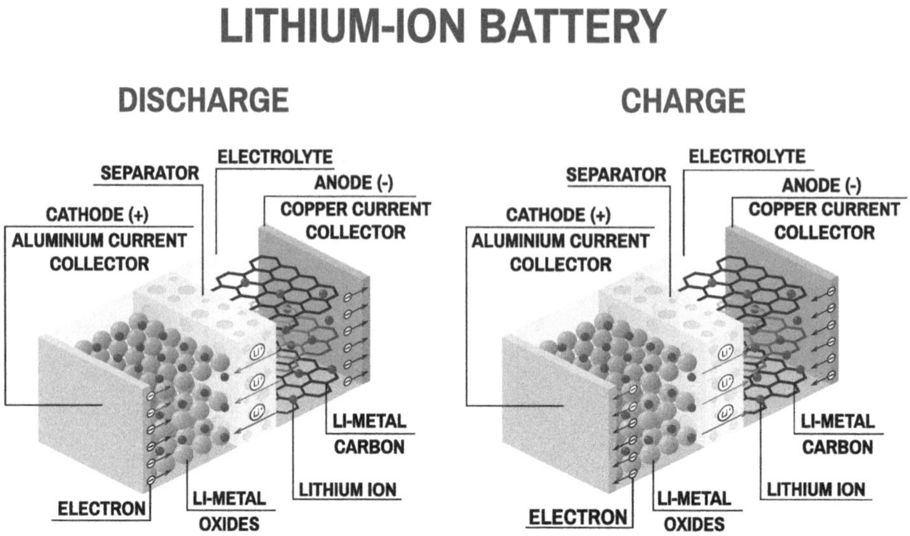

Figure 4-13. Instead of chemical reactions, the lithium ions flow back and forth between the node and the cathode.

electrochemical potential; therefore, Li-ion batteries are compact and small in volume and produce high voltages. In this type of battery, lithium is combined with a transition metal (cobalt, nickel, manganese, or iron) and oxygen to form the cathode. When a voltage is applied during recharging, the lithium-ion (which has a strong electrochemical driving force to be oxidized) migrates from the cathode to the graphite anode and becomes a Li-ion again, giving up its electron to the cobalt ion (Figure 4-13). The lithium ions in the electrolyte solution migrate through a porous polymer separator to the anode. The movement of electrons in the circuit generates usable current.

In rare circumstances, if a Li-ion battery short circuits, it can overheat inside the cell and ignite the liquid or rapidly raise its vapor pressure until the cell bursts. Overcharging or exposing a Li-ion battery to too high of a voltage by using the wrong charger or a failed charger can bring the cathode to an unsafe oxidizing state. Under such conditions, the cathode can react with and decompose the electrolyte solution, generating heat and hydrocarbons that can react further, generating more heat and triggering an uncontrollable heating condition that can violently destroy a battery.

Such an accident occurred in January 2013 when a Boeing 787 Li-ion battery caught fire as the empty plane sat on the ground at Boston's Logan International Airport. Although such events are extremely rare, because billions of these batteries (ranging from very small cells to 500 kg units for electric vehicles or even larger devices) are produced yearly, safety is still a concern because of their flammable electrolyte solutions. Unlike other common types of batteries, in which the electrolytes are aqueous solutions of acid or base, the electrolyte in Li-ion cells consists of lithium salts in flammable organic solvents, such as ethylene carbonate and ethyl methyl carbonate.

In 2020, only Tesla had a lithium demand of about 8000 tons, while the global consumption was 56,000 tons, and as of yet, no technology exists for recycling lithium from used batteries. Li-ion battery usage is growing quickly because prices have dropped due to increased economies of scale. However, Li-ion batteries cannot solve all of the challenges of electric power storage. Li-ion batteries are well positioned to meet the demand for the next 5 to 10 years, but in the long run, other storage technologies will be needed for long-term energy storage and large-scale applications. Right now, the most likely candidate for that is hydrogen.

4.4.4.1 Flow Batteries

Redox flow batteries are electrochemical cells in which the chemical energy is provided by two chemical components dissolved in liquids separated by a membrane. The flow of electric current occurs through the membrane while both liquids circulate in their respective spaces (Figure 4-14). The cell voltages of these designs range from 1.0 to 2.2 V.

A flow battery's operation is similar to that of a fuel cell (the spent fuel is extracted and new fuel is added continuously) or a rechargeable battery (an external electric

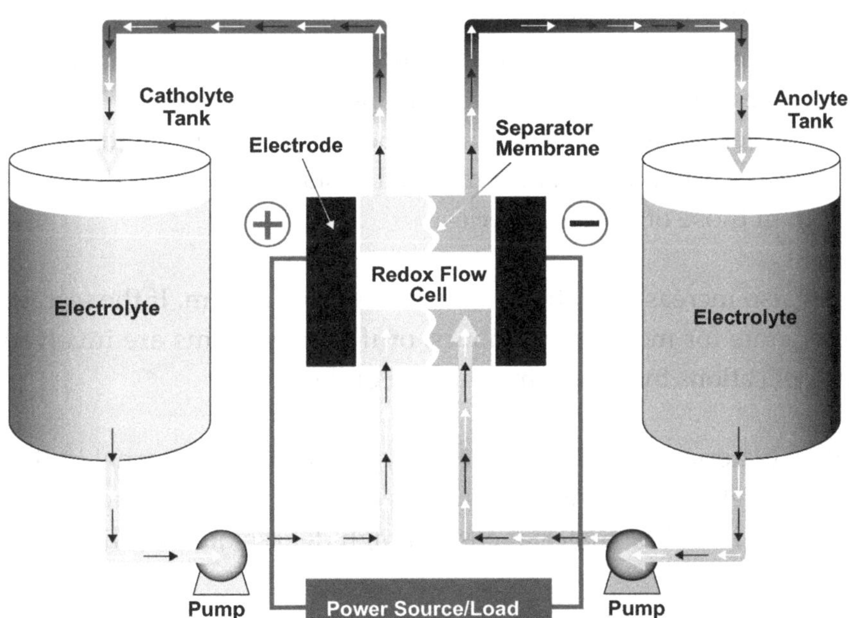

Figure 4-14. Flow batteries are normally considered for relatively large (1 kWh to 10 MWh) stationary applications.

Table 4-5. Energy and power densities of flow batteries.[36]

Battery Type	Volume-Based Energy Density (Wh/L)	Power Density (W/L)
Polysulfide–bromine	20 to 35	60
Vanadium–vanadium	20 to 35	60 to 100
Vanadium–bromine	20 to 35	50
Zinc–bromine	20 to 35	40
Zinc–cerium	20 to 35	50

Source: Bhaskar Garg, "Introduction to Flow Batteries: Theory and Applications," paper submitted as coursework for Introduction to the Physics of Energy, Stanford University, March 22, 2012, http://large.stanford.edu/courses/2011/ph240/garg1/.

power source is used to regenerate the unit). Although flow batteries have technical advantages over conventional rechargeable batteries, such as separate liquid tanks and nearly unlimited longevity, current implementations are comparatively less powerful and require more sophisticated electronics. One of the limitations of such batteries is their low power density (Table 4-5).[33]

For a flow battery, the energy capacity is a function of the electrolyte volume (amount of liquid electrolyte), and the power is a function of the surface area of the electrodes. These designs are under development for large-scale energy storage in which the electrolyte, anode, and cathode are all liquids. For portable applications, thin-film polymer battery designs with a flexible electrolyte made of nonflammable gel are being considered. Another development goal is to design batteries using abundant, inexpensive, and safe substances with the same commercial potential as popular lithium-based batteries. Flow batteries allow the decoupling of power and energy and can use different materials, including vanadium. Their material costs are lower, and their designs are less complex than those of Li-ion batteries.

Flow batteries also increase the life span of the battery system. If the storage system is expected to operate for more than 4 hours, or if large systems are involved, they can serve those applications by increasing the size of the tanks.

In zinc-air flow batteries (ZAFBs), the electrolyte is stored in external tanks and circulated through the cell. ZAFBs are promising as efficient energy storage systems for several reasons. They have low material costs, a high degree of flexibility, and the ability

33. M. R. Mohamed, S. M. Sharkh, and C. Walsh, "Redox Flow Batteries for Hybrid Electric Vehicles: Progress and Challenges" (paper presented at the IEEE Vehicle Power and Propulsion Conference VPPC '09, Dearborn, MI, September 7 to 11, 2009): TS04B6.

to stay at a zero state of charge with no degradation. They are most cost-effective if the storage system is expected to operate for more than 8 hours. On the other hand, they are not suited for operation at high power levels.[34]

4.4.4.2 Costs of Li-Ion Batteries

It seems to be a strange contradiction that the unit costs of lithium-ion battery packs are dropping at the same time that the cost of lithium itself is surging.

Let us first look at lithium. The global consumption of lithium in 2019 was estimated at 57,700 metric tons; by 2021, it had increased to 70,000 metric tons, and it reached 93,000 metric tons by 2022. In 2019, the average price of battery-grade lithium carbonate was under $13,000 per metric ton. According to S&P Global Platts data, as reported by Forbes,[35] lithium carbonate prices have gained 413% since then, and by the year 2021, the price reached $32,600 per metric ton. By March 2022, the price more than doubled

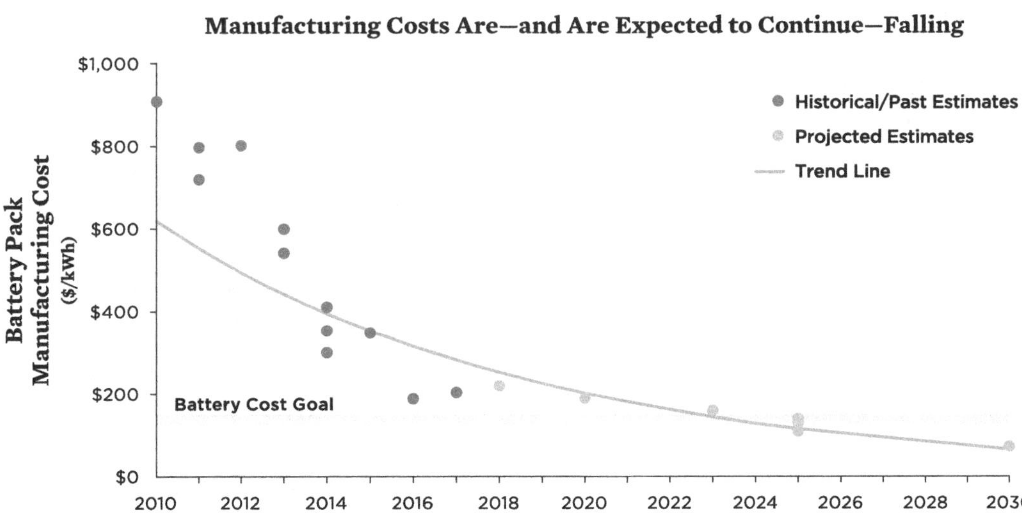

Figure 4-15. If battery costs continue to decline as EV production increases, they will drop to $200/kWh within a few years, making EVs competitive with conventional gasoline vehicles.

Source: Reprinted with permission from "Electric Vehicle Batteries: Materials, Cost, Lifespan," Union of Concerned Scientists, March 9, 2018, https://www.ucsusa.org/resources/ev-batteries.

---

34. Lisa Cohn, "Long-Term Energy Storage: What are the Options When Lithium-ion Falls Short?," Microgrid Knowledge website, March 17, 2017, https://microgridknowledge.com/long-term-energy-storage/.

35. Robert Rapier, "The Challenges Posed by Rising Lithium Prices," Forbes website, December 31, 2021, accessed November 21, 2022, https://www.forbes.com/sites/rrapier/2021/12/31/the-challenges-posed-by-rising-lithium-prices/?sh=150958833af9.

to $76,700 per metric ton. Lithium hydroxide prices have also climbed 254% to $31,900 per metric ton.[36]

As to storage costs in batteries, it is likely that in a couple of years, that cost will drop below the cost of storing energy in fossil fuels. The cost of Li-ion battery packs is falling (Figure 4-15) because of mass production and high competition and because they are still using lithium purchased at a lower price. As of 2022, the storage of each kWh costs about $250. The average EV stores 20 kWh to 60 kWh of electricity, so the battery pack costs range from $5000 to $15,000. Yet, according to some estimates, by 2030, there will be 200 million EVs on the roads, so the demand for lithium will rise even more, and therefore, it is probable that the battery pack costs will also rise.

According to other estimates, Li-ion battery costs for stationary applications[37] could fall to $200/kWh by 2030 for installed systems, other estimates anticipate even lower costs and markets around $10 billion.[38] Battery storage in stationary applications looks set to grow from only 2 GW worldwide in 2017 to over 175 GW in 2022. This rivals PSH applications, which is projected to reach 235 GW in 2030. In the meantime, lower installed costs, longer lifetimes, increased numbers of cycles (number of times a battery can be recharged), and improved performance will probably drive down the costs of stored electricity services.

4.4.4.3 Large, Grid-Stabilizing Installations

Batteries can be used alongside solar and wind power sources to compensate for the latter's intermittent or fluctuating power generation. Flow batteries are currently used in solar microgrid applications throughout the Caribbean. In load balancing applications, the batteries are connected to an electrical grid to store excess electrical power during off-peak hours and release electrical power during peak demand periods. In addition to storing energy from renewable sources such as wind and solar for discharge during periods of peak demand, they are used for peak shaving—managing demand to smooth out peak loads—by using power from the batteries to eliminate short-term demand spikes.

When the main supplies to the grid fail, the storage has to be large enough to supply the grid for several hours. The sudden change in large amounts of renewable energy

---

36. Alex Kimani, "Lithium Prices Have Nearly Doubled In 2022 Amid Insane Commodity Rally," Oilprice.com, March 21, 2022, accessed October 14, 2022, https://oilprice.com/Energy/Energy-General/Lithium-Prices-Have-Nearly-Doubled-In-2022-Amid-Insane-Commodity-Rally.html.

37. Tom Kenning, "Partnerships in Saudi and Australia plan future of long-duration flow battery deployment," Energy Storage News website, May 15, 2019, accessed November 21, 2022, https://www.energy-storage.news/partnerships-in-saudi-and-australia-plan-future-of-long-duration-flow-battery-deployment/.

38. Justin Gerdes, "Reducing battery cost is essential for a clean energy future," Energy Monitor website, October 5, 2020, https://www.energymonitor.ai/tech/energy-storage/reducing-battery-cost-is-essential-for-a-clean-energy-future/.

to the grid can cause grid instability. For example, when a solar farm is used, there can be instability in the morning when too much solar electricity suddenly comes online and in the evening when the amount of solar energy drops off.

An example of a large battery storage system for large solar and wind farm applications is Tesla's 100 MW/129 MWh Powerpack project in South Australia. It was started up in 2017 and cost approximately $66 million, but it reportedly made as much as $17 million within its first 6 months of operation.[39] According to financial adviser Lazard Ltd.,[40] building solar and wind farms has started to become cheaper than running aging coal and nuclear generators in parts of the United States. Building and operating a utility-scale wind farm costs $30 to $60 per megawatt-hour over its lifetime. In comparison, just keeping an existing plant running can cost $26 to $39/MWh for coal-burning power plants and $25 to $32/MWh for nuclear power plants.

A 20 MW Tesla energy storage project is currently online in Ontario, California. The installation consists of 396 Powerpacks and two dozen inverters and can supply 80 MWh to utilities. Also in California, AES Corporation started commercial operations of its 100 MW, 400 MWh Alamitos energy storage project in Long Beach on January 1, 2021. This system, one of the largest operating battery storage projects in the world, employs a 4-hour, lithium-ion battery system supplied by Fluence Energy LLC, an AES joint venture with Siemens AG.

4.4.4.4 Electric Vehicles

As of 2022, the population of our planet is about 7.9 billion, and the total number of cars in use is about 1.5 billion, consisting of 1.1 billion passenger cars and 0.4 billion commercial vehicles. This total number is projected to double by the end of the century.

In 2010, Tesla sold only a couple hundred BEVs. Yet, a decade later, the number of Tesla cars on the roads exceeded 3 million, and in a single quarter in 2022, they sold over a quarter million. The proportion of BEVs in 2022 exceeded 1% of the total. The share of EVs is still rising by about 7% a year, at which rate it is expected to reach 10% by 2030 and about 20% of new car sales. This increase in market share is likely to occur because the weight of battery packs will also be reduced, and efficiencies will rise. Nevertheless, the weight of EV battery packs (1200 to 1800 pounds today) will remain

---

39. Fred Lambert, "Tesla's massive Powerpack battery in Australia cost $66 million and already made up to ~$17 million," electrek website, September 24, 2018, accessed November 21, 2022, https://electrek.co/2018/09/24/tesla-powerpack-battery-australia-cost-revenue/.

40. Naureen S. Malik, "Renewables Are Starting to Crush Aging US Nukes, Coal Plants," *Bloomberg News*, November 2, 2017, accessed November 21, 2022, https://www.bloomberg.com/news/articles/2017-11-02/renewables-are-starting-to-crush-aging-u-s-nukes-coal-plants.

high, and the time required for recharging will remain long. Both factors make the increased use of hydrogen FCEVs likely in the long range.

Hydrogen FCEVs are at the point where BEVs were in the early 2010s. This is because as soon as hydrogen fuel stations become available, their refilling time will drop to minutes. BEVs require long recharging times and carry heavy batteries, FCEVs do not. A number of long-range R&D projects aimed at designing roads that would wirelessly recharge moving BEVs are in progress, but it is debatable as to whether developing this technology is either realistic or necessary.[41]

The cost of storing each kilowatt-hour of electricity was $250 in 2020; it is estimated to fall further to about $150/kWh by the end of this decade. In 2010, a 50 kWh battery pack in 2010 cost about $50,000. This cost has dropped to about $15,000 today. The power consumption of the average EV is 0.25 to 0.5 kWh/mile.

In a review of EV safety, the National Highway Traffic Safety Administration concluded that electric cars do not pose a greater fire risk than gasoline-powered vehicles. Currently, battery-operated electric cars are good for local driving when they can be charged during the night. They are convenient if the owners do not drive more than roughly 200 miles a day. In that case, they can be recharged from 240 V outlets overnight. If the owner's daily commute is only 20 miles, a 120 V outlet will suffice. With a 240 V system, each hour of charging can add enough electricity for 25 miles of additional driving.

As was mentioned earlier, the Chevrolet Bolt battery pack is estimated to cost about $205/kWh. When the first mass-market EVs were introduced in 2010, their battery packs cost an estimated $1,000/kWh. Today, Tesla's Model 3 battery pack costs $190/kWh. The downward trend in battery pack costs is shown in Figure 4-15, but because of the recent sharp increases in lithium costs, this trend is unlikely to continue.

Many governments support converting the transportation industry to electric in the shorter range and possibly to hydrogen in the longer range. For example, President Biden has announced a goal of having EVs constitute 50% of the new automobiles sold in the United States by 2030 and 100% by 2035. Battery weight represents a limitation for EVs because carrying 1200 to 1800 lb of batteries (the weight of up to 10 passengers), even when the car is empty, reduces driving efficiency. In the future, this factor will

---

41. Maurizio Di Paolo Emilio, "Wireless Charging Technology for EVs," Power Electronics News website, May 27, 2021, accessed November 21, 2022, https://www.powerelectronicsnews.com/wireless-charging-technology-for-evs/.

Figure 4-16. Home batteries for storing solar energy.

support a preference for hydrogen fuel cells, particularly for trucks, buses, and longer-distance driving. The power consumption of battery-fueled cars is usually between 0.25 and 0.5 kWh per mile.

The cost of battery storage for homes (Figure 4-16) has also dropped. Just a decade ago, the cost of a 7 kW battery, together with an inverter and installation, was greater than $15,000. However, because solar battery prices are falling, analysts predict that the prices will soon reach the point where using stored electricity will cost less than using grid-supplied electricity.[42]

I should emphasize that using batteries to store intermittent green energy is not a solution in many cases. This is because of the batteries' high weights, slow recharging times, and the carbon emissions produced when fossil-fuel power plants are used to generate the recharging electricity. There are other reasons. One is that batteries are stationary storage devices that are not suited for transporting large quantities of electricity; they require the grid to do that. Another is that lithium batteries require large quantities of lithium, the cost of which can become prohibitive.

The time required for recharging the average battery is usually an hour or more. For detailed information about the four levels of chargers for plug-in electric vehicle (PEV) batteries (Levels 1, 2, and 3 and superchargers), see Table 1-1 in Chapter 1.[43]

---

42. "Home solar batteries and energy storage solutions," *Australian Solar Quotes Blog*, June 26, 2018, accessed November 21, 2022, https://www.australiansolarquotes.com.au/2018/06/26/home-solar-batteries-energy-storage-solutions/.

43. Tom Moloughney, "What Are the Different Levels of Electric Vehicle Charging?," Forbes Wheels website, updated October 4, 2021, accessed November 21, 2022, https://www.forbes.com/wheels/advice/ev-charging-levels/.

Table 4-6. Approximate comparison among gasoline-powered, PEV, and FCEV cars in 2020.

Feature	Gasoline	PEV	FCEV
Make	Toyota Corolla	Tesla 3	Hyundai NEXO
Engine	170 hp	460 hp	160 hp
Initial cost	$21,000 to $28,000	$40,000 to $50,000	$60,000 to $65,000
Fuel cost per mile[a] (city/highway)	$0.12/$0.09	$0.06/$0.03	$0.20/$0.14[b]
Fuel efficiency[c] (city/highway)	31 MPG/40 MPG	115 MPGe/125 MPGe	58 MPGe/65 MPGe
Range per filling	600 miles	250 to 350 miles	400 miles

[a]Assumed fuel costs at the pump: gasoline, $3.50/gal; electricity, $0.20/kWh; hydrogen, $14/kg.
[b]To become competitive with gasoline at $3.50/gal, the cost of hydrogen must drop to $6/kg (today's cost of gray hydrogen, which is obtained from natural gas and produced from fossil fuels).
[c]The energy equivalent of a gallon of gasoline is about 34 kWh in the form of electricity and 0.85 kg in the form of hydrogen. Abbreviation: MPGe, miles per gallon of gasoline-equivalent.

In comparison, hydrogen fuel-cell-operated cars can be charged in about 5 minutes, have a driving distance of more than 400 miles, and have zero carbon impact on the environment; however, the electricity used to charge the batteries is often produced by fossil-fuel plants. In addition, the life of fuel cells is at least twice that of batteries, and their fuel capacity does not decrease with age. The weight of batteries is about 10 times higher than that of fuel cells because the mass-based energy densities of batteries are some 150 times lower than that of hydrogen. Table 4-6 compares the features of gasoline-fueled, PEV, and FCEV cars as of 2020.

The International Energy Agency's (IEA's) EV30@30 Campaign[44] encourages policies aimed at reaching a 30% market share for electric cars by 2030. In 2016, Bloomberg New Energy Finance (BNEF) forecasted that the total sale of EVs would reach 35% of global sales of new cars (41 million) by 2040. BNEF also forecasts that EV ownership costs will drop below those of conventional vehicles by 2025.[45]

4.4.4.5 Li-Ion versus NiMH Batteries

Although NiMH batteries were the primary batteries used in EVs for a long time, most EVs currently use Li-ion batteries because of some specific advantages.[46]

---

44. Sarah Gray, "125 Million Electric Vehicles Will Be on the Road by 2030, Agency Says," *Fortune Magazine*, May 31, 2018, accessed November 21, 2022, https://fortune.com/2018/05/31/electric-vehicles-international-energy-agency/.
45. "Electric vehicles to be 35% of global new car sales by 2040," February 25, 2016, accessed November 21, 2022, https://about.bnef.com/blog/electric-vehicles-to-be-35-of-global-new-car-sales-by-2040/?.
46. Kristen Hall-Geisler, "How can lithium-ion batteries improve hybrids?," HowStuffWorks website, accessed March 12, 2022, https://auto.howstuffworks.com/lithium-ion-batteries-improve-hybrids1.htm.

Li-ion batteries are made of carbon and lithium, a highly reactive metal, and can store a great deal of energy. NiMH batteries use hydrogen to store energy, and they control the flow of hydrogen ions with nickel and other metals (such as titanium). Currently, NiMH batteries are less expensive than Li-ion batteries, but this will probably change as the production volume of Li-ion batteries rises.

NiMH batteries are larger and heavier than Li-ion batteries and have lower energy densities (Table 4-4). Although Li-ion and NiMH batteries can store similar amounts of power, Li-ion cells can be charged and discharged more rapidly. Both types of batteries are durable and have been in use for years, but NiMH batteries last longer in hot climates. Manufacturers are working to resolve this issue and improve the chemistry of Li-ion batteries so they last longer. New designs of Li-ion batteries are also under development that will have 17% greater power densities because they will use silicon powder instead of graphite for their anodes.

4.4.5 Summary

When one talks about electricity storage, most people think of batteries. In the past few years, the cost of utility-size Li-ion storage systems has decreased dramatically, and larger-capacity battery farms are being built worldwide. For example, in 2020, Tesla's Powerpack facility in Jamestown, South Australia, was expanded from its original 100 MW/129 MWh capacity to 150 MW/193.5 MWh. Batteries are also used in private homes and electric cars. Still, they are heavy, their driving distances per charge[47] are relatively short, their recharging times are long, the time during which suppliers will replace degraded batteries under warranty is unresolved, and they are not practical for transporting large quantities of electricity.

Stored energy can provide nighttime electricity for the home, recharge electric cars, and be sent to the grid. The time when excess solar electricity is available usually coincides with the times when air-conditioning loads are high (peak electricity usage). This electricity is valuable to power companies because it reduces the need to increase production at a time when the plants are already operating near their full generating capacities. Optimizing processes for integrating stored electricity into the existing distribution system involves sending the stored electricity to the most profitable destination.

---

47. Jill McIntosh, "2018 PHEV and EV Range and Charging Times," Auto Trader Canada website, July 11, 2018, accessed November 21, 2022, https://www.autotrader.ca/newsfeatures/20180710/2018-phev-and-ev-range-and-charging-times/.

Battery performance has improved during the past few decades. The oldest and most mature battery technology is the PbA battery, which has a mass-based energy density of about 50 Wh/kg and a life expectancy of between 2 and 5 years. For example, a 6 V, 210 Ah golf-cart battery stores about 1.26 kWh (6 V × 210 Ah = 1.26 kWh) and weighs about 30 kg. Trailers and motor homes usually use only a single 12 V house battery. These batteries often use gel-cell or absorbed glass mat (AGM) designs instead of wet cells because these designs are suited for harsher environments, require less maintenance, and provide the greatest reserve capacity. The individual cells of PbA house batteries generate about 0.8 V per cell, and two 12 V batteries are needed to power 4 kW inverters.

The introduction of Li-ion significantly improved the capabilities of batteries (Table 4-7) because Li-ion batteries have relatively good energy densities, low weights, and shorter charging times (Figure 4-17).

Both the driving range and the economics of batteries are improving with time. Depending on its size, a regular electric car requires between 0.2 and 0.6 kWh of electric energy per mile of driving; hybrids naturally require less. The cost of today's batteries is dropping rapidly (Figure 4-15).

The issue of how electric car filling stations of the future will be designed is unresolved. Will they replace used battery blocks with full ones (which would take minutes), or will they operate on a plug-in recharge basis (which would take an hour or longer)? In any case, today's battery costs are still high, driving distances between recharges (up to 300 miles) are still short, and recharging times still limit their widespread use.

Table 4-7. Comparison of the energy densities of some rechargeable battery types.

Cell Chemistry	**Energy Density**	
	Mass-Based (MJ/kg) [Wh/kg]	**Volume-Based (MJ/L) [Wh/L]**
NiMH (also Ni–MH)	0.36 [100]	1.44 [401]
NiCd (also NiCad)	0.11 [30]	0.36 [100]
Li-ion[b]	0.36 to 0.875 [100 to 265]	0.90 to 2.43 [250 to 693]
PbA (SLA, VRLA)[a]	0.11 to 0.14 [30 to 40]	0.22 to 0.27 [60 to 75]

[a]Abbreviations: SLA, sealed lead-acid; VRLA, valve-regulated lead-acid.
[b]Lithium-ion data: "Lithium-ion battery," Wikipedia, accessed March 12, 2022, https://en.wikipedia.org/wiki/Lithium-ion_battery.
Source: "Comparison of commercial battery types," Wikipedia, accessed March 12, 2022, https://en.wikipedia.org/wiki/Comparison_of_commercial_battery_types.

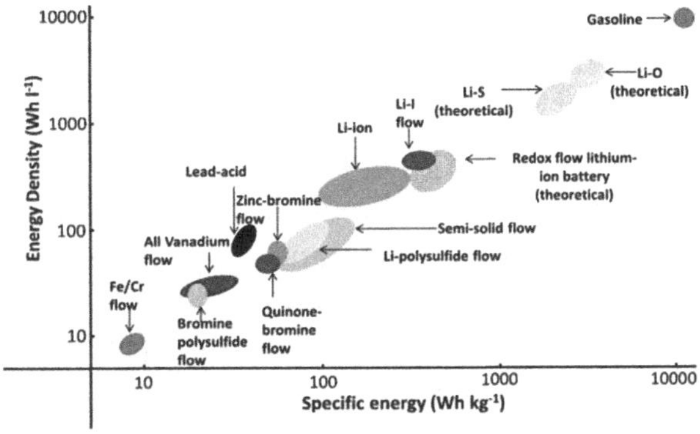

Figure 4-17. Comparison of the energy densities of battery types. (kWh = 1000 Wh).

Source: Reproduced with permission. Q. Huang, and Q. Wang, "Next-Generation, High-Energy-Density Redox Flow Batteries," *ChemPlusChem* 80, 2015: 312–322, https://doi.org/10.1002/cplu.201402099. Copyright Wiley-VCH GmbH.

Although the initial costs of EVs are still higher, the operating costs already favor EVs over internal combustion (IC) vehicles. The specific costs vary depending on many variables, including local conditions, oil prices, and other factors, as already mentioned in Table 4-6. Other data (as discussed in the next paragraph) will yield different results than the values in the table, but the net result is still that the operating costs of BEVs are lower than those of regular IC designs.

At a gasoline cost of \$3.50/gal and an average fleet mileage of 30 mpg, the cost of driving an IC car is about \$0.12 per mile. The energy content of a gallon of gasoline is about 34 kWh; therefore, the per-mile energy consumption of the average IC vehicle is about 1.1 kWh/mi (obtained simply by dividing 34 kWh/gal by 30 mpg = 1.1 kWh/mi). The cost of driving an electric car is significantly less than that because, at an electricity cost of \$0.20/kWh, the cost per mile corresponds to \$0.06, compared to \$0.12 per mile for the average IC car. Naturally, as the cost of gasoline rises and the cost of batteries falls (Table 4-6), the economic advantage of driving electric cars will further increase.

The race between fuel cells and batteries has not yet really started. Today, there are over a million electric cars on the road, but less than 10,000 fuel-cell-driven ones. As of this writing, the economics favor batteries, but the long-range outcome of this competition is undecided. As electric cars replace today's gasoline vehicles, the availability and cost of the materials needed for fuel cells and batteries will become important factors. The critical cost factor in the case of fuel cells is the cost and availability of the catalysts. The critical cost factor for batteries is the cost and availability of lithium, although other materials (nickel, manganese, antimony, and multiple miniature carbon terminals) and other designs (asymmetric supercapacitors, microelectromechanical systems, nanopowders, etc.) are also being considered.[48]

Automatic controls are also helping to improve the convenience of using electric cars because of the increased sophistication of battery charge controllers. These controllers are usually provided with digital interfaces and interactive displays that can be integrated with the controller or connected wirelessly to the user's PC. In either case, the human-machine interface allows the homeowner to check the charging voltage and the amount of electrical energy stored at any time and to modify the limit settings for each mode of operation.

4.5 The Hydrogen Age

The conversion from our fossil-nuclear energy economy to a solar-hydrogen energy economy will require a dramatic change in our culture, a tremendous global technological and economic effort, and decades of hard work. Here, I describe the steps that must be taken on this road of transition, including the required financial and land investments and additional R&D advances needed to complete the conversion.

In the coming "hydrogen age," most of the energy used will come from intermittently available, stationary, green energy sources that will be converted into hydrogen for continuous availability. The hydrogen will be transported and distributed to users in the form of either high-pressure gas or cryogenic liquid. The hydrogen "energy chain" will include the following main component blocks: solar collectors; electrolyzers; compressors; liquefaction plants; and storage, transportation, and distribution systems. These components are still under development today, their efficiencies are increasing, and their unit costs are dropping. Therefore, the performance data included in this list is expected to improve further.

- If the global average solar collector efficiency is about 30% and the average yearly global insolation is 1200 kWh/m^2, each square meter of solar collectors will generate about 400 kWh of electricity annually.

48. Microelectromechanical systems, or MEMSs, are miniaturized mechanical and electromechanical elements (i.e., devices and structures) that are made using microfabrication techniques and range in size from less than 1 µm to several millimeters. Bridget McCrea, "MEMS Are Having a Moment," January 26, 2018, accessed October 29, 2022, https://www.sourcetoday.com/news/article/21866997/mems-are-having-a-moment#.

- When producing hydrogen from water at low pressure, an electrolyzer[49] requires about 48 kWh of electricity to generate 1 kg of hydrogen.
- A compressor requires about 1.36 kWh of electric energy to compress one kilogram of hydrogen from 20 to750 bars (from 290 to 10,800 psi).
- If the hydrogen is to be used in liquid form, the energy required for liquefaction will range from 10 to 18 kWh/kg.
- The energy requirement for hydrogen storage, transportation, and distribution is about 5 kWh/kg.
- The efficiency range of burning hydrogen in boilers or fuel cells is between 40% and 80%.

Therefore, a conservative value for the energy needed to produce and use hydrogen as a fuel is about 180 kWh.

I do not know when we will complete the conversion to a green energy economy, but as a process control engineer, I know the steps we take in this decade will largely determine that date. As we saw in the previous chapters, the global temperature has already increased by more than 1.0°C (1.8°F), and it is still rising at an increasing rate. We also saw that GW is a self-accelerating process because of some large positive feedback loops (permafrost melting, ocean degassing, increased concentration of water vapor in the air, albedo effect, etc.).

In the past two centuries, humankind's energy demand has increased by a factor of 27, resulting in the emission of an extra 280 GT of carbon into the atmosphere and increasing the atmospheric carbon content from 600 to 880 GTC. We are increasing this quantity at a rate of about 10 GTC per year partially. Besides other causes, this emission is also getting worse because oil or gas wells are not always sealed when they are shut down. In short, we are dealing with CO_2 emission as an uncontrolled and runaway process.

Fortunately, we also know that this uncontrolled process can be brought under control, and we still have the time and the technology to do so. To start that discussion, in the next section, I provide an estimate of the surface area of the planet needed to collect the energy required to enter this solar-hydrogen age.

---

49. Green hydrogen is made from green energy (solar, wind, etc.), whereas gray hydrogen is made from natural gas.

4.5.1 Solar Collector Area Required for Conversion

Today, humankind's total annual energy consumption is about 600 Q (176 PWh), and our global annual electricity consumption is more than 85 Q (26 PWh). We can calculate the solar farm area required to meet these needs. Figure 4-18 shows the yearly quantities of solar energy received at different points on the planet in kilowatt-hours per square meter. Assuming that the global yearly average insolation at the collection sites amount to 1200 kWh/m^2 (1200 GWh/km^2) and that the average solar collector efficiency is about 30%, the approximate area required to meet this need with solar energy would be about 490,000 km^2, which is about the size of Spain[50] (498,980 km^2). This area is less than that of our roofs, roads, and parking lots. Given that the gravimetric energy density of hydrogen is 39.6 kWh/kg and assuming that the generation of one kilo of hydrogen requires 180 kWh, the generation of about one billion tons of hydrogen is needed.

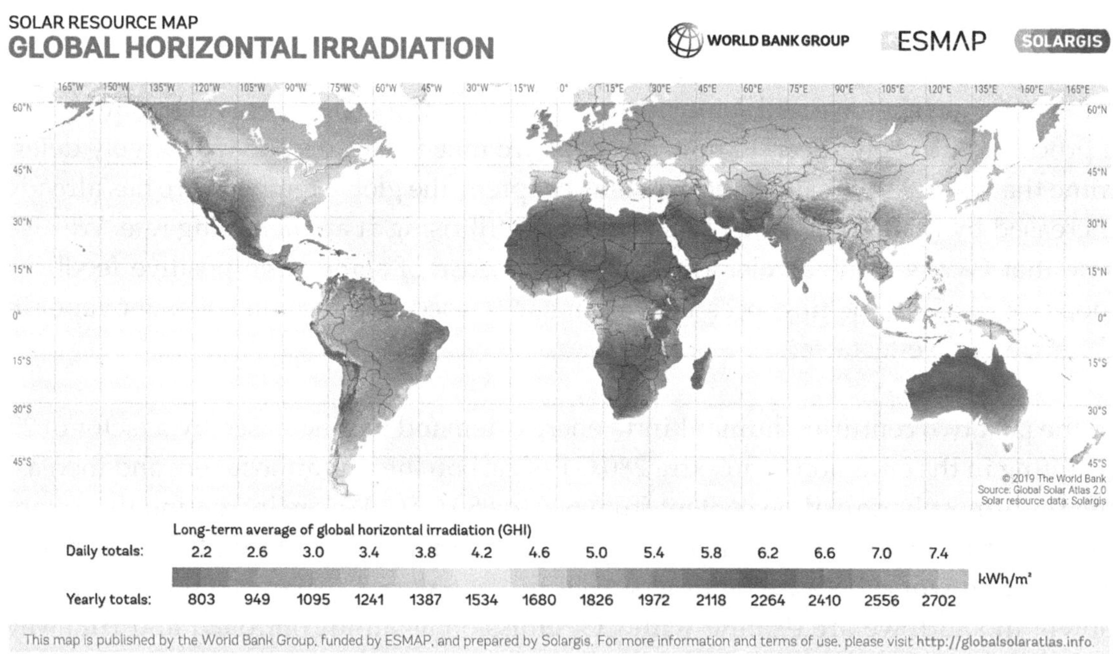

Figure 4-18. The yearly average global horizontal irradiation (GHI) received per unit area of Earth's surface ranges between 500 and 2500 kWh/m^2. (See color images after page 306.)

Source: Map obtained from the "Global Solar Atlas 2.0," a free, web-based application developed and operated by the company Solargis s.r.o. on behalf of the World Bank Group, utilizing Solargis data, with funding provided by the Energy Sector Management Assistance Program (ESMAP).[51]

50. "Countries Compared by Geography > Land area > Sq. km. International Statistics," NationMaster.com website, accessed July 6, 2022, https://www.nationmaster.com/country-info/stats/Geography/Land-area/Sq.-km.

51. For additional information, see: "World GHI Solar-resource-map GlobalSolarAtlas World-Bank-Esmap-Solargis.png" by Solargis, © The World Bank / CC BY 4.0, https://commons.wikimedia.org/wiki/File:World_GHI_Solar-resource-map_GlobalSolarAtlas_World-Bank-Esmap-Solargis.png.

Therefore, I would estimate that the area required to meet the global energy consumption of humankind by solar energy is about a half-million km^2 and the hydrogen requirement to meet that need is about one billion tons yearly. The numbers I used in this calculation are debatable. Still, I believe they support the conclusion that the energy needs of the coming hydrogen age can be met by solar energy alone, which is just one of the available green energy sources.

4.5.2 Hydrogen Age Economics

In a solar-hydrogen economy, the hydrogen will probably be distributed in the form of gas at a pressure of 700 bar (10,000 psig). The energy density of this hydrogen is 39.6 kWh/kg (Table 4-1). Therefore, if the yearly electricity consumption of a home is 1500 kWh, a 40 gal (150 L), 700 bar (10,000 psig) hydrogen tank would meet the home's electricity needs for about 6 weeks.

First, I discuss methods for improving solar collector and electrolyzer technology to serve both large solar farms and individual home collectors efficiently. Next, I discuss compressing and liquefying hydrogen, which are needed for storing and transporting this fuel. After that, I examine hydrogen distribution and utilization in terms of cost and efficiency.

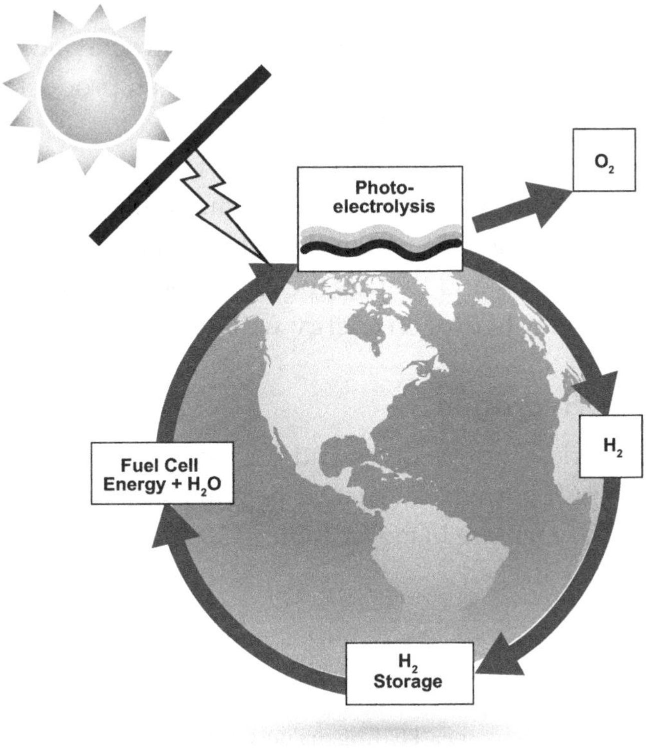

Figure 4-19. The operation of the clean, inexhaustible, and carbon-free solar-hydrogen energy economy.

The solar-hydrogen energy economy starts with splitting water into oxygen and hydrogen at a location where green energy is available. During this step, the oxygen is released into the atmosphere, and the hydrogen is stored, ready to be transported to users (Figure 4-19). This process is very similar to the vegetation process of splitting water, releasing the oxygen, and using the hydrogen to produce a fuel (food in the form of glucose, cellulose, etc.) to be stored, transported, and eventually consumed to provide an energy source for animals and humans.

In the coming sections, I focus on the process of splitting water to generate hydrogen, describe the properties and safety record of hydrogen, and discuss the methods of hydrogen storage, transportation, and distribution.

As of this writing, gray hydrogen is produced primarily from methane by *steam reforming*. In California today, the cost of one liter of gray hydrogen at the dispenser[52] is about $14. This makes the fuel cost of FCEVs nearly twice that of battery-operated EVs and 50% more than that of IC vehicles. Mass production and improving the efficiency of solar-hydrogen system components are needed before green hydrogen will be cost-competitive with other fuels.

The term *solar-to-hydrogen* (STH) describes the complete process of hydrogen production from water by electrolysis. Some believe that, unless hydrogen is subsidized or fossil fuels are taxed, STH conversion using photoelectrochemical (PEC) technology will not be commercially viable until the cost of hydrogen production comes down to meet the US DOE target of less than $2 per kilogram.[53]

I have no opinion on when this will happen or what efficiency solar-hydrogen technology will have to reach to get there. On the other hand, I am sure that the price of hydrogen will continue to fall and that, as the conversion to green energy accelerates, the competitiveness of solar-hydrogen energy will also improve.

4.5.3 Properties of Hydrogen

Hydrogen (H_2) is a colorless, odorless gas and the most abundant material on Earth. It is seldom found in its pure form because it easily combines with other elements. Some of the properties of hydrogen are listed in Table 4-8.

---

52. "Cost to refill," California Fuel Cell Partnership, accessed March 12, 2022, https://cafcp.org/content/cost-refill.

53. For more details on projections, see: "Hydrogen Fuel Cost vs. Gasoline," Hydrogen Energy Systems LLC website, accessed March 12, 2022, https://heshydrogen.com/hydrogen-fuel-cost-vs-gasoline/.

Table 4-8. Properties of hydrogen.

Temperatures	
Boiling point	–252.88°C (–423.18°F, 20.271 K)
Melting point	–259.16°C (–434.49°F, 13.99 K
Autoignition point	571°C (1060°F, 844 K)
Explosive Limits in Air	
Lower explosive limit (LEL)	4.0%
Upper explosive limit (UEL)	74.2%
Heat[a]	
Latent heat of evaporation at boiling point	0.124 kWh/kg (192 Btu/lb)
Density	
Gas (at STP[53])	0.08988 kg/m^3 (0.00523 lb/ft^3)
Liquid at the melting point	70.85 kg/m^3 (4.432 lb/ft^3)
Gas at 700 bar (10,000 psia)	42 kg/m^3 (2.62 lb/ft^3) or 5 kg per 125 L tank
Solid	76.3 kg/m^3 (4.76 lb/ft^3)

[a]The mass-based energy density of hydrogen is more than three times that of gasoline (39.6 kWh/kg vs. 12.7 kWh/kg). On the other hand, hydrogen's volume-based energy density is more than six times lower than that of gasoline (1.55 kWh/L versus about 10 kWh/L for gasoline).

Source: "Hydrogen," Wikipedia, accessed August 8, 2022, https://en.wikipedia.org/wiki/Hydrogen.

As a gas, hydrogen has a molecular weight of 2.01 at 0°C and atmosphere pressure. Its relative density, compared with that of the air, is 0.0695. Hydrogen can be used not only as a fuel but also to store and transport energy. Table 4-9 lists the approximate energy densities of different fuels, including hydrogen, and other energy storage devices (batteries) in both mass and volumetric units.[54]

From the values in Table 4-9, one can see that gasoline has the highest volumetric energy density; therefore, it requires the smallest-volume tank to carry the same amount of energy. Hydrogen compressed to 700 bar (10,000 psia) requires about six times that volume. On the other hand, in terms of weight, hydrogen can provide four times the energy of gasoline per unit of weight. Therefore, it is a light and extremely powerful rocket propellant that burns with extreme intensity (approximately 3050°C, or 5500°F). In combination with an oxidizer, such as liquid oxygen, liquid hydrogen yields the highest specific impulse of any known rocket propellant, or the highest

54. Standard temperature and pressure (STP) is defined as a temperature of 273.15 K (0°C, 32°F) and an absolute pressure of exactly 10^5 Pa (100 kPa, 1 bar).

Table 4-9. Energy densities of hydrogen and other energy sources.

Energy Source	Energy Density	
	Mass-Based (kWh/kg)	Volume-Based (kWh/L)
Hydrogen liquid	39.6	2.8
High-pressure hydrogen gas at 700 bar (10,000 psia)	39.6	1.55
Liquefied natural gas (LNG)	12.7 to 14.7	6.1
High-pressure natural gas at 250 bar (3500 psia)	12.7 to 14.7	3.05
Gasoline or diesel fuel	12.7 to 13	9.7 to 11.6
Battery (Li-ion)	0.1 to 0.27[a]	0.25 to 0.67[a]

[a]Power density.

efficiency in terms of propellant consumed. This is why hydrogen is the primary fuel of space exploration.

All fuels provide much higher volumetric energy densities than batteries, so batteries require more space than any fuel. Batteries also require more frequent refueling and take more time to refuel. All of the listed energy sources generate carbon emissions. Naturally, fossil fuels generate the most emissions, but batteries are also charged with electricity—generated primarily by fossil-fuel-burning power plants—and even hydrogen (gray hydrogen) also has a carbon footprint if it is made from hydrocarbons.

Naturally, in the future, if hydrogen is made by splitting water using green energy, it will have no carbon emissions and the green hydrogen will be a zero-carbon energy source.

4.5.4 Safety of Hydrogen

Hydrogen gas forms explosive mixtures with air in concentrations from 4% to 74%[55] and with chlorine in concentrations from 5% to 95%. The explosive reactions may be triggered by sparks, heat, or sunlight. The autoignition temperature (temperature of spontaneous ignition in air) of hydrogen is 540°C (1004°F). Although hydrogen is explosive over a wide range of concentrations in air, it is important to remember that gasoline can present a greater danger than hydrogen because its potential for explosion starts at even lower concentrations, namely, 1.1% to 3.3%. The enthalpy of the combustion of hydrogen is 286 kJ/mol.

---

55. "Hydrogen: Combustion," Wikipedia, accessed March 12, 2022, https://en.wikipedia.org/wiki/Hydrogen#Combustion.

Hydrogen ignites automatically at a temperature of 560°C (1040°F) and burns violently in air. Pure hydrogen-oxygen flames burn in the ultraviolet color range, and therefore, burning hydrogen leaks can only be seen in daylight but can be continuously detected by flame detectors. Another characteristic of a hydrogen fire is that the flames tend to rise rapidly in the air, causing less damage than hydrocarbon fires. Another consideration is that hydrogen reacts directly with other oxidizing elements. For example, it reacts spontaneously at room temperature with chlorine and fluorine.

Interestingly, after decades of successful and safe use of hydrogen as fuel in space exploration and various industrial and fuel-cell applications, its public image is still influenced by the *Hindenburg* accident. For this reason, I describe that accident in the next section in full detail.

I am not saying that hydrogen is inherently safe. Hydrogen is the most flammable of all known substances and it needs only a small amount of energy to ignite. It burns cleanly, and when it is burned with oxygen, the only byproducts are heat and water. Also, many metals, such as steel, absorb hydrogen, which can result in brittleness and can lead to damage.

Hydrogen is used as a fuel because it burns into harmless water and has a higher energy density per unit weight than fossil fuels. Another advantage of hydrogen is that it disperses very quickly, meaning that if hydrogen leaks outdoors, its concentration quickly drops to incombustible levels. Because hydrogen in the air rises very quickly, it does not accumulate on the ground, as do fossil fuels, and therefore, it is less of a safety concern or threat. Whereas hydrogen has a high burning velocity, its explosive power is less than those of other fuel-air mixtures.

Hydrogen is odorless and tasteless. Although its flame is visible under daylight conditions, it is much less so than other flames because the hydrogen flame contains no soot. A hydrogen flame can most easily be identified by the mirage-like effect on the air over and around the flame, it does not produce significant heat radiation.

Hydrogen has many characteristics that warrant handling it with great care, such as the fact that it is more easily ignited than other fuels. To be stored in liquid form, hydrogen must be cooled below its boiling temperature of only 20 K (–253°C, –423°F). This causes the fuel to boil off very quickly when spilled, creating only a narrow window for ignition because it cannot be ignited either at high concentrations or at concentrations below 4%. On the other hand, extremely cold liquid hydrogen can cause serious cold burn damage to people and metallic components to break due to embrittlement. Also, as was seen at Fukushima, in enclosed and unvented buildings, leaking hydrogen will accumulate near the roof, creating pressures that—if not released by safety valves—will tear apart the vessel or building.

Hydrogen is very reactive at high temperatures, but it is not reactive at ambient temperatures. Although it reacts with oxygen to form water, this reaction is extraordinarily slow at ambient temperatures. However, the reaction can be fast and explosive when accelerated by a catalyst such as platinum or an electric spark.

In the United States, nine million tons of hydrogen are produced each year[56] for use in chemical production, petroleum refining, and electrical applications, as well as for rocket fuel. As of this writing, NASA is the primary user of hydrogen for fueling space rockets. According to advocates for a hydrogen economy, hydrogen should be used as an energy carrier or fuel because decades of wide usage have demonstrated that it can be safely stored and transported. Also, the use of hydrogen fuel would improve air quality.

The average gasoline tank holds three to four times the energy of a hydrogen tank of the same size; therefore, hydrogen tanks need to be larger in volume (Figure 4-20). Tanks for hydrogen FCEVs are sized in kilograms. A 4 kg (8.8 lb) hydrogen tank holds the energy equivalent of 4 gallons of gasoline. Today's hydrogen fuel stations can fill such a tank in 3 to 5 minutes.

Hydrogen is about as safe as, or safer than, gasoline. Hydrogen is 14 times lighter than air; therefore, it rises quickly. When its concentration drops below 4%, it can no longer be ignited. Hydrogen vapors do not accumulate on the ground as do vapors that are heavier than air; therefore, hydrogen does not present as great a fire or explosive danger as gasoline when it is stored above the ground in well-ventilated areas.

In upcoming sections, I discuss how industry has succeeded in obtaining a safety record for hydrogen that is better than other fuels.

4.5.4.1 NASA and DOE Experience

Because changing the public perception of the safety of hydrogen is essential, I now discuss the safety record of hydrogen during the past several decades, including its successful and safe use in general industry and space exploration and as a fuel for fuel cells.

It is essential to understand two key facts about hydrogen fuel. First, hydrogen is safer than fossil and nuclear fuels, meaning that converting to a solar-hydrogen economy is

---

56. Michelle Fung, "Energy Density of Hydrogen," in *The Physics Factbook*, 2005, accessed November 21, 2022, https://hypertextbook.com/facts/2005/MichelleFung.shtml.

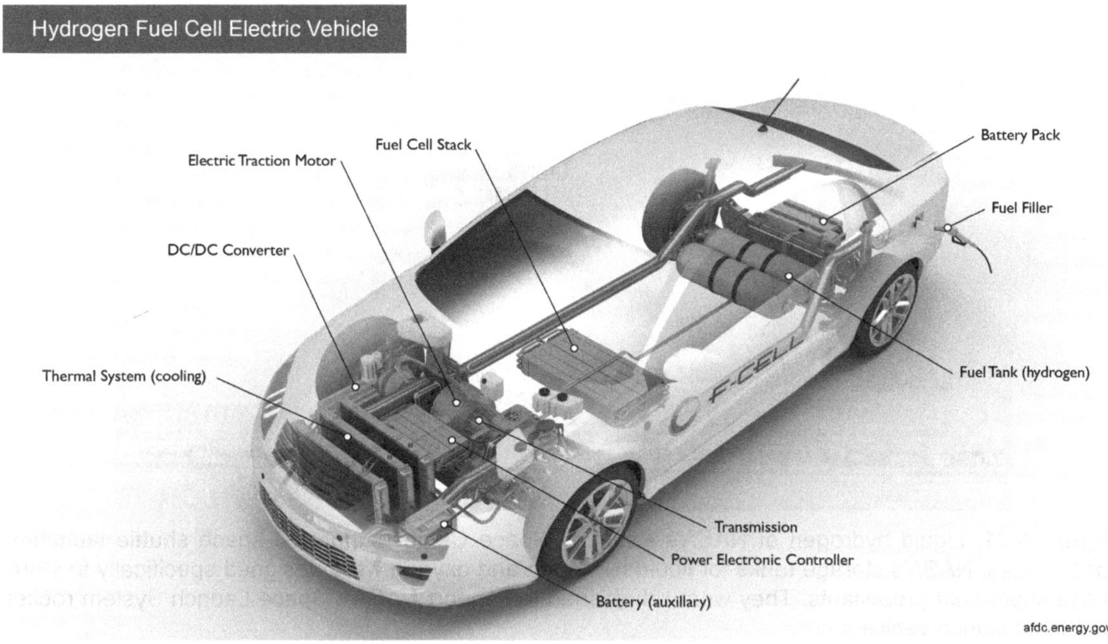

Figure 4-20. The hydrogen tank on a typical fuel-cell-powered electric car is larger than the average gasoline tank and can nearly double the driving distance between refills, which take minutes. Abbreviation: DC, direct current.

Source: "How Do Fuel Cell Electric Vehicles Work Using Hydrogen?" Alternative Fuels Data Center, https://afdc.energy.gov/vehicles/how-do-fuel-cell-electric-cars-work.

safe. Second, the technology to transport hydrogen from where it is produced to where it will be used already exists.

The safety record of using hydrogen fuel in space exploration is excellent. Despite criticism and early technical failures, the taming of liquid hydrogen proved to be one of NASA's most significant technical accomplishments. Hydrogen, a light and extremely powerful rocket propellant, has the lowest molecular weight of any known substance and burns with extreme intensity (approximately 3050°C [5500°F]). In combination with an oxidizer such as liquid oxygen, liquid hydrogen yields the highest specific impulse, or efficiency—in terms of the amount of propellant consumed—of any known rocket propellant (see Figure 4-21).

NASA summarizes the use of hydrogen and fuel cells in space applications as follows:[57]

> For decades, NASA has relied upon hydrogen gas as rocket fuel to deliver crew and cargo to space. With the Centaur, Apollo, and space shuttle vehicles, NASA has developed extensive

57. "Summary: Space Applications of Hydrogen and Fuel Cells," NASA, last updated July 29, 2010, accessed November 21, 2022, https://www.nasa.gov/topics/technology/hydrogen/hydrogen_2009.html.

Figure 4-21. Liquid hydrogen at NASA's Kennedy Space Center supported space shuttle launches for 30 years. NASA's storage tanks for liquid hydrogen and oxygen were designed specifically to store these super-cold propellants. They were refurbished to support NASA's Space Launch System rocket and other launch vehicles.

Source: "Liquid Hydrogen—the Fuel of Choice for Space Exploration," Kim Shiflett/NASA, October 20, 2015, https://www.nasa.gov/content/liquid-hydrogen-the-fuel-of-choice-for-space-exploration.

experience in the safe and effective handling of hydrogen. For example, the rocket engines of each shuttle flight burn about 500,000 gallons of cold liquid hydrogen; another 239,000 gallons are depleted by storage boil-off and transfer operations.

Experts at Glenn Research Center, Kennedy Space Center, Marshall Space Flight Center, Stennis Space Center, and White Sands Test Facility are proficient with hydrogen propellant transportation, storage, system design, training, safety standards, hazard analysis, testing, vehicle demonstration, technology transfer, and outreach.

With the recent focus on human missions to the Moon and eventually Mars, hydrogen will continue to be innovatively stored, measured, processed, and employed. Experience with fuel cell power and International Space Station life support systems provide an additional basis for future exploration. On the space station, water can be split into oxygen for breathing and hydrogen. Generating and recycling hydrogen in space will decrease the cost and complexity of remote missions by reducing the need for supplies delivered from Earth.

NASA is also pursuing a variety of alternative power sources for aerospace vehicles. One of those alternatives is fuel cells. To achieve improved performance, NASA is building on decades of success with Gemini, Apollo, and Shuttle alkaline fuel cell power system development and operations. The agency has also pursued other types of fuel cells, such as methane, metal-air, and hydrogen peroxide.

In recent years, the Department of Energy (DOE) and private industry have made significant advances in the development of proton exchange membrane (PEM) fuel cells using hydrogen and air as the fuel and oxidant for ground-transportation applications. Additionally, the DOE and the ground power industries are working on emerging solid oxide fuel cells (SOFC) for ground-based power generation. NASA is building upon these PEM and SOFC developments to dramatically advance fuel cell technologies, providing reliable, compact, and high-energy renewable power sources for aerospace applications.

With the aid of industry and universities, NASA is addressing aerospace challenges that include reduced gravity, low or no atmospheric pressure, extreme temperatures, dynamic vibration, shock loads, and extended-duration operations. For airless space applications, the focus is on closed-cycle regeneratable PEM technology. The targeted space systems feature power outputs of 1 to 10 kilowatt systems (eventually scalable up to 100 kilowatts), compact sizing (250 to 350 watts per kilogram), and high reliability for long lives (10,000 hours). With the appropriate investments and collaborations, NASA intends to revolutionize aerospace power generation to enable new capabilities. Experts at Glenn, the Jet Propulsion Laboratory, the Johnson Space Center, and Kennedy are leading these fuel cell efforts.

4.5.4.2 The *Hindenburg* Disaster

Some people still think of the Hindenburg accident when hearing about hydrogen and, based on what they heard about it, think that hydrogen is a highly dangerous gas. It is for that reason I describe the accident in some detail.

The German passenger airship *Hindenburg* arrived in the United States from Germany on May 6, 1937. While landing in Manchester Township, New Jersey, it was destroyed during its attempt to dock at Naval Air Station Lakehurst. On board were 97 people (36 passengers and 61 crew); there were 36 fatalities (13 passengers, 22 crew members, and one worker on the ground).

The cause of the accident has never been firmly resolved, although the widely held view is that the airship was leaking hydrogen that was ignited by an electrostatic discharge caused by a thunderstorm at the time of the landing. Numerous studies and books have advanced dozens of theories about the cause of this disaster (sabotage, lightning, static spark igniting incendiary coating on the fabric skin, damage caused by sudden turning of the zeppelin, sparks from an engine backfire, puncture of hydrogen cells, etc.). Regardless of the cause, however, the incident created the public perception that hydrogen is very dangerous (Figure 4-22).

One often-heard argument is that the hydrogen onboard the Hindenburg caused the disaster. It is true that the hydrogen burned, but it did not cause the disaster. Instead, the cotton fabric skin of the *Hindenburg* was covered in aluminum powder[58] to reflect

58. Josh Clark, "Is Hydrogen Fuel Dangerous?: Hydrogen Fuel Safety," HowStuffWorks website, accessed March 12, 2022, https://auto.howstuffworks.com/fuel-efficiency/alternative-fuels/dangerous-hydrogen-fuel2.htm.

Figure 4-22. The *Hindenburg* disaster: the *Hindenburg* on fire at the mooring mast of Naval Air Station Lakehurst on May 6, 1937.

Source: Sam Shere, "Hindenburg disaster.jpg," Public Domain, May 6, 1937, https://commons.wikimedia.org/wiki/File:Hindenburg_disaster.jpg.

the sunlight, and it was waterproofed with highly flammable acetate.[59] The aluminum powder, in a form equivalent to rocket fuel, and the acetate waterproofing on the skin likely caused the disaster. As evidence that hydrogen was not the cause, proponents point out that the flames burned upward, away from the carrier containing the passengers, because hydrogen is so lightweight. The proponents also point out that the 35 passengers and crew members who died did not die because of the hydrogen explosion but because they jumped from the airship, whereas everyone who stayed on board survived the accident.

We know that a spark ignited the flammable cotton outer skin, which was covered by a finish of plasticized lacquer that provided stiffness; protection; and a lightweight, airtight seal to the woven fabric. Because this type of coating is highly flammable in its liquid form, proponents of a hypothesis suggest that a static spark occurred when the mooring line touched the ground and ignited the coating on the skin. The volume of hydrogen in the *Hindenburg* was 200,000 m^3 (7,062,000 ft^3), which, because of its low density, created a fiercely and vertically upward-moving blue or invisible flame for a minute or so. The cameras did not show this. The hydrogen concentration would have had to decrease before it became explosive, which would suggest that the hydrogen would have had to rise above the zeppelin before exploding. According to observers,

59. "Hindenburg & Hydrogen," ABC Science website, published February 26, 2004, accessed November 21, 2022, http://www.abc.net.au/science/articles/2004/02/26/1052864.htm?site=science/greatmomentsinscience.

the color of the flames was that of the burning aluminum superstructure and the diesel fuel, which burned for several hours on the ground.

Some believe that the *Hindenburg* would have burned the same if it had been filled with inert helium gas because the main cause of the accident was not the bags of hydrogen that provided the lifting force for the zeppelin but a combination of dark iron oxide and reflective aluminum paint that coated the surface of the ship. These components are extremely flammable and burn at a very high rate once ignited.

The loss of 36 lives is a tragedy, but the public perception lasting three generations—that hydrogen is a dangerous substance and technology has not made it safe—is a mistake. One of the reasons the *Hindenburg* disaster is still present in the public mind is because the media of the whole world was present when it occurred. The sinking of the *Ocean Ranger* rig near Newfoundland, Canada that killed its crew of 84; the Texas City ship explosion that killed 578 people; the Oppau fertilizer explosion in Germany that killed 500 to 600 people, and many other mining disasters, dam and building collapses, and fire accidents did not result in the public perception that shipping, oil drilling, or fertilizer manufacturing were safety hazards. Even though these accidents and hundreds of others occurred more recently, few of us remember them.

4.5.4.3 Transportation Safety Record

Hydrogen is one means of storing solar energy in chemical form (as a liquid or gas compressed to pressures of 350 to 700 bar [5000 to 10,000 psig]) in which it can be used as a fuel.

On a mass basis, the energy density of hydrogen is more than three times that of gasoline (39.6 kWh/kg [135,000 Btu/kg] for hydrogen versus 13.0 kWh/kg [44,300 Btu/kg] for gasoline). On the other hand, hydrogen's volume-based energy density is more than six times lower than that of gasoline (1.55 kWh/L [5290 Btu/L] for hydrogen versus about 10 kWh/L [34,100 Btu/L] for gasoline).

Liquid hydrogen weighs 0.59 lb (0.27 kg) per gallon, and the energy content of 1 gal of gasoline equals the heating value of 3.58 gal of liquid hydrogen (Table 4-9). On a volume basis, hydrogen requires more than three times the volume of gasoline to store the same amount of energy. Hydrogen can also be stored in solids. These reversible solid storage processes are probably the safest, but their development is still in the experimental stage. Today, solid forms are capable of storing only small amounts of energy.

According to a report from Edmunds:[60]

> The University of California at Irvine has operated a public hydrogen station for a dozen years without incident, says Carl Baust, alternative energy projects specialist for the Orange County Fire Authority. Several other hydrogen stations opened in Southern California and the Sacramento area to fuel test the vehicles that have been plying the state's highways since 2002. So far, they also have been incident-free.
>
> The vehicles themselves have arrays of hydrogen sensors that sound alarms and seal valves and fuel lines in case of a hydrogen leak. Additionally, the pressurized tanks that hold the hydrogen have been tested repeatedly and found to be safe in collisions.
>
> One test involves firing live ammunition into the tanks to pierce them. Pierced tanks typically vent their hydrogen so quickly that even if the gas catches fire, it is out of the tank before explosive pressure could build up.
>
> In 2001, researchers at the University of Miami's College of Engineering set fire to the hydrogen in a tank mounted in an SUV and later punctured the fuel line on a conventional gasoline-powered vehicle and set the leaking gas on fire. The burning hydrogen versus gasoline test showed that flames caused "severe" damage to the gasoline vehicle, whereas the hydrogen vehicle was undamaged because the burning hydrogen gas vented up and away from the vehicle.

To my knowledge, hydrogen has never caused an accident in any space exploration projects or any fuel-cell-powered electrical cars. If I am accurate, this is quite a record considering the number of gasoline-related accidents and the fact that there are more than 10,000 fuel-cell-powered electric cars just on the roads of California. Naturally, there have been accidents in nontransportation cases (see Section 4.5.4.5, "Industrial Accidents"), such as in industrial use, storage, and distribution of hydrogen. Still, to my knowledge, none were caused by hydrogen explosions in FCEVs. Today, the number of traditional electric cars on the roads exceeds a million. The main reasons for the difference in numbers are the relatively low-energy efficiency, high costs, and the lack of a refueling infrastructure for hydrogen FCEVs.

The two prime concerns regarding fuel-cell- and hydrogen-powered vehicles are the danger of electrical shock and the flammability of the fuel. Fuel cells power vehicles by electrochemically combining hydrogen and oxygen gases from the surrounding air into water and electrical energy. The electrical energy is then used to power the electric motors that propel the vehicle and electrical devices such as the radio, lights, and air-conditioning. A notable difference between current and new technology vehicles is that the voltage needed to power the electric motors is much higher in new vehicles than can be accommodated by the current standard voltage of a 14 V system. The

60. "8 Things You Need To Know About Hydrogen Fuel-Cell Cars," Edmunds website, May 8, 2015, accessed November 21, 2022, https://www.edmunds.com/fuel-economy/8-things-you-need-to-know-about-hydrogen-fuel-cell-cars.html.

automobile industry is moving to a new standard of a 42 V system, which was chosen as an industry standard partly for safety reasons because anything greater than 50 V can stop a human heart. On the other hand, some FCEV motors run on voltages exceeding 350 V. With such high voltages, the danger of electric shock is substantial.[61]

The second concern lies in the fuels used to power this future generation of vehicles. Even though hydrogen is expected to be the primary fuel used in FCEVs, it is not the only possible new fuel for them. Potential fuels include methanol, ethanol, and methane. In addition, the hydrogen used to power a vehicle does not necessarily have to be stored in the form of hydrogen. Reforming such materials as alcohols, methane, propane, and even regular gasoline can produce gaseous hydrogen in the vehicle.

Even though all current fuel-cell and hydrogen vehicles are experimental and are only used in low numbers on public roadways, several states have mandated zero-emission vehicles (ZEVs) to be introduced in the near future. As fuel cells and hydrogen are new technologies, any safety concerns must be addressed. And, because some of the concerns raised are similar to those encountered with substances such as compressed natural gas and liquefied natural gas, they warrant special attention.

As exhibited by sales and lease volumes, the number of FCEVs continues to grow. As of 2019, the United States had 8000, China had 6000, and Japan had 3600 FCEVs on their roads. Globally there were 40,000 FCEVs on the roads[62] in 2022. By 2030, Japan is aiming for nearly a million and the Netherlands for 300,000 FCEVs.[63]

4.5.4.4 Industrial Safety

Hydrogen is extremely flammable; however, this disadvantage is mitigated by the fact that hydrogen rapidly rises in the atmosphere and often disperses before ignition unless it escapes into an enclosed, unventilated area. Demonstrations have shown that a fire in a hydrogen-powered vehicle can burn out completely with little damage to the vehicle, in contrast to the expected result in a gasoline-fueled vehicle.

---

61. "Safety issues regarding fuel cell vehicles and hydrogen fueled vehicles," International Consortium for Fire Safety, Health & the Environment website, accessed March 12, 2022, https://dps.mn.gov/divisions/sfm/programs-services/Documents/Responder%20Safety/Alternative%20Fuels/FuelCellHydrogenFuelVehicleSafety.pdf.

62. Marcus Baram, "Hydrogen fuel cells are not as 'dumb' as Elon Musk says. Here's how they could improve the EV market," *Fortune*, August 15, 2022, accessed October 29, 2022, https://fortune.com/2022/08/15/hydrogen-fuel-cells-elon-musk-toyota-renault-hyundai-cummins/.

63. Katharina Buchholz, "Japan Aims for the Sky in Fuel Cell Car Adoption," Statista website, December 9, 2020, https://www.statista.com/chart/23735/number-of-fuel-cell-hydrogen-cars-and-targets-selected-countries/.

Hydrogen gas burns with incredible speed and can quickly produce large explosive forces. However, because it has a lower volumetric energy density than gasoline, hydrogen is very useful for applications that require vast amounts of instantaneous power but do not have highly restrictive constraints on storage volume, such as rocketry and spaceflight.

Industry has used hydrogen in vast quantities during the past 50 years as an industrial chemical and fuel for space exploration. Over these decades, the safety challenges of its production, storage, transportation, and use have been solved. Such efforts have been so successful that, today, hydrogen is often safer than traditional fuels. One reason for this unexpected result is that when traditional fuels burn, they produce hot ash, whereas hydrogen produces no ash and very little radiant heat. In short, hydrogen is already a promising energy carrier that could help decarbonize industry and transportation.

One industrial use of hydrogen is in the synthesis of ammonia. Huge quantities of hydrogen are consumed in the catalytic hydrogenation of organic chemicals, such as in producing solid fat from unsaturated vegetable oils.

Hydrogen is the fuel most frequently used in fuel cells and space rockets, and it can be burned in IC engines. In the case of automobiles, one of the major challenges lies in designing hydrogen fuel tanks that are strong enough to withstand the impact of car accidents. Steel tanks can withstand such impacts without suffering a puncture or rupture, but steel becomes brittle when exposed to hydrogen for a long time. Another problem is that hydrogen tanks would have to be more than three times larger in volume than gasoline tanks, making the use of steel prohibitively heavy. Composite materials such as polyethylene are used because they are lightweight and can be shaped to fit a car. Another option is to make the storage tanks out of materials that can trap hydrogen molecules and store them safely until the metal is heated to a point where they are released. The required heat can be obtained, for example, from the waste heat generated by hydrogen fuel cells.[64]

An example of present-day designs is Toyota's Mirai fuel tank (Figure 4-23). Each tank contains multiple walls, starting with a plastic liner inside, followed by a carbon-fiber-reinforced plastic shell surrounded by a glass-fiber-reinforced plastic outer layer. The

64. For further information, see: Clark, "Hydrogen fuel safety."

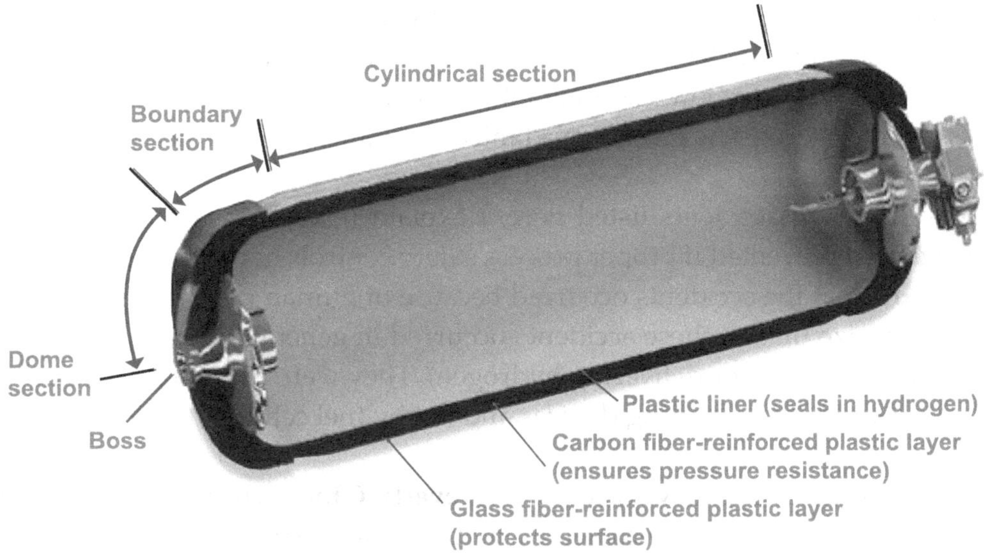

Figure 4-23. Toyota Mirai fuel tank design.[65]

Source: Reprinted with permission from "Toyota Mirai: The Safety Facts," *Toyota Blog*, August 14, 2015, accessed January 29, 2021, https://blog.toyota.co.uk/toyota-mirai-safety-facts.

glass-fiber layer serves to protect the carbon-fiber layer from abrasion. The tanks are meant to withstand at least 225% of their nominal working pressure.

Trace hydrogen sensors detect very low concentrations of hydrogen and not only sound an alarm but also automatically shut the safety valves and turn the car off. In addition to these safety precautions, the hydrogen compartment is separated from the passenger cabin to prevent hydrogen from entering the cabin. This design is to ensure that any leaking hydrogen will be dispersed into the atmosphere.

4.5.4.5 Industrial Accidents

In this section, I list some industrial hydrogen accidents to show their nature and discuss their prevention. A key aspect of hydrogen's behavior is that, in air, it rises at a speed of almost 45 mph (20.1 m/s, 65.6 ft/s). Therefore, unless a roof, poorly ventilated room, or some other structure prevents the gas from leaving, any leaking hydrogen will quickly escape. Simply stated, to become a fire hazard, hydrogen must first be confined.

Hydrogen combustion produces no carbon, only heat and water. The amount of heat produced is significantly less than that produced in a hydrocarbon fire. As a result, hydrogen

65. For further information, see: "Toyota Mirai: the safety facts," *Toyota Blog*, August 14, 2015, accessed November 21, 2022, https://blog.toyota.co.uk/toyota-mirai-safety-facts.

flames usually do not cause secondary fires. Hydrogen has a wide flammability range[66] (4% to 75% in air and up to 95% in pure oxygen), and the energy required to ignite hydrogen (0.02 mJ)[67] can be very low, except at low concentrations, where the energy required is higher. If only hydrogen is present in a tank or container, an explosion cannot occur.[68]

For the well-known accidents listed here, I explain in each case how the accidents could have been prevented if proper process safety controls had been applied. In other words, I show that the accidents occurred because of human carelessness and error. I should emphasize that all these accidents occurred in general industry (most of them while a truck is loading or delivering hydrogen). They were most often caused by the failure of pressure safety venting devices and not by fuel cells.

- **Muskingum River Power Plant near Beverly, Ohio** – The turbines of coal-fired supercritical Unit 5 were cooled by hydrogen for years. On January 8, 2007, the high-pressure relief valve on the delivery truck failed to open, and the explosion killed one employee. *Note: Proper process control design would have included both a high-pressure alarm and a backup relief valve. If they had existed, the accident could not have occurred.*
- **Fukushima Daiichi Nuclear Power Plant in Ōkuma, Fukushima Prefecture, Japan** – Three reactor buildings were damaged by hydrogen explosions during the Fukushima nuclear accident (March 11, 2011). The hydrogen was formed as the exposed zirconium-cladded fuel rods became very hot and reacted with steam, releasing hydrogen. The reactor containment vessels were filled with inert nitrogen, which prevented the hydrogen from burning. However, the hydrogen leaked from the containment vessels into the reactor building, where it mixed with air and exploded. *Note: Proper process control design would have located the pressure relief rupture disk (RD) connection on the top of the primary containment vessel (PCV) and would not have inserted shutoff valves between the PCV and the RD. If this had been done, the accident could not have occurred.*
- **Diamond Bar, California** – In February 2018, a truck carrying high-pressure gas cylinders caught fire on its way to a hydrogen fuel-cell refueling station. The fire caused the hydrogen pressure in the tanks to rise, and because the pressure relief devices on the tanks were set too low (~400 bar [5844 psig] instead

66. "Hydrogen Flames," Hydrogen Tools website, accessed November 29, 2022, https://h2tools.org/bestpractices/hydrogen-flames.

67. "Lessons Learned from a Hydrogen Explosion," *POWER* Magazine, May 1, 2009, accessed November 21, 2022, https://www.powermag.com/lessons-learned-from-a-hydrogen-explosion/.

68. "Will Your Hydrogen Tank Explode?" Toyota of Laredo website, April 4, 2016, accessed October 13, 2022, https://www.toyotaoflaredo.com/blog/2016/april/4/will-your-hydrogen-tank-explode.htm.

of the proper setting of ~690 bar [10,000 psig]) and their vent lines released the hydrogen below the top of the tank, the hydrogen ignited. No injuries occurred because the hydrogen quickly rose into the air and exploded in a fireball well above the truck when mixed with the air. *Note: Proper process control design would have prevented this accident by not releasing the hydrogen until its pressure reached ~690 bar (10,000 psig) and by locating the pressure relief valves on the top of the hydrogen tanks.*

- **Veridiam, Inc., in El Cajon, California** – In August 2018, a delivery truck carrying 2000 gal of liquid hydrogen experienced a flash fire when a vent shutoff valve was left open. *Note: Proper process control design would have prevented this accident by providing the truck driver with a hydrogen detector alarm to remind him to close the valve.*

- **Gangwon Technopark in Gangneung, South Korea** – In May 2019, a hydrogen storage tank exploded, killing two and injuring six. The cause of the explosion was the failure of the pressure relief valve to open when the pressure inside the vessel reached the tank's design pressure. *Note: Proper process control design would have included both a high-pressure alarm and a backup relief valve. If they had existed, the accident could not have occurred.*

- **Santa Clara, California** – On June 2, 2019, at a hydrogen manufacturing plant belonging to Air Products & Chemicals, Inc., an improper procedure led to the release of hydrogen while loading a tanker truck. The accumulated hydrogen mixed with air, resulting in an explosion, after which the hydrogen leak was stopped by manually closing a shutoff valve. No major injuries occurred. *Note: Proper process control design would have included a hydrogen detector switch and an automatic shutoff valve. If they had existed, the accident could not have occurred.*

- **Norway** – On June 10, 2019, there was a hydrogen explosion at an Uno-X hydrogen fueling station. The cause of the accident was a plug that was leaking high-pressure hydrogen. There was no hydrogen sensor on the truck; therefore, the leak was not noticed until the hydrogen concentration in the air reached the explosion limit and an explosion followed. No injuries occurred. *Note: Proper process control design would have included a hydrogen detector and an alarm to signal and help locate any leaks. If they had been provided, the accident could not have occurred.*

From the preceding list, we can note that there were relatively few hydrogen accidents (compared to gasoline ones) during the noted period, and none of them were caused by fuel cells or hydrogen leaks inside cars. In addition, all these accidents could have been prevented by the most basic process safety controls. Finally, these accidents demonstrate how sensitive the public is to hydrogen-related accidents. For example, the explosion at the hydrogen fueling station in Norway (which caused no casualties) prompted Toyota and Hyundai to freeze the sales of their hydrogen FCEVs there. Compare that response to the probable response to a gasoline tank accident that caused no casualties.

4.5.5 Green Hydrogen Production

Although there are many methods for making hydrogen, in this section, I focus on only those processes that produce green hydrogen by splitting water and have zero carbon emissions. Before I begin, I must mention that, currently, the United States is using some 10 billion cubic feet of gray hydrogen per day in the petrochemical, food, and rocket propulsion industries, 97% of which is produced by reforming natural gas with steam. This process is carried out by reacting methane with water vapor over a catalyst to form carbon monoxide (CO) and hydrogen. Most of the currently operating hydrogen production plants depend on reforming natural gas. This process emits carbon dioxide (CO_2) while consuming a nonrenewable fossil fuel. Hydrogen can also be made from ethanol (alcohol), biomass, fossil fuels, and organic wastes through the process of *reforming*. I do not discuss these processes here. Instead, I focus on electrolysis-based water splitting fueled by green energy that produces no carbon emissions.

Although electrolysis currently produces only about 3% of our hydrogen consumption, I consider it to be the process of the future because it is "natural" in that it is similar to what plants do. Vegetation uses sunshine as the energy source to break down water from the soil into hydrogen and oxygen. After that, this green hydrogen reacts with CO_2 in the air to produce glucose and cellulose and emit oxygen. Animals consume glucose as their energy source and exhale CO_2. When the ratio of plant and animal life on the planet is balanced (plants take up the CO_2 that animals exhale), the atmospheric concentration of CO_2 remains constant. In the following sections, I discuss how this balance can be reestablished by supplementing the energy production of plant life with solar-hydrogen energy.

4.5.5.1 Green Hydrogen by Water Electrolysis

In 1820, Michael Faraday discovered electrolysis by passing electricity through water (Figure 4-24), thereby generating hydrogen at the negative electrode (cathode) and oxygen at the positive electrode (anode).

The production of pure hydrogen is compelling because hydrogen is a clean fuel that can be stored easily. This makes it a potential alternative to fossil and nuclear fuels and to batteries, which have a low storage capacity, create high amounts of waste materials,

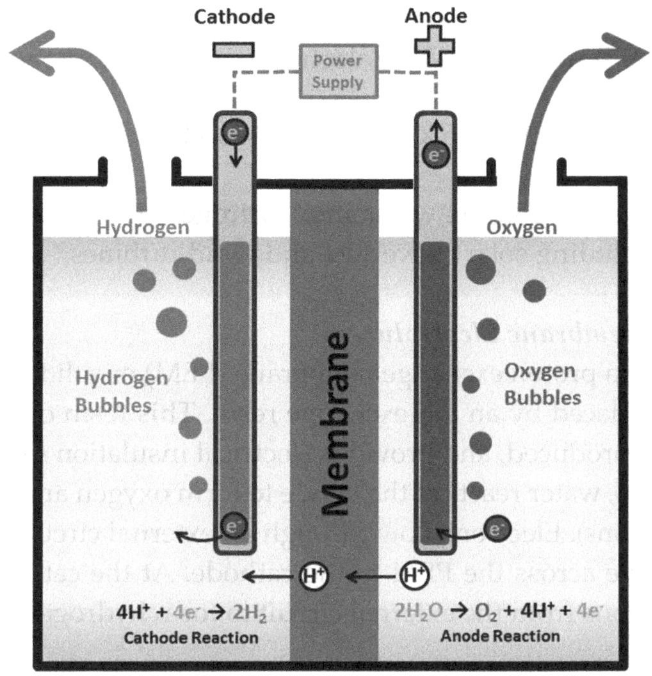

Figure 4-24. Schematic of the hydrogen generation process by electrolysis.

Source: "Hydrogen Production: Electrolysis," Office of Energy Efficiency & Renewable Energy, US DOE, Public Domain, accessed March 12, 2022, https://www.energy.gov/eere/fuelcells/hydrogen-production-electrolysis.

and are usually recharged using fossil-fuel-produced electricity. Electrolysis is currently the most promising method of producing hydrogen from water because of its high conversion efficiency and relatively low required energy input as compared to thermochemical and photocatalytic production methods.

Electrolysis is an endothermic reaction (requires energy), whereas hydrogen combustion is exothermic (generates energy). Electrolysis requires 0.066 kWh (224.79 Btu) of electricity to synthesize one mole of hydrogen. The weight of one mole of hydrogen molecules is about 2 g; therefore, the electricity needed to obtain 1 kg of hydrogen is about 33 kWh (112,600 Btu). This is about equal to the energy content of one gallon of gasoline (33.5 kWh [114,100 Btu]).

The same amount of energy is required to split water into hydrogen and oxygen as is obtained when hydrogen is oxidized into water. The only difference between the two operations (the processes that take place in fuel cells and electrolyzers) is that electrolysis increases entropy and, therefore, not all of the energy needs to be supplied in the form of electricity because the environment contributes about 17% of the total in the form of thermal energy (heat). Therefore, only 0.066 kWh of electrical energy is needed to produce one mole of hydrogen from water, whereas, in a fuel cell, one mole of hydrogen will generate 0.077 kWh of electricity (at 100% efficiency).

$$H_2O + (286\ kJ^{69}\ \text{of energy}) \leftrightarrow H_2 + {}^1/{}_2O_2$$

The electricity needed to convert water into hydrogen and oxygen can be obtained from any source, including solar converters and wind turbines.

Proton Exchange Membrane Electrolyzers

In electrolysis using a proton exchange membrane (PEM) or solid polymer electrolyte, the electrolyte is replaced by an ion-exchange resin. This resin conducts the protons, separates the gases produced, and provides electrical insulation for the electrodes. As shown in Figure 4-25, water reacts at the anode to form oxygen and positively charged hydrogen ions (protons). Electrons flow through an external circuit, and the hydrogen ions selectively move across the PEM to the cathode. At the cathode, hydrogen ions combine with electrons from the external circuit to form hydrogen gas.

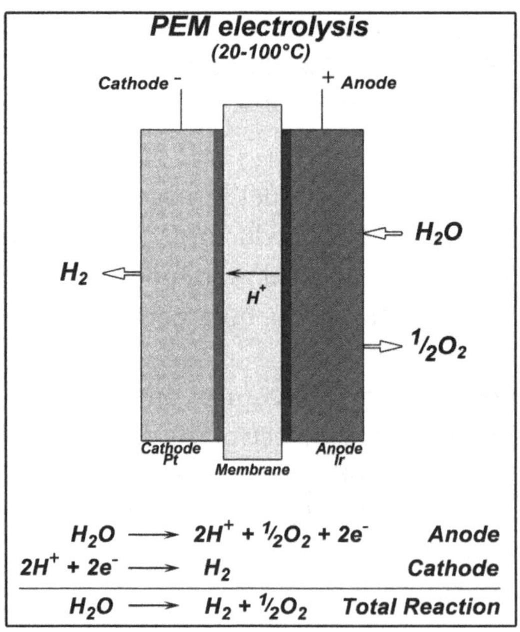

Figure 4-25. Design of a PEM electrolyzer.[70]

Source: "PEMelectrolysis.jpg" by Davidlfritz, May 16, 2013, is licensed under CC BY-SA 3.0, https://commons.wikimedia.org/w/index.php?curid=26111847.

69. 1 kWh = 3600 kJ = 3492 Btu.

70. For further information, see: "Polymer electrolyte membrane electrolysis," Wikipedia, March 12, 2022, https://en.wikipedia.org/wiki/Polymer_electrolyte_membrane_electrolysis.

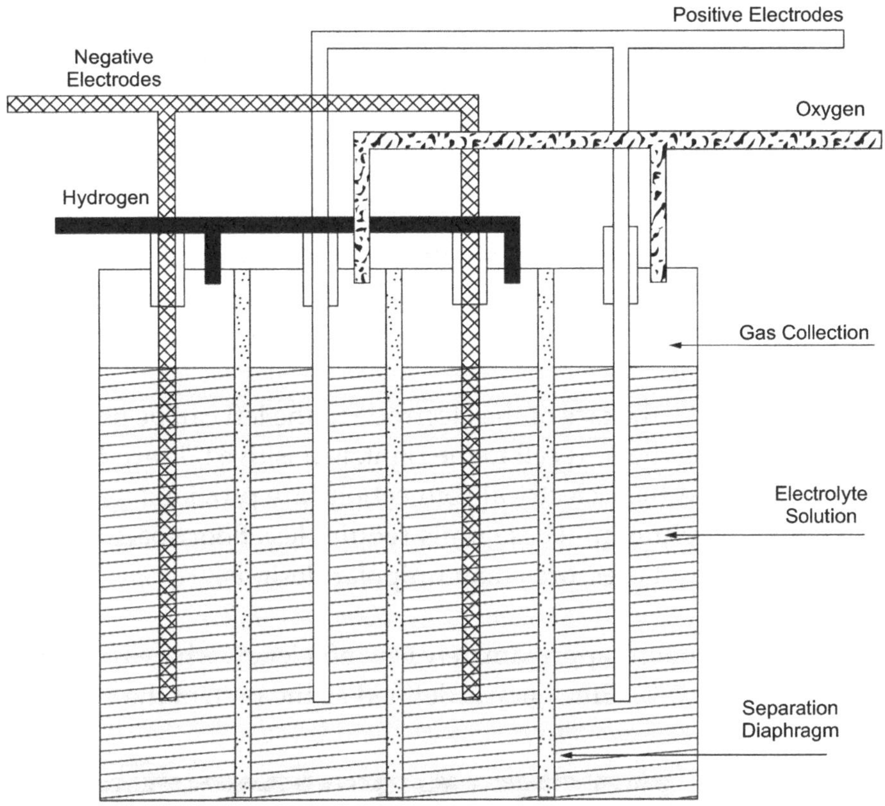

Figure 4-26. Unipolar alkaline electrolyzer design.

Source: Brian Williams, "Electrolysis of Water," *Global Warming Causes Blog*, last updated April 29, 2020, https://www.briangwilliams.us/solar-hydrogen-generation/electrolysis-of-water-1.html.

$$\text{Anode reaction: } 2H_2O \rightarrow O_2 + 4H^+ + 4e^-$$

$$\text{Cathode reaction: } 4H^+ + 4e^- \rightarrow 2H_2$$

The PEM electrolyzer was introduced to overcome the problems of partial loads, low current densities, and low-pressure operation of alkaline electrolyzers. These units are compact and provide high current densities, but they are more expensive and require special construction materials because of the corrosive nature of the electrolyte.

With fast dynamic response times, large operating ranges, and high efficiencies, water electrolysis is a promising technology for converting intermittently available green energy into hydrogen, which can be stored and then used as a continuously available energy source.

Alkaline Electrolyzers

Alkaline electrolyzer designs can be unipolar or bipolar (Figure 4-26). In the *unipolar alkaline electrolyzer design*, the anode and cathode are separated by a semipermeable diaphragm membrane and immersed in an alkaline solution. The diaphragm allows only salts to pass through and prevents the gases from mixing. The produced hydrogen and oxygen are dried downstream of the electrolyzer to remove water vapor. These units operate at low current densities and temperatures.

Electrolysis requires minerals to be present in the solution. Tap, well, and groundwater contain various minerals, some are alkaline and others are acidic. Electrolysis can occur only if the water is one or the other. The requirement is that there must be ions in the water to conduct electricity for the water electrolysis process to occur.

In *bipolar electrolyzer stacks*, the face of the electrode to the left can be negative, in which case the other side, facing the next cell to the right, is positive (Figure 4-27). These electrodes are separated, and their electrical connections are provided by a metal separator plate (separation diaphragm). These units require less floor space and can operate at higher temperatures and pressures than unipolar designs.

Today, electrolyzers operate at approximately 66% efficiency, whereas the ultimate target set by the US DOE is 75%. Efficiency is defined as the ratio of the energy content of

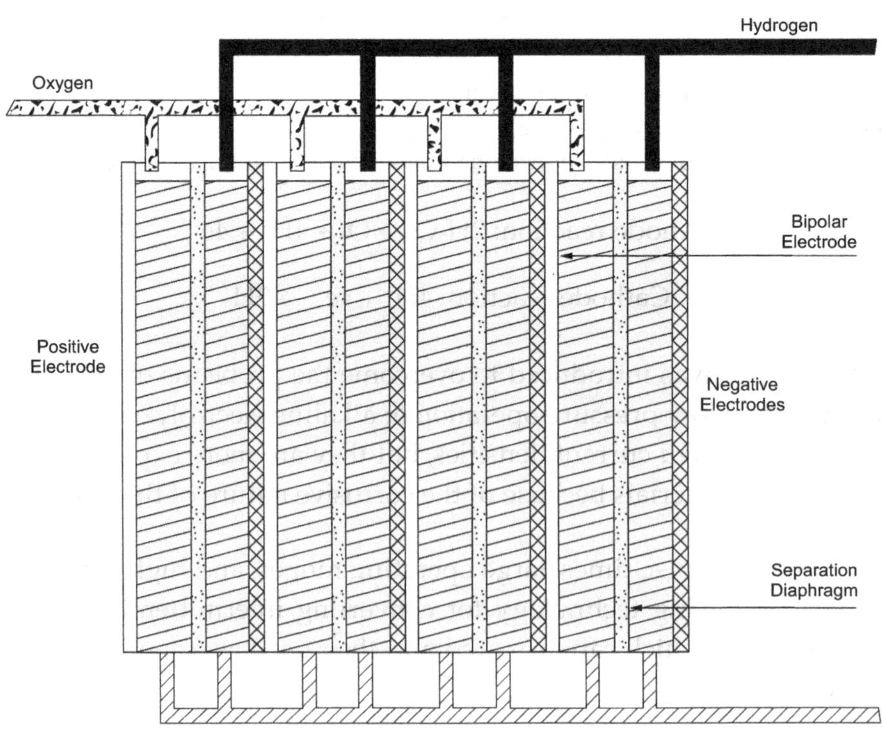

Figure 4-27. Bipolar alkaline electrolyzer design.

Source: Brian Williams, "Electrolysis of Water," *Global Warming Causes Blog*, last updated April 29, 2020, https://www.briangwilliams.us/solar-hydrogen-generation/electrolysis-of-water-1.html.

the generated hydrogen divided by the energy required to produce that hydrogen. The DOE efficiency target has already been reached in the laboratory using new, porous electrode materials. Another goal of research and development efforts is to replace the expensive platinum electrodes with electrodes made of less expensive material.

4.5.5.2 New Developments

Solid Oxide Electrolyzer Cells

A solid oxide electrolyzer cell (SOEC) is a fuel cell that runs in regenerative mode to achieve the electrolysis of water using a solid oxide, or ceramic, electrolyte to produce hydrogen gas and oxygen. SOECs use electrolytes made of a solid ceramic material that selectively conducts negatively charged oxygen ions (O_2^-) at elevated temperatures. Water at the cathode combines with electrons from the external circuit to form hydrogen gas and negatively charged oxygen ions. The oxygen ions pass through the solid ceramic membrane, react at the anode to form oxygen gas, and generate electrons for the external circuit (Figure 4-28).

Figure 4-28. A 60-cell SOEC stack.[71]

Source: "Solid oxide electrolyser cell prefab.jpg" by EERE, Public Domain, March 12, 2022, https://commons.wikimedia.org/w/index.php?curid=5403677.

71. For further information, see: "Solid oxide electrolyzer cell," Wikipedia, March 12, 2022, https://en.wikipedia.org/wiki/Solid_oxide_electrolyser_cell.

SOECs must operate at temperatures high enough for the solid oxide membranes to function properly (about 700°C to 800°C [1290°F to 1470°F]. In comparison, PEM electrolyzers operate at 70°C to 90°C (158°F to 194°F) and commercial alkaline electrolyzers operate at 100°C to 150°C [212°F to 302°F]). SOECs, therefore, require heat to operate.

The advantages of solid-oxide-based regenerative fuel cells include high efficiencies, long-term stability, fuel flexibility, low emissions, and low operating costs. Their greatest disadvantage is their high operating temperatures, which result in long start-up and break-in times. The high operating temperatures also lead to thermal expansion mismatch and chemical stability issues, such as diffusion between layers of different materials in the cell.

Photoelectrochemical Methods

Photoelectrochemical (PEC) methods of water splitting are being studied by the US National Renewable Energy Laboratory (NREL). Such methods use semiconductor materials to convert solar energy directly into chemical energy in the form of hydrogen. The semiconductor materials used in the PEC process are similar to those used in photovoltaic (PV) solar electricity generation. However, for PEC applications, the semiconductor is immersed in a water-based electrolyte where sunlight energizes the water splitting process.

Microbial Electrolysis Cells

Microbial electrolysis cells are devices that harness the energy and protons produced by microbes breaking down organic matter, combined with an additional small electric current, to produce hydrogen. This technology is very new, and researchers are working on improving many aspects of the system, from finding lower-cost materials to identifying the most effective type of microbes to use.

Photobiological Hydrogen Production

According to the US DOE Office of Energy Efficiency & Renewable Energy, "...photobiological hydrogen production uses microorganisms and sunlight to turn water, and sometimes organic matter, into hydrogen. This is a longer-term technology pathway in the early stages of research that has long-term potential for sustainable hydrogen production with low environmental impact."[72] One of the research teams working in this area is at the Hungarian Academy of Science in Szeged, Hungary.

---

72. "Hydrogen Production: Photobiological," DOE Hydrogen and Fuel Cell Technologies Office website, accessed March 12, 2022, https://www.energy.gov/eere/fuelcells/hydrogen-production-photobiological.

Artificial Photosynthesis

Artificial photosynthesis represents an effort to mimic the natural process of plants capturing energy from sunlight and storing it in chemical bonds. Plants harvest about 6% of the solar energy they receive, whereas PV solar cells harvest 18% to 30%. Artificial photosynthesis research is focusing on maximizing the efficiency of photon absorption and developing materials that are easy and inexpensive to produce. Metal oxides are not the only candidates for light harvesters, also called *photon antennas*. Both organic and synthetic polymers can also act in that capacity. Rare earth and other metals are also being used to extract useful amounts of energy from sunlight.

For example, at the Massachusetts Institute of Technology, Daniel Nocera and colleagues produced an artificial leaf made primarily of silicon, nickel, and cobalt. The leaf is the size of a playing card, is about 10 times more efficient (about 60%) than a natural leaf, and can operate continuously for about 45 hours.

When the layered leaf is exposed to light and water simultaneously, a silicon junction captures incoming photons and energizes the cobalt-based, oxygen-evolving complex at one end and a nickel-molybdenum-zinc alloy at the other. Together, they split the water touching the device and produce hydrogen that can be stored. Oxygen is released from the cobalt side of the leaf, and hydrogen gas is released from the nickel–molybdenum–zinc side. Previously, platinum was the catalyst of choice for creating the electrolytic effect, but the nickel-molybdenum-zinc alloy performs the task more than adequately while being much less expensive.

Thermochemical Water Splitting

This process involves using solar-driven, high-temperature thermochemical water splitting cycles to produce hydrogen with near-zero greenhouse gas emissions. Additional research is still needed because this process is not yet economical. On the other hand, exciting progress continues in this field, taking maximum advantage of interactions with concentrated solar power technologies and emerging solar-fuel production technologies.

4.5.5.3 Larger Electrolyzer Systems

When larger quantities of hydrogen are produced at solar farms or solar power plants, larger-capacity electrolyzer stations are required. As of 2022, the largest electrolyzer units in the world include a 150 MW alkaline electrolyzer system in China;[73] a PEM

73. Leigh Collins, "Record breaker | World's largest green hydrogen project, with 150MW electrolyzer, brought on line in China," updated April 27, 2022, accessed October 14, 2022. https://www.rechargenews.com/energy-transition/record-breaker-world-s-largest-green-hydrogen-project-with-150mw-electrolyser-brought-on-line-in-china/2-1-1160799.

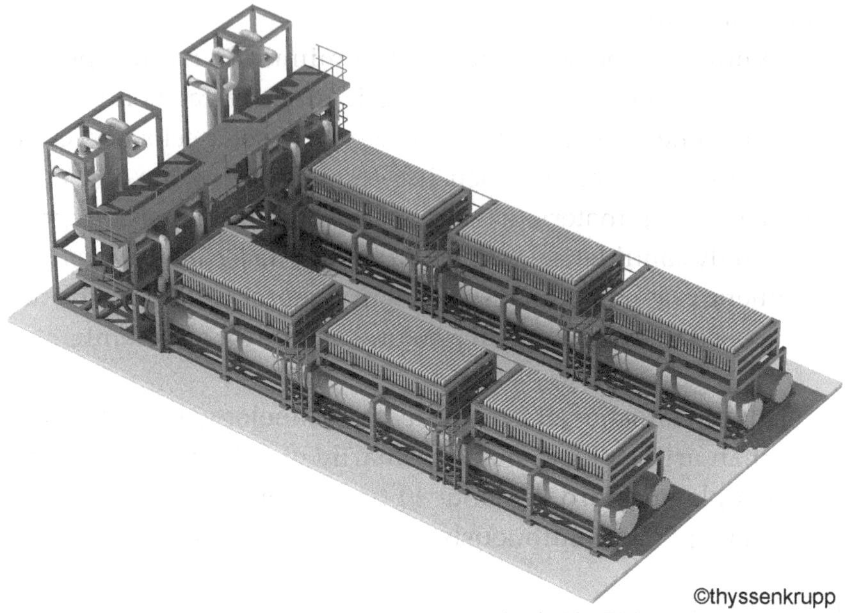

Figure 4-29. The thyssenkrupp 20 MW electrolyzer unit in Donaldsonville, Louisiana, will start up in 2023. It will operate at 82% efficiency and produce hydrogen at a rate of 4000 Nm^3/h at a power consumption of 49 kWh per kilogram of hydrogen produced.

Source: Reprinted with permission from "thyssenkrupp offering large-scale water electrolysis," Green Car Congress website, July 27, 2018, https://www.greencarcongress.com/2018/07/20180728-tk.html.

unit in Bécancour, Canada, producing 8.2 tons of hydrogen a day;[74] a number of 20 MW and 10 MW units in Zhangjiakou, Hebei Province, China;[75] and one that will start up in 2023 in Donaldsonville, Louisiana.[76] All will generate green hydrogen. The Donaldsonville system is shown in Figure 4-29. This system, manufactured by thyssenkrupp, will consist of large electrochemical cells that provide very high efficiency in industrial-scale hydrogen production. thyssenkrupp's multimegawatt hydrogen installations operate at efficiencies of about 82%, requiring only 49 kWh to produce 1 kg of hydrogen. A 20 MW module, such as that being built in Louisiana, can produce hydrogen at a rate of 4000 Nm^3/h with a power consumption rate of 4.3 kWh/Nm^3. (The normal cubic meter, Nm^3, is defined as a volume of 1 m^3 under standard conditions of 1.013 bar [14.7 psi] and 0°C [32°F]). The resulting hydrogen product is at a pressure of about 300 mbar (4.4 psi) and a temperature of up to 90°C (194°F).

---

74. David Sear, "World's largest PEM electrolyzer installed in Canada," October 18, 2021, accessed October 14, 2022, https://hydrogentechworld.com/worlds-largest-pem-electrolyzer-installed-in-canada.

75. Anmar Frangoul, "Shell says one of the largest hydrogen electrolyzers in the world is now up and running in China," January 28, 2022, accessed October 14, 2022, https://www.cnbc.com/2022/01/28/green-hydrogen-one-of-planets-largest-electrolyzers-up-and-running.html.

76. "thyssenkrupp to supply 20 MW electrolysis plant to CF Industries for green hydrogen for green am monia production," Green Car Congress website, April 23, 2021, accessed November 7, 2022, https://www.greencarcongress.com/2021/04/20210423-thussenkrupp.html.

As explained in an article on the Green Car Congress website, the thyssenkrupp water electrolysis system "features a well-proven cell design paired with an especially large active cell area of 2.7 m^2. By further optimizing the proven 'Zero-Gap' electrolysis technology (leaving virtually no gap between membrane and electrodes), very high efficiencies of more than 82% are achieved. To make deployment of large hydrogen projects as easy as possible, the thyssenkrupp technology is available in pre-fabricated, skid-mounted standard modules. They easily add up to the desired project size, potentially into the hundreds of megawatt range."[77]

4.5.6 Hydrogen Compression

If a solar farm is far from where the hydrogen is to be used, the hydrogen must be compressed or liquefied for transportation. In fuel-cell-operated electric cars, hydrogen is usually stored at 350 or 700 bar (about 5000 or 10,000 psig). The energy cost of compression varies with the discharge pressure required, the compressor size, and the number of compression stages.

If a single-stage compressor is used, the energy required to compress the hydrogen is about 16% of the higher heating value (HHV) of the compressed hydrogen. Multistage units with intercoolers are more efficient, consuming only 8% to 12% of HHV. Compressors with both large[78] and small[79] capacities are available to compress the gas. For example, the energy required to compress hydrogen from 1 to 200 bar (2900 psig) at a rate of 1000 kg/h in a five-stage compressor is 7.2% of the HHV of the gas. If hydrogen is compressed from 1 to 800 bar (14.5 to 11,600 psig), the energy cost is about 10% of the HHV of hydrogen. In terms of electricity consumption, compressing gas from 7 to 470 bar (from 100 to 7000 psig) requires about 3 kWh per kilogram of hydrogen. For these high-pressure applications, diaphragm compressors are most often used (see Figure 4-30).

Polymer-coated metal diaphragm compressors are used to compress high-purity gases to eliminate leakage. In these designs, the metal diaphragm isolates the process gas from the hydraulic fluid to ensure purity. A self-energizing static O-ring seals the head assembly and provides a positive seal and low bolt torque. Any hydrogen that leaks due to seal or diaphragm failure is retained in the head assembly and can be automatically detected to initiate maintenance.

---

77. "thyssenkrupp offering large-scale water electrolysis," Green Car Congress website, July 27, 2018, accessed October 14, 2022, https://www.greencarcongress.com/2018/07/20180728-tk.html.

78. Suppliers of large units: Ariel, Dresser-Rand, Neuman-Esser, and Sulzer Burckhardt.

79. Suppliers of smaller (fuel-station-size) units: CompAir, Fluitron, Greenfield, Hydro-Pac, Neuman-Esser, PDC, and Rix.

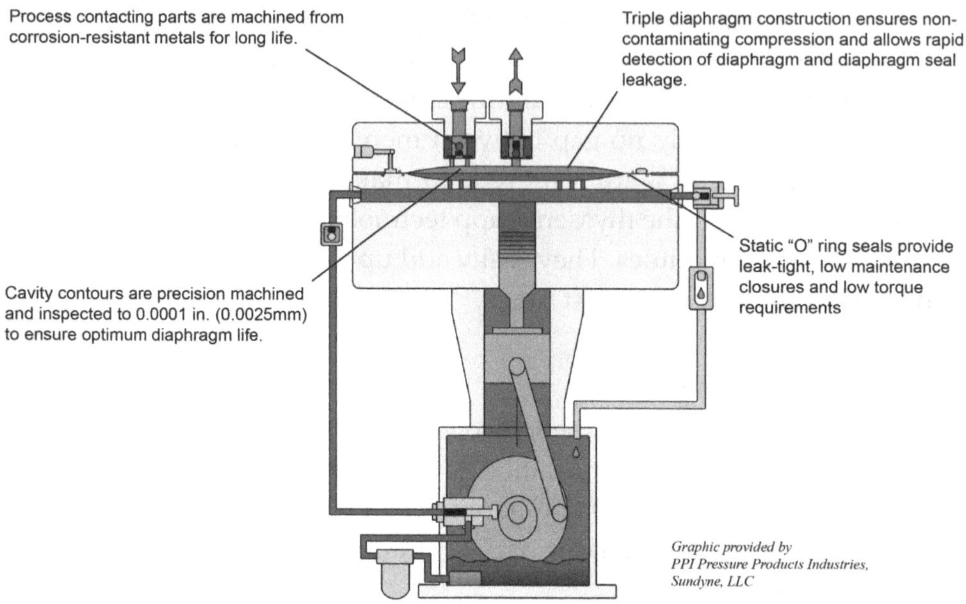

Figure 4-30. Metal diaphragm compressor manufactured by Pressure Products Industries (PPI).[80]
Source: Reprinted with permission from Sundyne, LLC.

These metal diaphragm compressors are available with capacities of up to 680 m^3/h (400 ft^3/min) and discharge pressures of up to 2000 bar (30,000 psi). Their drive motor sizes range from 1 to 200 hp (0.75 to 150 kW). A 1000 bar (15,000 psi) unit with a 5 hp motor drive will compress 14 m^3/h (8 ft^3/min) of hydrogen, a 10 hp unit will compress 85 m^3/h (50 ft^3/min), and a 200 hp unit will compress 680 m^3/h (400 ft^3/min).

In high-pressure applications (70 to 8250 bar, or 1000 to 120,000 psi) with a capacity range from 3 to 1200 kg/day (1 to 350 standard cubic feet per minute—SCFM), intensifiers are often added to the compressor. In these oil-free, nonlubricated gas pistons, the pressure of a hydraulic fluid moves the piston as it compresses the hydrogen gas. The hydraulic drive controls the flow and the discharge pressure of the hydrogen. This way, the rate at which the electrolyzer generates the hydrogen is matched to the hydrogen flow in the compressor.

Figure 4-31 illustrates the double-ended piston design of the intensifier, which can boost the pressure of the hydrogen from 5 bar (70 psi) to 1000 bar (15,000 psi). This pressure increase is standard for filling hydrogen-gas storage tanks or transport vehicles. In a single-stage machine, the high-pressure hydraulic fluid moves the motive

80. Sundyne PPI compressors are used in petrochemical plants and hydrogen fueling stations worldwide, including Hyundai in South Korea, Powertech in Canada, Senju in Japan, GM in the United States, and Hydrogenics in Canada.

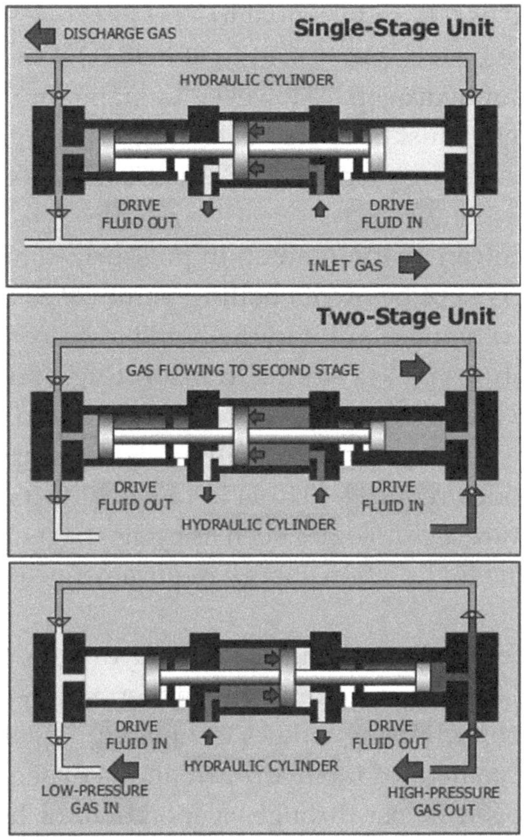

Figure 4-31. An intensifier with a double-ended piston can boost the pressure by a ratio of 8:1 in a single-stage unit and a ratio of 64:1 in a two-stage unit.

Source: Reprinted with permission from Hydro-Pac Inc. Accessed August 12, 2021. http://www.hydropac.com/how-intensifiers-work.html.

cylinder to the left, compressing the hydrogen. The compression ratio for a single-stage machine is limited to approximately 8:1.

In two-stage machines, the principle of operation is similar except that the diameter of the second-stage gas cylinder is smaller than that of the first-stage gas cylinder. The interstage cooling that occurs between stages allows for higher overall compression ratios. The compression ratio of a two-stage machine is limited to approximately 64:1. The controls on hydrogen compressor systems include inlet and outlet pressure controls, motor starters, temperature and flow controls for the cooling water, and sequencing controls.

4.5.7 Hydrogen Liquefaction

Hydrogen liquefaction is a very energy-demanding process because it takes a great deal of cooling to condense hydrogen gas. The lowest possible temperature, known as absolute zero, is denoted as 0 K and is equal to –273.15°C (–469.67°F). When hydrogen gas is condensed, it must be brought to within at least 33 K of absolute zero, which is the critical temperature of hydrogen.[81] However, to maintain the liquid state at atmospheric pressure, hydrogen must be cooled below the critical point, specifically below its atmospheric-pressure boiling point of 20.28 K (–252.88°C [–423.18°F]).

Once liquefied, hydrogen can be maintained in pressurized and thermally insulated containers as long as it is kept below its boiling point (20.28 K). Although hydrogen must be kept pressurized to maintain it in the liquid state, continuously maintaining sufficiently low temperatures makes storing, transporting, and using liquid hydrogen expensive. It is assumed that these costs will be much reduced over time, as hydrogen production costs are expected to continue falling. For example, whereas producing 1 kg of pure hydrogen today requires 10 to 14 kWh (equal to 33% to 42% of hydrogen's lower heating value), future technologies such as cyclic magnetization may ultimately be capable of reducing the liquefaction energy requirements to perhaps 7 kWh/kg.

After solar energy has been collected and stored as hydrogen, the stored energy must be transported to users. The hydrogen must be either a high-pressure gas or a cryogenic liquid to be transported. Using today's technology, liquefaction consumes more than 30% of the energy content of the hydrogen and is expensive. In addition, some amount of stored hydrogen is lost through evaporation, or boil-off, of the liquefied hydrogen, especially in small tanks with large surface-to-volume ratios. The gravimetric energy density of hydrogen is 39.6 kWh/kg (see Table 4-1) if it is in the form of a liquid or a high-pressure gas (700 bar [10,000 psia]). Next, I briefly discuss the liquefaction process.

At atmospheric pressure, hydrogen condenses at –253°C (–423°F). As the pressure increases, the condensation temperature also increases. The liquefaction process uses a number of heat pumps in series to reduce the hydrogen temperature from ambient to liquefaction temperature. The compression-condensation process efficiency ranges from 35% to 60%, depending on the system size and the refrigerant used in the heat pumps. One of the most efficient refrigerants is a mixture of helium and neon. The energy consumption of liquefiers theoretically ranges from 5 to 15 kWh per kilogram of liquefied hydrogen, but the practical range is more like 10 to 18 kWh/kg, depending on the liquefier size (decreasing as size increases).

---

81. In thermodynamics, a critical point (or critical state) is the end point of a phase equilibrium curve. The most prominent example is the liquid-vapor critical point, the end point of the pressure-temperature curve that designates conditions under which a liquid and its vapor can coexist.

Liquid hydrogen suppliers in the United States include Linde, Air Products, Praxair, BOC, and Air Liquide. The daily capacities of liquefiers range from 5400 to 32,000 kg (6 to 35 tons). The total cost of hydrogen liquefaction is the sum of the capital costs (about 65%), the energy costs of production (about 30%, corresponding to about 12 kWh/kg), and operation and maintenance costs (5%). The proportion of capital costs falls as the plant size increases. Energy efficiency can be improved by using expanders, high-speed centrifugal compressors, and newer technologies such as magnetic and acoustic refrigeration.

For a plant to achieve a liquid hydrogen production capacity of 100 kg/h, the electrical energy requirement is about 17 kWh/kg (60 MJ/kg). The required energy input decreases[82] as the plant size increases to a minimum of about 11 kWh/kg.[83] As the HHV of hydrogen is 39.3 kWh/kg, the energy cost of liquefaction at a 100 kg/h capacity is about 42%. In large plants (1000 kg/h capacity), it can be approximately 35%, whereas the energy cost of production in small units can exceed 100% (of the HHV of the hydrogen).

In the process of liquefaction, one must also consider the inversion temperature of hydrogen because the behavior of this gas changes (inverts) at that temperature. Below the inversion temperature, the hydrogen temperature decreases as the pressure is reduced. Above that temperature, the opposite occurs: a decrease in pressure causes an increase in temperature. Therefore, in the process of liquefaction, the hydrogen first has to be cooled below its inversion temperature using a process such as cooling with liquid nitrogen.

Figure 4-32 depicts a hydrogen liquefaction process developed by Praxair. This particular process uses both external refrigeration and liquid nitrogen precooling. It has three sets of heat exchangers, two compressors, two expander turbines (for energy recovery), two accumulators, and a number of control and letdown valves.[84]

---

82. The US DOE target for 2012 was to improve the energy cost of hydrogen liquefaction for plants with capacities of 30 and 300 tons/day from 14 kWh/kg (2005 status) to 11 kWh/kg (2012 goal), a 22% improvement. I could not find more recent data. *Source:* Joe Schwartz, "Advanced Hydrogen Liquefaction Process," Praxair website, May 10, 2011, https://www.hydrogen.energy.gov/pdfs/review11/pd018_schwartz_2011_p.pdf.

83. Some estimates suggest that, theoretically, this energy consumption could be reduced to 5 kWh/kg, but I have not seen proof of these claims.

84. Other liquefaction process configurations have been developed under names such as hydrogen Claude, helium Brayton, combined reverse-Brayton Joule–Thompson (CRBJT), neon Brayton, and neon with cold pump. They can use more exchangers in series or more expanders and compressors, and they naturally have different performance characteristics.

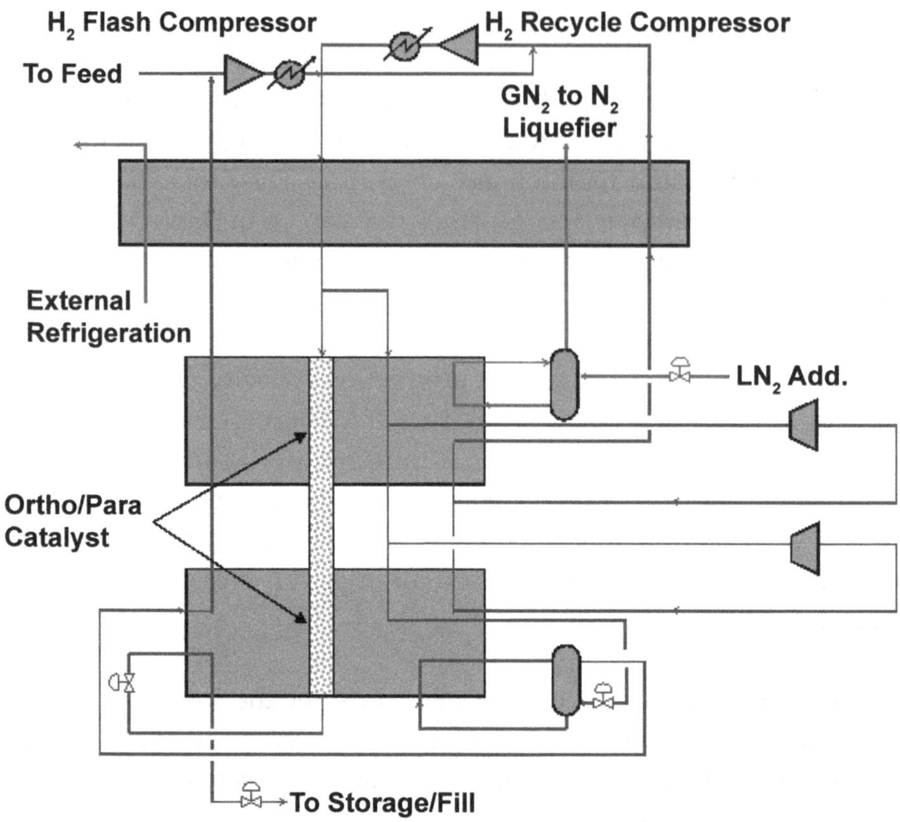

Figure 4-32. Hydrogen liquefaction process developed by Praxair. Abbreviations: GN_2, gaseous nitrogen; LN_2, liquid nitrogen.

Source: Reprinted with permission from the US Department of Energy Hydrogen and Fuel Cell Technologies Office. Joe Schwartz, "Advanced Hydrogen Liquefaction Process," contract no. DE-FG36-08GO18063, presented at the DOE Annual Merit Review Meeting, May 10, 2011, *https://www.hydrogen.energy.gov/pdfs/review11/pd018_schwartz_2011_p.pdf.*

4.5.8 Hydrogen Storage

Once the electrolyzers have converted the solar energy to hydrogen in the form of either pressurized gas or cryogenic liquid, it must be stored for transportation to users.

4.5.8.1 Liquid Storage

At atmospheric pressure (1.013 bar [14.7 psi]), hydrogen gas condenses into a liquid when its temperature is below –252.88°C (–423.18°F), and liquid hydrogen has a density of close to 71 kg/m^3. The weight density of hydrogen under these conditions is 42 kg/m^3 (2.32 lb/ft^3). At this pressure, 5 kg of hydrogen can be stored in a 75 L tank. To maintain hydrogen in the liquid state at this temperature, these tanks must be perfectly insulated to prevent boil-off. (This boil-off problem caused car manufacturers to select high-pressure gas storage for fuel-cell-driven automobiles.)

Currently, hydrogen is stored in liquid form only in high-tech applications, such as NASA's space exploration. For example, the storage tanks on the Ariane rocket launcher (designed and manufactured by Air Liquide) contain 28 T of liquid hydrogen. When empty, these tanks weigh only 5.5 T; their casings are only 1.3 mm thick.

NASA's Space Launch System (SLS) will have a tank that is 61 m (200 ft) tall and 8.4 m (27.6 ft) in diameter to store cryogenic and a similar size one for liquid oxygen. This disposable tank has a storage capacity of 537,000 gallons of hydrogen, which will be kept at a temperature of –253°C (–423°F).

4.5.8.2 Compressed Gas Storage

At 700 bar (10,000 psia), hydrogen has a density of about 40 kg/m³ (Figure 4-33), compared with 0.090 kg/m³ under atmospheric pressure and temperature conditions. At this high pressure, 5 kg of hydrogen can be stored in a 125 L tank.

Table 4-10 lists the approximate weights of pressurized hydrogen at pressures of 350 and 700 bar (5000 and 10,000 psia) in fuel tanks of various volumes.

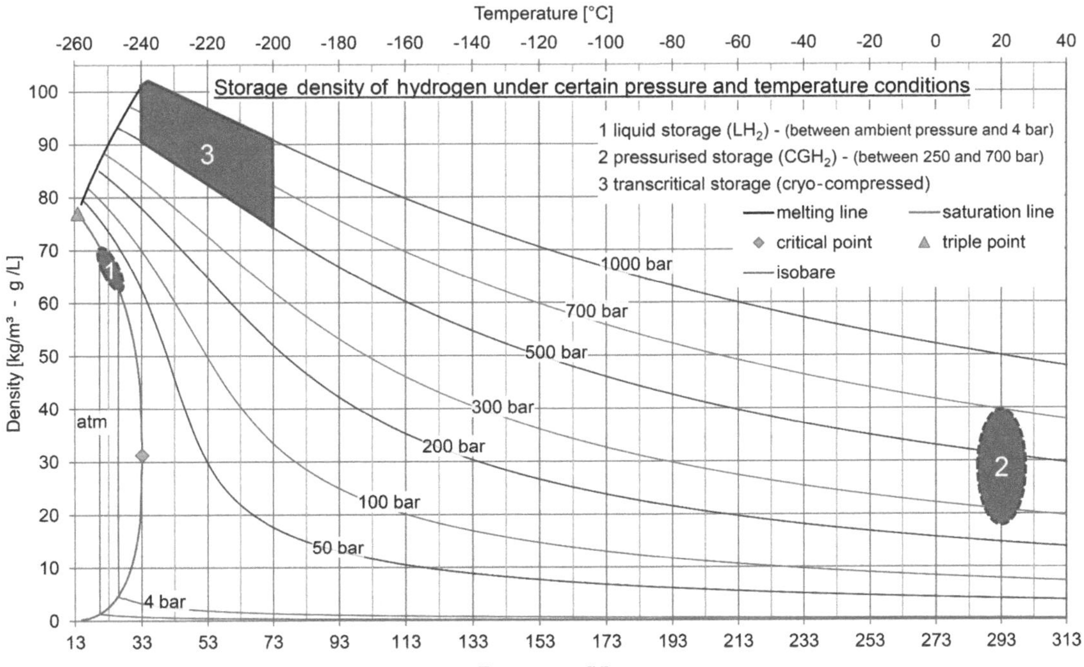

Figure 4-33. The volumetric density of hydrogen gas at various pressures and temperatures. Abbreviations: CGH_2, compressed gaseous hydrogen; LH_2, liquid hydrogen.

Source: "Storage Density of Hydrogen.jpg" by Moritz Kuhn, ILK Dresden, July 9, 2015, is licensed under CC BY-SA 4.0, https://fr.wikipedia.org/wiki/Fichier:Storage_Density_of_Hydrogen.jpg.

Table 4-10. Weights of pressurized hydrogen in fuel tanks of various sizes.

Tank Volume (L)	Full Weight (kg)	
	350 bar (5000 psi)	700 bar (10,000 psi)
100	2.3	3.5
200	4.5	7.1
300	6.8	10.6
400	9.1	14.1
500	11.4	17.6

Today, most car manufacturers have opted for storing hydrogen in gaseous form at high pressures. Using high-pressure gas as fuel allows for storing enough hydrogen to drive a distance of 500 to 600 km (310 to 370 mi) between refills.[85]

Fuel tanks for compressed hydrogen are usually made of composites of carbon fibers and metal alloys. The inner liner is a high-molecular-weight polymer that serves as a hydrogen-gas permeation barrier. A carbon-fiber, epoxy-resin composite shell is placed over the liner and constitutes the gas-pressure-load-bearing component of the tank. Finally, an outer shell is placed on the tank to provide impact and damage resistance. These tanks are made strong enough to resist the penetration of rifle bullets. The pressure regulator for the 700 bar (10,000 psi) tank is located in the interior of the tank. There is also an in-tank sensor to monitor the tank temperature during the gas-filling process, which is when tank heating occurs.

FCEVs weigh substantially less than BEVs (Figure 4-34). Among BEVs, the Li-ion battery design is the lightest, the NiMH design is heavier, and the PbA design is the heaviest and is seldom used anymore.

4.5.8.3 Underground Hydrogen Storage

Hydrogen can be stored in underground salt caverns, depleted oil/gas reservoirs, or aquifers. Oil or gas reservoirs are the most convenient to use, as they have already been proven safe for operation and are provided with the required infrastructure. Hydrogen is also injected into operating natural gas reservoirs, resulting in a mixture containing about 10% hydrogen; the mixture is used in the same way as natural gas.

---

85. For details for a number of fuel cell cars, see: "Compare Fuel Cell Vehicles," Office of Energy Efficiency & Renewable Energy, US DOE, updated January 7, 2021, accessed November 23, 2022, https://fueleconomy.gov/feg/fcv_sbs.shtml.

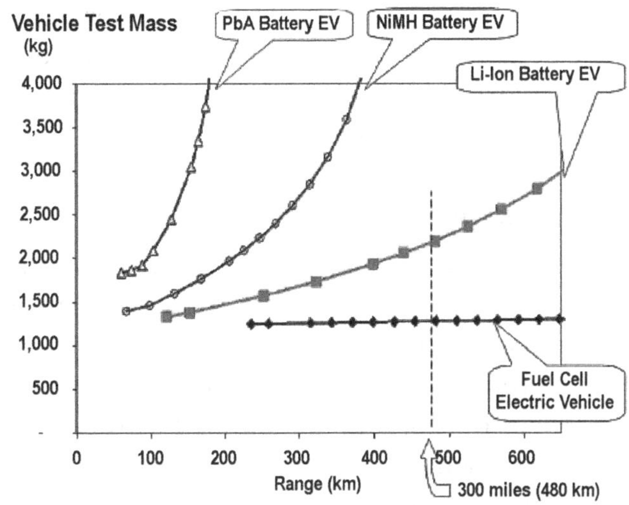

Figure 4-34. Weights of BEVs and FCEVs as a function of vehicle range.

Source: Reprinted with permission from C. E. Thomas, "Hydrogen-Powered Fuel Cell Electric Vehicles Compared to the Alternatives," AZOCleantech website, July 1, 2010, https://www.azocleantech.com/article.aspx?ArticleId=214.

Storing hydrogen in salt caverns is not a new technology. Hydrogen has been stored in three salt caverns at Teesside in the UK since 1972 and two at the US Gulf Coast in Texas since 1983. The experience of storing hydrogen in these caverns proved that hydrogen could be stored safely for long time periods.[86] As the conversion from fossil fuels to intermittently available renewable energy is progressing, storing solar and wind energy is gaining importance. One means of storing these intermittent energies is to convert them to hydrogen and store the hydrogen in salt caverns.

Most caverns operate at up to 200 bars and are located at a depth of 500 to 2500 meters. These caverns usually have diameters of up to 100 meters and heights from 100 to 500 meters. About 30% of their volume is taken up by buffer gas to maintain the cavern's integrity, and the other 70% is used for storage. The energy density of hydrogen is 39.6 kWh/kg; therefore, if the energy content of the stored hydrogen is recovered by 60% efficient fuel cells, each cavern can supply about 300 GWh of electricity.

Underground hydrogen storage caverns are in operation in several locations in the world. In the United States, the Linde salt cavern supports a 340-mile (545 km) hydrogen pipeline that serves more than 50 refineries and chemical plants from Sweeny,

86. Radoslaw Tarkowski, "Underground hydrogen storage: Characteristics and prospects," *Renewable and Sustainable Energy Reviews* 105 (2019): 86–94, https://www.sciencedirect.com/science/article/abs/pii/S1364032119300528.

Texas, to Lake Charles, Louisiana.[87] Similarly, in Europe, Germany identified salt caverns as a feasible and flexible solution for hydrogen storage. It is estimated that Europe has the potential to inject hydrogen in bedded salt deposits and salt domes, having the equivalent electricity storage capacity of 84.8 PWh.[88]

4.5.9 Hydrogen Transportation and Distribution

After the compressed or liquefied hydrogen is stored at a solar farm, it is ready to be transported to users. Some of the necessary infrastructure is already in place because hydrogen has long been used in industrial applications. Today, hydrogen is transported from the point of production to the point of use by pipeline, cryogenic liquid tanker trucks, gaseous tube trailers, rail, and barge. Hydrogen used in portable and stationary applications can be delivered by truck to a storage facility; in cylinders, like the propane used for gas grills; or in cartridges that resemble batteries.

Hydrogen used in FCEVs is dispensed the same way as gasoline in IC automobiles. Drivers pull into a filling station, connect a dispenser to the vehicle, fill, disconnect, pay, and then drive away with a full tank. Refueling an FCEV takes approximately the same amount of time as refueling a gasoline-powered car, 3 to 5 minutes.

4.5.9.1 Hydrogen Transportation as a Liquid

In the absence of pipelines, hydrogen is most commonly transported and delivered as a liquid when high-volume transport is needed. Trucks transporting liquid hydrogen are referred to as liquid tankers. For longer distances, hydrogen is transported as a liquid in super-insulated cryogenic tanker trucks (Figure 4-35) to distribution sites, where it is vaporized into a high-pressure gaseous product for dispensing. Over long distances, trucking liquid hydrogen is more economical than trucking gaseous hydrogen because a liquid tanker truck can hold a much larger mass of hydrogen than a gaseous-hydrogen tube trailer. However, transporting liquid has the potential for boil-off during delivery.

At a density of 70.85 kg/m^3, about 3500 kg of liquid hydrogen can be carried at a loading volume of 50 m^3. The tankers have a range of approximately 4000 km. Over the journey time, the cryogenic hydrogen heats up, causing the pressure in the container to rise.

---

87. "Storing Hydrogen in Underground Salt Caverns," Linde website, accessed October 13, 2022, https://www.lindehydrogen.com/technology/hydrogen-storage.
88. Emiliano Bellini, "Hydrogen storage in salt caverns," June 16, 2020, accessed October 13, 2022, https://www.pv-magazine.com/2020/06/16/hydrogen-storage-in-salt-caverns/.

Figure 4-35. Liquid hydrogen tanker.

Source: "Liquid Hydrogen Delivery," Office of Energy Efficiency & Renewable Energy, US DOE, accessed March 12, 2022, https://www.energy.gov/eere/fuelcells/liquid-hydrogen-delivery.

Transporting hydrogen is similar to, but safer than, transporting liquefied natural gas. If leakage occurs, hydrogen will quickly rise into the atmosphere, whereas liquefied natural gas, propane, and gasoline accumulate on the ground, pool, and soak into clothing or the ground, where they can be ignited. In addition, when gasoline burns, glowing hot soot particles carry the heat and can become secondary ignition sources. In contrast, hydrogen burns cleanly without a residue.

Liquid hydrogen can also be transported by ship or rail, provided that suitable waterways, railway lines, and loading terminals are available. Kawasaki launched a liquid hydrogen carrier ship to transport large volumes of liquid hydrogen in 2019.[89] In 2022, the 380-foot-long Suiso Frontier set sail from Australia, bound for Japan with liquid hydrogen in its insulated hold.[90]

Norway is also working on liquid hydrogen shipping technology. The country has provided grants to leading maritime companies to conduct feasibility studies into the various aspects of this emerging technology sector. For example, a grant was issued through the state-sponsored research organization Innovation Norway to design a

89. Loz Blain, "Kawasaki launches the world's first liquid hydrogen transport ship," New Atlas website, December 15, 2019, accessed October 13, 2022, https://newatlas.com/marine/kawasaki-worlds-first-liquid-hydrogen-transport-ship/.

90. Amy Gunia, "Australia Just Exported Its First Batch of Fuel That Doesn't Emit CO_2. There's Just One Catch," January 28, 2022, accessed October 13, 2022, https://time.com/6139820/liquid-hdyrogen-shipment-hesc/.

carrier to transport liquid hydrogen in bulk having a 9000 m^3 capacity that is expected to serve an integral role in an emerging hydrogen economy. Central to these efforts is the question of how hydrogen will be transported from central storage to end users in the coming decades.

4.5.9.2 Hydrogen Transportation as a Gas

Generating high-pressure hydrogen gas requires about 50% less energy than liquid hydrogen production. Long-term storage of liquid hydrogen is a source of further losses. In conventional automotive applications without special seals, leakage losses can amount to about 30% per month, but special seals can significantly reduce that. The higher production costs and leakage are disadvantages of liquid hydrogen transportation, but on the other hand, liquid hydrogen is less expensive to store and transport than hydrogen gas. Transporting compressed hydrogen gas by trucks is inefficient because, for a constant volume, a truck can carry much less fuel (lower mass of hydrogen) when carrying pressurized gas than when transporting liquid hydrogen.

Because of the decades of experience using liquid hydrogen in space exploration, the storage, transportation, and handling were found to be safe; the design of hydrogen tankers and trucks is similar to those of liquefied natural gas storage vessels and tankers. Hydrogen is usually stored safely in double-walled vessels, in which the space between the walls is filled with superinsulation or evacuated.

Railcars, ships, and barges can also be used to deliver hydrogen as a compressed gas in tanks certified by the International Organization for Standardization to carry large volumes of high-pressure hydrogen gas over rail or water.

BMW tested its hydrogen tanks in a series of collisions that caused both fire and tank ruptures. In most of these tests, hydrogen was found to be safer than gasoline and diesel fuels. Even though the simulated 88 km/h (55 mph) crash tests left the cars totaled, the newly designed hydrogen tanks remained intact. The potential for pressurized hydrogen tanks to explode, although possible, would require enormous impact forces; therefore, such tanks are generally considered to be safe.

Small quantities of hydrogen can be transported by truck in compressed gas cylinders. Larger amounts are transported in steel cylinders that are bundled and secured within a frame in so-called tube trailers (Figure 4-36).[91] The hydrogen gas is com-

91. Evgenios Zogopoulos, "Hydrogen: Is this the future? (Part II)," Energy Industry Review website, updated March 30, 2021, accessed November 23, 2022, https://energyindustryreview.com/analysis/hydrogen-is-this-the-future-part-ii/.

Figure 4-36. Composite tube trailers are typically capable of delivering 550 kg of hydrogen at 250 bar (3600 psig) pressure.

Source: "Hydrogen Tube Trailers," Office of Energy Efficiency & Renewable Energy, US DOE, accessed March 12, 2022, https://www.energy.gov/eere/fuelcells/hydrogen-tube-trailers.

pressed to about 180 bar (~2600 psig) or higher in the cylinders. The net weight of the cylinders can be an issue and lead to transport restrictions. For example, tube trailers are currently limited to pressures up to 250 bar (3600 psig) by the US Department of Transportation (DOT) regulations. Combined with the DOT weight limitations for on-road vehicles, this results in a limited carrying capacity of approximately 280 kg of compressed hydrogen gas for steel tube trailers. In contrast, the newest pressurized storage systems use lighter composite storage containers for truck transport. These containers can deliver 560 to 720 kg of hydrogen per trailer and are within the DOT's height, width, and weight requirements. Such tube trailers are currently being used to deliver compressed natural gas in other countries.

The low density of hydrogen also has an impact on its transport: Under standard conditions (1.013 bar [14.7 psig] and 0°C [32°F]), hydrogen has a density of 0.0899 kg/m^3. If hydrogen is compressed to 200 bar (2900 psig), the density under standard conditions increases to 15.6 kg/m^3; at 500 bar (7300 psig), it reaches 33 kg/m^3.

Single tube trailers carry approximately 500 kg of hydrogen, depending on the pressure and the container material used. According to energy analyst Evgenios Zogopoulos, "the largest tank volumes for gaseous hydrogen transport are currently 26 cubic meters. Taking account of the low hydrogen density factor at 500 bar [7300 psi], this results in a load of around 1,100 kg [of] hydrogen per lorry. This figure extrapolates to approximately 12,000 normal cubic meters of hydrogen. At 250 bar [3600 psi], both the weight of hydrogen and its transport volume in Nm^3 would be roughly halved."[92]

92. Evgenios Zogopoulos, "Hydrogen: Is this the future? (Part II)."

4.5.9.3 Hydrogen Transportation by Pipeline

A pipeline network would be the best option for the comprehensive and large-scale distribution of hydrogen as an energy source. However, pipelines require high levels of initial investment, which can be justified only if the user demand is high. Although hydrogen pipelines are in operation worldwide, their use is limited. In 2016, only 4500 km of pipelines were in use globally. The longest pipelines are operated in the United States, in Louisiana and Texas, followed by Belgium and Germany. Micro-networks, pipelines for hydrogen distribution in local areas, can be developed and later combined into transregional networks.

The high initial capital costs of new pipeline construction constitute a major limitation to expanding the hydrogen pipeline delivery infrastructure. Transporting gaseous hydrogen through existing pipelines would be a low-cost solution for delivering large volumes of hydrogen, but this solution is limited because of embrittlement and corrosion problems. Hydrogen pipes must resist corrosion and embrittlement of the steel and the welds used to fabricate the pipelines because hydrogen can easily migrate into the crystal structures of most metals.[93] For process metal piping at pressures up to 480 bar (7000 psig), high-purity stainless steel piping with a maximum hardness of 80 on the Rockwell B scale[94] is preferred but expensive.

Research is in progress to overcome the limitations, including the need to control hydrogen permeation and leaks, reduce costs, and develop more reliable and durable hydrogen compression technology. Potential solutions include using fiber-reinforced polymer (FRP) pipelines. Installation costs are about 20% lower for FRP pipelines than for steel pipelines because the FRP can be obtained in much longer sections than steel, reducing the welding labor requirements.

One possibility for rapidly expanding the hydrogen delivery infrastructure would be to adapt part of the natural gas delivery infrastructure to accommodate hydrogen. Converting natural gas pipelines to carry a blend of natural gas and hydrogen (up to about 15% hydrogen) would require only modest modifications, but converting them

---

93. The design of steels and components that resist hydrogen embrittlement by reducing the intake of hydrogen or rendering it innocuous when it penetrates the steel is beyond the scope of this book. For further information, see: Harshad Kumar Dharamshi Hansraj Bhadeshia, "Prevention of Hydrogen Embrittlement in Steels," *ISIJ International* 56, 1 (2016), 24–36, https://www.jstage.jst.go.jp/article/isijinternational/56/1/56_ISIJINT-2015-430/_pdf/-char/en.

94. Hardness is very important for metal casting, heat treatment, and machining processes. Rockwell hardness (HRC and HRB) and Brinell hardness (HB or BHN) are most commonly used for steel and iron castings.

to deliver pure hydrogen would require substantial modifications. Research evaluations are in progress to resolve this problem.

Because of Russia's 2022 natural gas "trade war" against Europe, Germany is considering using their existing natural gas pipe grid for future hydrogen distribution.[95] However, the current natural gas infrastructure is not suitable in its present state to transport hydrogen, so conversion is required. Converting the existing natural gas infrastructure to transport hydrogen is generally much cheaper than building a new network from scratch. In 2020, German gas grid operators said that an initial hydrogen grid could be established at a "justifiable cost" of 660 million euros, largely by converting current pipelines.

4.5.10 The RFC

In the solar-hydrogen energy economy, it could become possible to minimize or eliminate our dependence on the electric grid. In addition to eliminating the need for grid maintenance, this would eliminate resistance losses and the risks and consequences of cyberattacks on the grid.

The fuel cell and the RFC would play important roles in this fully distributed energy supply structure. As already discussed, the fuel cell itself is a device that generates electricity through reverse electrolysis, a chemical reaction in which hydrogen and oxygen react to produce water, heat, and electricity. The fuel for these cells is hydrogen obtained by water electrolysis (see Section 4.5.5.1, "Green Hydrogen by Water Electrolysis") and oxygen taken from the atmosphere. The electricity, distilled water, and heat produced by the fuel cell can all be used in a home. A fuel cell can be compared to a flow battery because it is an electrochemical cell that converts the chemical energy of a fuel (hydrogen) and an oxidizing agent (oxygen) into electricity through a pair of redox reactions.[96] Fuel cells can produce electricity continuously for as long as fuel and oxygen are supplied.

There are many types of fuel cells, but they all consist of an anode, a cathode, and an electrolyte that allows ions—often positively charged hydrogen ions (protons)—to move between the two sides of the cell. At the anode, a catalyst causes the fuel to undergo oxidation reactions that generate positively charged hydrogen ions and

---

95. Julian Wettengel, "German regulator paves way for converting natural gas infrastructure to hydrogen," March 22, 2021, accessed October 13, 2022, https://www.cleanenergywire.org/news/german-regulator-paves-way-converting-natural-gas-infrastructure-hydrogen.

96. A reduction-oxidation (or redox) reaction is a chemical reaction characterized by the transfer of electrons between chemical species. Most often, the reduction agent is oxidized (loses electrons) and the oxidizing agent is reduced (gains electrons).

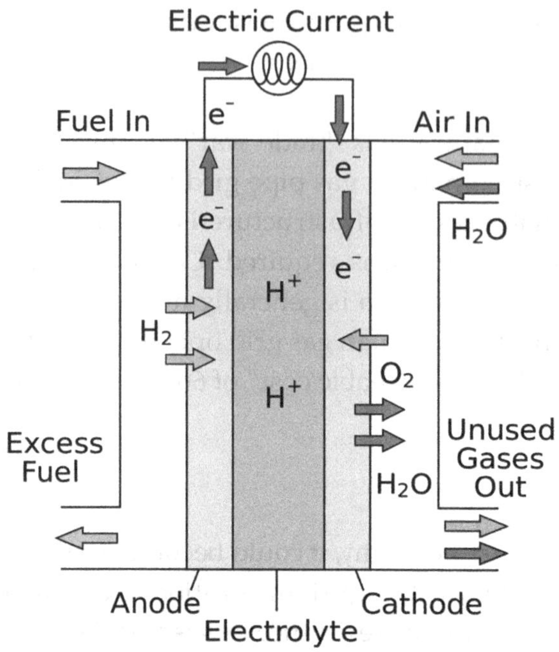

Figure 4-37. A fuel cell generates electricity by moving ions from the cathode to the anode.[97]

Source: "Solid oxide fuel cell protonic.svg" by R. Dervisoglu, May 2012, Public Domain, https://commons.wikimedia.org/w/index.php?curid=19314043.

electrons. The ions move from the cathode to the anode through the electrolyte. The electrons travel from the anode to the cathode through an external circuit, which produces a direct current (Figure 4-37). At the cathode, another catalyst causes ions, electrons, and oxygen to react, forming water.

Individual fuel cells produce relatively small electrical potentials, about 0.7 V per cell, so cells are stacked, or placed in series, to create sufficient voltage to meet application requirements. The energy efficiency of a fuel cell is about 60%; however, if the waste heat is captured, efficiencies up to 85% can be obtained.

In principle, because of the inherent reversibility of chemical reactions, the process of any fuel cell could be reversed. However, a given fuel cell is usually optimized for operating in only one direction. Such fuel cells are either unsuitable for being operated backward or inefficient in that mode of operation (whereas other devices, such

97. For further information, see: "Polymer electrolyte membrane electrolysis," Wikipedia, https://en.wikipedia.org/wiki/Polymer_electrolyte_membrane_electrolysis. "Fuel cell," Wikipedia, accessed March 12, 2022, https://en.wikipedia.org/wiki/Fuel_cell.

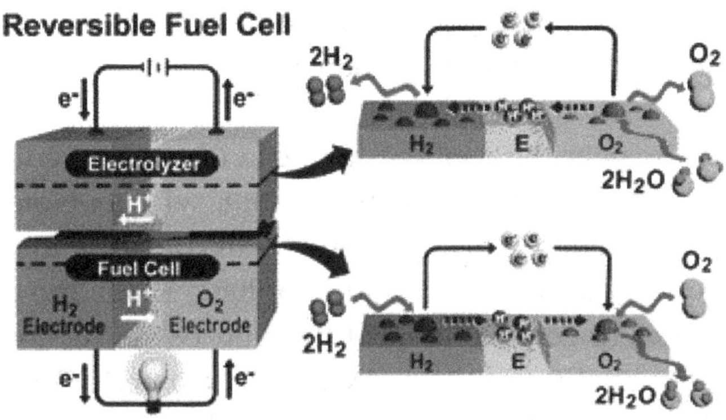

Figure 4-38. The RFC, when operated in one direction, produces hydrogen from water and, when reversed, uses hydrogen to generate electricity and water.

Source: Sehkyu Park, Yuyan Shao, Jun Liu, and Yong Wang, "Oxygen Electrocatalysts for Water Electrolyzers and Reversible Fuel Cells: Status and Perspective," *Energy & Environmental Science 5*, no. 11 (2012): 9331–9344, https://doi.org/10.1039/C2EE22554A. Reproduced with permission from the Royal Society of Chemistry.

as high-pressure electrolyzers, can be constructed for operation in both directions, Figure 4-38).

Work is in progress at the US DOE Hydrogen Program National Science and Technology Council[98] and elsewhere to develop commercially available reversible electrolysis modules. For example, the German company sunfire GmbH has developed a commercial reversible electrolysis module based on a reversible solid oxide cell (RSOC). This unit is in operation at a US Navy microgrid test facility in California. The fully integrated system converts surplus renewable energy from wind or solar PV arrays into hydrogen, which is stored at high pressure. In electrolysis mode, the system yields 42 m^3/h of hydrogen at up to 85% efficiency. When switched to fuel cell mode, it delivers 50 kW of electric power at up to 60% efficiency.[99]

4.5.10.1 The RFC Could Be Part of Homes of the Future

In 50 years or so, the abbreviation of RFC will probably be as familiar as PC is today because—just as oil and propane tanks are today—an RFC and its hydrogen tank could be sitting next to each home. This will likely occur because, at the end of the fossil-nuclear age, we will have to completely convert our energy economy to using green energy, the source of which will mostly be the Sun, which is an inexhaustible,

98. "Space Applications of Hydrogen and Fuel Cells," NASA website, last updated May 27, 2021, accessed November 23, 2022, https://www.nasa.gov/content/space-applications-of-hydrogen-and-fuel-cells.

99. "sunfire Supplies Boeing with Reversible Solid Oxide Electrolyser/Fuel Cell System," *Fuel Cells Bulletin* 2016, no. 2, February 29, 2016, accessed November 23, 2022, https://www.sciencedirect.com/science/article/abs/pii/S1464285916700022.

free, and clean energy source. This will be done by converting and storing the intermittent solar energy into continuously available hydrogen.

In the next century, I expect the roofs of homes will be covered by solar-collecting roof tiles and shingles, which, when the Sun is out, will generate electricity for the home and send the excess energy to an RFC that will convert it to hydrogen and store it (electrolyzer mode). (Today, because we do not have RFCs, the excess electricity is either stored in batteries or sent to the grid through two-directional electric meters, as described in Chapter 3, Section 3.2.1, "My Solar Home.") Once RFCs are widely available, there will no longer be a need for the grid because the RFCs will convert the excess electricity into hydrogen and store it in a tank. If the solar collector surface area is large enough, storing the hydrogen will eliminate the need for electricity from the grid and the associated transmission system or the power plants that serve it today. This would eliminate all costs associated with transmission and distribution and thereby substantially cut the cost of electricity.

Let us examine the operation of an RFC in a little more detail. When the Sun is out (red in Figure 4-39), the RFC acts as an electrolyzer. It receives the excess solar energy that the home does not need, converts it into hydrogen, and stores it in a pressurized tank (labeled H_2 in the Figure). During the night or whenever the Sun is not out (blue in Figure 4-39), the RFC switches to fuel cell mode and supplies the home with electricity generated from the hydrogen in the storage tank.

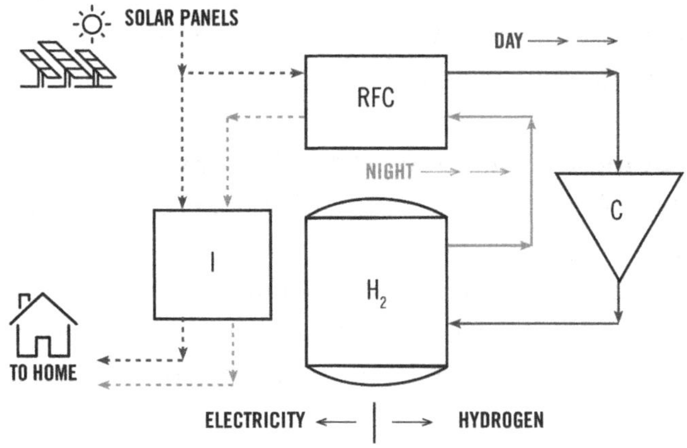

Figure 4-39. When the Sun is out (red), the RFC operates in electrolyzer mode and stores the excess solar energy as hydrogen. During the night (blue), it operates in fuel cell mode and supplies electricity to the home. Legend: C, compressor; H_2, hydrogen storage tank; I, inverter; RFC, reversible fuel cell.

Source: Courtesy of *Control* magazine. Béla Lipták, "Controlling the Future, Part 4," September 21, 2021, https://www.controlglobal.com/articles/2021/controlling-the-future-part-4/.

The storage tank size will vary depending on the location of the home and its roof area. As with today's propane tanks, when the hydrogen pressure in the tank falls below a certain level, the tank will be refilled, whereas when the hydrogen is in excess, that excess hydrogen will be trucked away. A hydrogen pressure detector in the tank will signal the trucking company when the tank needs refilled or the excess hydrogen needs to be removed. In other words, when the hydrogen in the storage tank drops to its low limit or when it approaches its maximum limit, the hydrogen supplier will be automatically alerted by a wireless pressure detector on the tank to keep the pressure within those limits.

In this arrangement, once the system is installed, the solar electricity for the home (including that needed for charging electric cars, drones, etc.) will be free. If the roof is large enough, the yearly excess energy can be used to generate heat or income for the home. This system can also be designed to act as a backup for local grids because the electricity from the RFCs in a neighborhood can provide backup without the need for diesel generators.

In parts of the world where the yearly insolation equals or exceeds 1500 kWh/m^2 (Figure 4-18), the roofs are likely to be large enough for the insolation to meet the total electricity needs of the average home. Later, as the needed technologies become available, other building surfaces (such as windows) can also become solar collectors. Naturally, hydrogen tanks serving homes with high air-conditioning loads, several electric cars, and/or large swimming pool pumps will probably need more collector area than just the roof, or they will need to be refilled periodically with hydrogen delivered by truck. The larger the tanks, the less frequent the refilling will be.

4.5.10.2 Controlling the RFC

Fully developing the RFC would represent a significant step toward a carbon-free energy future, and the contributions of the process control profession will be an essential part of this success. RFC packages do not exist today (although, as already mentioned, their development is in progress[100]). It will probably take a decade or more for them to be available and mass-produced. However, we have electrolyzers and fuel cells, and we could assemble them into a system similar to the one in Figure 4-40.

To be a viable option, the RFC would have to become an inexpensive, mass-produced, miniaturized, completely reliable, and fully automatic device in a couple of decades. The schematic layout in Figure 4-40 shows that this control system will be quite

100 "sunfire Supplies Boeing," Science Direct website.

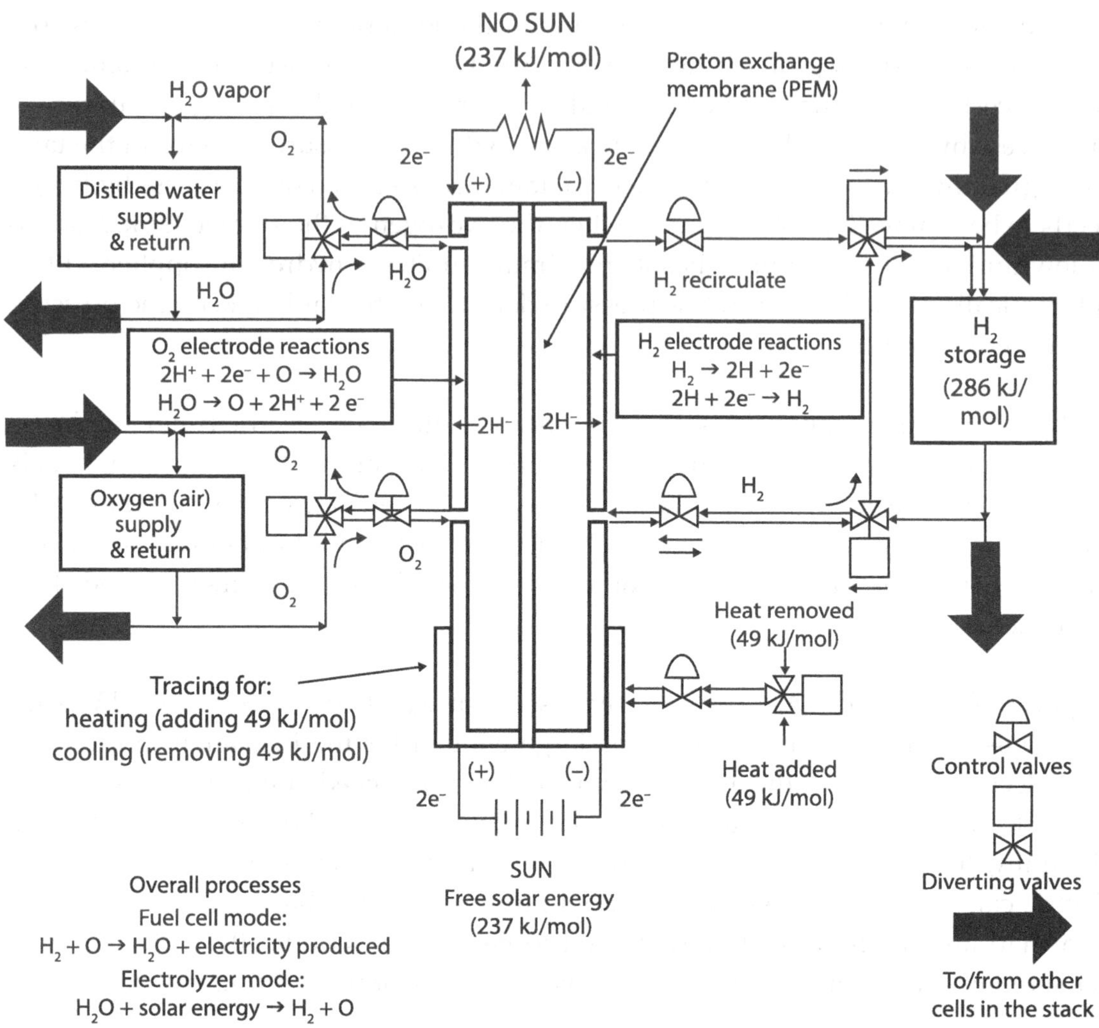

Figure 4-40. Developing a fully automated, optimized, and mass-produced RFC package will take a decade or more to finalize, and once finalized, it might look like this. In this sketch, the blue lines show flow directions for daytime mode (hydrogen generation), and the red lines show the flow directions for nighttime mode (hydrogen consumption). (See color images after page 306.)

Source: Courtesy of *Control* magazine. Béla Lipták, "Automating the reversible fuel cell," May 13, 2016, Figure 2, https://www.controlglobal.com/articles/2016/automating-the-reversible-fuel-cell/.

complex. There will be dozens of sensors and communication links to serve simple tasks, such as informing the hydrogen supply company when the hydrogen in the storage tank is approaching its upper or lower limits, as well as more elaborate tasks, such as controlling switching between daytime and nighttime operation and operating the system as a grid backup or EV charger.

Developing both the hardware and software for the RFC package will face many challenges. On the hardware side, absolute safety will be the number one goal. Nevertheless, no matter how good the safety design is, the RFC must remain outdoors (similar to a propane tank) so that, if leaking occurs, the hydrogen can safely escape. Also, the sensors will be configured in a voting system configuration, using three sensors to measure all critical variables. Whenever one of the three sensors disagrees with the readings of the other two, it will be automatically disregarded, and the RFC installer will be contacted automatically if maintenance is required.

On the software side, there will be an evolution similar to the one currently in progress with self-driving cars. This evolution will include redundant multisensors and self-learning, artificial-intelligence algorithms to maximize system efficiency, minimize operation costs, and facilitate wireless communications between the homeowners and the hydrogen supplier.

4.6 Summary

The goal of this chapter was to show that intermittently available and free green energy can be converted to become an uninterrupted, continuous supply that is inexhaustible and causes no climate change. Being able to store the energy received from intermittent energy sources is a prerequisite for eliminating humankind's carbon footprint. We know how to store, transport, and distribute solar energy to achieve that goal. We also know that we have no option but to implement the coming solar-hydrogen energy economy because the competition for fossil resources is starting wars, climate change is causing disasters, and if we burn all the fossil fuels, the planet could become unlivable. Therefore, we should learn from our Stone Age ancestors who stopped using stone not because they had run out of it but because bronze was better.

In the remaining chapters, I summarize the conclusions of my analysis, including the timeframe for effective corrective action and the steps we must take to move from the fossil-nuclear age to the age of hydrogen.

5 Summary of the Process Control Analysis

It was already apparent when I started this analysis that GW was endangering our physical environment and that the uncontrolled development of AI was threatening our cultural environment. As I continued, it became more and more obvious that this decade is the critical one, that if we do not take the drastic steps needed by the end of this decade, our later efforts will come too late to be able to protect against the coming disaster. And then came the pandemic, inflation, cyber terrorism, and the energy war, which combined to increase carbon emissions and diminished both the resources and the attention needed.

In this chapter, I provide a summary of the process control analysis presented in the previous chapters. In Chapter 6, I summarize the many types of data used in or developed by this analysis. Finally, Chapter 7 contains my conclusions regarding where the AI and GW processes are leading, where they will reach if left uncontrolled, what steps are needed to get them under control, and how the fossil-free and green energy economy might function in the twenty-second century.

When the heat input from the Sun and the heat loss to outer space are the same, GW is zero, the proportions of the three phases of water (vapor, liquid, and ice) are relatively constant, and Earth's temperature range is conducive to life.

When the heat input from the Sun is less than the heat loss to outer space, the global temperature drops, and the anomaly becomes negative. Each degree of anomaly change has tremendous consequences because of the immense heat capacity of our planet. For example, when the negative anomaly was –2.0°C (–3.6°F), the Little Ice Age occurred, and when it increased to –5.0°C (–9.0°F), North America was covered by an ice layer a mile thick.

This should give us a hint of what it means when the Paris Agreement talks about a positive anomaly (GW) of +1.5°C or +2.0°C (+2.7°F or +3.6°F). When the anomaly is positive, the heat received from the Sun exceeds the heat loss of the planet to outer space, and this has enormous consequences. These consequences first show their symptoms in the air (climate) because air mass is relatively small. The masses of the oceans and the land are hundreds of times greater than that of the air. This is why my process control analysis concluded that the impact of the oceans is delayed by a decade or so because the dead time of the oceanic processes is about 10 years longer than the atmospheric ones.

By applying the rules of process control, we have also learned that, after the dead time passes, we will shortly reach several "tipping points" that will be extremely difficult to reverse. Why? It is because, at those points, self-sustaining and accelerating feedback processes are triggered that cause the GW to rise, which further accelerates the feedback effect. At a positive anomaly exceeding +3.0°C (+5.4°F), the United Nations (UN) and the scientific community agree that about one billion people will have to migrate toward the poles because their present locations will become unlivable, and human civilization will be at risk.

The process control analysis has also shown that increasing the anomaly increases the rate of water circulation on the planet. In other words, as the evaporation of the oceans rises, the atmosphere contains more and more water vapor. As a result, more rain falls over colder land masses as the water vapor condenses, and droughts increased on land masses where water evaporates from the soil. Concurrently with the increased evaporation, there is a reduction in the amount of the solid phase of water (ice). As the ice melts, its reflective surface is replaced with heat-absorbing water and land surfaces, further accelerating GW.

Our analysis also shows how the atmosphere serves to sustain life. If Earth had no atmosphere (like the Moon), its surface temperature during the day would rise to several hundred degrees, and at night, it would drop to near absolute zero. Earth's atmosphere not only provides thermal insulation; it also helps to balance plant and animal life through the carbon cycle. When the thermal anomaly is zero (the temperature is stable), the amount of carbon emitted into the air by animals and humans and the amount of carbon absorbed by the plants and oceans are in balance. It became out of balance when humans put extra carbon into the air by burning fossil fuels to meet the 30-fold increase in their energy needs, while they were also destroying carbon-absorbing vegetation. This resulted in an increase in our planet's "thermal insulation" and a rise in the temperature anomaly (GW).

From a heat-balance perspective, this is perfectly normal because the only way the planet can reestablish its own heat balance is by increasing the global temperature until the heat reflected back into space (cooling) rises sufficiently to match the rise in the incoming heat. The global average temperature was first measured in 1880 at 13.7°C (56.7°F). Today's average global temperature is about 15°C (59°F). The average global temperature during the past million years (see Figure 2-3 in Chapter 2) remained within the band from 6°C to 18°C (42.8°F to 64.4°F). Throughout the past million years, ice ages alternated with warm periods, but parts of the planet were still able to support life. This tells us that Earth's temperature is self-regulating: it remains within the mentioned temperature band, and despite its up-and-down fluctuations, it does not destroy life. The Earth's marvelous design protects life, but it can destroy or relocate species. In our case, it can destroy human civilization because Earth's current warming can make most of the planet unlivable for humans.

So, what is the solution? Process control suggests the obvious: Reduce the carbon content of the air (don't just stabilize it, reduce it) and meet our energy needs through the inexhaustible and free energy supply from the Sun (a nuclear reactor that is a safe distance from Earth). This can be done by converting solar energy into a substance that does not release carbon when burned and can be stored, distributed, and (using fuel cells) converted back into electric energy while emitting nothing other than distilled water. This substance is hydrogen, and one of the main goals of the preceding chapters of this book was to describe the process of generating, storing, and distributing it.

In the previous chapters, I analyzed GW and artificial intelligence (AI) as if they were just processes to be brought under control. I showed that both of these processes are out of control and that, if the basic laws of process control are followed and implemented, they can still be brought under control. I analyzed these processes together because they interact: AI can support the successful limiting of GW.

5.1 Nonindustrial Processes

Before one can control a process, one must fully understand it. I hope that the content of the previous four chapters provided that understanding. We understand that both the AI development and GW processes are out of control, and we must look at them as control loops to bring them under control,. In other words, we must determine their set points (the goals we want to reach), measure their errors (the difference between the present conditions and our goals), and identify the manipulated variables (the variables we must change to eliminate the errors).

In a traditional temperature control loop, this determination is easy because we can measure the difference between the existing temperature and the desired one and cool or heat the process until that difference is reduced to zero. In the case of the GW and AI processes, we can also identify their "personalities" (time constants, rates of responses, capacities, inertias, etc.), measure their existing conditions, and compare them to the desired ones. The rest of the loop is different because the controller and the controlled are not separated, they are the same. They both are humankind itself! The controller output is a function of humankind's understanding of what needs to be achieved, and the manipulated variables are the means that can influence human behavior (education, taxation, etc.).

In the previous chapters, I identified the natures of these processes, the reasons they are out of control, the consequences of them remaining out of control, and the tipping points at which they might become irreversible. I have approximated the length of time we still have to bring these processes under control, and I have shown that we have the technology to do so.

GW and AI are also interacting processes: They can reinforce or conflict with each other. AI can help address GW by educating society about climate change and about the corrective steps needed. Simultaneously, GW can benefit the AI evolution process by providing job openings to replace the ones that robotics will eliminate. In fact, building the green infrastructure will require more (and better educated) human resources than the one replaced by robotization.

First, I briefly describe what will happen if these processes remain uncontrolled.

Suppose GW and AI processes are left uncontrolled. In the wrong hands, AI could be used to discredit the scientific findings about GW and introduce the culture of the "alternative truth," a culture in which spreading false propaganda is unchecked and tolerated, and thereby a "truthless, lie-tolerant" age evolves. If that happens, AI can, for example, help the fossil-fuel lobbies in their propaganda campaigns.

The AI and GW processes differ from industrial processes because their time constants are much longer, sometimes decades or generations. For humans, this is a significant difference because the mental time frames of even our well-intentioned leaders are too short to make them good captains of the ship of human evolution. It is hard for political leaders to serve the needs of future generations when that conflicts with the immediate needs of their present constituents. It seems that we need to establish global institutions and enforcement structures that would establish such rules—as does the US Constitution—so the acts of today's generation consider the needs of future generations.

5.2 Evolution of Artificial Intelligence

Regarding the evolution of AI, we must fully understand that, today, this evolution is out of control and must be changed. The controls needed are very simple, but their implementation is not. All we need is to keep this evolution on the road toward serving what it should and not what it can. We must realize that AI is just a tool, like a knife, the use of which must be controlled. A knife is useful in the hands of a cook but a danger in the hands of a bank robber. In other words, although we must make sure that AI "stays in the kitchen," we must also realize that we should support the advance of its capabilities while curtailing its applications.

For example, AI should be allowed to serve education and communications, but it must not be allowed to help spread false information or propaganda. AI must not be allowed to become an antiscience propaganda tool; it must not be allowed to become a means of destroying our privacy, spreading false information, brainwashing, or even destroying democracy itself.

As seen in the Ukraine war of 2022, AI-based cyberterrorism in the wrong hands can make us vulnerable to cybercriminals attacking our critical infrastructure, including electric grids, energy pipelines, water supplies, mass transit systems, and nuclear power plants. The 1000 hackers at the Russian Foreign Intelligence Service (SVR) spread false propaganda to attack democracies around the world, to create a truthless environment, and in the case of the 2021 SolarWinds attack, break into government and private servers. In addition, cybercriminals have used AI in a wide range of illegal activities, such as obtaining millions of dollars in ransom payments (as in the 2021 DarkSide attack on the Colonial Pipeline) and attacking individuals using autonomous AI drones. In short, it seems to me that the time has arrived for a digital "Geneva Convention" to establish rules and enforcement mechanisms to limit the allowable applications of AI.

In the first chapter of this book, I showed that AI can be very effective in modeling the dynamics of GW, determining the scope and speed at which its consequences will evolve if inaction continues. AI can also serve in the design and production of the components of the new technological infrastructure that must be built. AI can educate the public about scientific facts and help change our culture. It can show that the impact of humans on our planet exceeds what Earth can tolerate and that globalized action is needed because it is an ethical necessity to care not only about our individual or national self-interests but also those of future generations. Finally, AI can provide sophisticated model-based control (MBC) for operating our new technology and free up the labor force needed to implement the conversion to green energy.

5.3 The GW Process

During the past century, the global average temperature has increased by 7% (from 14.0°C [57.2°F] in about 1880 to approximately 15.2°C [59.4°F] in 2022). GW affects the oceans and the atmosphere, which respond at different rates. This is because the mass of the oceans is a couple of hundred times greater than that of the atmosphere. As a result, the response of atmospheric processes is faster than that of the oceanic ones when GW changes the global heat balance. Consequently, in locations such as the Arctic, the atmosphere has already warmed by some 4 to 5°C (7 to 9°F). GW is causing melting in the Arctic, while elsewhere severe droughts, wildfires, heat waves, water shortages, hurricanes, flooding, and the destruction of wildlife are in progress. In contrast, the changing of oceanic processes, such as the increased thermal expansion of ocean waters or alterations of ocean currents, El Niño cycles, or the Pacific Decadal Oscillation (PDO), is just beginning.

One hundred years ago, the gain (rate of rise) of the GW process was about 0.15°C (0.27°F) per decade; in 2021, it was about 0.25°C (0.45°F) per decade; and if inaction continues, it might reach 0.5°C (0.9°F) per decade by 2030. The inertia of this process is very large because of the immense masses involved and the greenhouse effect of the gases that have already accumulated in the atmosphere. Therefore, even after net-zero conditions are reached (when the carbon dioxide [CO_2] concentration in the air is no longer increasing but has not yet started to fall), still more heat will be reaching the planet than leaving it.

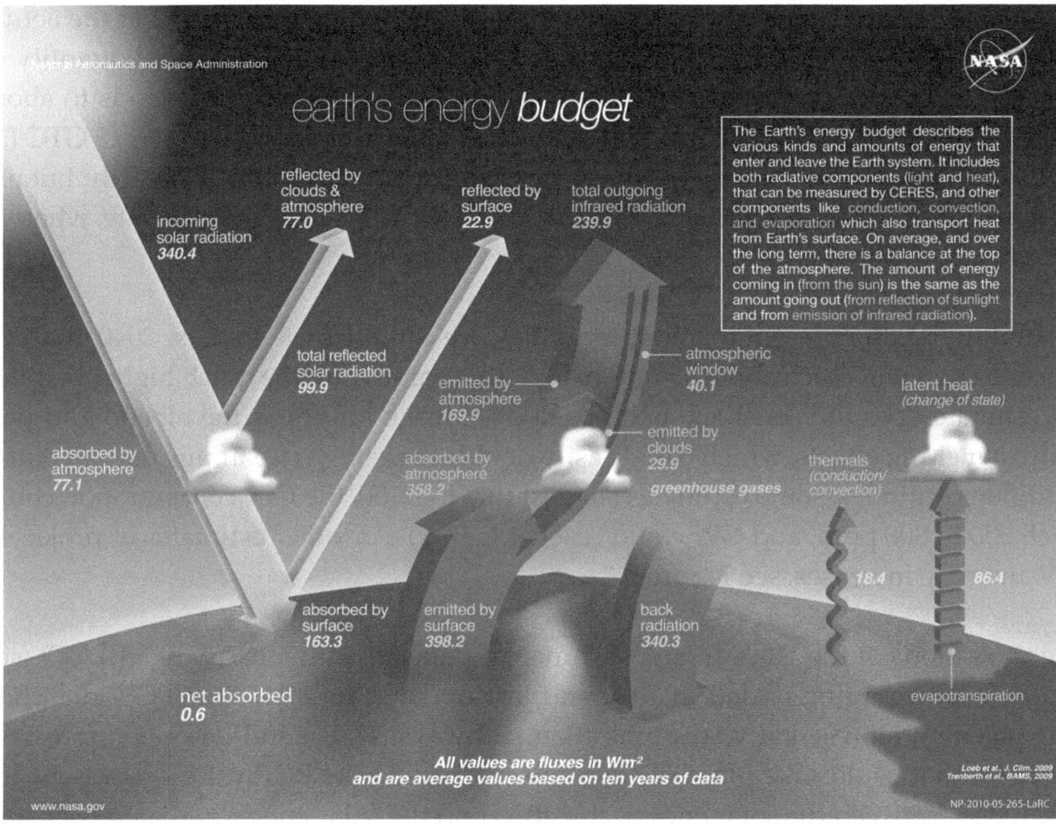

Figure 5-1. Earth's energy budget.

Source: "The-NASA-Earth's-Energy-Budget-Poster-Radiant-Energy-System-satellite-infrared-radiation-fluxes.jpg" by NASA, Public Domain, April 20, 2014, https://commons.wikimedia.org/w/index.php?curid=32285340.

Changes in Earth's thermal equilibrium cause temperatures to rise or fall. This imbalance between the heat received and the heat leaving is called climate forcing or radiative forcing (RF). Positive RF means that Earth receives more incoming energy from sunlight than it radiates back into space (Figure 5-1). This net gain of energy causes GW. Today, the total radiation energy received from the Sun is more than 340 W/m^2, and the RF caused by human activity is nearly 1.7 W/m^2.

During the industrial age, climate forcing increased, and it will continue to exist for decades, even when humankind has fully converted to an energy economy with net-zero carbon emissions. This is because, even though a net-zero carbon energy economy does not increase the greenhouse gas (GHG) concentration in the atmosphere, it does not reduce the amount of GHGs that are already there. After considering the time constants of the GW process, let us now examine the rising trend in CO_2 concentration. As of this writing, the CO_2 concentration in the air exceeds 410 ppm. The present rate at which it is rising is over 2.0 ppm/year.[1,2] (Some reports indicate that the actual increase between 2018 and 2019 was as high as 3.14 ppm/year.[3]) The total quantity of carbon in the air is about 880 gigatons of carbon (GTC) (which corresponds to about 3,260 gigatons of CO_2 [$GTCO_2$]), and it is increasing every year by close to 10 GTC (37 $GTCO_2$). (Because of the pandemic, this rate of yearly increase fell temporarily, but not for long.) Of this increase, about 50% is absorbed by the oceans and plant life, whereas the rest remains in the air.

Assuming that the present inaction continues, we can estimate that, by 2040, the CO_2 concentration will reach about 450 ppm. Still, this value could be much higher because the estimate does not consider such positive feedback effects as the albedo and permafrost melting effects, which could easily double the rate of rise. If this out-of-control condition continues, some estimates suggest that by 2100, the CO_2 concentration could reach 600 to 800 ppm, and GW could reach 4°C (7°F). For other estimates of projected GW anomaly increases, see Figure 5-2a.

The projections of the UN's Intergovernmental Panel on Climate Change (IPCC) (see Figure 5-2c) were first issued in 2000 and last updated in 2014. The IPCC identified four potential representative concentration pathways (RCPs) and described four GW scenarios that could take place during this century. These RCP values represent the differences between the solar energy absorbed and the solar energy radiated back into space per square meter of the Earth's surface in 2100. They are based on assumed combinations of future population growth, per capita energy use, energy source mix, and other trends.[4] The next update of these projections is expected in 2022.

---

1. Over the past decade, however, the growth rate has been closer to 2.3 ppm per year. *Source:* Rebecca Lindsey, "Climate Change: Atmospheric Carbon Dioxide," NOAA Climate.gov website, August 14, 2020, https://www.climate.gov/news-features/understanding-climate/climate-change-atmospheric-carbon-dioxide.

2. Overall, the annual average CO_2 concentration for 2021 was estimated to be 416.3 ppm (±0.6). With an annual increase of about 2.5 ppm—even in 2020, when emissions fell sharply because of the economic impacts of the COVID-19 pandemic—2022 will be the first year in which the annual average CO_2 concentration will be 50% above preindustrial levels. Richard Betts, "Met Office: Atmospheric CO_2 now hitting 50% higher than preindustrial levels," World Economic Forum website, March 22, 2021, https://www.weforum.org/agenda/2021/03/met-office-atmospheric-co2-industrial-levels-environment-climate-change/.

3. "COP 21 | 2015 Paris Climate Talks," CO_2 Earth website, last updated December 20, 2015, accessed November 1, 2022, https://www.co2.earth/cop-21-2015-paris-climate-talks.

4. For more details, see the following web pages: "Representative Concentration Pathway," Wikipedia, https://en.wikipedia.org/wiki/Representative_Concentration_Pathway. David Furphy, "What on earth is an RCP?," September 29, 2013, accessed November 1, 2022, https://medium.com/@davidfurphy/what-on-earth-is-an-rcp-bbb206ddee26.

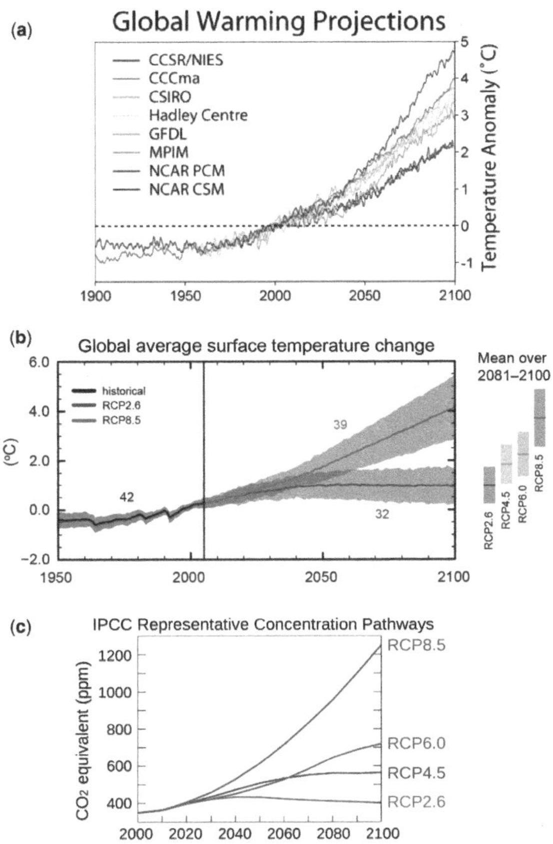

Figure 5-2. GW trend projections for RCP scenarios for different RF values (2.6, 4.5, 6.0, and 8.5 W/m²) that could be reached by 2100 under different conditions of population growth, economic growth, climate policies, and so on. The projected parameters include (a) the temperature anomaly (total GW), (b) the average global surface temperature anomaly, and (c) the CO_2 equivalent concentration of all GHGs. For the identities of the research organizations in Figure 5-2a, see footnote.[5]

Source of Figure 5-2a: "Global Warming Predictions.png" by Robert A. Rohde for the Global Warming Art Project is licensed under CC BY-SA 3.0, April 10, 2006, https://commons.wikimedia.org/w/index.php?curid=694325.

Source of Figure 5-2b: Figure SPM.7 (Panel a) from IPCC, 2013: Summary for Policymakers, in Climate Change 2013: The Physical Science Basis. Contribution of Working Group I to the Fifth Assessment Report of the Intergovernmental Panel on Climate Change, T. F. Stocker, D. Qin, G.-K. Plattner, M. Tignor, S. K. Allen, J. Boschung, A. Nauels, Y. Xia, V. Bex and P. M. Midgley (eds.). Cambridge University Press, Cambridge, United Kingdom, 1–30, DOI:10.1017/CBO9781107415324.004.

Source of Figure 5-2c: "All forcing agents CO_2 equivalent concentration.svg" by Efbrazil, March 7, 2020, is licensed under CC BY-SA 4.0, https://commons.wikimedia.org/w/index.php?curid=87801257.

5. Names of research organizations and models cited in Figure 5-2a: CCSR/NIES,Centre for Climate System Research/National Institute for Environmental Studies (Japan); CCCma, Canadian Centre for Climate Modelling and Analysis (Canada); CSIRO, Commonwealth Scientific and Industrial Research Organisation (Australia); Hadley Centre, Hadley Centre for Climate Prediction and Research (United Kingdom); GFDL, Geophysical Fluid Dynamics Laboratory (United States); MPIM, Max Planck Institute für Meteorologie; NCAR, National Center for Atmospheric Research (United States); PCM, parallel climate model; and CSM, climate system model.

The four curves in Figure 5-2c are based on four projected RF values that could exist at the end of this century, namely, 2.6, 4.5, 6.0, and 8.5 W/m². The lower values are based on more optimistic assumptions about the future, whereas the higher values are based on less optimistic assumptions. For example, RCP2.6 assumes that RF will start dropping in 2025, which seems over-optimistic.

The higher the RF values are, the higher the GW is projected to be by 2100. For example, scenario RCP8.5 is based on the assumption that RF will reach 8.5 W/m² by 2100, which would make the planet uninhabitable. (Remember that, during the past 65 million years, only the asteroid impact in the Yucatán Peninsula caused such conditions!) Such scenarios are unthinkable, as they would result in the destruction of human civilization.

Figure 5-3 shows the amount of RF caused by the increasing concentration of GHGs in Earth's atmosphere as reported by the US Environmental Protection Agency (EPA). This forcing represents the size of the energy imbalance in the atmosphere. For example, CO_2 contributes an average RF value of 1.66 W/m² with a range between 1.49 and 1.83 W/m², and methane has an average RF value of 0.48 W/m² with a range between 0.43 and 0.53 W/m². Still, recent measurements indicate that the contribution of methane could be much higher.

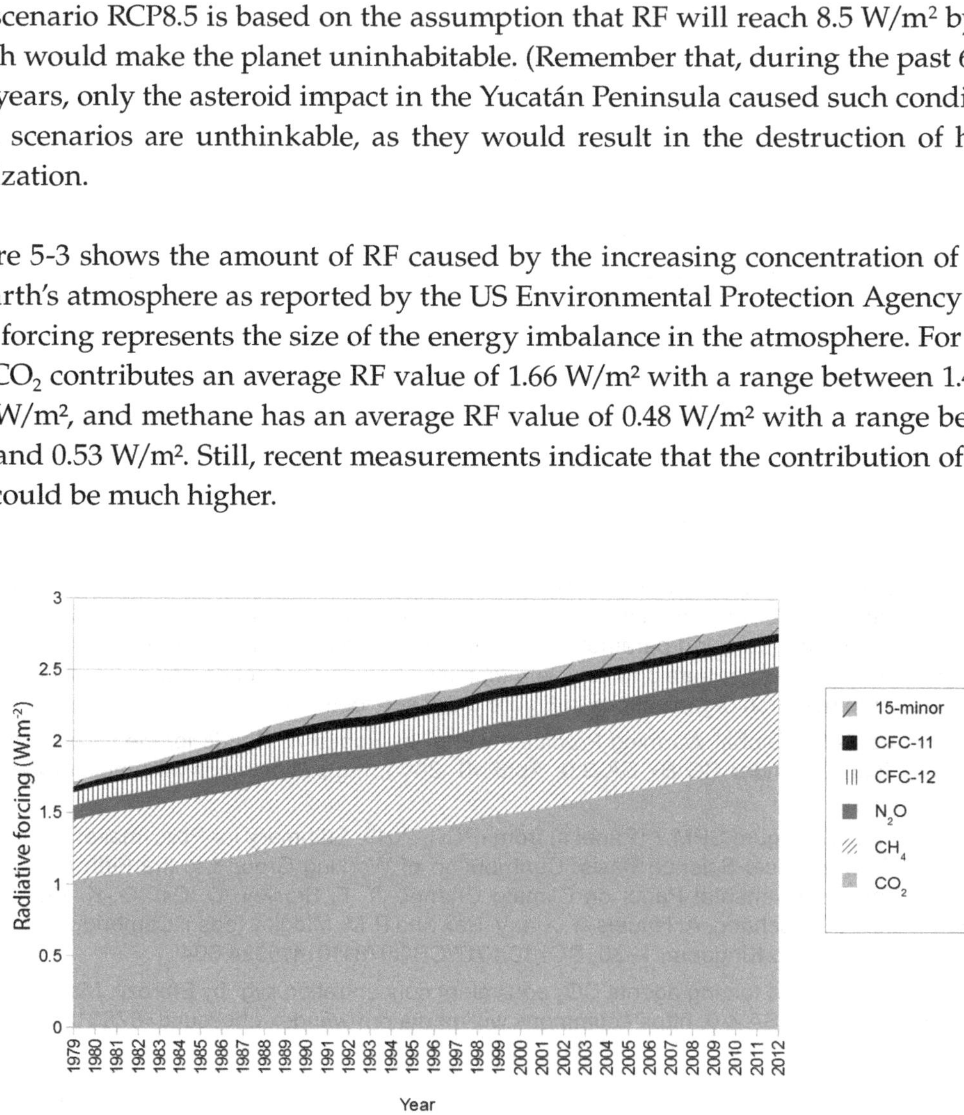

Figure 5-3. Changes in the RF of long-lived GHGs between 1979 and 2012.

Source: "Changes in radiative forcing of long-lived greenhouse gases between 1979-2012.png" by Enescot, August 4, 2013, is licensed under Creative Commons CC0 1.0 Universal Public Domain, https://commons.wikimedia.org/w/index.php?curid=28685086.

The estimated total of proven fossil energy reserves on Earth is some 750 GTC. If we continue burning these reserves at the present rate of about 10 GTyC—50% of which stays in the atmosphere and 50% of which is dissolved in oceans or absorbed by vegetation—they could run out in about a century. The estimates of the size of these reserves vary, as do the projected rates of their consumption. Most estimates indicate that if consumption remained at today's rate, oil and gas reserves could suffice for 30 to 40 years, whereas coal could last for 100 to 150 years.

5.3.1 Reducing Heating, Increasing Cooling

In addition to having the means to reduce warming, we also have the means to increase cooling on Earth. One well-understood method of reducing heating is to reduce energy consumption, which decreases the emissions of GHGs. For example, we saw how the economic slowdown caused by the coronavirus outbreak temporarily reduced carbon emissions. We also know that the same effect results from using more bicycles and public transport, converting to electric cars, insulating our homes, and using smart thermostats and appliances. In addition, we can use drought-resistant seeds; reduce food waste and meat consumption (the carbon emissions associated with the production of 1 kg of pork meat amount to 200 kg of CO_2); add solar-radiation-collecting surfaces; introduce carbon taxes (not a cap-and-trade system, but taxes); and eventually, convert our energy economy from a fossil-nuclear-based system to a carbon-free system.

Although we do not currently have the technology to remove the carbon that has already accumulated in the atmosphere, we can remove some of it by planting trees. However, that approach takes time and would not produce a significant change because trees absorb CO_2 only when growing and release it when they die and rot. Another much more effective approach is to make our roofs and road surfaces from more reflective (lighter in color) material so that they will reflect more of the solar radiation back into space. We can also reduce the human footprint by making agricultural surfaces lighter or revising agricultural practices such as tillage and plowing.[6]

The speed of conversion will be a function of both market forces and government support. Where both are present, the conversion will be faster (for example, in California today, green electricity use is already 30% of the total). In contrast, where only the

---

6. It has been estimated that, to date, approximately 55 GTC has been lost from the soil. (Remember that the yearly total carbon emissions today amount to about 10 GTC.)

marketplace is driving the conversion, it will be much slower (about 15% in the United States overall).

Naturally, even after we have balanced our emissions and increased the removal of GHGs from the atmosphere, we will still not have returned the planet to its preindustrial state because, even after this tremendous technological and political transformation effort, we will still have the already-accumulated carbon in the air. As long as that carbon remains, the planet will continue to warm until a new heat balance is reached. GW is a self-stabilizing process in that, as the planet warms, its heat loss to outer space also increases. Eventually, this loss rebalances the incoming and outgoing quantities of energy. Unfortunately, this is not an option for us because the new balance would occur at a temperature that is too high for human civilization to survive.

As already mentioned, in addition to reducing the heating of the planet, we can also cool it. Cooling can be increased by using more reflective roofs, roads, and agriculture.[7] As of today, this option is not yet being exploited. In fact, instead of rising, the global reflectivity is dropping because the highly reflective snow and ice are melting and are being replaced by highly heat-absorbing water and land. As of this writing, the Earth's reflectivity is dropping, which is a self-accelerating process because melting increases GW and GW further increases melting.

5.3.2 Warming and RF

The average of the total surface temperature (land and oceans combined) of the planet between 1951 and 1980 was about 14.0°C (57.2°F). As of May 21, 2021, the US National Oceanic and Atmospheric Administration (NOAA) reported the global average temperature to be[8] 15.6°C (60.1°F). (Other sources reported somewhat lower values.) Today, the average land surface temperature is about 10°C (50°F), and the average temperature of the ocean surfaces is approximately 17°C (62.6°F). The land surface temperature is rising at about 0.3°C (0.5°F) per decade; the oceans' surface temperature is rising at about 0.1°C (0.18°F) per decade. Currently, the average rise of the Earth's total surface is about 0.25°C (0.45°F) per decade, and the increase is accelerating (because of the albedo, permafrost, forest clearing, peat and forest fires, and ocean degassing).

---

7. If it is correct that humankind can influence the reflectivity of 5% of the planet's surface and if we increased the reflectivity of that 5% from about 30% to about 32%, that would remove enough heat to eliminate GW.

8. "May 2021 Global Climate Report," NOAA National Centers for Environmental Information, June 2021, accessed November 1, 2022, https://www.ncdc.noaa.gov/sotc/global/202105.

Therefore, I estimate that the global temperature rise will reach 1.5°C (2.7°F) by 2030 or sooner.

About half of the carbon emitted by humankind stays in the atmosphere; the other half is absorbed by the oceans and by the growth of plants and forests. During the industrial age, the amount of carbon in the air increased from about 600 to 880 GTC, and we are adding another 10 GTC to it every year. This corresponds to *daily* carbon emissions of about 3.5 kg (~10 lb) per person on the planet. In addition, the global energy demand is rising as the economic catch-up of developing countries continues. The EIA predicts a 39% increase in the use of fossil fuels by 2040.[9] The CO_2 content of the air (which for a million years did not exceed 280 ppm) is approaching 420 ppm today[10] and is still rising.

According to the fourth report of the IPCC,[11] since 1975, the RF (or climate forcing) has been about 1.6 W/m^2, meaning that 1.6 W more solar energy was absorbed per square meter on Earth's surface than was radiated back into space. Other sources use even higher values (Figure 5-4). We do not yet know exactly what the RF was in 2022, but we know that about two-thirds of the RF is caused by the increased concentration of CO_2 (Figure 5.2c). We also know that this concentration is rising. Based on these and other facts discussed in the previous chapters, my conclusion is that the 1.5°C limit of the Paris Agreement cannot be kept.

5.3.3 Dynamic Characteristics of Warming

I find the treaties and future projection scenarios to be little more than guesses and, as such, rather useless. Therefore, I focus on the factual determinations already made in the previous chapters. My goal is to identify the time constants, dead times, interactions, and other characteristics of this lag-dominant and high-capacity process because that is what we need to approximate the future global temperature values. Specifically, I estimate the values this temperature would reach at various times in the future if left uncontrolled and describe the steps that must be taken to limit GW to various maximum values.

---

9. Stephen Eule, "Future Global Energy Demand: Up, Up, and Away!," Global Energy Institute website, May 18, 2016, accessed November 1, 2022, https://www.globalenergyinstitute.org/future-global-energy-demand-and-away.

10. "Trends in Atmospheric Carbon Dioxide," NOAA Global Monitoring Laboratory website, accessed November 1, 2022, https://gml.noaa.gov/ccgg/trends/.

11. S. Solomon, D. Qin, M. Manning, Z. Chen, M. Marquis, K. B. Averyt, M. Tignor, and H. L. Miller, eds., *Climate Change 2007: The Physical Science Basis.* Contribution of Working Group I to the *Fourth Assessment Report* of the Intergovernmental Panel on Climate Change (Cambridge, UK: Cambridge University Press, 2007).

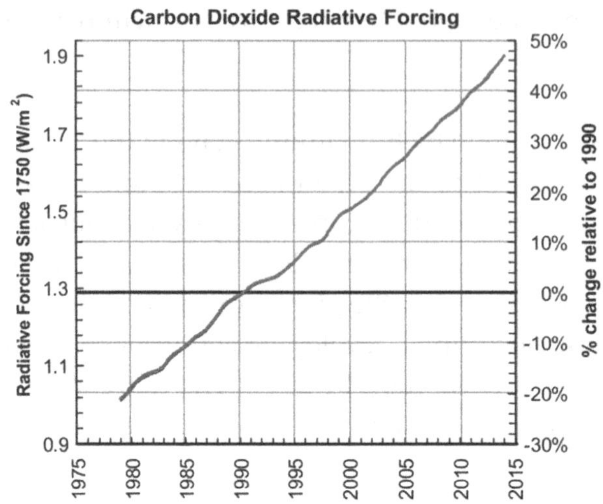

Figure 5-4. According to the IPCC, climate forcing between 1980 and 2015 increased by about 0.8 W/m^2 and reached a total of 1.9 W/m^2 in 2015. This is a 50% change from January 1, 1990, to 2015. During the past 5 years, these values increased further.[12]

Source: "Carbon Dioxide radiative forcing.png" by James Butler, NOAA, Public Domain, 2014, https://en.wikipedia.org/w/index.php?curid=39537281.

Let me first illustrate how slow the response of GW is to changes in RF.[13] If we use the RF in 1750 as the zero reference (RF = 0), according to the Annual Greenhouse Gas Index (AGGI)[14] between 1979 and 2014, the global RF caused by CO_2 alone was 1.027 W/m^2 in 1979 and 1.985 W/m^2 in 2014.

The total forcing for 2018 was 3.101 W/m^2, which corresponds to 2.48°C (4.46°F) if we multiply it by the commonly accepted climate sensitivity parameter[15] of 0.8. This value is much greater than the observed increase, which was about 1.2°C (2.2°F). Part of the reason for this difference is probably the lag (dead time) that exists in the response of the global temperature to the new value of the forcing.

---

12. Antero Ollila, "Challenging the Greenhouse Effect Specification and the Climate Sensitivity of the IPCC," *Physical Science International Journal*, 22, no. 2 (2019): article PSIJ.49095, Figure 6, https://www.researchgate.net/figure/Radiative-Forcing-RF-curves-of-carbon-dioxide-according-to-different-research-studies_fig4_333355598.
13. For more information, see: "Earth's energy budget," Wikipedia, accessed November 1, 2022, https://en.wikipedia.org/wiki/Earth%27s_energy_budget.
14. For more information, see: "Radiative forcing," Wikipedia, accessed November 1, 2022, https://en.wikipedia.org/wiki/Radiative_forcing.
15. "Radiative forcing," Wikipedia.

5.3.3.1 Variable-Gain Process

The dynamic response of the atmospheric temperature to the same rise in RF is quite different from the response of the temperature of the oceans. These reaction curves differ because, in the case of the atmosphere, the dead times are shorter and the gains are greater due to the relatively small mass of the atmosphere compared to that of the oceans. This explains why we already see the symptoms of GW in atmospheric processes (e.g., hurricanes, storm surges, heat waves, wildfires, and droughts).

The gains of the atmospheric and oceanic processes rise when self-intensifying positive feedback processes, such as ocean degassing, permafrost melting, and albedo changes, are triggered. Today, the GW process is uncontrolled. Its gain (rate of rise) increases and accelerates when the tipping points of the positive feedback subprocesses are triggered. The relationship between the GHG concentration and its impact on RF is shown in Figure 5-5. Based on that relationship, we can approximate that each doubling of the GHG concentration causes an increase of about 3.05 W/m^2 in the RF.

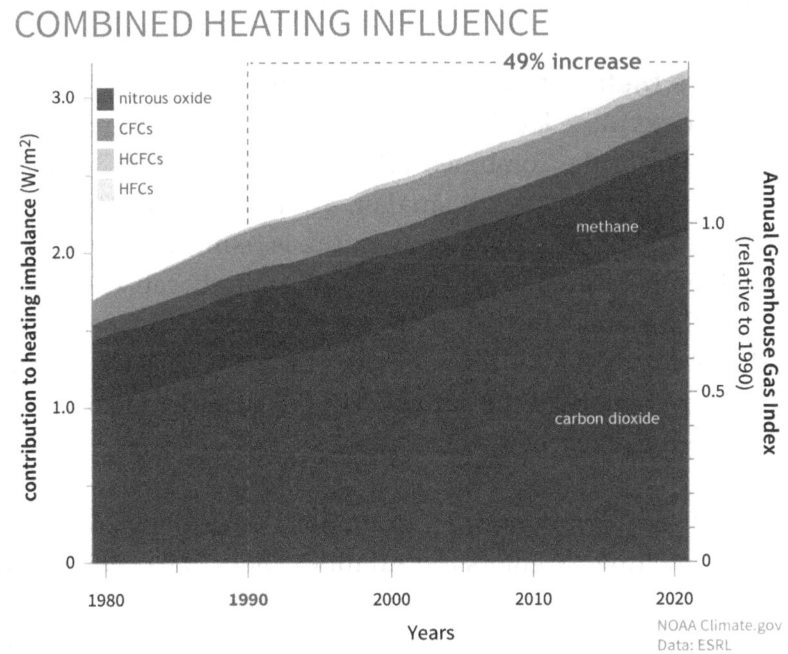

Figure 5-5. During the 40 years between 1980 and 2020, a 49% increase in GHG concentrations caused approximately a 1.4 W/m^2 increase in RF. Legend: Carbon dioxide (gray), methane (dark purple), nitrous oxide (medium purple), chlorofluorocarbons (CFCs, lavender), hydrochlorofluorocarbons (HCFCs, blue), and hydrofluorocarbons (HFCs, light blue).

Source: Rebecca Lindsey, "Climate Change: Annual greenhouse gas index," June 17, 2022, https://www.climate.gov/news-features/understanding-climate/climate-change-annual-greenhouse-gas-index.

The gain of the GW process is the ratio of the change in the global temperature and the change in CO_2 concentration that caused it. Figure 2-7 (Chapter 2) shows that GW reached about 0.9°C (1.6°F) between 1960 and 2010. As I explained elsewhere,[16] that rise was caused by an increase in the atmospheric CO_2 concentration from 310 to 390 ppm. Thus, the average rate of increase in GW was 1.12°C (2.02°F) per 100 ppm. In terms of temperature, the rate of increase was about 0.17°C (0.31°F) per decade. Today, that rate (the gain of the GW process) is about 0.25°C (0.45°F) per decade, and some estimates suggest that it is still rising.

5.3.3.2 Error Accumulation (Integral)

If we consider the error in the GW loop as the extra carbon accumulation in the air during the industrial age, that quantity is about 280 GT and it is still rising because we are adding 10 GT to it yearly. From a process control perspective, we would call this "error accumulation." Even if we reach a carbon-neutral lifestyle and the quantity of carbon in the air stops rising, the extra carbon that accumulated in the atmosphere during the industrial age will remain, and both the corresponding RF and the global temperature will continue to increase, only at a lower rate.

5.3.3.3 Temperature Distribution and Ice Melting

Local temperatures determine local conditions. Today, the average GW is about 1.2°C (2.2°F) which results from the combined effects of about a 2.0°C (3.6°F) temperature rise on the land surfaces and a much smaller, around 0.5°C (0.9°F) rise in the ocean surface temperature. Today, the global temperature is rising at about 0.25°C (0.45°F) per decade. Still, local temperatures vary more dramatically because of phenomena such as the reduction in the albedo, the degassing of the oceans, the melting of the permafrost, forest and peat fire effects, and so on. The average GW in the Arctic has reached about 4°C (7.2°F) or even more in the summer. In the summer of 2020, at Verkhoyansk in Siberia, the temperature reached 38°C (100°F).

Another example of local temperature variations is Greenland, where the relatively cold water from the melting ice enters the Atlantic. There, the ocean is cooling, which, according to some theories, could be slowing down the Gulf Current. Globally, the increase in sea levels is slow, 3 to 4 mm yearly (probably not more than 1 m by the end of this century). However, hurricane-caused storm surges and rainstorms cause coastal flooding. In other words, the impact of the melting rate on the level of the

16. Béla Lipták, "Compressor rod position detection. And how quickly we need to bring global warming under control," *Ask the Experts Blog*, February 22, 2019, https://www.controlglobal.com/measure/level/article/11303634/compressor-rod-position-detection.

oceans is small, but if, for example, all the ice on Greenland (2.5 million cubic kilometers) melted, the ocean levels could rise by about 7.2 m (24 ft).[17]

5.3.4 Set Point of the GW Control Loop

If we look at the task of limiting GW as if we were configuring a control loop, the first feature we would need to define would be the set point of that loop. Actually, this controller would have two set points. The first would be used during the transition from the present fossil-nuclear energy economy to a green energy economy. This set point would be used until the conversion was 100% complete. The second set point would be used after that point, when the carbon content of the air was no longer rising but GW was still rising because of the accumulated carbon in the air and the albedo effect of the melting permafrost.

Regarding the first set point, the time available to convert to a net-zero-carbon energy economy is shorter than the various models suggest, and the efforts made do not even aim for that goal. Discussions focus on target dates to reduce emissions by different percentages or to reach net zero by specific dates, but there are no global commitments. According to the UN,[18] to limit GW to 1.5°C (2.7°F), GHG emissions need to be cut to reach 55% of net-zero conditions by 2030. The IPCC indicated that global CO_2 emissions should reach net zero by 2050 to limit warming to 1.5°C (2.7°F), but the European Union leaders did not commit[19] to that goal. In fact, instead of eliminating or reducing the burning of fossil fuels, the EIA predicts an *increase* in fossil fuel use. That prediction is reinforced by the impact of Russia's energy war, which is forcing even Germany to open coal power plants temporarily.

In addition, instead of falling, carbon emissions reached new records in 2019,[20] mainly because of the unprecedented burning of forests and peat, which not only dumped large amounts of CO_2 and heat into the air but also reduced the size of these carbon-absorbing "lungs" of the planet (Figure 5-6). In 2020, because of the pandemic, CO_2

---

17. For further information, see: "Greenland ice sheet," Wikipedia, accessed November 1, 2022, https://en.wikipedia.org/wiki/Greenland_ice_sheet. Lorin Hancock, "Why are glaciers and sea ice melting?," World Wildlife Fund website, https://www.worldwildlife.org/pages/why-are-glaciers-and-sea-ice-melting.

18. Eillie Anzilotti, "Global emissions must drop 55% by 2030 to meet climate goals," Fast Company website, November 27, 2018, accessed November 1, 2022, https://www.fastcompany.com/90272330/global-emissions-must-drop-55-by-2030-to-meet-climate-goals.

19. Martina Larkin, "How can the EU hit net-zero emissions?," World Economic Forum website, July 18, 2019, https://www.weforum.org/agenda/2019/07/how-can-the-eu-hit-net-zero-emissions/.

20. Climate research institute CICERO based in Oslo, Norway, compiles a report each year on the amount of heat-trapping GHGs emitted into the planet's atmosphere by humans. In 2019, CICERO's *Global Carbon Budget* report indicated that CO_2 emissions continued growing during that year, although by a lower-than-usual amount of about 0.6%, reaching a new record high.

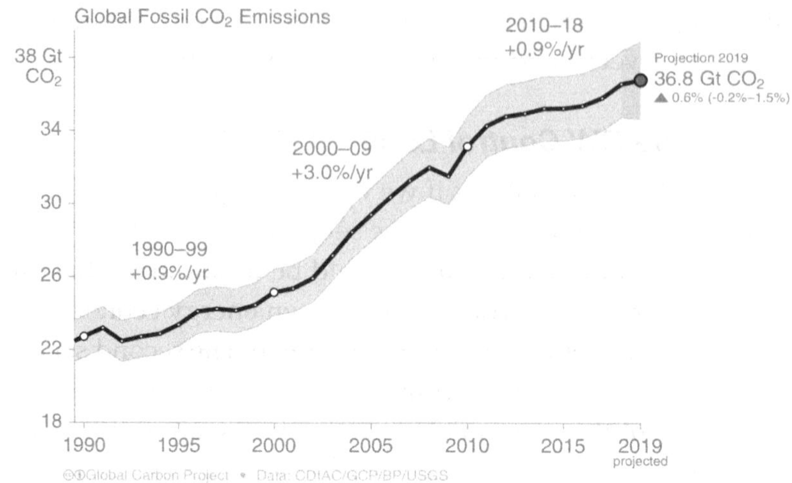

Figure 5-6. CO_2 emissions continued to rise in 2019. The data here is in gigatons of CO_2 per year ($GTyCO_2$). One GTC is equal to 3.7 $GTCO_2$ because of the molecular weight ratio of CO_2 to carbon (44/12 = 3.7). Emissions fell by 5% in 2020 because of the pandemic but rebounded in 2021 and 2022.

Source: Global Carbon Project, Global Carbon Budget 2019, Slide 11, https://folk.universitetetioslo.no/roberan/img/GCB2019/PNG/s11_2019_FossilFuel_and_Cement_emissions_1990.png. Used with permission of the Global Carbon Project under the Creative Commons Attribution 4.0 International license.[21]

emissions fell by approximately 5%[22] (mainly because of reduced fossil energy use in transportation), but by 2022 they continued to rise.

5.3.5 Manipulated Variables of the GW Loop

On February 28, 2022, the contribution of Working Group II to the IPCC *Sixth Assessment Report* was released. Among its findings are that "nearly half the world population [is] living 'in contexts that are highly vulnerable to climate change'... [and] that climate change has exacerbated food and water insecurity, extreme weather disasters, declines in people's physical and mental health, premature deaths, species loss and extinctions, and vector-borne diseases in regions around the world."[23]

Let us consider what it would take to achieve even the small emission cuts that are contemplated. It would require a change in human behavior, and achieving that requires

---

21. "Global Energy Review: CO_2 Emissions in 2020," IEA, March 2, 2021, https://www.iea.org/articles/global-energy-review-co2-emissions-in-2020.

22. For additional information, see: Mark Kaufman, "World's carbon emissions grew in 2019 to their highest levels ever," Mashable Science website, December 3, 2019, accessed July 27, 2022, https://mashable.com/article/carbon-emissions-grow-2019-climate-change/.

23. Dana Nuccitelli, "New IPCC report highlights urgency of climate change impacts," *Yale Climate Connections*, February 28, 2022, accessed March 4, 2022, https://yaleclimateconnections.org/2022/02/new-ipcc-report-highlights-urgency-of-climate-change-impacts/.

changing our old tools—such as education, science, and cultural and ethical arguments—because they have all failed. Mother Nature is already telling us, "Enough!" Yet, all we are doing is adapting to the symptoms of GW instead of eliminating its cause. We are moving roads, drinking water pipelines, electric cables, and entire communities from places that are threatened by wildfires or floods as if adapting to them would solve anything. Living with these disasters is not a solution, and hoping that humankind will respond to empty words with no action is also unrealistic.

The best way to influence human behavior is through our pocketbooks. The market forces themselves can motivate the conversion to a green energy economy if doing so is profitable. As of this writing, the opposite is the case; the market forces are slowing progress because Russia's energy war simultaneously increased the cost and the demand for fossil fuels. I believe this harm will eventually help because the high energy prices will make using more green energy more profitable. Naturally, carbon taxes could speed up this transition even more.

These taxes should include cross-border import taxes and motivations for industries to cut their emissions by allowing them to sell their cuts to companies that exceed their allowed limits. All these taxes could go directly into a dedicated fund that can be used only to reward emission reductions, subsidize green conversion, and finance the reeducation of the workforce so that, as fossil-fuel-related jobs disappear, the unemployed can be hired to carry out the immense transformation that is coming. Naturally, this effort will require the total support of society, but that support can be gained if it is in our financial interest. The Marshall Plan at the end of World War II rebuilt Europe and created an economic boom for decades. The Green Marshall Plan can do even more; it can create the longest global economic boom ever.

The costs of both action and inaction are in the trillions of dollars, but whereas action to solve the problem of GW brings jobs and a better future, inaction brings disaster. Regarding the costs of solving the GW problem, the investment required to reach net-zero carbon emissions is estimated to range between 1.5% and 4% of GDP yearly.[24] In comparison, Moody's Analytics has estimated the global cost of GW by 2100 to be $69 trillion (in 2019, the entire prepandemic world economy was $88 trillion). Moreover, if the current warming continues, other studies estimate that the global GDP will fall by 7% to 23% by the end of this century.

---

24. Julia Rosen, "The Science of Climate Change Explained: Facts, Evidence and Proof," *New York Times*, published April 19, 2021, updated November 6, 2021, accessed November 1, 2022, https://www.nytimes.com/article/climate-change-global-warming-faq.html.

In contrast to inaction, converting our energy economy will create hundreds of millions of jobs and require a tremendous retraining effort during this decade. During this same period, AI-supported robotization will free the required workforce that will be needed to fill the jobs created by the conversion.

After this first phase of action (after reaching the state of zero carbon emissions), a second phase must be initiated to decrease the amount of carbon in the atmosphere to near preindustrial levels and to stabilize the global temperature low enough to reestablish acceptable living conditions on Earth. Today, we do not have the technology to remove gigantic quantities of CO_2 from the air, but we do have the means of cooling the planet by increasing its albedo. This can be done by making our roofs, roads, and agriculture more reflective (lighter in color to reflect more solar radiation into space).

5.4 We Can Fix Things

There are many ways to modify our lifestyle to decrease our per capita emissions. I have already mentioned some of these steps, such as using mass transit, changing agricultural practices (reducing tillage, increasing the use of cover crops), reducing food waste and meat consumption, improving building insulation, and adding solar-collecting rolls to the outside surfaces of buildings. Figure 5-7 shows the contributions of the different economic sectors to total GHG emissions in terms of both direct and indirect (through fossil-fuel-based electricity or heating) emissions.

Even though GW is currently out of control, 20 years is a long time, and human ingenuity, science, and technology can achieve wonders similar to those created in past decades. I am optimistic not only because an estimated 1.6 million students from 125

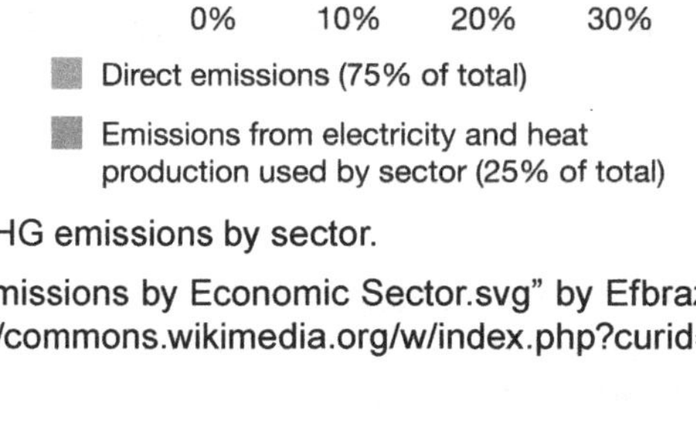

Figure 5-7. Percentage of GHG emissions by sector.

Source: "Greenhouse Gas Emissions by Economic Sector.svg" by Efbrazil, March 9, 2020, is licensed under CC BY-SA 4.0, https://commons.wikimedia.org/w/index.php?curid=87864359. Accessed March 4, 2022.

countries rallied in March 2019 to protest the present inaction but also because our scientists have amassed data that convincingly and accurately determines the size of the threat and the steps we must take to save our planet.

Being able to identify the future consequences of our present actions was the greatest achievement of modern science. In the coming decades, it will also be scientific technology that will double the efficiency of our solar collectors by developing perovskite cells,[25] designing hydrogen-fuel-cell-operated electric cars, and developing reversible fuel cells to serve our energy-independent homes. Most importantly, our technology will convert today's only intermittently available green energy sources into continuously available sources and will design the entire infrastructure for storing, transporting, and distributing them to all the users on the planet.

Today, the global population is increasing by 25 million each year, three billion people go to bed hungry every day, and a similar number would like to migrate from regions that are becoming unlivable. Close to 90% of humankind is inhaling unhealthy, toxic, and sometimes radioactive air. Our per capita daily carbon emissions amount to 3.6 kg, resulting in a global heat input equivalent to five Hiroshima bombs exploding every second (not every minute or every day, but every *second*). Our forests are disappearing, and soon, our oceans will contain more plastic than fish. Still, instead of building global institutions to face our global problems, we build walls, barbed wire fences, and nuclear bombs to protect the status quo.

Once we agree that the steps described in this book and elsewhere must be taken, two additional agreements are needed. One is to agree on our future energy source (I believe that solar-hydrogen energy is the logical option), and the other is to determine how society can be motivated to take the steps needed (I believe that the free market forces, possibly assisted by a carbon tax, can do that). After the conversion is completed, our energy sources will be free, inexhaustible, and clean so that the life of the coming generations will be easier.

5.4.1 Education Is Needed

One cause of the present inaction is the uncertainty about GW-related facts caused by the successful propaganda from the fossil-fuel and nuclear industries. A segment of society is not yet convinced that the need to fight GW is beyond politics and that climate change is real and in progress. Books like this one aim to change this by providing indisputable facts.

25. For more information, see: "Perovskite solar cell," Wikipedia.

Such facts provided in the previous chapters included that the CO_2 concentration in our atmosphere was about 280 ppm at the start of the Industrial Revolution, and today, it is over 410 ppm (an increase of more than 130 ppm). Compared to the preindustrial age, we have accumulated about 280 GT of "extra" carbon in the air (1.0 ppm = 2.15 GTC). As of May 2021,[26] this concentration reached 419.1 ppm.

Model simulations by the IPCC estimate (Figure 5-8) that Earth will warm by between 2°C and 4°C (between 3.6°F and 7.2°F) this century, depending on the growth rate of the economy.[27] Depending on the economic growth rates assumed, these values range from less than 400 ppm to more than 1200 ppm (Figure 5-2c).

During the half-century between 1860 and 2010, the temperature rose at a rate of 0.17°C (0.31°F) per decade, the CO_2 concentration increased at a rate of about 16 ppm per decade, and each 100 ppm increase in CO_2 concentration caused the temperature to

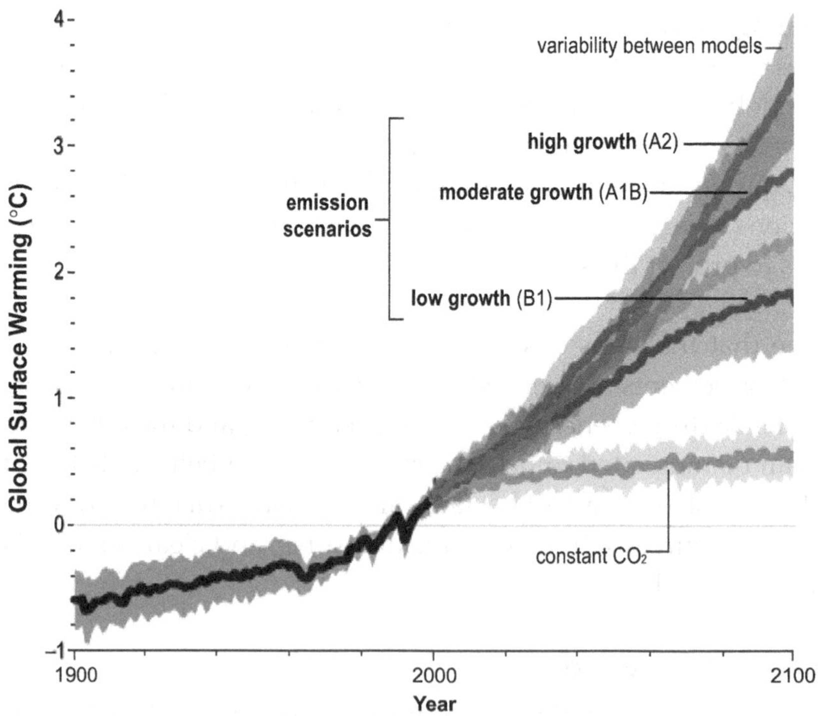

Figure 5-8. The IPCC projections of the increase in GW during the twenty-first century.

Source: Holli Riebeek, "Global Warming," NASA Earth Observatory website, June 3, 2010, https://earthobservatory.nasa.gov/features/GlobalWarming/page1.php.

26. "Trends in Atmospheric Carbon Dioxide," Global Monitoring Laboratory website, accessed February 28, 2022, https://gml.noaa.gov/ccgg/trends/.

27. "How Much More Will Earth Warm?," NASA Earth Observatory website, published June 3, 2010, accessed November 1, 2022, https://earthobservatory.nasa.gov/features/GlobalWarming/page5.php.

rise by about 1.1°C (2.0°F). We know that if GW exceeds 2.0°C (3.6°F), large sections of the planet will become uninhabitable. Yet, today, GW is rising at a rate of about 0.25°C (0.45°F) per decade, and I believe that if the present inaction continues, it might reach 3.0°C (5.4°F) by the end of this century.

Therefore, to stop the warming, additional cooling is needed. That cooling can be obtained by reflecting more solar energy back into space. Earth's average albedo is about 0.3 (30% reflectivity). This albedo is decreasing because the highly reflective snow and ice are melting and being replaced by much less reflective surfaces such as ocean water and permafrost (Figure 5-9).

All of this information should be taught in schools. For example, students could discuss the ways humankind could increase the global albedo by making the human

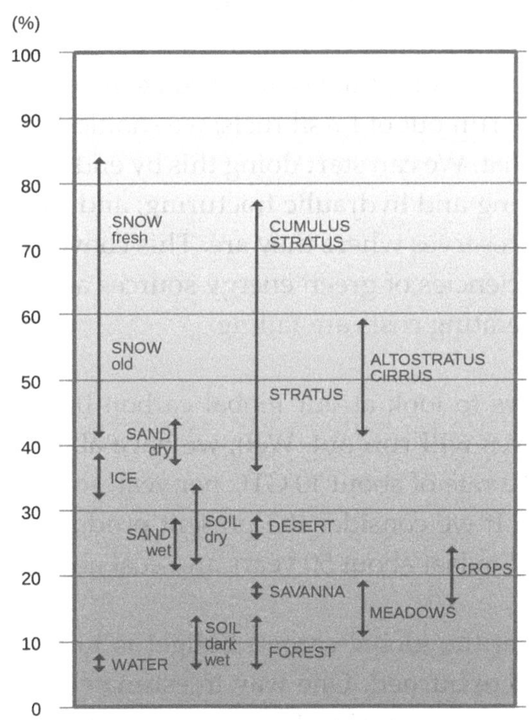

Figure 5-9. Percentages of diffusely reflected sunlight (albedo) as a function of the type of the reflecting surface.[28]

Source: "Albedo-e hg.svg," Hannes Grobe, August 16, 2006, is licensed under CC BY-SA 2.5, https://commons.wikimedia.org/w/index.php?curid=1060378.

---

28. For more information, see the following web pages:
- Albedo," Wikipedia, accessed November 1, 2022, https://en.wikipedia.org/wiki/Albedo.
- Diffuse reflection," Wikipedia, accessed November 1, 2022, https://en.wikipedia.org/wiki/Diffuse_reflection.
- Sunlight," Wikipedia, accessed November 1, 2022, https://en.wikipedia.org/wiki/Sunlight.

footprint "whiter." They could be given homework to calculate how much the albedo of the planet would have to increase to eliminate GW. They could study how the albedo of agriculture could be increased. They could estimate the amount of albedo increase that could be achieved by making our roofs, parking lots, and road surfaces whiter. They could study reports that suggest that the increase in our albedo could fully compensate for the warming effect of the GHGs.[29] In addition to costing less than emissions reductions, such "whitening" can also be done faster, and its equivalent carbon reduction value can be calculated and rewarded by tax rebates.

5.4.2 The Politics of GW

I have touched on the role of politics throughout this book, but because of its importance, I devote a few more paragraphs to it here. Understandably, the fossil-fuel lobby opposes converting to green energy sources. After all, such a conversion would leave $40 to $60 trillion worth of fossil fuels in the ground and under the ocean. But we also know that our Stone Age ancestors switched from using stone to using bronze, not because they had run out of stone, but because bronze made better tools. Similarly, we should not wait until we run out of fossil fuels; we should have the wisdom to switch to green energy before that. We can start doing this by ending fossil-fuel subsidies; terminating deep-sea drilling and hydraulic fracturing; and leaving the Arctic oil, shale gas, tar sands, and coal reserves where they are. This conversion is becoming increasingly feasible as the efficiencies of green energy sources are improving, and their initial investments and operating costs are falling.

There are only two ways to look at our global carbon budget. One is to determine when our proven reserves will run out. Well, we have about 750 GTC in the ground and are consuming it at a rate of about 10 GTC per year, so one might say that we have approximately 75 years. If we consider the present production-to-reserve ratio (P/R), gas and oil are projected to last about 50 years and coal about 125 years.

The other way to look at the global carbon budget is to see what would happen if all the fossil reserves were burned. One way to estimate this is to look at the excess carbon accumulation in the air during the industrial age (880 GTC – 600 GTC = 280 GTC), which has already caused more than 1.0°C (1.8°F) of warming, and compare that temperature rise with what would happen if we emitted another 750 GTC more. The rise would be 750/280 = 2.7°C (4.9°F) on top of the already existing warming of 1.0°C (1.8°F), indicating that GW would reach a minimum of 3.7°C (6.7°F), which would make a large part of the planet either flooded or unlivable because of lack of water.

---

29. "How much will Earth warm?," NASA Earth Observatory website.

Today, the only method contemplated to fight GW is reducing carbon emissions. However, neither the size nor the speed of the planned steps are being met, and even if they were, they are insufficient. The United Nations, the European Union, and the United States are all aiming for carbon neutrality by some date, such as 2040 or 2050 (with the dates for others, such as China and some third-world countries, being even later). They hope to get halfway there by 2030.

A net-zero or carbon-neutral emission rate is reached when we put only as much carbon into the air as is removed by the absorption of the oceans and vegetation. In other words, net-zero emissions are not the same as zero emissions; net-zero emissions are only a state in which the quantity of carbon in the atmosphere is no longer rising. Therefore, in a carbon-neutral state, all the excess carbon that accumulated during the industrial age remains in the air, and GW will keep increasing.

Moreover, while waiting for the effects of these cuts to evolve, other processes will be triggered that will increase (not cut!) emissions further. These positive feedbacks include CO_2 release due to burning vegetation in wildfires, CO_2 release due to the oceans warming, polar melting causing a drop in global albedo, and methane release due to the melting permafrost. These sources of added emissions will more than offset the impact of the contemplated cuts.

The fossil-fuel industry is also arguing that fossil fuels can continue to be burned because their CO_2 emissions can be much reduced by injecting CO_2 into underground caverns. Much of the public and some of our political leaders accept this logic and are willing to support the development of this technology using public funds. What is wrong with that? Injecting CO_2 into partially emptied oil wells not only gets rid of the CO_2 but it also dislodges more crude oil. This provides a double advantage to the oil companies because, on the one hand, they gain more oil, and on the other hand, they receive financial support for capturing carbon. Oil companies hope to receive as much as $35 to $50 per ton of CO_2 captured in this way through tax breaks.[30]

What is wrong with that? First, carbon capture technology still emits some carbon into the air (about 10% of what the fossil power plant would have emitted otherwise). Second, it is a temporary means of reducing emissions because space in oil wells or other caverns is limited and will fill up quickly. Third, the financial support of carbon capture will extend the life of the fossil-fuel industries. Instead of such subsidies, the

30. Minnkota Power Cooperative in North Dakota plans to inject five million tons of CO_2 underground per year, at a cost of $1 billion. Enchant Energy in New Mexico plans to capture six million tons of CO_2 per year, at a cost of $1.3 billion.

money should be directed toward developing and converting to the permanent solution of using green energy.

As of today, our emissions (particularly methane emissions due to the increased use of natural gas) are still rising, and our efforts are focused not on cutting emissions but on adapting to living with the symptoms of GW. For example, we spend our resources on moving people out of flood-prone or drought-stricken areas, rerouting roads and water pipes, burying electric grid cables, and even sending specialized drones into clouds to trigger rain. We are not reducing GW; we are just trying to live with its consequences.

In addition to focusing on eliminating the causes of GW, we should also consider which nations should carry most of the burden. It is only fair that the nations emitting the most GHGs and those with higher per capita incomes should contribute the largest shares of the emission reductions and the financial burden (Figure 5-10).

To summarize, it is my view that the goal of keeping GW under 1.5°C (2.7°F) cannot be reached, and even staying under the target of 2.0°C (3.6°F) is unlikely. My evaluation suggests that, if our inaction continues, in a decade or two, the melting of the permafrost, the reduction in the albedo of the planet, the degassing of the oceans, and other greenhouse effects could bring us to a tipping point where the GW process becomes

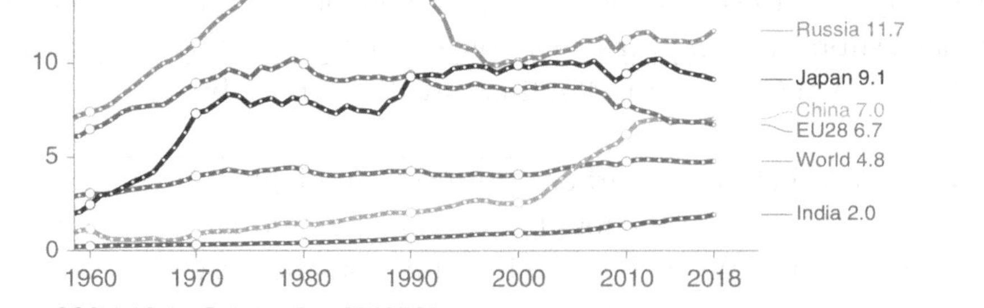

Figure 5-10. Annual per capita CO_2-equivalent emissions of different nations between 1960 and 2018. The values listed on the right are the emission rates in 2018. Note that the global average per capita emission in 2018 was 4.8 TCO_2e/y (equivalent to every person on the planet emitting 13 kg/day).

Source: Global Carbon Project, Global Carbon Budget 2019, Slide 16, https://folk.universitetetioslo.no/roberan/img/GCB2019/PNG/s16_2019_Top_FF_Emitters_percapita.png. Used with permission of the Global Carbon Project under the Creative Commons Attribution 4.0 International license.

self-accelerating, meaning that these triggered feedback processes will increase GW and the increased GW will further increase their impact. Therefore, my analysis projects that we have only about 20 years to convert our energy economy into a green (solar-hydrogen-based) one. This is the bad news. The good news is that we have the technology and the financial and labor resources to carry out this conversion.

Like GW, AI development is also approaching a tipping point. Where Nature directed human development in the past, today, humans are beginning to influence their own evolution by changing their physical and cultural environments. For a process control engineer, the conclusion of this analysis is obvious: With our past lifestyle, we have exceeded Earth's natural limits, yet we want to maintain our comfortable lifestyle, including having abundant water and food, cheap energy, biodiversity, and an unpolluted planet. Therefore, we must change our energy source before Nature tells us we have overstayed our welcome. We can change and bring the AI and GW processes under control; we not only can but must do that. Otherwise, our children will pay for our inaction.

6
Data Summary

Table of Contents

Those reading this book from cover to cover can skip this chapter because its contents have already been discussed in the previous chapters. This chapter is included only to serve the convenience of those readers who want quick access to specific GW-related information.[1]

The information in this chapter is based on respected sources[1] and on estimates and predictions that are no more than that: estimates. For the reader's convenience, the listing begins with information on metric prefixes and conversions between units that might not be familiar. To assist readers, the topics are in alphabetical order starting with "Albedo" and ending with "Wind Energy."

6.1 Prefixes

When reporting data about global processes such as climate change, the numbers are often very large or very small. To make writing these numbers easier, prefixes that represent corresponding multipliers are inserted in front of the unit names and symbols. Those for large numbers are listed in Table 6-1.

---

1. For additional data, refer to the following sources (all accessed in March 2022).
   - "Data and Statistics," International Energy Agency (IEA) website, https://www.iea.org/data-and-statistics.
     "Climate Change 2022: Impacts, Adaptation and Vulnerability," Working Group II contribution to the IPCC *Sixth Assessment Report*, Intergovernmental Panel on Climate Change (IPCC) website, https://www.ipcc.ch/report/ar6/wg2/.
   - "Climate change: Related reports," IEA website, https://www.iea.org/topics/climate-change#reports.
   - "Global Greenhouse Gas Emissions Data," US Environmental Protection Agency website, https://www.epa.gov/ghgemissions/global-greenhouse-gas-emissions-data.
   - "Fifth National Climate Assessment Update: August 2021," US Global Change Research Program website, https://www.globalchange.gov/news/fifth-national-climate-assessment-update-august-2021.

Table 6-1. Metric prefixes.

Prefix	Symbol	Multiplier
Deka	da	10
Hecto	h	10^2
Kilo	k	10^3
Mega	M	10^6
Giga	G	10^9
Tera	T	10^{12}
Peta	P	10^{15}
Exa	E	10^{18}
Zetta	Z	10^{21}
Yotta	Y	10^{24}

6.2 Units and Conversions

To assist the reader in quickly converting energy and power units, some of the most commonly used units are provided in Table 6-2. For the values of the prefixes used, see Table 6-1.

Note that, according to the International Energy Agency (IEA), the total energy consumption of humankind in 2020 was about 600 Q (176 PWh).[2]

6.3 Albedo

Albedo is the portion of the incoming solar radiation that is reflected back into space. Zero reflection (full absorption) is denoted by 0.0, and 100% reflection is denoted by 1.0 (Table 6-3). The planet's average albedo is about 0.31 (31% reflection), which is dropping because the highly reflective ice and snow at the poles are melting and being replaced by much less reflective water and land.

6.4 Areas on Earth

The total surface area of the planet is 5.1×10^8 km^2. The areas of some oceans and land areas are listed below:

- Antarctica: 1.42×10^7 km^2 (2.8%)
- Arctic Ocean: 1.40×10^7 km^2 (2.75%)

---

2. *International Energy Outlook 2021*, released October 2021, Table A1, https://www.eia.gov/outlooks/ieo/data/pdf/hm/A01_hm.pdf.

Table 6-2. Conversions among energy and power units.

Energy	
British thermal unit (Btu)	Btu = 1054 J = 253 cal = 0.293 Wh
Calorie (cal)	cal = 4.16 kJ
Joule (J) Exajoule (EJ) Megajoule (MJ)	J = 1.0 W.s EJ = 10^{18} J = 0.95 Q MJ = 948 Btu = 0.277 kWh
Horsepower hour (hp · h)	hp · h = 2511 Btu
Kilowatt hour (kWh) Megawatt hour (MWh) Gigawatt hour (GWh) Terawatt hour (TWh) Petawatt hour (PWh)	kWh = 3.41×10^3 Btu = 3.41×10^{-12} Q MWh = 3.41×10^6 Btu = 3.41×10^{-9} Q GWh = 3.41×10^9 Btu = 3.41×10^{-6} Q = 86 toe TWh = 3.41×10^{12} Btu = 3.41×10^{-3} Q = 8.6×10^4 toe PWh = 3.41×10^{15} Btu = 3.41 Q
Quad (Q)	Q = 1.0×10^{15} Btu = 293 TWh = 0.293 PWh = 2.51×10^8 toe = 1.055×10^{18} J
Ton of carbon equivalent (TOC)	TOC = 2.93×10^{10} Btu
Ton of oil equivalent (toe)	1 toe = 11.6 MWh = 39.6×10^6 Btu
Power	
Horsepower (hp)	hp = 746 W = 0.746 kW
Watt (W)	W = 1.0 J/s = 9.47×10^{-4} Btu = 0.24 cal/s
Kilowatt (kW)	kW= 10^3 W = 1.34 hp
Megawatt (MW)	MW = 10^6 W = 1340 hp
Gigawatt (GW)	GW = 10^9 W = 1.34×10^6 hp
Ton of refrigeration (TR)	TR = 1.2×10^4 Btu/h = 200 Btu/min = 3.5 kW

- Greenland: 2.17×10^6 km^2 (0.4%)
- Ice-covered land (glaciers): 1.6×10^7 km^2 (3.1%)
- Ice-free land: 1.3×10^8 km^2 (25.5%)
- Oceans: 3.6×10^8 km^2 (71.4%)
- Permafrost: 1.9×10^7 km^2 (3.7%)
- Sahara: 8.6×10^6 km^2 (0.17%)

6.5 Carbon Dioxide-Related Data

During the past 4 million years, the concentration of CO_2 in the atmosphere never exceeded 280 ppm (parts per million), but when humans began burning fossil fuels,

Table 6-3. Albedo values and reflection percentages of different surfaces on Earth.

Type of Surface	Albedo	Reflection (%)
Fresh snow	0.84 to 0.90	84% to 90%
Cumulus clouds[a]	0.7	70%
Bare ice	0.65	65%
Stratus clouds[b]	0.5	50%
Old snow	0.5	50%
Fog	0.4	40%
Dry sand	0.4	40%
Ice	0.35	35%
Earth's overall surface[c]	0.31	31%
Dry soil (desert)	0.3	30%
Wet sand	0.25	25%
Crops	0.20	20%
Permafrost[d]	0.14	14%
Forest or dark/wet soil	0.1	10%
Water	0.07	7%

[a]Cumulus clouds are low-level, flat-based clouds that appear fluffy.

[b]Stratus clouds are low-altitude, featureless layers of clouds that can range in color from dark gray to nearly white and are often associated with light precipitation.

[c]"Measuring Earth's Albedo," NASA Earth Observatory website, October 13, 2014, https://earthobservatory.nasa.gov/images/84499/measuring-earths-albedo

[d]Permafrost is land that is frozen year-round. This land is assumed to contain 1.5 trillion tons of carbon, in the form of methane and other hydrocarbons, that, if released, would have an impact twice that of the amount of carbon already in the atmosphere. Methane GHG is 25 times more powerful than CO_2.

atmospheric CO_2 quickly began accumulating. Today, its concentration is greater than 421 ppm[3] and still rising. In 2020, because of the pandemic, CO_2 emissions dropped by approximately 5%[4] (mainly because of reduced fossil energy use in transportation). In 2021, they began to rise again, and as of 2022, the atmospheric CO_2 concentration had increased by another 2 ppm.

3. "Carbon dioxide now more than 50% higher than preindustrial levels," NOAA website, June 3, 2022, https://www.noaa.gov/news-release/carbon-dioxide-now-more-than-50-higher-than-pre-industrial-levels.

4. "Global Energy Review: CO_2 Emissions in 2020," IEA, March 2, 2021, https://www.iea.org/articles/global-energy-review-co2-emissions-in-2020.

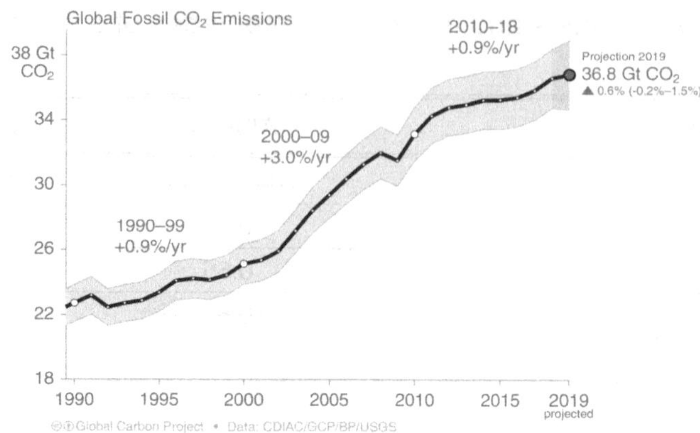

Figure 6-1. CO_2 emissions data from 1990 through 2018.

Source: Global Carbon Project, Global Carbon Budget 2019, Slide 11, https://folk.universitetetioslo.no/roberan/img/GCB2019/PNG/s11_2019_FossilFuel_and_Cement_emissions_1990.png. Used with permission of the Global Carbon Project under the Creative Commons Attribution 4.0 International license.[5]

The yearly emission rate has increased over time. For emission data between 1990 and 2019 in units of gigatons of CO_2 equivalent per year (GTyCO_2e), see Figure 6-1.

In June 2022, the US National Oceanic and Atmospheric Administration (NOAA) reported the post-pandemic emission rate as 36.3 GTyCO_2e (gigatons of CO_2 equivalent per year), approximately the same as it was in 2019. As of today, the total amount of extra CO_2 equivalent GHG that has accumulated in the atmosphere during the industrial age is about 1600 GT (1.6 trillion tons).

Table 6-4 lists the main greenhouse gases (except water vapor) and the percentages of their contributions to the total GHG effect. The second column of the table describes the relative effectiveness of these gases, gives the amount of energy one ton of each gas would absorb over its lifetime in the air, and compares that to the energy that the same amount of CO_2 would absorb over its lifetime, which "effectiveness" is 1. This shows why even small quantities of certain GHGs contribute large percentages to the total greenhouse effect. The last column lists the number of years each of these gases stays in the atmosphere.

6.5.1 Abbreviations and Terminology

Abbreviations associated with carbon include:

5. For additional information, see: Mark Kaufman, "World's carbon emissions grew in 2019 to their highest levels ever," Mashable Science website, December 3, 2019, https://mashable.com/article/carbon-emissions-grow-2019-climate-change/.

Table 6-4. GHGs in the atmosphere.

GHG	Contribution to the Total Greenhouse Effect	Energy Absorbed by One Ton of Gas Over Its Lifetime in the Air	Lifetime in the Atmosphere (years)
Carbon dioxide (CO_2)	76%	1	>100
Methane (CH_4)	16%	25 to 36	~10
Nitrous oxide (N_2O)	6%	250 to 300	100
Chlorofluorocarbons (CFCs)	2%	>10,000	>1000

- GTC: gigatons of carbon
- GTyC: gigatons of carbon per year
- TOC: ton of carbon
- toe: tons of oil equivalent

Abbreviations associated with carbon dioxide include:

- CO_2e: CO_2 equivalent (used to indicate the greenhouse effect of all GHG combined)
- $GTyCO_2$: gigatons per year of CO_2
- $GTyCO_2e$: CO_2-equivalent gigatons per year of all GHGs
- RCP: representative concentration pathway

Terminology associated with carbon include:

- *Carbon-neutral* and *net-zero* emissions are used interchangeably to describe processes that do not change the atmospheric concentration of carbon-containing compounds. Such processes are used also called traditional renewable processes. (Use of common language is needed.)
- *Carbon-free* and *zero-carbon* (sometimes inaccurately called *modern renewable*) describe green energy processes that emit no carbon compounds.
- *Carbon sequestration* and *CO_2 sequestration* describe CO_2 capturing and storing processes.

- RCPs were introduced by the IPCC in 2014 to describe four RF (radiative forcing) ranges that would result by 2100 if they took place in the future. The numbers following the letters RCP (2.6, 4.5, 6, and 8.5) refer to the RF level (the difference between incoming and reflected solar energy in W/m^2) that would exist in 2100.

6.5.2 CO_2 Accumulation in the Air

- Preindustrial (1880) total: 2200 $GTCO_2$ (equivalent to about 600 GTC)
- Current (2022) total: 3220 $GTCO_2$ (equivalent to about 880 GTC)
- Amount that can be added to the atmosphere before the GW limit of 1.5°C is reached: 550 $GTCO_2$ (the equivalent of about 150 GTC)
- Amount that can be added to the atmosphere before the GW limit of 2.0°C is reached: 920 $GTCO_2$ (equivalent to about 250 GTC)
- Increase during the industrial age to date: 1025 $GTCO_2$ (280 GTC)
- Increase in the quantity of CO_2 in the atmosphere resulting from each 1 ppm rise in concentration: ≈8.3 $GTCO_2$ (2.26 GTC)

6.5.3 Evolution of Concern About GW

Although humankind's concerns related to carbon emissions date back to World War I, even today, they remain little more than just that: a "concern." In 1995, the first Conference of the Parties (COP) was held in Berlin, Germany, initiating a series of global summit meetings under the United Nations (UN) Framework Convention on Climate Change (UNFCCC). In addition, the Intergovernmental Panel on Climate Change (IPCC) has been issuing climate change assessment reports every 6 to 7 years. These reports provide temperature rise projections for the rest of this century, corresponding to four possible future scenarios estimating four RCPs[6] for the coming decades (Figure 6-2).

These pathways describe different climate futures, all of which are considered possible depending on the volume of GHGs emitted in the years to come. Higher GHG emissions, in turn, increase the difference between the energies entering and leaving Earth's atmosphere, called RF, resulting in greater GW. The RCPs—originally RCP2.6, RCP4.5, RCP6, and RCP8.5—are labeled according to the RF values projected for the year 2100. Naturally, the higher the RF becomes, the more irresponsible human behavior is assumed to be, and the less avoidable the coming climate disaster will be.

---

6. An RCP is a trajectory of atmospheric GHG concentrations (not emissions) adopted by the IPCC. Four pathways were developed for climate modeling and research for the IPCC *Fifth Assessment Report* (AR5) in 2014.

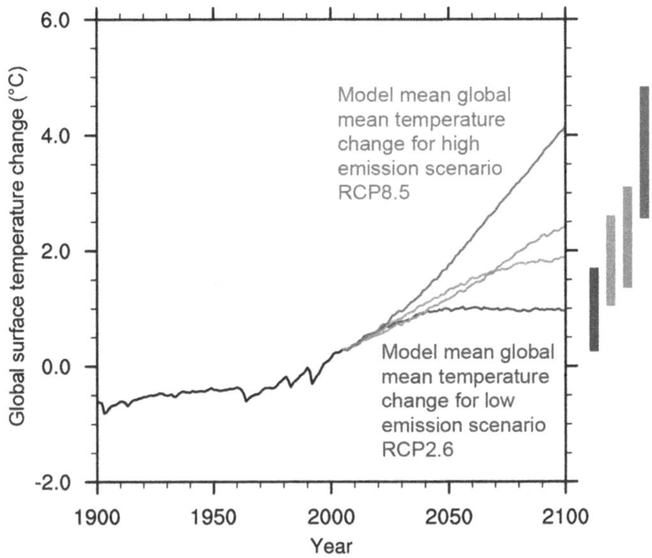

Figure 6-2. Scenarios for future GW trends under four emissions trajectories, called RCPs,[7] where the vertical colored bars show the ranges of uncertainty. (See color images after page 306.)

Source: Reproduced with permission. IPCC, *Climate Change 2013: The Physical Science Basis*,[8] FAQ 12.1, Figure 1, http://www.climatechange2013.org/report/reports-graphic/ch12-graphics/.

6.5.4 CO_2 Concentrations and Rates of Rise

There being a wide variety of CO_2 concentration scales in use, which differ by zero references, measurement locations, timing (winter or summer), etc., my recommendation is to use the I WMO X2019 scale from NOAA, which combines some 19 scales. The atmospheric CO_2 concentrations listed below are either from the NOAA or similar scales:

- 1880 to Preindustrial period: 280 ppmv
- End of 2019: 410 ppmv
- End of 2020: 412 ppmv
- July 2021: 413.45 ppmv

7. RCPs (representative concentration pathways) were introduced by the IPCC in 2014 to describe four RF (radiative forcing) ranges that would result by 2100 if they took place in the future. The numbers following the letters RCP (2.6, 4.5, 6, and 8.5) refer to the radiative forcing (RF) level (the difference between incoming and reflected solar energy in W/m^2) that would exist in 2100.

8. M. Collins, R. Knutti, J. Arblaster, J. L. Dufresne, T. Fichefet, P. Friedlingstein, X. Gao, W. J. Gutowski, T. Johns, G. Krinner, M. Shongwe, C. Tebaldi, A. J. Weaver, and M. Wehner, "Long-term Climate Change: Projections, Commitments and Irreversibility," in *Climate Change 2013: The Physical Science Basis. Contribution of Working Group I to the Fifth Assessment Report of the Intergovernmental Panel on Climate Change*, edited by T. F. Stocker, D. Qin, G. K. Plattner, M. Tignor, S. K. Allen, J. Boschung, A. Nauels, Y. Xia, V. Bex, and P. M. Midgley (Cambridge, UK: Cambridge University Press, 2013): chapter 12, 1029–1136.

- July 2022: 415.58 ppmv[9]
- Rise during the industrial age to date: 135 ppm
- Rate of rise during the past 50 years: 2.0 ppm/year
- Rate of rise during 2022: ~2.75 ppm/year

6.5.5 Actual and Projected Yearly CO_2 Emissions

Following the 1992 Earth Summit in Rio de Janeiro, Brazil, that established the UNFCCC, the United Nations has organized 26 global conferences on climate change (COPs); the most recent was held in 2021 in Glasgow, Scotland. In addition, the UN's IPCC has issued six assessment reports.[10] (The final portion of the *Sixth Assessment Report* [AR6], the *Synthesis Report*, is scheduled to be released in late 2022 or early 2023.) Despite this organized effort, warming has continued, and according to AR6, Earth is the hottest it has been in the last 125,000 years.

Table 6-5 provides global and national data on CO_2-equivalent emissions of GHGs. The table also shows that, even if the Glasgow pledges are met, GW will likely exceed 1.5°C by 2030. Table 6-5 includes actual emission data for 2005 and 2020 and probable emission values for the rest of the decade. It is clear that if the present trends are continued, global emissions will reach 48.8 $GTyCO_2e$, whereas global emissions would have to drop to 21.0 $GTyCO_2e$ to limit GW to 1.5°C.

NASA measured the total global emissions in 2019 to be much higher (i.e., 59.1 $GTyCO_2e$) than given in the table, and although the pandemic caused a drop of about 6.5 $GTyCO_2e$, by 2021, it had risen back to about 60 $GTyCO_2e$ as of this writing. The reason the data is lower than the NASA measurements is probably that NASA measures the total increase in the GHG concentration (to a precision of 0.3 ppm), which includes ocean- and land-based degassing.

I am focusing here on the Glasgow conference because it took place before the pandemic and the Ukraine war. Data released at the Glasgow conference and the later events show that we are far from achieving the goals of the Paris Agreement, although they were supposed to be legally binding. The United States abstained from the Glasgow accord; China, Japan, and South Korea did not sign the agreement to stop financing the building of fossil-fuel-based power plants. On top of that, the global energy crisis caused by Russia's aggression in Ukraine made the future even more unpredictable.

9. "Trends in Atmospheric Carbon Dioxide," National Oceanic & Atmospheric Administration (NOAA), https://gml.noaa.gov/ccgg/trends/global.html.

10. Intergovernmental Panel on Climate Change (IPCC), Reports web page, https://www.ipcc.ch/reports/.

Table 6-5. Past and projected future global and national GHG emissions quantities in yearly CO_2-equivalent units ($GTyCO_2e$).

Region/Nation	2005	2020	2030 - If Current Policies Continue	2030 - If Glasgow Pledges Are Met	To Keep GW < 1.5°C
Global total	35.6	43.5	48.8	42.1	21.0
China	8.1	13.8	14.3	14.3	8.0
United States	7.5	6.0	6.0	4.0	3.0
European Union	4.6	3.6	3.4	2.1	1.0
India	1.9	3.6	4.0	4.0	2.5
Indonesia	1.9	2.3	2.8	1.8	1.0
Brazil	1.7	2.3	2.5	1.4	1.0
Russia	2.0	2.0	2.0	2.5	1.1
Japan	1.4	1.2	1.0	0.8	0.5
Iran	0.7	1.0	1.1	5.0	0.5
Canada	0.7	0.7	0.7	0.5	0.4
Other nations	5.1	7.0	11.0	10.0	5.0

Therefore, it is no surprise that global coal consumption is increasing by over 6.0% this year. Table 6-5 also shows that the emissions of China, India, and Iran are projected to increase most dramatically. It also shows that the 2030 emissions of all nations will exceed the limits that would have to be met for GW to be kept below 1.5°C. In fact, only the United States, the European Union, Japan, and Canada have pledged to emit less GHG in 2030 than they would have emitted if they had made no pledges at all.

The reality is that elections are still decided at the gas stations and by the money that flows into the reelection coffers of our political leaders—who care far more about the next election than the next generation. "Actions speak louder than words," and for once, the parties on the left and right— the globalized international corporations on the left and the oil-rich autocrats on the right—find that inaction on climate change is profitable.

Data on the anthropogenic (caused directly by human activities) emissions rates of the past in units of yearly gigatons of $GTyCO_2$ (0.7 GTyC) are as follows:

- 1905: 2.5 $GTyCO_2$ (0.7 GTyC)
- 2000: 25 $GTyCO_2$ (6.75 GTyC)
- 2018: 32 $GTyCO_2$ (8.75 GTyC)
- 2019: 33$GTyCO_2$ (9 GTyC)

- 2019 from forest fires: 22 GTyCO_2 (6 GTyC)
- 2019 from anthropogenic sources: ~43 GTyCO_2 (11.6 GTyC)
- 2019 from land-use practices: 6 GTyCO_2 (1.6 GTyC)
- 2020: 31GTyCO_2 (8.5 GTyC)
- 2021: 37GTyCO_2 (10.0 GTyC)
- Per person daily emission: 13.3 kg of CO_2 (3.6 kg of carbon[11])

6.6 Energy

During the industrial age, the planet's population increased 7 to 8 times, and the per capita energy consumption 5 to 6 times. Nature finds this 40- to 50-fold increase in the human "energy footprint" too much and is telling us to stop meeting our needs by burning fossil and uranium fuels.

Earth has two continuous energy sources: solar radiation (about 4 to 5 million Q) and internal heat release (less than 1000 Q). The immense energy supply from the Sun keeps the planet warm and provides energy for the wind, ocean currents, and life on Earth.

Energy-related and global-heat-balance related data:

- Solar energy received on Earth: ~4 to 6 in $\times 10^6$ Q/year
- Heat required to melt all of the ice on Earth: ~10×10^6 Q
- Excess heat received causing GW: ~28×10^3 Q/year
- Total energy captured by plants: ~4×10^3 Q/year
- 2020 total used by humans: 553 Q (14 Gtoe or 162 PWh)
- 2021 total used by humans: ~600 Q/year[12] (15.2 Gtoe or 176 PWh)
- In 2040 (IEA projection): 820 Q (20.8 Gtoe or 240 PWh)

11. Given the current global population of about 8 billion, this quantity equals the weight of 100,000 Empire State Buildings.

12. According to the EIA, the world consumed a projected total of 558.7 quads in 2014; 160 of those quads came from burning coal. The EIA also estimates that by 2040 coal-energy will grow to 219.5 quads (China: 121.5 Q, India: 22.4 Q, U.S. : 20.4 Q), *Source:* Thomas W. Fuller, 3000 Quads website, blog, February 21, 2015, accessed October 29, 2022, https://3000quads.com/2015/02/21/our-global-energy-future/.

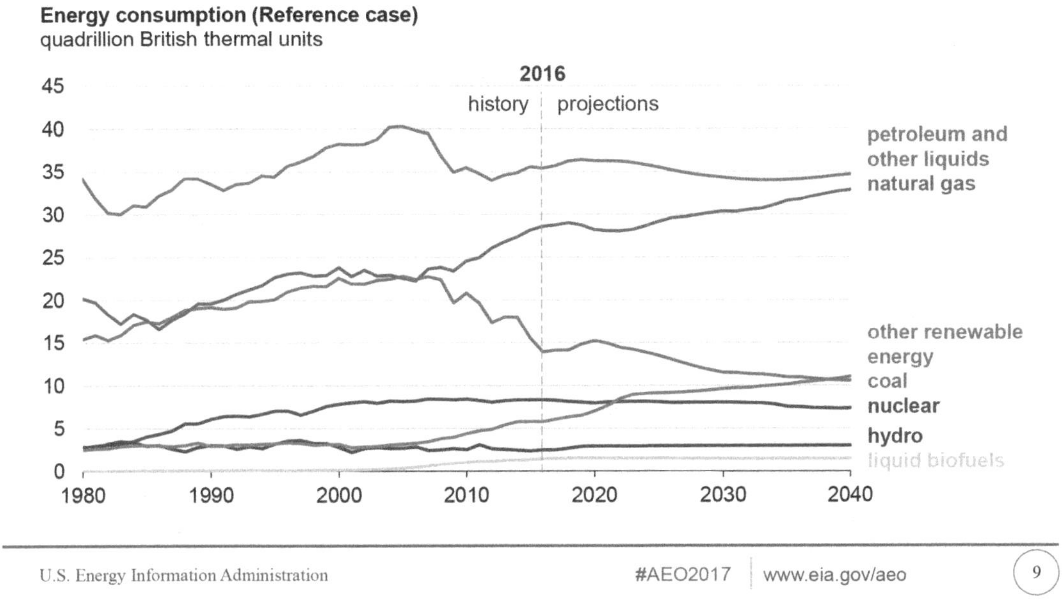

Figure 6-3. In *International Energy Outlook 2017*, the US Energy Information Administration (EIA) projects that world energy consumption will grow by 28% between 2015 and 2040. (See color images after page 306.)

Source: International Energy Outlook 2017, EIA, https://www.eia.gov/outlooks/ieo/pdf/0484(2017).pdf.

6.6.1 Consumption (Present and Future)

The energy consumption data in Figure 6-3 was prepared before the pandemic and the energy war in Ukraine. The pandemic caused a temporary drop in consumption, and the war caused a rise in energy costs. Still, in the longer range, the increase in the cost of fossil energy will probably make green energy less expensive and increase its share of the total consumption.

6.6.2 Energy Densities of Fuels and Batteries

Table 6-6 lists the gravimetric and volumetric energy densities of several energy sources, including fuels and batteries. The energy densities of batteries vary with their 1design, but the ranges of both their mass-based and volume-based energy densities fall below those of all fuels.

6.6.3 Cost of Energy Source Conversion

We can appreciate the cost of conversion only if I compare it to some of the expenses of the international community:

- Global GDP in 2019: $85 trillion
- US GDP in 2019: $21 trillion
- Global military budget in 2019: $2 trillion

Table 6-6. Energy densities of fuels and batteries.

	Energy Density	
Energy Source	**Mass-Based (kWh/kg)**	**Volume-Based (kWh/L)**
Liquid hydrogen	39.6	2.8
High-pressure hydrogen gas at 700 bar (10,000 psia)	39.6	1.55
Liquefied natural gas (LNG)	12.7 to 14.7	6.1
High-pressure natural gas at 250 bar (3500 psia)	12.7 to 14.7	3.05
Gasoline or diesel fuel	12.7 to 13	9.7 to 11.6
Batteries	~0.21 (present)[a] to 1.0[b,c]	0.25 to 1.1[d]

[a]The energy density of Li-ion batteries must increase from the current 210 Wh/kg^{-1} to 500 to 700 Wh/kg^{-1} for use in electrical vehicles. *Source:* Chen-Xi Zu and Hong Li, "Thermodynamic analysis on energy densities of batteries," *Energy & Environmental Science* 8, 2011, https://pubs.rsc.org/en/content/articlelanding/2011/ee/c0ee00777c#!

[b]Mark Kane, "Innolith Claims It's on Path to 1,000 Wh/kg Battery Energy Density," Inside EVs website, April 4, 2019, accessed October 21, 2022, https://insideevs.com/news/343771/innolith-claims-its-on-path-to-1000-wh-kg-battery-energy-density/.

[c]Innolith has withdrawn its claim of hoping to achieve this energy density for batteries. The company has registered new fuel cell patents and produced full cells in 2021 delivering 300 Wh/kg (>800 Wh/l). The company hopes to achieve an energy density of 400 Wh/kg (>1000 Wh/l) by 2023. *Source:* Innolith website, https://innolith.com/faq/.

[d]"Battery Energy – What Battery Provides More?," AllAboutBatteries website, https://www.allaboutbatteries.com/battery-energy.html.

- **Estimated Costs of Converting to the Green Energy Economy**
    - Total cost of reaching net-zero GHG emission by 2050:[13] $50 trillion
    - Yearly costs of reaching net-zero carbon emissions by 2050: $4 trillion (about 3% of global GDP)
    - Total cost of limiting GW to 1.5°C by 2050: more than $50 trillion
    - Total cost of limiting GW to 2.0°C:[14] could reach $69 trillion
    - Cost to reach one billion electric vehicles on the roads by 2050: $11 trillion

13. Sergei Klebnikov "Stopping Global Warming Will Cost $50 Trillion: Morgan Stanley Report," Forbes website, October 24, 2019, accessed October 21, 2022, https://www.forbes.com/sites/sergeiklebnikov/2019/10/24/stopping-global-warming-will-cost-50-trillion-morgan-stanley-report/?sh=47932c8851e2.

14. Wal van Lierop, "Our Carbon Problem Is a Multibillion-Dollar Opportunity," Forbes website, February 27, 2020, accessed October 21, 2022, https://www.forbes.com/sites/walvanlierop/2020/02/27/our-carbon-problem-is-a-multibillion-dollar-opportunity/?sh=1bb5a3a93649.

- **Counterproductive Investments**
  - o Global subsidies to the fossil-fuel industries:[15] about $6 trillion
  - o Expected value of such subsidies in 2025:[16] 7.4% of global GDP
- **Consequences of Inaction**
  - o IPCC's estimated cost of damages from allowing GW to rise to 2°C:[17] $54 trillion
  - o Morgan Stanley estimated cost of GW-related disasters incurred during the past 3 years:[18] $0.65 trillion
  - o Decrease in global GDP in 30 years[19]
    - o If GW is allowed to rise to 3.2°C: 18%
    - o If GW is allowed to rise to 2.6°C: 14%
    - o If GW is allowed to rise to 2.0°C: 11%

6.7 Electricity

It was 200 years ago that Michael Faraday[20] and Ányos István Jedlik[21] invented electric motors that, in a way, started the Industrial Revolution. Today the total global electricity consumption is rising faster than the population, although this increase is tempered by efficiency improvements in electricity-consuming devices. As a result, per capita consumption has recently increased more slowly, or even decreased, in high-income nations. However, this trend will probably reverse as the mobility, comfort,

15. "Fossil Fuels Received $5.9 Trillion in Subsidies in 2020, Report Finds," YaleEnvironment360 website, October 6, 2021, https://e360.yale.edu/digest/fossil-fuels-received-5-9-trillion-in-subsidies-in-2020-report-finds.
16. Ian Parry, Simon Black, and Nate Vernon, " Still Not Getting Energy Prices Right: A Global and Country Update of Fossil Fuel Subsidies," IMF Working Paper WP/21/236, International Monetary Fund, September 24, 2021, https://www.imf.org/en/Publications/WP/Issues/2021/09/23/Still-Not-Getting-Energy-Prices-Right-A-Global-and-Country-Update-of-Fossil-Fuel-Subsidies-466004.
17. Katherine J. Wu, "The World Was Just Issued 12-Year Ultimatum on Climate Change," *Smithsonian Magazine*, October 8, 2018, https://www.smithsonianmag.com/smart-news/world-was-just-issued-12-year-ultimatum-climate-change-180970489/.
18. Tom DiChristopher, "Climate disasters cost the world $650 billion over 3 years — Americans are bearing the brunt: Morgan Stanley," CNBC website, February 14, 2019, https://www.cnbc.com/2019/02/14/climate-disasters-cost-650-billion-over-3-years-morgan-stanley.html.
19. "World economy set to lose up to 18% GDP from climate change if no action taken, reveals Swiss Re Institute's stress-test analysis," Swiss Re Group website, April 22, 2021, https://www.swissre.com/media/press-release/nr-20210422-economics-of-climate-change-risks.html.
20. "Michael Faraday," Wikipedia, accessed November 23, 2022, https://en.wikipedia.org/wiki/Michael_Faraday.
21. "Ányos Jedlik," Wikipedia, accessed November 23, 2022, https://en.wikipedia.org/wiki/%C3%81nyos_Jedlik.

and electrification of transportation increase and as manual agricultural and some industrial jobs are taken over by robots.

6.7.1 Global Electricity Generation Capacity and Consumption

Electricity generating capacity is the maximum power output in TW. This is the amount of electricity that generators can produce when running at full blast. The global electricity generating capacity in 2020 (Table 6-7) was about 7.7 TW, representing an increase of about 0.5 TW from 2019 (Table 6-8).

Table 6-8 provides the global electricity generating capacities in GW of the various electricity sources in 2019[22] (except as indicated).

Global consumption in 2019 was 27.0 PWh;[23] it dropped to 26.8 PWh in 2020 because of the pandemic.[24] In 2020, contributions to the global consumption of electricity from green sources were as follows:

- Biofuels: 1000 TWh
- Hydroelectric: 4200 TWh
- Wind: 1500 TWh

Table 6-7. GW capacities of the various electricity generation types.

Energy Source	Electricity Generation Capacities in 2020
Fossil Fuels	4415
Hydraulic	1162
Wind	736
Solar	716
Nuclear	394
Pumped hydraulic	168
Biomass and waste	136
Geothermal	14
Wave and fuel cell	2
Total	**7743**

---

22. Sönnichsen, "Installed electricity capacity worldwide in 2019, by source."
23. "Generation of electricity worldwide from 1990 to 2019, by energy source," Statista website, accessed November 1, 2022, https://www.statista.com/statistics/273273/world-electricity-generation-by-energy-source/.
24. "List of countries by electricity production," Wikipedia, accessed November 1, 2022, https://en.wikipedia.org/wiki/List_of_countries_by_electricity_production#Total_production.

Table 6-8. Global capacities of electricity generating sources in 2019 (GW).

Fossil fuels[a]	4213
Green and renewables	2497
Hydroelectricity (conventional)	1140
Wind	622
Solar	584
Share of solar in China	35%
Share of solar in the European Union	22%
Share of solar in the United States	12%
Nuclear	369
Hydroelectric (pumped storage)	168
Biomass and waste	134
Geothermal	14
Tide, wave, and fuel cell	2
Global total in 2019	**7246**
US total in 2020[b]	**1200**

[a]2018 data.

[b]"America's Electricity Generating Capacity," American Public Power Association website, accessed July 25, 2022, https://www.publicpower.org/resource/americas-electricity-generating-capacity.

- Solar: 900 TWh
- Other: 700 TWh
- Total: 8300 TWh

6.7.2 Global Electricity Costs

Because of the energy war of 2022, the cost of electricity has become erratic and widely differs in various parts of the world. Since the start of the Ukraine war the prices varied wildly. After the war they will probably stabilize, but not knowing what those prices will be, Table 6-8 lists only the prewar values. Even before the war, the cost of electricity varied depending on the distance between the source and the user. Although the cost of electricity leaving the power plants is usually between $0.05 and $0.07 per kilowatt-hour, transmission, maintenance, and other fees can cause the cost to triple by the time it reaches users.

The cost of electricity also depends on local fuel costs and pricing regulations. In 2021, US Energy Information Administration (EIA)[25] reported extreme prices ranging

25. "Electricity explained: Factors affecting electricity prices," EIA website, last updated April 20, 2022, https://www.eia.gov/energyexplained/electricity/prices-and-factors-affecting-prices.php.

from 30.35¢/kWh in Hawaii to 8.17¢/kWh in Idaho. The levelized cost of electricity[26] (LCOE) can be determined by first taking the net present value of the buildings and other power generating assets and then dividing that number by the total electricity generation over the plant's lifetime. Other terms used when discussing electricity costs include levelized avoided cost of electricity (LACE) and levelized cost of storage (LCOS). Electricity costs are also affected by various financial motivations, such as carbon emission taxes, when they are introduced. For these reasons, the electricity cost data reported herein is outdated and is provided only for general orientation purposes.

6.7.2.1 Household Electricity Costs in Some Nations

Household electricity costs[27] vary among nations, as their governments use different rules that influence user costs. The household cost of 1 kWh of electricity in the 2019 prepandemic world were as follows:

- Cuba: $0.01
- Russia and India: $0.06
- China: $0.08
- United States (average): $0.14
- France: $0.21
- Connecticut, United States: $0.22
- Denmark: $0.26
- Germany: $0.39

6.7.2.2 Power Plant Costs

Power plant costs can be expressed on the basis of the initial cost per unit kilowatt production capacity or the lifetime average cost of a certain amount (kilowatt-hour or megawatt-hour) of electric energy produced. In terms of initial costs of generating capacity as of 2019, some approximate values were as follows:

- Solar photovoltaics (PV) (fixed, utility): $1,060/kW
- Solar PV (tracking, utility): $1,130/kW
- Onshore wind: $1,600/kW

---

26. LCOE represents the cost of building and operating a power plant over its lifetime. It is a measure of a power source that allows for the comparison of different methods of electricity generation on a consistent basis. It can also be regarded as the minimum constant price at which electricity must be sold to break even over the lifetime of the project.

27. "Electricity prices," GlobalPetrolPrices.com website, accessed July 25, 2022, https://www.globalpetrolprices.com/electricity prices/.

- Battery storage: \$2,000/kW
- Conventional hydropower: \$2,680/kW
- Geothermal: \$2,800/kW
- Coal (with SO_2 and NO_x controls): \$3,500/kW
- Advanced nuclear: \$6,000/kW
- Offshore wind: \$6,500/kW
- Fuel cells: \$7,200/kW

The lifetime costs of power plants vary depending on the energy source,[28] size, and design. Figure 6-4 shows the global lifetime average LCOE per megawatt-hour of electricity produced by various plants, as estimated by three different studies.[29]

6.7.2.3 Green Energy Costs

Of the total global electricity generating capacity, which in 2020 was about 7.6 TW, the zero-carbon green energy segment, including hydraulic, amounted to about 2.6 TW (hydraulic, 1.3 TW: wind, 0.60 TW; solar, 0.6 TW; geothermal, etc., 0.1 TW) or 34%. Despite the pandemic, global green energy consumption increased from 2019 to 2020 by 1.5%.[30] As fossil-fuel prices rise and shortages evolve, the costs of green energies are becoming more competitive, as shown in Figure 6-5.

6.8 Hydrogen Age

During the present transitional period, while we still depend on fossil and nuclear energy sources, the rate of increasing carbon emissions and the resulting rate that GW rises can be limited by increasing energy efficiency and by "whitening the human footprint" (increasing our albedo). Our ultimate goal should be complete conversion into the solar-hydrogen or green hydrogen energy economy.

---

28. For detailed data on capital and production costs (LCOE, LACE, and LCOS), see: "Levelized Costs of New Generation Resources in the *Annual Energy Outlook 2021*," Additional Annual Energy Outlook (AEO) Documentation, US Energy Information Administration, February 8, 2021, https://www.eia.gov/outlooks/aeo/pdf/electricity_generation.pdf.

29. Jon Moore and Nat Bullard, "BNEF Executive Factbook," BloombergNEF website, March 2, 2021, retrieved March 3, 2021, https://assets.bbhub.io/professional/sites/24/BNEF-2021-Executive-Factbook.pdf. "Levelized Cost of Energy, Levelized Cost of Storage, and Levelized Cost of Hydrogen 2020," Lazard website, October 19, 2020, retrieved October 24, 2020, https://www.lazard.com/perspective/lcoe2020. *Renewable Power Generation Costs in 2019* (Abu Dhabi: International Renewable Energy Agency [IRENA], June 2020), retrieved June 6, 2020.

30. International Energy Agency (IEA), *Global Energy Review 2020* (Paris: IEA Publications, April 2020), https://www.iea.org/reports/global-energy-review-2020.

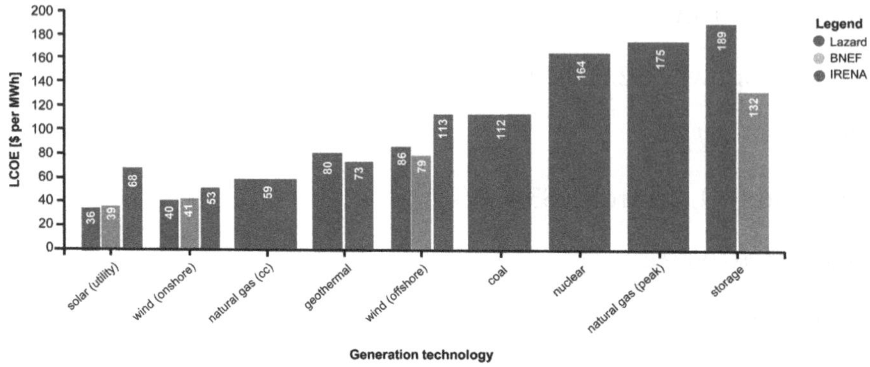

Global levelized cost of generation (US$ per MWh)

Source	Solar (utility)	Wind onshore	Gas CC	Geothermal	Wind offshore	Coal	Nuclear	Gas peaker	Storage (1:4)
BNEF	39	41			79				132
Lazard	36	40	59	80	86	112	164	175	189
IRENA	68	53		73	113				
Lazard (ranges)	29-42	26-54	44-73	59-101	86	65-159	129-198	151-198	132-245

Figure 6-4. Global LCOE cost ($/MWh) of electricity. Sources of information: Bloomberg New Energy Finance (BNEF), 2021; Lazard and the International Renewable Energy Agency (IRENA), 2020. (See color images after page 306.)

Source: "Cost of electricity by source," Wikipedia, accessed March 22, 2021, https://en.wikipedia.org/wiki/Cost_of_electricity_by_source#Global_studies

Today gray hydrogen is produced by steam reforming[31] natural gas,[32] and it supplies more than 99% of the current hydrogen market. It is called gray because its production involves large quantities of carbon emissions. Even so, it is widely used in space exploration[33] (each shuttle flight to the space station burns 500,000 gallons of liquid hydrogen) and in fuel cells for applications such as long-distance transportation, buses, trucks, trains, and ships.

This section focuses on green hydrogen, which is produced by the electrolysis of water using green energy (solar, wind, etc.) as an energy source without any carbon being emitted. One can regard green hydrogen as playing a role similar to that of vegetation in balancing the energy flows on the planet. Just as plants use solar energy to split groundwater, release oxygen into the air, and use hydrogen to provide energy (food) for humans, green hydrogen uses solar energy to split water, release oxygen, and generate energy (hydrogen) for humans.

---

31. "Steam reforming," Wikipedia, accessed November 1, 2022, https://en.wikipedia.org/wiki/Steam_reforming.
32. "Natural gas," Wikipedia, accessed November 1, 2022, https://en.wikipedia.org/wiki/Natural_gas.
33. William Harwood, "Test firing of NASA moon rocket sets stage for maiden flight," Yahoo! News website, March 18, 2021, accessed October 21, 2022, https://news.yahoo.com/test-firing-nasa-moon-rocket-133735951.html.

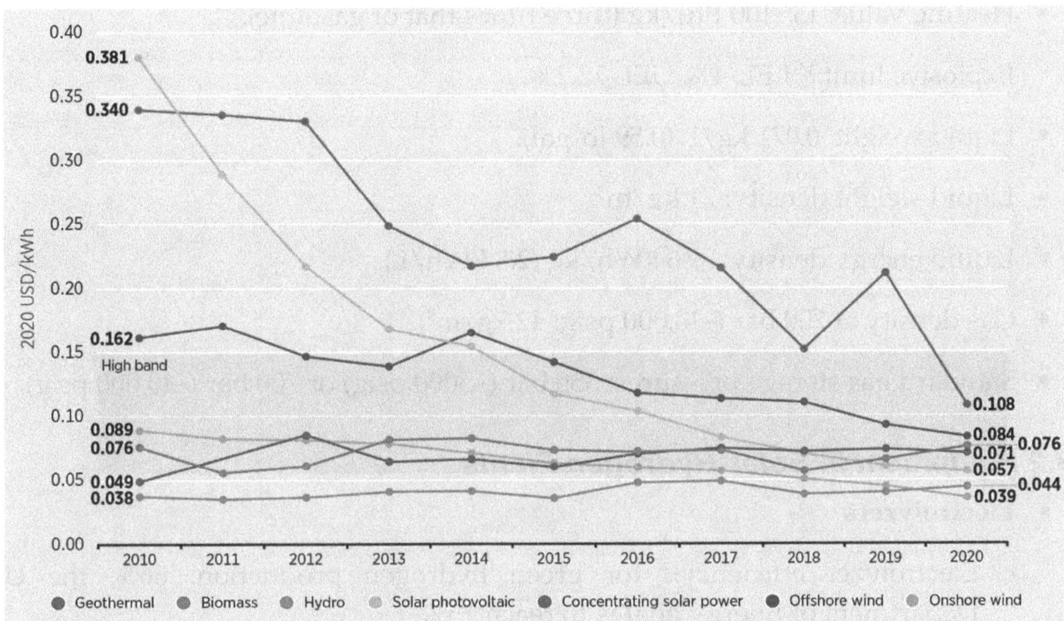

Source: IRENA Renewable Cost Database

Figure 6-5. IRENA estimate of green electricity costs. (See color images after page 306.)

Source: Renewable Power Generation Costs in 2020, Figure 1.6, IRENA, https://web.archive.org/web/20210622210701/https://www.irena.org/publications/2021/Jun/Renewable-Power-Costs-in-2020.

Today's high production cost[34] is the main reason for the low use of green hydrogen. However, this is projected to change. Some sources predict that the prices of green and gray hydrogen will be similar by as soon as 2025. Once the conversion to a green hydrogen energy economy is completed, the free and inexhaustible, but only intermittently available, green energies (solar and wind) will be collected, stored, transported, and distributed in the form of hydrogen, and the global heat balance will be reestablished.

6.8.1 Properties of Hydrogen[35]

- Energy density: Mass-based – 39.6 kWh/kg, volume-based – 1.55 kWh/L at 700 bars (mass-based density is three times higher than that of gasoline, volume-based density is six times lower). One kg holds roughly the same energy as 4 gallons of gasoline.
- Boiling point of liquid: –253°C (–423°F).

---

34. "Manufacturing cost," Wikipedia, accessed November 1, 2022, https://en.wikipedia.org/wiki/Manufacturing_cost.

35. "Hydrogen," Wikipedia, accessed October 21, 2022, *https://en.wikipedia.org/wiki/Hydrogen.*

- Heating value: 135,100 Btu/kg (three times that of gasoline).
- Explosive limit:[36] LEL-4%, UEL-72.2%.
- Liquid weight: 0.071 kg/L (0.59 lb/gal).
- Liquid weight density: 71 kg/m^3.
- Liquid energy density: 39.6 kWh/kg (2.8 kWh/L).
- Gas density at 700 bar (~10,000 psig: 42 kg/m^3).
- Standard gas storage pressures: 350 bar (~5000 psig) or 700 bar (~10,000 psig).

6.8.2 Equipment in Solar-Hydrogen Plants

- **Electrolyzers**
  - Electrolyzer efficiency[37] for green hydrogen production: 66%, the US Department of Energy goal is to reach 75%
  - Maximum electrolyzer sizes[38] in 2020: 10 MW,[39] 20 MW[40]
  - Amount produced by a 20 MW electrolyzer: 3000 tons/year
  - Total energy needed to produce 1 kg by electrolysis: 33 kWh/kg
  - Used by NASA to generate oxygen from water in space travel
- **Storage**
  - NASA's Ariana rocket launcher carried 28 tons of hydrogen.
  - To store it in liquid form at atmospheric pressure, its temperature has to be below: 20 K (–253°C, –423°F).

---

36. "Hydrogen: Combustion," Wikipedia, accessed March 12, 2022, https://en.wikipedia.org/wiki/Hydrogen#Combustion.
37. "Hydrogen Energy Storage – an overview," ScienceDirect website, https://www.sciencedirect.com/topics/engineering/hydrogen-energy-storage.
38. As of 2022, the largest electrolyzer units in the world include a 150 MW alkaline electrolyzer system in China;[43] a PEM unit in Bécancour, Canada, producing 8.2 tons of hydrogen a day;[43] and a number of 20 MW and 10 MW units in Zhangjiakou, Hebei Province, China.[43]
39. The Refhyne PEM electrolyzer at the Rhineland chemicals and energy plant is the first industrial-scale application of a PEM electrolyzer. See: Brent Haight, "Europe's Largest PEM Green Hydrogen Electrolyzer Begins Operation," *Gas Compression Magazine*, July 14, 2021, accessed October 29, 2022, https://gascompressionmagazine.com/2021/07/14/europes-largest-pem-green-hydrogen-electrolyzer-begins-operation/.
40. "thyssenkrupp to supply 20 MW electrolysis plant to CF Industries for green hydrogen for green ammonia production," Green Car Congress website, April 23, 2021, accessed November 3, 2022, https://www.greencarcongress.com/2021/04/20210423-thussenkrupp.html.

- Large gas quantities under high pressure can be stored in salt caverns.[41,42]
- In 2022, the 380-foot-long Suiso Frontier set sail from Australia, bound for Japan with liquid hydrogen in its insulated hold.[43]
- For trucks, 16 kg can be stored in a 0.4 m^3 (130 gal) tank at 700 bar (~10,000 psig).
- For small gas quantities for home backup or car transport, a 125 L (33 gal) tank holds 5 kg at pressures of up to 700 bar (~10,000 psig). See Table 4-10 for details.

- **Compression**
  - Energy required to compress 1 kg from 20 to 700 bar [from 290 to 10,000 psig]: 1.36 kWh/kg
- **Liquefaction**
  - Energy required to liquefy 1 kg of hydrogen gas: 10 to 18 kWh/kg
  - Energy required to produce each kg of LH_2 in a liquefaction plant that is producing at a rate of 100 kg/h: 17 kWh
- **Fuel Cell**
  - Fuel cell efficiency range:[44] 40% to 80% (internal combustion engine efficiency is 20%). NASA uses them in space exploration[45] and is also developing new fuel cells using proton exchange membranes (PEMs), solid oxide (SOFC), and many other materials with the goal of achieving power densities of 0.25 to 0.35 kW/kg.
- **Costs**[46]
  - Production cost of gray hydrogen in 2021: $4.0/kg to $6.5/kg
  - Gray hydrogen retail cost at fuel stations in California in 2021: $16/kg
  - Expected production cost of green hydrogen in 2030: $2/kg to $3/kg

---

41. "Storing Hydrogen in Underground Salt Caverns," Linde website, accessed October 29, 2022, https://www.lindehydrogen.com/technology/hydrogen-storage.
42. Hydrogen has previously been stored in three salt caverns at Teesside in the UK since 1972 and in two at the US Gulf Coast in Texas since 1983.
43. Amy Gunia, "Australia Just Exported Its First Batch of Fuel That Doesn't Emit CO_2. There's Just One Catch," *Time*, January 28, 2022, accessed November 4, 2022, https://time.com/6139820/liquid-hdyrogen-shipment-hesc/.
44. "Fuel Cell Efficiency – an overview," ScienceDirect website, https://www.sciencedirect.com/topics/engineering/fuel-cell-efficiency.
45. "Space Applications of Hydrogen and Fuel Cells," NASA website, accessed November 4, 2022. https://www.nasa.gov/topics/technology/hydrogen/hydrogen_2009.html.
46. Stanley Reed and Jack Ewing, "Hydrogen Is One Answer to Climate Change. Getting It Is the Hard Part," *New York Times*, July 13, 2021, https://www.nytimes.com/2021/07/13/business/hydrogen-climate-change.html.

- **Global Generation**
  - The generating capacity in 2021 was 1.34 TW, which was about 17.5% of the total global electricity generating capacity of about 7.65 TW.

6.9 Ice

Here information is provided concerning ice properties and the potential ocean level rises if certain bodies of ice were to melt. In addition, Table 6-9 provides some data for the mass, volume, and melting rates of ice in different locations on Earth.

- **Ice Properties**
  - Ice volume on Earth: 33×10^6 km^3 (30.4×10^{16}t)
  - Specific weight of ice: 920 kg/m^3
  - Latent heat:[47] 334 kJ/kg (334 MJ/t)
- **Potential Ocean Levels**
  - Current yearly rate of rise: ~5 mm/year
  - IPCC-projected ocean level rise by 2100 from ice melting: 65 cm (2.1 ft)
  - Ocean level rise if all of the ice on Greenland melts: 7.7 m (25.3 ft)
  - Ocean level rise if all of the ice on Antarctica melts: 61 m (200 ft)
  - Ocean level rise if all of the ice on Earth melts: 66 m (216 ft)
  - Heat required to melt all of the ice on Earth: 10×10^6 Q (11×10^6 EJ)

6.10 Oceans

The oceans are gigantic carbon sinks; they contain 50 times more CO_2 than the atmosphere. One-third of the extra carbon emitted into the atmosphere during the industrial age has been reabsorbed by the oceans. GW increases the water temperature which also increases the movement of CO_2 from the oceans back into the atmosphere (degassing) and leads to even higher temperatures.[48]

On the other hand, the higher the CO_2 concentration in the atmosphere, the more CO_2 the oceans will absorb. Oceanic processes are much slower than atmospheric processes but have higher inertia. Atmospheric effects (drought, wildfires, hurricanes, flooding,

47. The latent heat of a solid (e.g., ice) is the amount of heat required to melt a unit weight of that solid.
48. Holli Riebeek, "The Ocean's Carbon Balance," NASA Earth Observatory website, November 29, 2007, accessed October 21, 2022, https://earthobservatory.nasa.gov/features/OceanCarbon.

Table 6-9. Distribution of ice mass, volume, area, and melting rate on Earth.

Location	Weight (PT)	Volume (10^6 km^3)	Area (10^6 km^2) [Percentage of Earth's Area]	Melting Rate in 2019 (GTy)[a]
Antarctica[b]	28	30	14.2 [2.8%]	250
Greenland	2.6	2.8	0.2 [0.4%]	350 to 500
Other glaciers	1.0	1.0 to 2.0	1.6 [0.3%]	350
Arctic Ocean (floating ice)	Changes by season	Thin ice	Summer: 3.6 [0.6%] Winter: 15 [3.0%]	Area decreases by 50,000 km^2/year
Total land ice	32	35	16 [3.5%]	>1000[c]

PT = petaton

[a]Total yearly melting: about 1.0 trillion tons/year (1090 km^3/year).

[b]Madison Dapcevich, "200 Billion Tonnes of Antarctic Ice Is Melting into the Sea Each Year and It's Expected to Increase," IFL Science website, June 14, 2018, https://www.iflscience.com/environment/200-billion-tonnes-of-antarctic-ice-is-melting-into-the-sea-each-year-and-its-expected-to-increase/.

[c]Bob Berwyn, "Greenland's Nearing a Climate Tipping Point. How Long Warming Lasts Will Decide Its Fate, Study Says," Inside Climate News website, December 23, 2019, https://insideclimatenews.org/news/23122019/greenland-ice-sheet-climate-tipping-point-temperature-duration-sea-level-rise-pnas-study.

etc.) have already been observed, whereas the effects of changes in ocean currents, degassing, and acidification are just beginning.

The primary role of the oceans in GW is absorbing the extra heat from the atmosphere, thereby delaying its full impact. The top few meters of the ocean store as much heat as the entire atmosphere, and as the planet warms, the oceans receive about 90% of the energy. The oceans cover almost 70% of Earth's surface and contain 97% of Earth's water. Global ocean conditions profoundly affect the weather, environment, marine ecosystems, and coastal regions where a large fraction of the human population lives. Several national and global institutions focus on accurately monitoring and understanding the roles that the oceans play in the weather, seasonal variations, interannual variations (such as El Niño), and the process of climate change.

6.10.1 Ocean-Related Data

- **Size and Carbon Content**
  - Area of Earth's oceans (relative to its total surface area of Earth): 70%
  - Mass of the oceans:[49] 1.35×10^{21} kg (200 to 300 times the mass of the atmosphere)

49. Matt Williams, "What percent of Earth is water?," Phys.org website, December 2, 2014, accessed October 29, 2022, https://phys.org/news/2014-12-percent-earth.html.

- Total amount of carbon in the oceans:[50] 39,000 GTC
- Yearly amount of carbon absorbed by the oceans: 2.0 to 3.0 GTC

- **Temperature**
  - Surface temperature (average): 17°C (62.6°F)
  - Surface temperature (tropics): up to 28°C (82.4°F)
  - Surface temperature (poles): down to –2°C (28.4°F)
  - Temperature increase during the industrial age: 0.7°C (1.26°F)
  - Rate of temperature rise today: about 0.20°C (0.36°F) per decade
  - Average GW of the ocean surface as of January 2022:[51] 0.67°C (1.21°F)
- **Acidification, Coral Damage, and Plastics**
  - Current pH resulting from acidification: pH 8.1 (lowest value in 2 million years).
  - Potential ocean pH resulting from acidification by 2100: pH 7.8.
  - Coral reef loss today: 25% (projected to rise to 90% by 2100). Coral reefs dissipate 97% of the ocean's wave energy thereby reducing coastal flooding and erosion; they also produce half of the oxygen in the atmosphere.
  - Amount of plastics in the oceans: 7 GT (could exceed the weight of fish in the oceans by 2030).

6.10.2 Ocean Levels

Ocean level rise is caused by the melting of the glaciers and by direct thermal expansion due to GW. This rise is small. It is not widely understood that flooding is caused mainly by coastal storm surges and by storms caused by increased water circulation that is caused by increased evaporation of the oceans and land near the Equator. Over land, this increased circulation causes water shortages and drought, whereas over the oceans, the increased concentration of water in the air intensifies rainstorms and hurricanes.[52]

---

50. "Ocean storage of carbon dioxide: Dilute carbon dioxide injection and storage," Wikipedia, accessed July 27, 2022, https://en.wikipedia.org/wiki/Ocean_storage_of_carbon_dioxide#Dilute_carbon_dioxide_injection_and_storage.
51. "State of the Climate: Monthly Global Climate Report for January 2022," NOAA National Centers for Environmental Information, published online February 2022, accessed June 27, 2022, https://www.ncei.noaa.gov/access/monitoring/monthly-report/global/202201.
52. To date, the response has mostly been to increase insurance rates, move population centers away from the shores, and improve building codes instead of treating the cause of GW, which will worsen significantly with time.

GW has resulted in a shift in the ratio of the water phases on the planet, reducing the solid phase (ice) and increasing both the liquid and vapor phases. It has also increased the rate of water circulation (ocean evaporation and condensation over land). Figure 6-6 shows the ocean level rise during the past 25 years and its causes.

- **Ocean Level Rise**
  - Average yearly ocean level rise: 3 to 4 mm (~5 mm in 2022)
  - Level of the Pacific Ocean compared to the Atlantic Ocean: +20 cm (+8 in)
  - Estimated rise in the last 25 years: 10 cm
  - Estimated rise on the US coast by 2050:[53] 0.3 m (1.0 ft)
  - Estimated ocean level rise by 2100: 0.3 to 1.8 m
  - Ocean level rise projected by 2100 due to Greenland melting alone: 65 cm
  - Ocean level rise caused by a 1°C GW temperature increase: 2.3 m (4.2 ft/1°F)
  - Estimated ocean level rise per century:[54] 30 cm (12 in)
- **Ocean Level Rise if All Ice Melts**
  - On Earth: 66 m (220 ft)
  - On Antarctica: 61 m (210 ft)
  - On Greenland: 7.7 m (25.6 ft)
- **Ocean Level Rise Since 1990**
  - Total = 8 cm
  - Due to melting = 4 cm
  - Due to thermal expansion = 4 cm

---

53. W. V. Sweet, B. D. Hamlington, R. E. Kopp, C. P. Weaver, P. L. Barnard, D. Bekaert, W. Brooks, M. Craghan, G. Dusek, T. Frederikse, G. Garner, A. S. Genz, J. P. Krasting, E. Larour, D. Marcy, J. J. Marra, J. Obeysekera, M. Osler, M. Pendleton, D. Roman, L. Schmied, W. Veatch, K. D. White, and C. Zuzak, *Global and Regional Sea Level Rise Scenarios for the United States: Updated Mean Projections and Extreme Water Level Probabilities Along U.S. Coastlines*, NOAA Technical Report NOS 01 (Silver Spring, MD: National Ocean Service, NOAA, 2022), https://aambpublicoceanservice.blob.core.windows.net/oceanserviceprod/hazards/sealevelrise/noaa-nos-techrpt01-global-regional-SLR-scenarios-US.pdf.

54. "Understanding Sea Level Change," NASA EarthData website, accessed November 4, 2022, https://sealevel.nasa.gov/understanding-sea-level/global-sea-level/ice-melt.

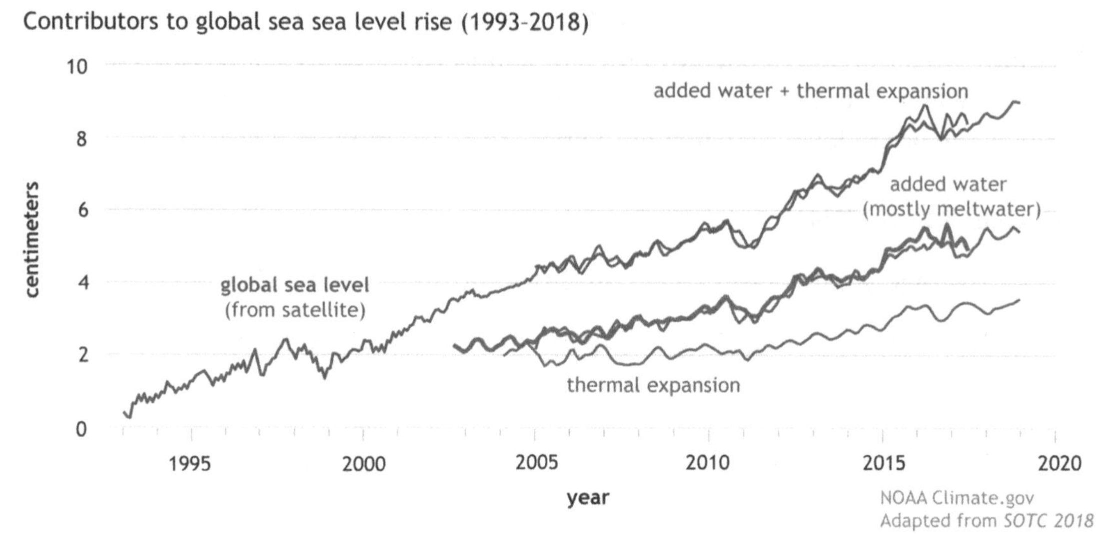

Figure 6-6. Contributions to ocean level rise during the past quarter century. (See color images after page 306.)

Source: Rebecca Lindsey, "Climate Change: Global Sea Level," Climate.gov website, January 25, 2021, https://www.climate.gov/news-features/understanding-climate/climate-change-global-sea-level.

6.11 Solar Energy

The first phase of the energy economy on Earth depended on the direct use of solar energy with little storage and maintained a stable temperature on Earth. The second phase of this evolutionary process arrived with the steam engine. During this industrial phase, the human population and its per capita energy use exploded, and with the discovery of electricity, human muscle power was replaced by machines. During this phase, as the need for energy increased, humans began to burn the fossil fuels that had accumulated underground over many millennia. This burning increased the emission of greenhouse gases which upset the global heat balance and triggered climate change due to global warming.

Today, we are entering the third phase in the evolution of our energy use. We are beginning to realize that further climate change cannot be tolerated. We also know that the population and the per capita energy consumption will increase, so a new energy source is needed. Because we cannot consider nuclear power for safety reasons, global heat balance can only be reestablished by leaving the accumulated fossil fuels where they are and switching to the inexhaustible and emission-free age in which solar-hydrogen will be our energy source.

6.11.1 Global Energy Budget

The total amount of solar energy received on Earth is about 5.0×10^6 Q/year, of which about 30% is reflected back into space by clouds, aerosols, and the planet's surface. The temperature is stable when the incoming and outgoing energy are the same. When the incoming energy exceeds the outgoing energy, the difference is called radiative forcing (RF). As mentioned previously, RF measures the difference between the entering and leaving energies, which is increasing primarily due to anthropogenic causes such as growing GHG emissions. The rate at which solar energy is received at the top of the atmosphere is estimated to be slightly more than 340 W/m^2 (Figure 6-7). As solar energy travels down toward Earth's surface, it is reflected or absorbed. Some of the reflected energy travels back into outer space, and the rest is reflected back to Earth by the greenhouse effect.

Modeling these heat flows is a complex task not only because they involve positive feedbacks between processes of dramatically different masses and inertias but because of the continuous changes in solar intensity, global volcanic activity, atmospheric GHG concentrations, and the reflectivity of Earth's surface.

6.11.2 Solar Energy Data

GW is caused by RF, the energy flux caused by natural and anthropogenic effects, and measured in watts per square meter. The RF of our planet is increasing as the CO_2 concentration in the atmosphere rises (Figure 6-8), and that rise causes temperature anomalies (GW).

- **Global Energy Flows**
    - Total incoming global solar radiation:[55] 340 to 342 W/m^2
    - Total yearly global solar energy received: about 5.0×10^6 Q/year
    - Percentage of the total solar energy absorbed by:
        - Oceans: 93%
        - Atmosphere: 2%
        - Land: 3%
        - Ice melting: 2%
- **Radiative Forcing**
    - RF caused by CO_2 in the atmosphere today:[56] 1.6 to 1.7 W/m^2
    - Global yearly heat accumulation due to total RF: 24,000 to 28,000 Q/year

55. According to most sources, the total global solar radiation has been falling during the past century and is about 340.4 W/m^2 today.

56. Global temperature has increased by about 1.6°C above the zero reference temperature that existed in 1750. This increase was caused by the increase in CO_2 concentrations from 278 ppm in 1750 to 405 ppm today, which caused a radiative forcing of about 2.0 W/m^2. If the CO_2 concentration reaches 560 ppm, it is predicted that this will cause a further warming of 1.4°C above present temperatures. See: "Radiative forcing: Radiation balance," Wikipedia, accessed July 27, 2022, https://en.wikipedia.org/wiki/Radiative_forcing#Radiation_balance.

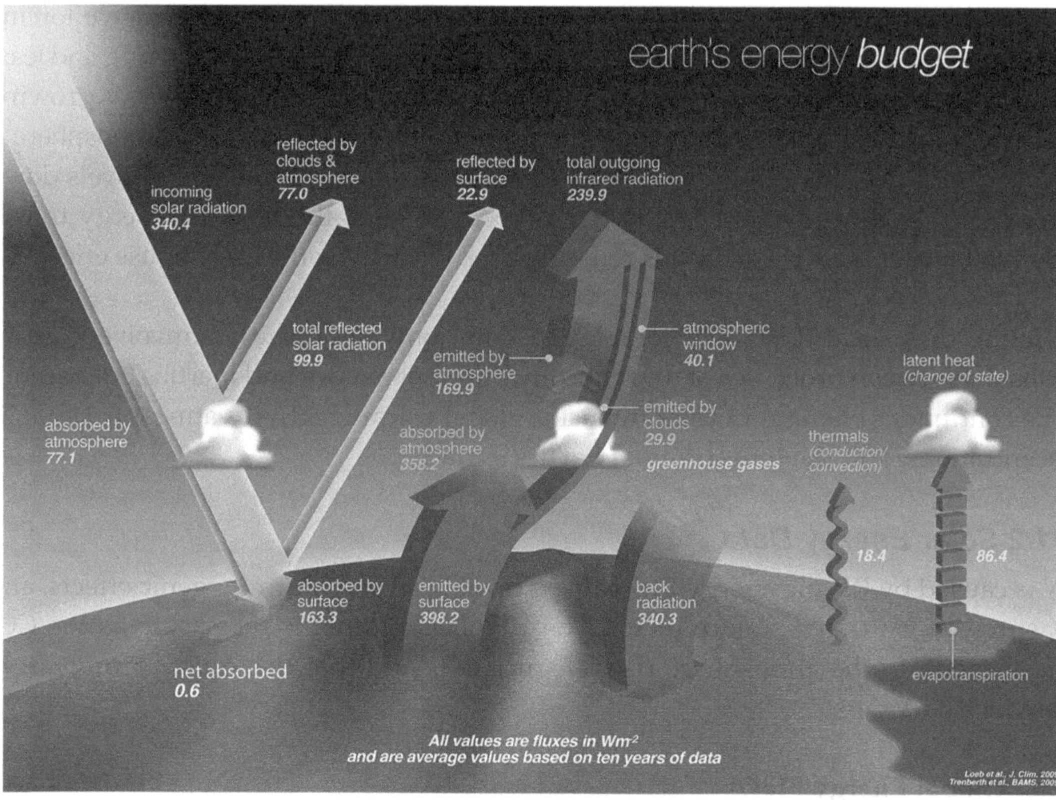

Figure 6-7. Earth's energy balance shows values averaged over a decade.

Source: Eric Gillard, "Keeping an Eye on Earth's Energy Budget," NASA Global Climate Change website, October 24, 2017, https://climate.nasa.gov/news/2641/keeping-an-eye-on-earths-energy-budget/.

6.11.3 Generation, Consumption, Efficiencies, and Costs

- **Global Solar Electricity Generation Capacity**
  - In 2017: about 404 GW
  - In 2019: 534 GW
  - In 2020: 716 GW
  - In 2021: 872 GW
  - In 2022: 1072
  - By 2030: 3000 GW is estimated by the IEA
  - Increase from 2019 to 2020: 182 GW
  - 2020 shares of global generating capacity

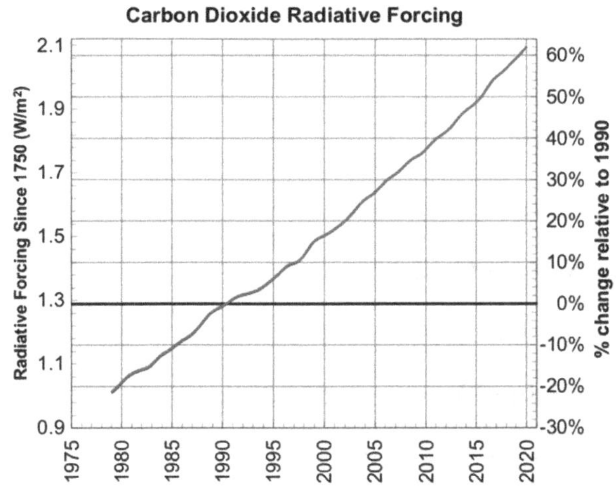

Figure 6-8. Increase in RF since 1750 (left) and percentage change in RF relative to 1990 (right). The RF values shown are only the portion of the increase caused by humans.

Source: "Carbon Dioxide radiative forcing.png" by James Butler, NOAA, Public Domain, 2014, https://en.wikipedia.org/w/index.php?curid=39537281.

- China: 35%
- European Union: 22%
- United States: 12%

- **Global Solar Electricity Consumption**
  - In 2019:[57] about 720 TWh
  - In 2020: about 900 TWh
  - In 2020 3.6% of the total global electricity consumption of 26.8 PWh
- **Solar Energy Received**
  - Depending on the location, 0 to more than 8.5 kWh/(m$^2\cdot$ d), which equals 0 to 3100 kWh/[m$^2\cdot$ year] (see Figure 3-9 for details)
- **Solar Collector Efficiencies**
  - Trees, plants: 5% to 6%
  - Thin-film PV: 5% to 15%

---

57. Heymi Bahar, *Solar PV* (Paris: IEA, June 2020), https://www.iea.org/reports/solar-pv. License: CC BY 4.0.

- Single-junction (standard) PV: 10% to 18% (maximum theoretical,[58] 33.16%)
- Multijunction PV: 20% to 25% (maximum, 40%)
- Bifacial solar:[59] 30%
- Multijunction concentrator: 46% maximum

- **Solar Collector System Costs**[60]
  - Utility-size fixed PV system: $1,100/kW
  - Utility-size tracking PV system: $1,200/kW
  - Household installation cost of a 10 kW system (without subsidy): $30,000
  - Solar panel (18% efficient, 300 W capacity, 3 ft × 5 ft): $300
  - Cost of uninstalled PV panels: about $2/$W_p$[61]
  - Cost of small unsubsidized residential system: about $3,000/kW

6.12 Temperature and GW

Earth has been cooling since its early molten age. Today, it has a 40 km thick crust. When cooling cracks the crust, immense amounts of dust and GHGs are released into the air. This cycle is about 100,000 years (Figure 6-9). During periods of high volcanic activity, the resulting greenhouse effect cools the planet. When the volcanic activity stops, the GHGs build and slowly rewarm the planet.

What is remarkable about this self-controlled cycling process is that it occurs within a narrow temperature band of just a few degrees, even when Earth's position changes relative to the Sun. Because volcanic activity, meteorite impacts, and tectonic shifts of Earth's crust can cause great heat-balance changes, it is remarkable that, during the past 1 million years, the GW (anomaly) of the planet never exceeded 3°C.

---

58. "Solar power in US.pdf," ELECTRICAL ME205 Power, Pakistan Institute of Engineering & Applied Sciences, Islamabad, Course Hero website, https://www.coursehero.com/file/p17n8ct/Efficiency-The-maximum-theoretical-efficiency-of-a-single-solar-cell-is-3316/.
59. "Bifacial solar panels: Innovative, eye-catching, and best of all, more efficient," SolarReviews website, updated January 29, 2021, https://www.solarreviews.com/blog/bifacial-solar-panels.
60. There are locations where solar energy is already the least expensive energy source. George McAtee, "Yes, America, There Is (Some) Hope for the Environment," *New York Times*, March 15, 2021, https://www.nytimes.com/2021/03/15/opinion/environment-climate-technology.html.
61. The unit of the watt peak (W_p), which is used to express the capacity of solar systems, reflects the maximum electrical capacity that a solar cell can yield under optimal conditions.

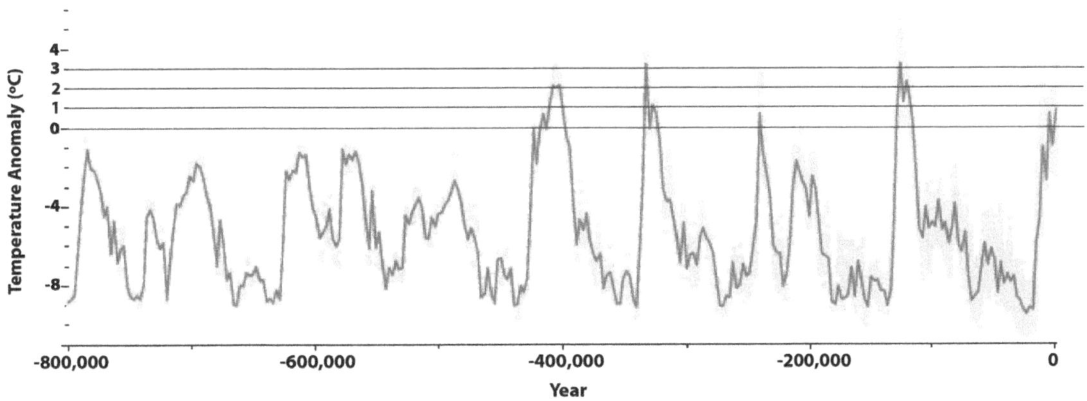

Figure 6-9. GHGs in the atmosphere kept the global temperature anomaly stable and under 3°C during the past 800,000 years.

In the early 1900s, Serbian scientist Milutin Milanković (also spelled Milankovitch) studied the effect of Earth's orbital motion on its climate. He found that changes in Earth's position relative to the Sun are responsible for variations in the amount of solar energy reaching the Northern Hemisphere and, thereby, for cycling the glaciation periods (ice ages). He also demonstrated that these planetary motions (with cycle periods of about 100,000 years) could not be the cause of GW because it evolved over a couple of centuries at a rate that is much faster than the rates caused by the Milanković cycles. Therefore, the cause of GW must be anthropogenic.

6.12.1 Global Warming

GW (i.e., the temperature anomaly) is the difference between past temperatures as the zero reference values and the actual temperature (Figures 6-10 and 2-8). The most often used zero reference is the global average surface air temperature from 1951 to 1980, which was about 14.0°C (57.2°F).[62]

6.12.2 Data on Temperature and Tipping Points

These temperature-related data are estimates based on a variety of variables including measurement location (albedo) and zero reference.

- **Global Temperatures at Different Times**
    - Twentieth-century global average:[63] 13.9°C (57.0°F)

---

62. "World of Change: Global Temperatures," accessed June 27, 2022, https://earthobservatory.nasa.gov/world-of-change/decadaltemp.php.

63. Rebecca Lindsey and Luann Dahlman, "Climate Change: Global Temperature," published June 28, 2022, accessed November 4,2022, , https://www.climate.gov/news-features/understanding-climate/climate-change-global-temperature.

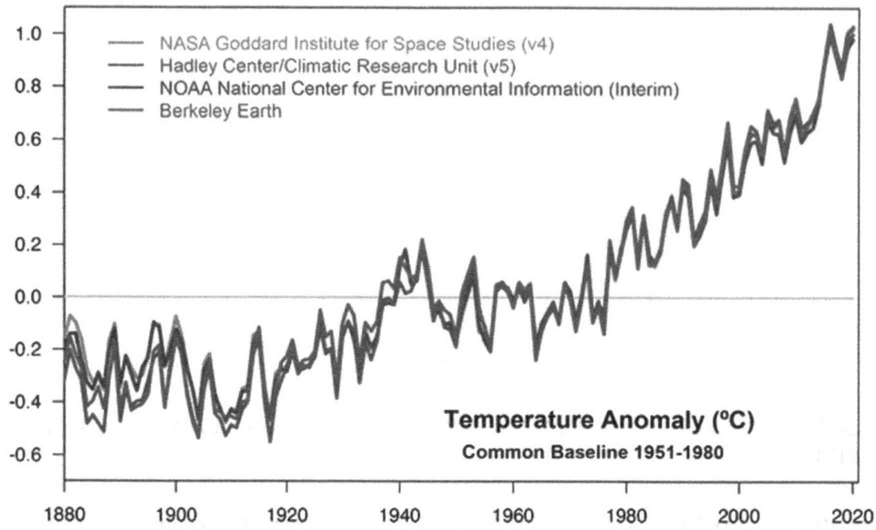

Figure 6-10. Comparison of four independent methods for estimating global temperature anomalies. (See color images after page 306.)

Source: NASA, https://climate.nasa.gov/internal_resources/2337/.

- 1951 to 1980 global average (reference temperature for GW): 14.0°C (57.2°F)[64]
- 2010 to 2020 global average: 15.1°C (59.2°F)
- 2021 global average on land: 10°C (50°F)
- 2021 global average on ocean surfaces: 17°C (62.6°F)
- May 2021 global average:[65] 15.6°C (60.1°F)
- July 2021 global average (hottest on record):[66] 16.73°C (62.07°F)

- **Global Temperatures at Different Locations**
  - North Pole (average) in Summer: 0°C
  - North Pole in Winter: –40°C
  - South Pole (average) in Summer: –28°C
  - South Pole in Winter: –60°C
  - Siberia maximum on June 20, 2020: 38°C (100°F)

64. "World of Change: Global Temperatures," accessed June 27, 2022, https://earthobservatory.nasa.gov/world-of-change/decadaltemp.php.

65. "State of the Climate: Monthly Global Climate Report for May 2021," NOAA National Centers for Environmental Information, published online June 2021, accessed June 27, 2022, https://www.ncdc.noaa.gov/sotc/global/202105.

66. "It's official: July was Earth's hottest month on record," NOAA website, August 13, 2021, accessed October 29, 2022, https://www.noaa.gov/news/its-official-july-2021-was-earths-hottest-month-on-record.

- Death Valley maximum set in 2020: 54.4°C (130°F)

- **Global Warming**
  - Average global GW in 2021: 1.1°C (2.0°F)
  - Average GW of ocean surface temperatures in 2021: 0.7°C (1.26°F)
  - Average GW rise per decade during the industrial age: 0.15°C (0.27°F)
  - Average GW rise per decade during the past 50 years: 0.18°C (0.32°F)
  - Average GW rise per decade during the past 5 years: 0.2°C (0.36°F)
  - Ocean-surface GW rise during the last decade: 0.25°C (0.5°F)
  - Land-surface GW rise during the past decade: 0.1°C (0.18°F)
  - Projected total GW rate of rise per decade in 2025: 0.25°C (0.5°F)
  - IPCC projection for 2100 according to the global IPCC scenario:[67,68] 2.0°C (3.6°F)
  - IPCC projection for 2100 according to global IPCC scenario A2:[69] 4.0°C (7.3°F)
  - MIT prediction for 2100: 5.4°C (9.4°F)

Tipping points[70] are thresholds beyond which positive feedback subprocesses are activated and start to accelerate GW. Even when carbon emissions are stopped, the carbon concentration in the air will only stop increasing, therefore, the planet will continue warming at a slower rate. Some of the effects tipping points affecting GW are the following:

---

67. The International Panel on Climate Change (IPCC) in 2000 issued four possible future scenarios (A1, A2, B1, B2) that assume future population growth, economic development, available primary energy sources, and advances in technology. These hypothetical evolutions are called "scenario families." A1 is a rapid economic growth scenario that assumes that the global population will continue to increase until 2050 and will begin decrease after that. For more details, see: https://www.ipcc.ch/report/emissions-scenarios/ and the 2022 update: https://www.ipcc.ch/report/sixth-assessment-report-working-group-3/.

68. Glenn Sherer pointed out that the IPCC projections were under estimated because they "...haven't included likely feedbacks such as large-scale melting of Arctic permafrost and subsequent release of large quantities of CO_2 and methane, a greenhouse gas 20 times more potent, albeit shorter lived, in the atmosphere than carbon dioxide." Glenn Scherer and DailyClimate.org, "How the IPCC Underestimated Climate Change," Scientific American website, December 6, 2012, accessed November 4, 2022, https://www.scientificamerican.com/article/how-the-ipcc-underestimated-climate-change/.

69. Scenario A2 assumes that a steady 3% yearly rise in carbon emissions continues to about 28 GTC by 2100.

70. Robert McSweeney, "Explainer: Nine 'tipping points' that could be triggered by climate change," CarbonBrief website, October 2, 2020, https://www.carbonbrief.org/explainer-nine-tipping-points-that-could-be-triggered-by-climate-change.

- Ice melting reduces albedo by replacing the highly reflective ice with heat-absorbing land and water surfaces.
- Permafrost melting releases immense amounts of GHGs, including methane.
- Wildfires and forest cutting increase CO_2 emissions.
- Ocean warming decreases CO_2 solubility, causing degassing.
- Ocean evaporation increases humidity, which increases the greenhouse effect.
- Melting ice reduces global albedo.

6.13 Wind Energy

Wind energy is similar to solar in that it has no GHG emission, and it is an intermittent energy source that requires storage. Wind turbines are usually colored white for day-time visibility by aircraft and birds; the lengths of their blades range from 20 to 80 m (66 to 260 ft). The sizes and heights of turbines increase year by year. Offshore wind turbines built today have capacities of up to 8 MW and blade lengths of up to 80 m (260 ft). Currently, the world's largest wind turbine is the GE-designed offshore Haliade-X, which has a power generating capacity of up to 14 MW.

- **Global Generating Capacity**
  - In 2005: 72 GW
  - In 2016: 460 GW
  - In 2018: 564 GW
  - In 2019: 654 GW
  - In 2020: 750 GW (700 GW onshore, 50 GW offshore)
  - Expected for 2050: 4040 GW
- **Global Consumption**
  - In 2020: 1500 TWh
  - In 2021 in the United States: 19.2 % of electricity generated (IEA)
- **Costs**
  - Electricity production: between $0.04/kWh and $0.06/kWh
  - Average size (2 to 3 MW) commercial units onshore: $2.5 to $4 million or $1.3 million per megawatt

- Capital cost onshore: $1,600/kW
- Capital cost offshore: $6,500/kW
- LCOE cost onshore 2021: $53/MWh
- LCOE onshore projected for 2027: $38.0/MWh[71]
- LCOE offshore in 2021: $53/MWh

71. Martin Armstrong, "The Falling Cost of Renewable Energy," Statista website, October 29, 2021, accessed November 3, 2022, https://www.statista.com/chart/26085/price-per-megawatt-hour-of-electricity-by-source/.

7

Conclusions

Table of Contents

To my knowledge, this is the first book that analyzes nonindustrial processes using the tools and know-how accumulated during the past century of industrial process control. It is also a book that does not analyze just one aspect of the evolution of the artificial intelligence (AI) and global warming (GW) processes but looks at their totality as they affect our physical and cultural environments and, as such, human evolution. I intentionally left out all equations and scientific jargon and attempted to describe complex processes using everyday language.

Until now, Nature has guided our evolution. Today, we are beginning to direct it ourselves; we are building our own future. In that sense, we have reached a fork in the road of human evolution, and it is important to understand where the two roads lead. The sign on one road reads "Better future for our children." The sign on the other road reads "Dead End." The first sign shows the direction toward protecting the only home we have and caring about the future of our children. We must not wait to find out where the second sign leads. Nature is also telling us to be careful about how we continue, that we are beginning to harm the only home we have, and if we don't stop, Nature will do it for us! We must shake off complacency, stop living in delusion, and face the threat that is the size of life itself.

We all are pessimistic about the future and have good reasons to be. Besides comfort and quick access to useless information, the digital age also brought us a "truthless" and sometimes immoral culture in which many spend their time clicking buttons as if we were the species of "clicksters." Our pessimism is also justified because our attention and resources are distracted from acting on the critically important tasks ahead by inflation, energy wars, pandemics, and political turmoil. On the one hand, the purpose of this book is to show where our attention should be focused, and on the other hand, it is also to show that we know what needs to be done and how to do it.

In this chapter, I repeat the key elements of my analysis and conclusions detailed in the previous chapters as they relate to these four topics: the present status, the cultural environment, the physical environment, and the future we must build.

7.1 The Present

My original plan was to write a summary of the various conferences, goals, and achievements of the past decades. I decided they were not worth mentioning because, most of the time, they only generated hot air and unkept promises while the out-of-control evolution of both AI and GW processes continued. Then came Russia's war and energy extortion, and the situation changed. At first, the soaring energy prices changed the situation for the worse, but now there are signs that, in the long range, this shock might wake up the world and speed up the fight against GW.

The first response to Russia's natural gas extortion[1] by some countries was to increase the burning of coal and the speed of building natural gas pipelines and floating liquefied natural gas (LNG) platforms, but many nations also responded to the soaring energy prices by becoming serious about converting to green energy sources. Japan started its Green Transformation Program, the United States Congress approved its Inflation Reduction Act (which includes $370 billion for GW reduction), and it is estimated that about $1.3 trillion will be spent globally on fighting climate change this year.

Yet, while the present energy crisis might bring a boom to the effort to convert to green energy, it is unclear whether that boom can be sustained. For example, the United Nation's *Emission Gap Report 2022*[2] states that even if the nations live up to their net zero target dates (which the report does *not* consider credible), GW will reach 1.8°C. Similarly, the International Energy Agency's (IEA) *World Energy Output 2022* report projects that GW will reach 2.5°C by 2025 if the present energy policies are maintained.[3] I believe it would take $2 to $4 trillion a year to reach net zero by 2040.

Today, we live in an age of pessimism and uncertainty, and we anticipate that the future will get worse. On the one hand, the main conclusion of this book is that we are

---

1. Europe is succeeding in its response to Russia's energy extortion by building up its natural gas (NG) stock. Prior to the war in Ukraine, 45% of their NG came from Russia. Today it is under 10%. It is estimated that Europe's NG storage is already sufficient to survive a mild winter. Other steps the European Union has taken are increasing the burning of coal by 10%; building 18 offshore terminals to receive liquefied NG, mostly from the United States and Qatar; delaying closing gas fields in the Netherlands; and searching for new gas fields.

The conservation effort is also substantial. Governments are lowering temperatures in government buildings, closing municipal swimming pools, and curbing street lighting. They are also making voluntary conservation recommendations to the public, including the "down a degree plan," using stairs instead of elevators, using bikes and public transportation, taking shorter showers, using electric devices (washing machines, dryers, dishwashers, etc.) only at night, and—what could have been unthinkable for past generations—asking Finns, to lower the temperature of their three to four million saunas from 212°F to 185°F.

The positive side effects of the energy crunch include the relative drop of the cost of green energy, causing an increase in the production of wind (192 TWh) and solar (153 TWh) generated electricity (between March and September 2022 according to think tanks E3G and Ember). Another valuable side effect is the increased public understanding that converting from fossil fuel to a green hydrogen economy is essential (blue hydrogen is made from NG and is undesirable) and that this conversion will result in an inexpensive and carbon-free energy economy.

2. United Nations Environment Programme (2022), *Emissions Gap Report 2022: The Closing Window — Climate crisis calls for rapid transformation of societies* (Nairobi), https://www.unep.org/emissions-gap-report-2022.

3. International Energy Agency (IEA), *World Energy Outlook 2022*, (Paris: IEA Publications, October 2022), https://www.iea.org/reports/world-energy-outlook-2022.

right to feel this way because if inaction is continued, the situation is rather hopeless. However, my analysis also shows that while the available time to fix things is short, we still have time to act and the financial and technical capability to do it.

7.2 AI and the Cultural Environment

In this book, I have analyzed the nonindustrial processes of GW and the development of AI by applying the scientific tools of the industrial process control profession. In this chapter, I make a few concluding comments on each, starting with AI.

In Chapter 1, we saw specific examples of the many useful contributions of AI to human development. For example, I described how, if correctly applied, AI can become an essential tool for converting to a green energy economy and optimizing the many components and subprocesses that will be used in the green infrastructure.

But just as a knife in the wrong hands can become a tool of crime, we have also seen examples of the harm AI can do. We have seen both physical and cultural examples of that damage. We saw that AI can become a tool to question the existence of GW, to convert this existential threat to all of humanity into a political issue, and to delay effective action until it is too late. On a broader scale, AI and its use on the Internet can confuse people to the point where they do not know who is telling the truth, where they cannot tell facts from lies, and where their confidence in their own institutions and even in their constitutional democracy is shaken. Similarly, AI's military applications must always be dissected by the repository of our moral values and must not be allowed to "advance" to the point that autonomous machines are allowed to make lethal decisions without human control.

The events of 2021 and 2022 clearly indicate that nationwide and global sets of rules and regulations are needed to forbid the harmful or criminal use of AI for cyberterrorism and the like. The time has arrived for a digital "Geneva Convention" to set the legal limits for the allowable use and applications of this new and powerful tool.

7.3 Physical Environment (Climate and GW)

Nature is telling us that we are overstaying our welcome and that if we are unwilling to control the consequences of our actions, it will. In this age of wildfires, droughts, flooding, heat waves, and the destruction of both types of rainforests (the ones on land and the coral reefs in the oceans), Mother Nature is beginning to say, "Enough!"

In the late 1970s, when I worked as an automation consultant on one of 14 US government projects aimed to develop green alternatives to fossil fuels, we made good progress. The first solar panels were installed on the roof of the White House in 1979, and we were on our way to a green transformation. But a few years later, politics changed, the 14 R&D teams were disbanded, the solar panels disappeared from the roof of the White House, and an eight-year oil war started in the Middle East. Part of the reason this occurred is that GW was insignificant in 1979; it had no manifest symptoms such as wildfires and droughts. Only the scientists knew that it was coming, and who believed the scientists?

The scientists did. They held the first climate conference in 1979 in Geneva. That conference was followed by many others, including the well-discussed Paris conference in 2015, the 2021 conference in Glasgow, and the last one in 2022 in Sharm el-Sheikh, South Sinai, Egypt. They all generated a lot of goals, promises, and hot air, while the CO_2 concentration in the atmosphere reached the highest level ever (421 ppmv[4]). In short, wildfires, coastal storm surges, heat waves, floods, malnutrition, water and food scarcity, and the devastation of plant and animal species are all signs of the coming environmental catastrophe, and they have already arrived.

Due to Russia's energy extortion, fossil-fuel prices are skyrocketing, oil pipelines are being built, Europe's natural gas storage is running low, and coal power plants are reopening. In short, we are—I hope temporarily—moving backward. The World Resource Institute reported in 2022 that if the current promises of the various nations are kept, global emissions will be reduced by 7% by 2030 (compared to that of 2019), while if we were to limit GW to 1.5°C, that reduction would be 43%. One could cite many numbers to illustrate the prevailing irresponsibility of most nations. For example, China, the world's largest emitter of greenhouse gases, announced that it intends to continue increasing the amount of coal burned and thereby increasing emissions until 2030.

7.3.1 The Role of Science and Religion

Today, science is colliding with a powerful enemy—money—the financial interests of a fossil-fuel industry, which for money, is willing to destroy the only home we have. Fortunately, this time, science is supported by the Church in the fight against greed and ignorance. Five hundred years ago, Pope Urban VIII honestly believed that Galileo was wrong and that the Inquisition was necessary to defeat the false teachings of science. Today, Pope Francis knows that science is right and climate change is not a hoax. Yet, leaving more than $40 trillion worth of fossil fuels in the ground is hard to achieve.

---

4. "Carbon dioxide now more than 50% higher than pre-industrial levels," NOAA website, June 3, 2022, https://www.noaa.gov/news-release/carbon-dioxide-now-more-than-50-higher-than-pre-industrial-levels.

Today's conflict is different from that of Galileo because 500 years ago, there was no urgency to prove that he was right; today, it is urgent to convince all that climate change is real. I am optimistic because, in December 2020, Pope Francis supported[5] the Secretary-General of the United Nations[6] in declaring that our number one task is to prevent the coming catastrophe. While most religious leaders also spoke up for action on climate change, a joint declaration from the leaders of Christian, Muslim, Hindu, Buddhist, Jewish, and other faiths would change climate change from a political topic to what it is, a moral and ethical challenge for all.

I am optimistic because the young generation understands what is at stake, as was shown in March 2019, when an estimated 1.6 million students from 125 countries protested the present inaction. The educated young not only understand what it will take to get rid of fossil fuels and convert to a solar-hydrogen energy economy but also know that they are like snowflakes who individually are powerless but jointly can create an avalanche. They are becoming educated and are starting to use their most powerful weapon, the ballot box! They are learning that process control and advanced automation can help stop climate change.

7.3.2 Denialist Arguments

In the very beginning of this book, I analyzed the heat-balance process of GW to determine whether this anomaly could have causes other than the greenhouse effect. I checked to see if the solar intensity had possibly increased during the industrial age and found that it had actually dropped. I investigated the possibility that the distance between the Earth and the Sun had decreased and found that the timescale of the processes that affect this distance is 25,000 to 100,000 years. This is 1000 times slower than the rate at which GW is rising today. This analysis confirmed that human activity can be the only cause of GW.

We know that the nature, or "personality," of the GW process is a self-stabilizing one. This is because, as Earth's temperature rises, the heat Earth radiates back into space also rises—which is the case with any warm object that radiates heat in a cold surrounding. When this increased cooling effect reaches the heat gain caused by GW, the global temperature will stop increasing and the planet's heat balance will be reestablished. GW will eventually stop rising, but that will be too late for us and human civilization.

---

5. Devin Watkins, "Pope to UN Climate Summit: 'Now is time to change direction'," Vatican News website, December 12, 2020, accessed March 8, 2022, https://www.vaticannews.va/en/pope/news/2020-12/pope-francis-videomessage-un-climate-summit-2020-vatican.html.

6. "Pope Francis & UN Secretary-General Meet and Record Message for World Peace and Harmony," Mercy International Association website, December 21, 2019, accessed November 7, 2022, https://www.mercyworld.org/newsroom/pope-francis-un-secretary-general-meet-and-record-message-for-peace/.

During the past million years, the GW on Earth never exceeded 3°C (5.4°F). Nevertheless, on September 9, 2020, the Secretary-General of the World Meteorological Organization (WMO), Petteri Taalas, reported that we are on our way to reaching GW of 3°C (5.4°F) compared to the temperatures in the late nineteenth century.[7] If the global temperature rises to that level or even higher, most projection models agree that one million plant and animal species will be decimated, humanity's food supply will be endangered, and a large part of Earth will no longer be able to support human life. In addition, disease and starvation will trigger a biblical-scale migration that could destroy human civilization. Clearly, the bottom line is that allowing the GW process to "self-stabilize" (leaving it uncontrolled) is not a sane option.

7.3.3 The Facts

The Earth's temperature anomaly cycled between ice ages and warm peaks. During the past 100,000 years, it never exceeded 1.0°C (1.8°F); during the past 800,000 years, it never exceeded 3.0°C (5.4°F) (Figure 7-1); and today, it has passed 1.1°C.[8] This is the case because the GHG concentration in the atmosphere is the highest in a million years. If inaction continues and the rise in temperature remains out of control, GW will reach 1.5°C (2.7°F) by 2030 or even sooner. On the other hand, if we convert half of our energy supplies to green fuels by 2030 and reach a net-zero-carbon emission state by 2035 to 2040, we might keep GW under 2°C (3.6°F). If we stop using fossil fuels completely by about 2045, GW will stop rising.

We have seen that the mass of the oceans is 200 times greater than the mass of the atmosphere; therefore, their time constants are also dramatically different. Whereas the atmospheric effects of GW are already being felt, the effects of the warming oceans are just beginning to evolve. Similarly, the consequences of the positive feedback processes (decreasing albedo, permafrost melting, ocean degassing and evaporation, etc.) have not yet triggered their tipping points. Even though these processes have not yet

---

7. Seth Borenstein, "UN report: Increased warming closing in on agreed upon limit," CTV News website, September 9, 2020, https://www.ctvnews.ca/sci-tech/un-report-increased-warming-closing-in-on-agreed-upon-limit-1.5097424. This statement was made during the launch of the report *United in Science 2020*, which was compiled by the WMO under the direction of the UN Secretary-General António Guterres to provide a "high-level compilation of the latest climate science information." See: *United in Science 2020: A multi-organization high-level compilation of the latest climate science information* (Geneva: WMO, 2020).

8. During the last 800,000 years, GW reached 3°C twice and 2°C once. Today, at nearly 1.2°C, it is the fourth time that GW has exceeded 1.0 °C. We do not know what happened on those first three occasions, but we do know that those GW values were not caused by humans. We know that our early ancestors survived those periods, but we do not know at what costs. According to my calculations, at those temperatures the ice caps at the poles were gone, the coasts and most islands must have been under water, and humans could probably survive only in caves. The overwhelming majority of scientific institutions agree that most of the GW is anthropogenic.

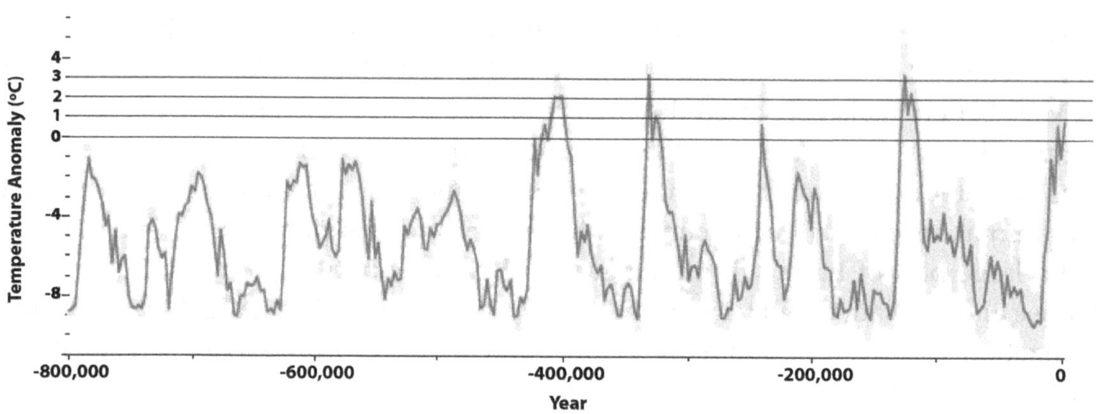

Figure 7-1. NASA prepared this 800,000-year global temperature anomaly record by analyzing air bubbles trapped in glacial ice.

Source: "How is Today's Warming Different from the Past?," NASA Earth Observatory web page, published June 3, 2010, https://earthobservatory.nasa.gov/features/GlobalWarming/page3.php. Modified by the author.

been triggered, the past decade was the hottest in about 100,000 years. We are already on our way to water shortages and droughts, to one billion people being exposed to life-threatening heat waves, and even to the existence of some island nations being threatened. This makes immediate and drastic action essential.

We have seen that climate change is no longer an abstract notion; its cascading effects are already apparent in our everyday lives. Not only do we see the consequences of devastating hurricanes and wildfires, but we are beginning to realize that better building codes, better-equipped firefighters, and bigger groundwater pumps cannot protect against the escalating effects of GW. We are beginning to understand that we must eliminate its causes; otherwise, not only will the hurricanes, droughts, and wildfires strengthen, but they will trigger other overlapping climate events.

7.3.4 Process Control Analysis

The Earth consists of four very different heat-absorbing bodies. The oceans absorb 94% of the total extra heat received through GW, whereas the atmosphere, the land, and the melting ice absorb only 2% each. From the perspective of process control analysis, we know that the speed of response (gain) of heat-transfer systems increases as the object's mass decreases. We also know that the mass of the oceans is 200 times greater than that of the atmosphere. Therefore, it is no surprise that the ocean surfaces are warming much more slowly (half as fast) than the atmosphere. (The fact that the deeper we go below the ocean's surface, the faster the temperature rises has nothing to do with the relative masses; it is caused by "upwelling.") Therefore, it is because of the drastically different masses of the oceans and the atmosphere that we are already experiencing the atmospheric effects of GW (drought, fires, etc.). In contrast, the oceanic responses (changes in ocean currents, degassing, coral deaths, etc.) are just beginning to be felt.

7.3.4.1 Positive Feedback Effects (Self-Acceleration)

Another fundamental difference between simple heat transfer and the climate change process is that GW is the sum of many subprocesses that are triggered at different times and, after being triggered, are self-accelerating. One of these is the reflectivity (albedo) drop of the planet, which is caused by the melting of the highly heat-reflecting ice, which is then replaced by highly heat-absorbing land and water surfaces. Today, 30% of the solar energy reaching Earth is being reflected into outer space, and if this amount decreased by only 1% (to 29%), that would add more heat to the planet than all of the GW caused by humans. This feedback process is accelerating because melting increases temperature, which further increases melting. The permafrost effect has the same characteristics as self-accelerating feedback because when permafrost melts, it releases immense amounts of methane which further increases GW and, thus, accelerates melting. Two other examples of self-accelerating positive feedback subprocesses are the degassing of the oceans (caused by the decrease in CO_2 solubility with increasing temperature) and wildfires.

7.3.4.2 CO_2 Accumulated in the Past Century

In "regular" heat-balance processes, if the heat input decreases, the process temperature is reduced. This is not the case with the GW process because more than 410 ppmv CO_2 equivalent (CO_2e) is already in the air, whereas the preindustrial stable concentration was 280 ppm CO_2e. This means that, even if we stop emitting CO_2, GW will continue (although at a lower rate) because the present 410 ppm concentration exceeds the preindustrial concentration by 130 ppm. It should be understood that when people talk about reaching "net-zero" conditions, they do not mean the elimination of all carbon emissions, only about half of that. Net zero means reducing emissions to the state at which the CO_2 concentration is no longer rising. GW is still increasing, only at a lower rate. Net-zero steps are not enough to stop GW and to start moving toward preindustrial conditions; cooling is needed as well. Cooling can be achieved by planting forests, converting our energy economy into a completely "green" one, or decreasing the human carbon footprint on the planet by increasing Earth's reflectivity.

7.3.5 Acceleration, Tipping Points, Albedo Effects

Over the past few centuries, human activity (particularly the burning of fossil fuels and changes in land use) has caused an increase in atmospheric GHG concentrations to levels that exceed any concentrations occurring during the past million years (Figure 7-2). Compared to the preindustrial age, when the CO_2 concentration was about 280 ppmv and we had about 600 GTC (2,220 $GTCO_2$) in the air, we reached about 950

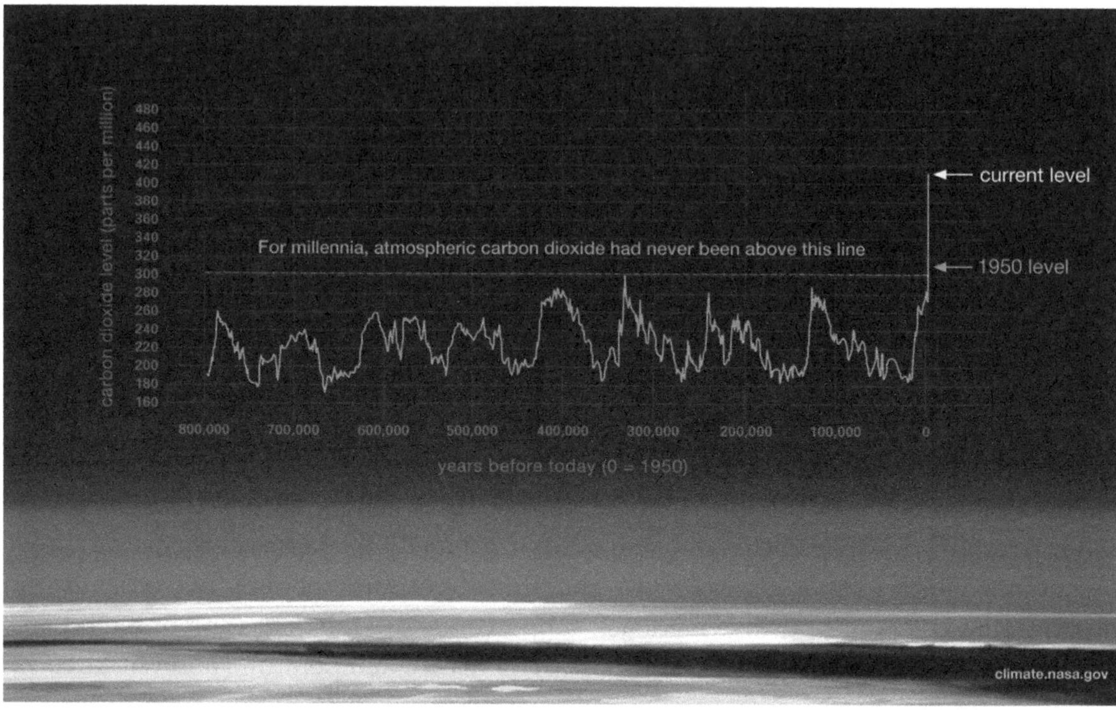

Figure 7-2. In the past, the CO_2 concentration in our atmosphere never exceeded 280 ppmv. Once the Industrial Revolution began, the concentration started to rise, and by the end of 2019, it had reached 410 ppmv. By May 2020, it had increased further to 417.3 ppmv, by May 2021, it reached 419.1 ppmv.

Source: "Climate Change: How Do We Know?," Global Climate Change website, accessed March 3, 2022, https://climate.nasa.gov/evidence/.

GTC (about 3,510 $GTCO_2$) by 2021. In May 2021,[9] the CO_2 concentration reached 419.1 ppmv (1.0 ppmv ~ 2.15 GTC), and according to some projections, it will rise another 16% by 2030. Today, we are emitting 9 to 10 GTyC (37.0 $GTyCO_2$). *At the beginning of the industrial age some 600 GTC (2,220 $GTCO_2$) was in the air; the accumulation during the industrial age was 350 GTC (1,290 $GTCO_2$), which is an increase of about 50%.* Therefore, whenever we manage to cut our carbon emissions to net zero—if we ever do—this extra 350 GTC (1,290 $GTCO_2$) will remain there and will keep increasing GW, only at a lower rate.

The GW process is the sum of many subprocesses. The nature of some of these processes is self-accelerating in the sense that they trigger positive feedback cycles; as they evolve, their evolution increases GW, and this rise further accelerates the subprocesses. When these positive feedback processes reach their tripping points, their impact on

9. "Trends in Atmospheric Carbon Dioxide," Earth System Research Laboratories website, accessed March 3, 2022, https://gml.noaa.gov/ccgg/trends/.

GW can become unstoppable. This book analyzed the time constants and gains of the following positive feedback subprocesses:

- The albedo effect occurs because glacial melting replaces the reflecting surface of ice with heat-absorbing water and land surfaces, which increases GW and further increases melting.
- Permafrost melting releases immense amounts of GHGs (mostly methane), which increases GW, which, in turn, further increases the melting of the permafrost.
- Warming of the oceans reduces the solubility of CO_2, resulting in the release of GHGs (called *degassing*), which further increases GW.
- GW increases the evaporation of ocean water. The increase in the amount of water vapor in the air further increases the greenhouse effect, which causes more evaporation.
- GW increases droughts and wildfires. Wildfires release heat and CO_2, causing deforestation[10] and drought, both of which further increase GW.
- Each 1°C of GW increases humidity by 7%, which increases the greenhouse effect, strengthens hurricanes and thunderstorms, and slows wind movement over land, which leads to giant rainstorms.

The analysis in this book also showed that there are negative feedback subprocesses that could be used to reduce GW. The one with the highest potential is increasing the amount of heat our planet radiates back into outer space. If net-zero-carbon emissions were reached, the carbon content in the atmosphere would stop rising; however, GW would continue (at a lower rate) because some 900 GTC has already accumulated in the atmosphere. Therefore, lowering emissions to the net-zero level is not enough to stop the rise of GW; the anomaly (the difference between incoming and outgoing heat) must be reduced. How can that be done?

The most obvious method is to increase cooling by increasing the albedo (reflectivity) of the planet. Today, Earth receives about 340 W/m^2 of solar power and, on average (depending on the cloud cover), reflects more than 30% of this amount back into

---

10. Deforestation can reduce rainfall over a wide region, even as it spurs increased rainfall in the immediate area where that deforestation took place. *Source:* Lela Nargi, "8 Things That Could Happen if the Amazon Rainforest Disappeared," updated June 2, 2022, accessed November 3, 2022, https://www.rd.com/article/what-could-happen-if-the-amazon-rainforest-disappeared/.

space. Earth's current (human-caused) radiative forcing (RF) is about 1.7 W/m^2; in other words, 1.7 W/m^2 more solar energy is received by Earth than is radiated back into space as heat.[11] If we could increase the albedo of the planet by only 1% (by "whitening" our roofs, roads, and agriculture), GHG could be reduced to the preindustrial state and GW would be eliminated.

7.3.6 Worth Repeating

Here I quickly list a few facts and recommendations that were discussed in the previous chapters:

- GW exists; its cause is anthropogenic, and it is presently out of control.
- Keeping GW under 1.5°C (2.7°F) can no longer be achieved; keeping it under 2.0°C (3.6°F) requires eliminating all humanmade CO_2 emissions by 2040.
- The mass of the oceans is 200 times greater than that of the atmosphere; consequently, most oceanic effects of GW are yet to evolve.
- The consequences of positive feedback processes (decreased albedo, permafrost melting, ocean degassing, etc.) will trigger their tipping points within five years.
- Resources spent on mitigating GW symptoms (e.g., building dams, improving firefighting equipment, planting drought-resistant crops, designing buildings to withstand hurricanes and floods, and relocating residents) should be secondary to converting to green energy.
- Estimates: According to the IPCC, if GW is kept under 1.5°C, the damages are estimated to cost $40 trillion; if 2°C is reached, the damage cost is estimated to be $54 trillion.[12]
- Industrialized, rich nations have caused most of the GW. It is their responsibility to mitigate the consequences by creating an international fund to compensate third-world and low-lying island nations for the climate-change-related damage they suffer.
- We have all we need to convert to green energy. We know how to collect, store, transport, and distribute the fuel of the future: solar hydrogen.

---

11. "Climate Forcing," Climate.gov website, accessed March 3, 2022, https://www.climate.gov/maps-data/primer/climate-forcing.

12. Katherine J. Wu, "The World Was Just Issued 12-Year Ultimatum on Climate Change," *Smithsonian Magazine*, October 8, 2018, accessed October 29, 2022, https://www.smithsonianmag.com/smart-news/world-was-just-issued-12-year-ultimatum-climate-change-180970489/.

- We can create good jobs for the workforce that would become unemployed due to robotization by training them to fill the jobs created by the transition to the green energy economy.
- Advanced nations should cut their net-zero-carbon emissions in half; less advanced nations by one-third.
- Stop building new coal and oil power plants.
- Stop developing new oil and gas fields.
- Stop all subsidies for the fossil-fuel industry.
- Stop making new gas- and diesel-fueled cars.
- Establish a methanol tax.
- Subsidize green electricity.
- If a carbon tax is introduced, spend the income to subsidize green energy.
- Set carbon limits for industries.
- Subsidize electric car purchases, the installation of home solar collectors, two-directional electric meters, and battery or hydrogen-based electricity storage.
- Require new buildings to include solar roofs, electricity storage, and higher-level means to recharge electric cars.
- Support switching from oil and propane fuels to locally made solar hydrogen.
- Add hydrogen outlets to electric charging stations.
- Support green hydrogen-fuel-cell-based transportation.
- Build green hydrogen-generating solar and wind farms.
- Develop the infrastructure to store and transport green hydrogen.
- Support increasing Earth's reflectivity by whitening roofs, roads, and agriculture.

7.4 Transition to Green Energy

The automation and process control profession will play a key role in the transition from the fossil-nuclear energy age to the hydrogen age by designing completely safe, AI-supported, multivariable, and autonomous control packages for the solar collector, electrolyzer, compressor storage, distribution, and fuel cell systems that compose the infrastructure. This development effort will include reducing the cost of these components through standardization and mass production.

The most effective transition method would probably be using carbon taxes, starting at about $50 per ton of CO_2 and increasing by 5% or so yearly. Cross-border import taxes can also be introduced and emission limits can be set, allowing industries that do better to sell their margin on the free market. Naturally, all of these taxes should go directly into a dedicated fund that can only be used for rewarding emissions reductions, subsidizing the conversion to green energy use, or financing the reeducation of employees who are impacted by this transformation so that they can fill the millions of new jobs that will be created.

Aside from the financial motivation through taxes and subsidies, the most important challenge is convincing society that spending to stop climate change is good for the economy. This is very important because the powerful fossil-fuel lobby is spreading the opposite view and a segment of our fellow citizens believe them. Therefore, all such denialists must be convinced of the truth that the green conversion will boost the economy (as did the Marshall Plan after World War II) and that we know how to do it and have the technology to do it.

Next, I discuss a few technologies that must be gradually discontinued during this period, such as the use of natural gas and nuclear energy. In addition, I discuss technologies that will have to mature during this transitional period, such as energy storage and electric transportation.

7.4.1 The Role of Natural Gas

The European Union has decided to temporarily treat nuclear energy and natural gas (NG) as acceptable fuels during Russia's energy war and possibly for a few years during the conversion from fossil fuels to green energy sources. For this reason, I briefly discuss both of these energy sources.

Because of the natural gas extortion by Russia, Europe has built floating peers to receive liquefied natural gas (LNG) from other sources. The LNG is gasified and distributed by the existing piping network that previously handled pressurized NG. They are also preparing for the future by installing pipe distribution systems that can handle both a mix of NG and hydrogen or poor hydrogen.

It is true burning natural gas emits only 53 kg of CO_2 per million Btu, whereas the emissions of oil and coal amount to 73 and 91 kg of CO_2 per million Btu, respectively.

However, CO_2 is not the only GHG emitted while burning natural gas, which is essentially methane gas. During the production and transportation of natural gas, methane leaks from the extraction and distribution infrastructure. For every cubic foot of natural gas drawn from underground reservoirs, about 1.4% escapes into the atmosphere. This leakage has an immense impact on GW because methane is more than 25 times as potent as CO_2 in trapping heat in the atmosphere. The environmental impact of the sum of the effects of burning and leaking methane can reach that of coal. Therefore, natural gas should only be considered for use as a transitional fuel as a last result and should not be referred to as a "transitional green" fuel.

During the past century, the methane concentration in the atmosphere has more than doubled. In 2020, the worldwide emissions of human-linked methane amounted to 400 million tons. Most of this methane was emitted by the Chinese coal production industry and the Russian natural gas pipelines and distribution system.

Carbon dioxide remains in the atmosphere for a few hundred years, which means that, even if the emissions were stopped or dramatically reduced, that would slow but not stop GW for a long time. On the other hand, it takes only a decade or two for methane to break down, so reducing methane emissions could reduce warming much more rapidly and significantly than lowering CO_2 emissions.

Methane is odorless, colorless, and flammable in concentrations between 4% and 17%. Sensors used to detect methane are either sampling-type spectrometers or remote-monitoring types. Ground-level detectors can be handheld portable or stationary units that monitor individual leaks. Drone-mounted sensors can view leaks up to a 100-m distance and are used in such applications as inspecting natural gas pipelines and wells. Satellite-based sensors can view most of the surface of the planet and detect the average methane concentration in the air column below.

Satellite-based monitoring of atmospheric gas concentrations has been used for decades. Its importance was demonstrated when the use of aerosols started to destroy the ozone layer, and such monitoring provided evidence that supported the passage of regulations that helped to solve the problem. The monitoring of methane in the atmosphere was advanced when the most reliable monitoring instrument package, the Tropospheric Monitoring Instrument, called TROPOMI, was launched on the Copernicus Sentinel-5 Precursor satellite.

As the satellite moves around the world, the instrument measures the sunlight that is backscattered by Earth's surface and partially absorbed by the atmosphere. It passes over an area approximately 2600 km long and 7 km wide and detects the reflected radiation within a resolution of 7 × 3.5 km, delivering up to 20 million spectra per day. The daily measurements are validated against data from well-maintained and calibrated ground-based spectroradiometers and broadband and multichannel radiometer measurements from 25 sites.

In February 2022, the IEA reported[13] that, according to recent satellite data, methane emissions are 70% higher than previously believed and are accelerating faster than the use of green energy. In short, because of its contributions to methane emissions, natural gas should not be treated as a transitional fuel; its use should be stopped as soon as possible.

7.4.2 Nuclear Energy

The fact that no carbon emissions are produced by nuclear power plants is a powerful and effective argument in favor of building more of them. For this reason, I decided to discuss them in this concluding chapter.

In my view, building new nuclear plants is undesirable and dangerous. It is undesirable for the following reasons:

- They are not needed (solar and wind energy are plentiful).
- The building cost is very high, and construction time is much longer than that of other energy production plants (often about five times that of solar or wind projects).
- The investment in nuclear power uses up financial resources that can be used to convert to a green energy economy.
- Most importantly, they are unsafe.

Therefore, in my view, new plants should not be built, and existing plants should be limited to 50 years of operation.

The presently operating nuclear power plants are of the fission type, obtaining their energy by splitting a radioactive isotope of uranium. I disapprove of their continued use for three safety reasons:

---

13. Lina Tran, "Satellites don't lie: Methane emissions are 70% higher than reported," Grist website, February 24, 2022, accessed March 7, 2022, https://grist.org/climate-energy/the-world-is-underestimating-methane-emissions-iea-report/.

1. I consider fission nuclear power plants to be potential nuclear bombs because they cannot be fully protected from cyberattacks.
2. I consider all presently operating designs unsafe because they will melt down if the energy sources (primary and backup) that serve their cooling fail.
3. No permanent disposal site exists anywhere in the world to store used fuel rods.

The other method of generating nuclear energy is fusing hydrogen into helium. This process has been investigated for a century, and in August 2021, the scientists at Lawrence Livermore National Laboratory (LLNL) managed to trigger this process for 100 trillionths of a second using giant lasers. The energy generated would have been enough to supply a 60 W lightbulb for five minutes. The net energy balance of the process was negative, meaning that it took more energy to operate the reactor than was obtained from the reaction.

In addition to the laser fusion design, doughnut-shaped tokamak reactors have also been tested, using magnetic fields to hold and compress the hydrogen fuel. This design has produced even less energy to date than the laser design of LLNL.

I am also against fusion-type nuclear plants for the following three reasons:

1. Fusion reactors operate at a minimum of 100 million degrees kelvin, a temperature at which all materials on Earth evaporate. Therefore, this process can only be contained by holding and compressing the hydrogen with giant magnets.
2. This process is not currently operational, and we cannot wait to find out if it ever will be because we must replace fossil-fuel energy now.
3. These reactors are prohibitively expensive and would drastically reduce the investment available to convert to solar-hydrogen energy.

7.4.3 Electricity Storage and Distribution

Once solar electricity has been generated, it powers the electrolyzer (Chapter 4) that splits the water into hydrogen and oxygen. The generated hydrogen is then compressed[14] to about 700 bar (10,000 psig) pressure and stored. At that point, the hydrogen

---

14. Monterey Gardiner, "Energy requirements for hydrogen gas compression and liquefaction as related vehicle storage needs," US Department of Energy Hydrogen Program, July 7, 2009, accessed October 29, 2022, https://www.hydrogen.energy.gov/pdfs/9013_energy_requirements_for_hydrogen_gas_compression.pdf.

is ready for continuous distribution. The technology is well established and is growing in both size and efficiency. The efficiencies of electrolyzers range from 64% to 75%. In 2022, the largest electrolyzer units in the world included a 150 MW alkaline electrolyzer system in China and a number of 20 MW and 10 MW units in other countries.

Converting to the use of solar electricity with storage will be as fundamental a change as the emergence of wireless telecommunication. As discussed in Chapter 4, the key to converting to green energy is storage that makes the energy continuously available. Today, the infrastructure for battery storage is more advanced than hydrogen storage, but in the long range, the overall economy will depend on hydrogen.

This conversion will probably reduce the role of centralized electric grids and convert them to more distributed, local, or even wireless systems. Today, one option is to update the existing grid or expand it into a gigantic, centralized network. On the other hand, in this age of cyberterrorism, it is safer and more economical to convert to a fully distributed system backed up by a fully distributed electricity storage network. I believe this second option is the better one, and it is feasible to gradually convert from the centralized to the distributed system.

Probably because of familiarity, the utility industries favor updating the old centralized grid system, which can transport electricity over long distances. I believe that investing our resources in local electricity storage and distribution systems would be less expensive and more efficient because the electricity will not need to be transported over long distances as it will be used close to where it is made. These local systems can be interconnected to provide backup energy for each other and the grid.

Local electricity distribution systems are faster to build and well-suited to store the intermittent energy from solar and wind sources, initially in batteries and then in hydrogen as the infrastructure becomes available. The hydrogen-based distribution systems would not necessarily replace the grid; they could provide backup for the grid because the hydrogen system in fuel cell mode can feed electricity back into the grid. Naturally, if a particular location depended totally on locally generated electricity from hydrogen, there would be no transmission losses because there would be no grid. In addition, the hydrogen-based electricity supply system has no central controls; therefore, it is safe from cyber terror attacks. If the hydrogen tanks are large enough, the system would not be subject to power outages. With the regular centralized grid, power outages often occur, such as the one caused by the deep freeze in Texas, another caused by a hurricane in Puerto Rico, and others caused by heat waves in various locations.

In the past, the electricity supply was a one-way operation: electricity flowed from the power plant to the users. Now, rooftop solar-fueled microgrids can generate and store electricity to satisfy the needs of homes and immediate neighborhoods; they can also help other grids and act as an infinite network of interconnected, two-directional mini-powerplants.

7.4.4 Electric Vehicles

GHG emissions by the transportation industry comprise 14% of total emissions globally and 28% in the United States. In 2022, the total number of cars and trucks on the roads was almost 1.5 billion,[15] and this number is expected to nearly double by 2035.

Ten years ago, only about 25,000 BEVs were on the roads. Today, they number 5.6 million, and by 2030, their number is projected to rise to about 200 million (10% of the total number of vehicles). Currently, the global number of fuel-cell electric vehicles (FCEVs) (including buses and trucks) is roughly 100,000, which is small but still larger than the number of BEVs a decade ago. The rate of increase in the number of BEVs is limited by infrastructure availability. For example, today there are about 125,000 regular gas stations in the United States, whereas the global number of BEV charging stations in 2019 was only 25,000, and the number of hydrogen fueling stations was about 500.

BEVs are superior to FCEVs in terms of costs and infrastructure availability. The advantages of FCEVs include that their energy source (solar energy) and fuel source (water) are essentially free and unlimited, they emit only water, and their driving range is higher. In addition, their weights, refueling times, and noise levels are all lower than those of BEVs.

7.4.4.1 Batteries

Today, lithium-ion batteries are the dominant designs, and the required battery capacity per vehicle averages about 60 kWh for regular BEVs and 12 kWh for plug-in hybrid EVs (PHEVs). Between 2020 and 2050, it is estimated that the demand for lithium and cobalt will increase 20-fold, and the demand for nickel will increase 30-fold. If the number of EVs on the roads is to increase to 200 million by 2030, keeping up with this rise in demand will probably cause an increase in battery costs and require battery recycling. Today, lithium costs are increasing quickly while battery costs are dropping because of advances and mass production.

---

15. "How Many Cars Are There in the World in 2022?," Hedges & Company website, accessed March 7, 2022, https://hedgescompany.com/blog/2021/06/how-many-cars-are-there-in-the-world/.

Batteries are about 12 times heavier than fuel cells. The mass-based energy densities of batteries range from 0.1 to 0.27 kWh/kg. (For comparison, gasoline has an energy density of 13, and hydrogen gas energy at a pressure of 700 bar is 39.6 kWh/kg.) Batteries consume between 0.24 and 0.87 kWh of electricity per mile (averaging about 0.33 kWh/mi). For standard cars, the capacity of battery packs ranges from 30 to 100 kWh, and their weight ranges from 250 to 700 kg (550 to 1500 lb).

In an average BEV, the weight of batteries is about one-third of the total weight of the vehicle, and the volume of the batteries limits free space in the car. Their charging time is a function of the level of the charger used. In 1 hour, a level 1 charger will supply enough electricity for 3 to 5 miles of driving; a level 2 charger, 20 to 80 miles; and a level 3 charger, up to 200 miles. A supercharger will supply enough electricity for 300 miles of driving in only 15 minutes. The problems with batteries have intensified the development and sales of fuel-cell-operated cars.

7.4.4.2 Hydrogen Fuel Cells

Electrolyzers have long been used in industrial processes, but their use for dedicated hydrogen production has not yet been widely adopted. Hydrogen produced by carbon-emitting processes is called *gray hydrogen*. Hydrogen produced by water electrolysis is called *green hydrogen*. The current production cost of green hydrogen is high compared to that of gray hydrogen, and closing this gap is a key goal of the conversion effort to reach a green energy economy. In Europe, the cost of driving 100 km using gray hydrogen is €9 to €10, whereas for BEVs, driving the same distance costs only €2 to €7, depending on the electricity prices in the particular country.

In 2019, the global demand for hydrogen was about 70 million metric tons, of which half was used to make ammonia and fertilizers and half was used in petrochemical refineries. Approximately 48% of hydrogen is produced from natural gas, 30% is a byproduct of petroleum refining, and 20% is from coal gasification (mainly in China). Compared to these numbers, the production and consumption of green hydrogen are negligible, but the Hydrogen Council expects green hydrogen production to reach nearly 550 million metric tons per year by 2050—a significant jump from the roughly 0.36 million metric tons produced in 2019.

Today, the number of FCEVs on the roads is only about 1% of all EVs, and we do not know if their numbers will rise as did the numbers of BEVs during the past decade. We know that Toyota and Hyundai have developed fuel cells that are ready for mass production, and in Japan, Panasonic is producing very small fuel cells for homes. FCEVs are already fueling buses, trucks, and trains, and there are plans to use hydrogen fuel for airplanes.

7.5 The Hydrogen Age

I hope that the fossil-nuclear age will be only a memory and that the transition to the hydrogen age will be completed by the twenty-second century. Assuming that it will be the case, let us look at what the energy economy will look like at that time.

I visualize a fully distributed and localized energy economy with zero-carbon, green energy sources made continuously available through hydrogen storage. I also visualize a largely wireless energy distribution system that provides hydrogen tanks to private homes, just as they are provided with oil and propane fuel tanks today. I visualize large, bifacial solar collector farms located in areas not used for any other purpose and with high insolation. I also visualize large hydrogen storage facilities (tanks or natural salt caverns) similar to the ones used to store high-pressure natural gas today. I see hydrogen being shipped in liquid form over long distances or transported by trucks, trains, or piped distribution networks over shorter distances. In short, I visualize an energy economy in which the fuel (sunlight) is unlimited and clean; the means of storage and distribution is hydrogen, which is produced from water at the solar power plant and used to generate electricity at the user end; and the only emission is distilled water.

Here, I describe the operation of the components of the hydrogen process in a little more detail. We know that plants use solar energy to split water from the ground into oxygen and hydrogen and then combine this hydrogen with CO_2 from the air to produce energy sources such as glucose for humans and animals. This process helped maintain a balanced amount of CO_2 in the atmosphere during the preindustrial age. Today, this balance is upset; plant life has diminished, and human energy consumption has increased nearly 50-fold. The solution is to supplement the energy from the remaining plant life with artificially obtained energy without emitting any GHGs into the atmosphere. We can do this by copying what plants and trees do. We can use solar energy to split water and generate hydrogen as our free and inexhaustible energy source. We already have the electrolyzer technology to do this. All it takes is more development (particularly in the area of automation and control) and mass production to reduce costs.

7.5.1 The Reversible Fuel Cell (RFC)

Reversible fuel cells (RFCs) do not exist as of this writing, but their two main components, the electrolyzer and the fuel cell, do. My concept is to combine the two so that the flow direction would reverse through it as a function of solar energy availability. This means the RFC would generate electricity for the home (fuel cell mode) when the Sun is not shining by using the stored hydrogen as fuel. When solar energy is available and the solar collectors generate more electricity than needed in the home, the RFC would switch to the opposite direction (electrolyzer mode), convert the excess energy into hydrogen, and store it (Figure 7-3).

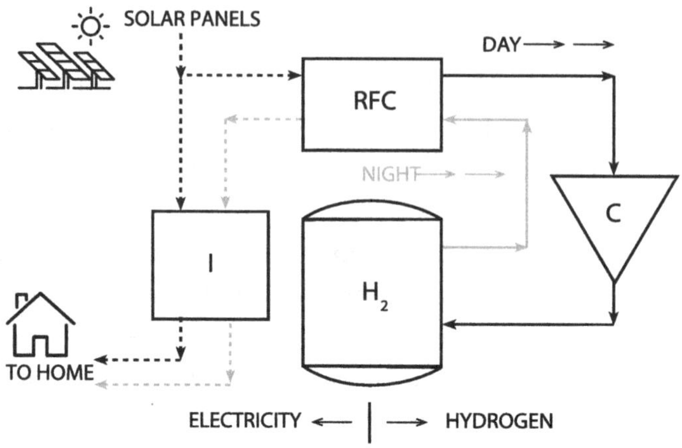

Figure 7-3. The main components and the operation of an RFC. When the Sun is out, the RFC operates in electrolyzer mode (red) and stores the excess solar energy as hydrogen in the storage tank (H_2). During the night, it operates in fuel-cell mode (blue) and supplies the electricity needed in the home. Legend: I, inverter; RFC, reversible fuel cell; H_2, hydrogen storage tank; C, compressor.

Source: Courtesy of *Control* magazine. Béla Lipták, "Controlling the future, part 4," September 21, 2021, https://www.controlglobal.com/articles/2021/controlling-the-future-part-4/.

Naturally, such a unit would require a substantial development effort, but if that is successful, the RFCs are mass-produced at a low cost, and their operation is fully automated, they could become the key component in a wireless and fully distributed electricity supply system serving the self-sustaining homes of the future. One advantage of such a system is that it has no centralized power supply grid, which means it could not be shut down by attacks like the one that stopped the electricity supply to part of Ukraine in 2022.

In this case, hydrogen storage tanks can be the same size as those used in FCEVs, and they can be serviced in the same way propane tanks are today. In other words, pressure detectors would detect the quantity of hydrogen gas in the tank, and the RFC package would automatically notify the supplier to refill the hydrogen if the pressure is low or to remove excess if the pressure is high. This way, the home of the twenty-second century would become fully self-sufficient—would not require the grid—and "RFC" would become a commonly used term just like "PC"[16] is today.

16. While IBM created the PC, it was Intel under Andy Grove that developed the microchip and triggered the digital revolution by reducing the brain of the PC into the small and cheap component that runs all smart digital devices today. It so happens that Andy was a classmate of mine. He was 19 and I was 20 when we emigrated together to the United States after Russian tanks crushed our Hungarian Revolution in 1956. I mention this in connection with the RFC not only because it would also be operated by a chip but because, if it could go through the same type of development that the computer did (from John von Neumann's vacuum tubes to Andy Grove's chip), it might just contribute to triggering the hydrogen age.

Using this system in remote areas makes the grid unnecessary. In areas where a grid exists, RFCs can provide backup energy for the grid during power outages. In this way, because the electricity is used where it is made, the transmission cost is eliminated, as are the infrastructure costs and the energy losses that occur in the supply cables. As a consequence, the cost of electricity drops sharply.

In private homes, batteries (or hydrogen storage) can eliminate the need for diesel generators because the storage can meet the home's electricity needs during power failures. In addition, the batteries in PEVs or the hydrogen in FCEVs can provide backup energy for the home or, if the home is connected to the grid, they can feed electricity to the grid when needed, such as during periods of peak demand, when otherwise, the cost of electricity would be high.

Likewise, green power plants can utilize large RFCs that generate green hydrogen for storage (red path in Figure 7-4) or continuously send green energy generated from the stored hydrogen (Figure 7-4 blue path) to users. Another important component of the green power plant is the underground storage cavern, which provides the space for storing large quantities of high-pressure hydrogen. The advantage of this storage method is that it has already been proven safe to use depleted natural gas caverns as reservoirs,[17] which is a benefit when considering reusing them to store hydrogen.

7.6 The Bottom Line

Humanity has reached the end of the fossil-fuel age and is at the beginning of the hydrogen age. At the same time, we are also moving from the analog to the digital age. Humanity is beginning to change its physical and cultural environment; therefore, we are beginning to influence our evolution. In a way, we are at a fork in our march into the future, and we have to select the road we will proceed on. The sign on one road reads "Controlled Future," while the sign on the other road reads "Dead End."

This book is unique because it uses the tools and know-how accumulated during the past century for analyzing and controlling industrial processes to analyze two nonindustrial processes: the evolution of GW and AI. To control processes, we must fully understand them. To fully understand the GW and AI processes, we must learn their dynamics, interactions, time constants, dead times, inertias, tipping points, and feedback subprocesses. We must also identify the means we have to influence their

---

17. S. Foh, M. Novil, E. Rockar, and P. Randolph, "Underground hydrogen storage. Final report. [Salt caverns, excavated caverns, aquifers, and depleted fields]," Technical Report BNL-51275, Brookhaven National Laboratory, December 1, 1979, accessed March 7, 2022, https://www.osti.gov/biblio/6536941.

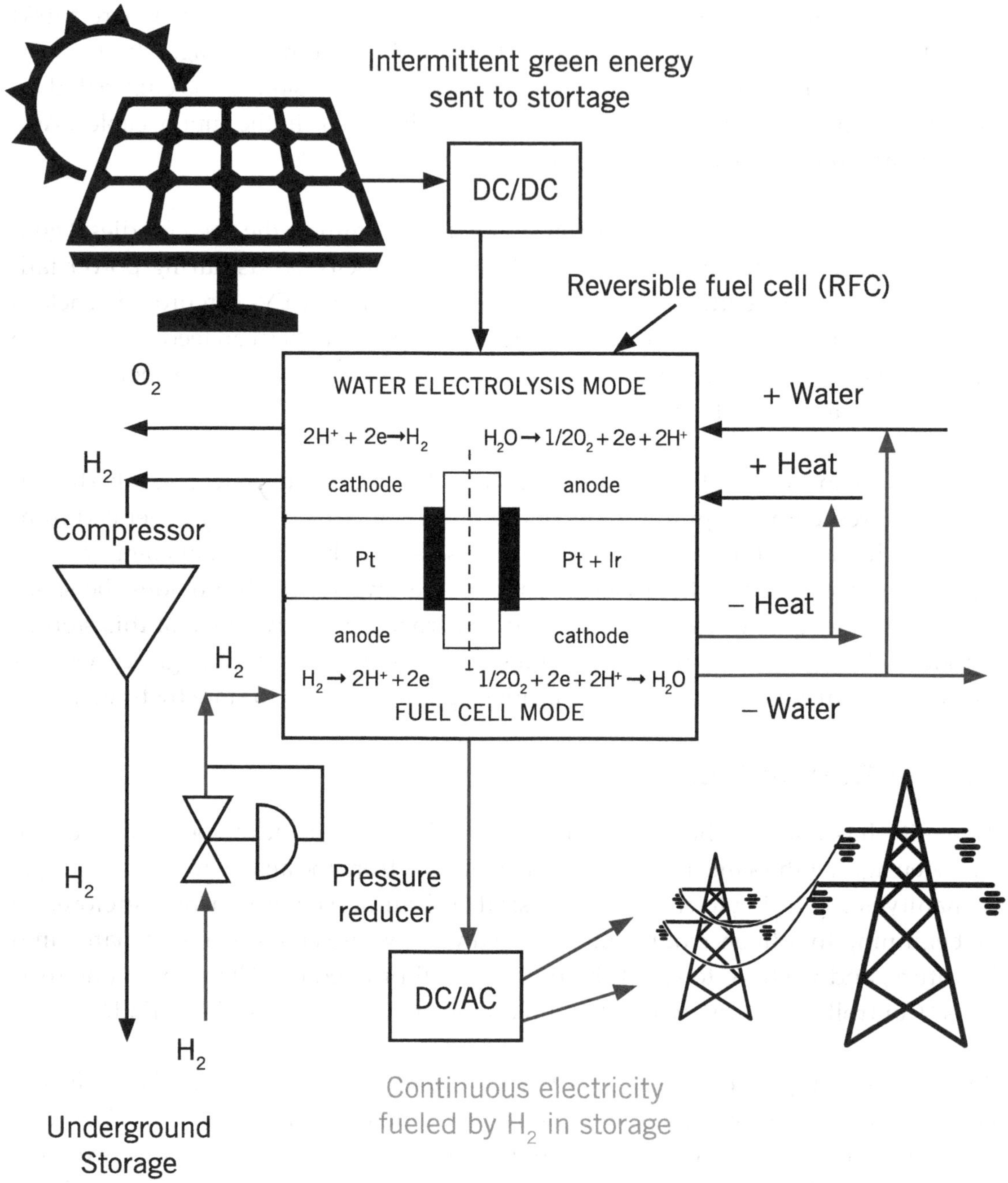

Figure 7-4. The large RFCs in solar-hydrogen power plants of the future[1] can generate and store hydrogen underground when the Sun is out and continuously supply electricity from that storage to the users whenever it is needed.

Source: Courtesy of *Control* magazine. Béla Lipták, "Hydrogen is key to sustainable green energy," January 24, 2022, https://www.controlglobal.com/articles/2022/hydrogen-is-key-to-sustainable-green-energy/.

behavior. Only after that can we estimate how fast they are progressing, how far they will get if left uncontrolled, and what we must do to get them under control.

7.6.1 Artificial Intelligence (AI)

Artificial intelligence is a dual-purpose tool. Just as a knife in the hands of a cook is very helpful, it can be dangerous in the hands of a gangster. AI can be very helpful in serving the conversion from the fossil to the hydrogen age; it can standardize and optimize the operation of the required new infrastructure and provide many good jobs for those workers whose jobs will be eliminated by robotization.

On the other hand, AI is just a tool. It performs according to the algorithms we load into the control chips. It has no sense of self and no moral or ethical standards; it just carries out what the algorithms dictate. Yes, AI can do wonders in manipulating data, speeding communications, processing and organizing information, and so on, but in the wrong hands, it can also brainwash people, spread lies and hate, operate weapons, carry out cyberattacks, and more.

Today AI development is out of control; it is allowed to do what it can instead of being limited to doing what it should. The out-of-control use of AI is impacting our cultural and moral environment because there are no standards or treaties to control its use. This lack of control enables the creation of an unethical, truthless cultural environment and the development of lethal tools such as fully autonomous weapons, killer robots that can make targeting decisions without human involvement, and face-recognition technology that can be used to carry out assassinations.

Yet, at the Sixth Review Conference of the Convention on Certain Conventional Weapons in December 2021, when some 120 nations agreed on the need to limit the development of killer robots, Russia and the United States vetoed the adoption of such a treaty. It seems to me that the time has come for a digital Geneva Convention to set the legal limits for the allowable use and applications of AI so that its development will be controlled. I hope the following quotations will reinforce my view that this is necessary.

"The rise of powerful AI will be either the best or the worst thing ever to happen to humanity. We do not yet know which."

– Stephen Hawking

"Whoever becomes the leader in this sphere will become the ruler of the world."

– Vladimir Putin

7.6.2 *Climate Change*

Our planet evolved into a marvelous home for all plant and animal species, including humans, when Nature directed its evolution. It protected all living creatures by keeping temperature relatively constant and maintaining water and carbon cycles so that plant and animal life could support each other.

Then came the Industrial Revolution. Humans upset the carbon cycle by sending it from underground into the atmosphere. They upset the water cycle by increasing the rate of ocean evaporation. And they increased global temperature to the point where the regions near the Equator could no longer support life, and the poles began melting, thus increasing ocean levels. In short, humans have upset the balance of their home, and the human "footprint" is becoming self-destructive.

During the industrial age, the global population increased 7-fold and is estimated to reach 12 billion by 2100; the per capita energy consumption increased 5- to 6-fold. The carbon content of the atmosphere nearly doubled (from 280 ppmv to 420 ppmv), and the weight of the extra CO_2 that accumulated in the air during the industrial age reached the weight of all buildings on the planet, causing GW to rise at a rate equaling the energy release of five Hiroshima bombs each second. In addition, the consumption of goods is increasing, as the global gross domestic product GDP increased from $50 trillion in 2000 to $100 trillion in 2020; it is estimated to reach $200 trillion by 2100.

Next, I will comment on the consequences of continuing inaction and how we can protect our physical and cultural environments in order to restabilize the planet.

7.6.2.1 Global Warming

GW can be reduced by reducing the heat entering our planet or increasing the heat leaving it. One reason that my analysis conclusions differ from most others is that they usually do not consider increasing the Earth's reflectivity to cool the planet, which can be done by making our roofs, roads, and agriculture lighter in color.

I consider my analysis to be more accurate because it takes into account the self-accelerating feedback effects of the many GW subprocesses and the differences between the speeds of the responses of the atmosphere and the oceans. Perhaps the most important difference in my analysis is in the evaluation of the tipping point effects, which have not been triggered yet. These effects—the albedo effect of the melting at the poles, the methane release of the melting permafrost, ocean degassing, the destruction of rain forests and coral reefs, and the effects of wildfires—will all occur suddenly, some irreversibly.

Today, GW is near 1.2°C, and my analysis shows that keeping it under 1.5°C is no longer possible. Staying under 2°C requires reaching net-zero emissions by 2035 to 2040, and if the present slow response to climate change continues, GW will reach 3°C by the end of the century. According to my analysis, the above numbers show that we have much less time for action than other estimates suggest. My conclusions are supported by the United Nation's Emission Gap 2022 report[18] which states that even if the nations live up to their net zero target dates (which the report does not consider to be likely or credible), GW will reach 1.8°C. Similarly, the IEA *World Energy Outlook 2022* report projects that GW will reach 2.5°C if the present energy policies are maintained.[19]

7.6.2.2 Consequences of Inaction

Last summer palm trees could grow in London and the weather in Europe felt like the continent was moved some 500 miles towards the Equator. Just in 2022, catastrophic floods killed 1,500 people in Pakistan, hurricane Ian caused $60 billion in insured losses in Florida, and there are many other examples. In fact, the IPCC estimates that 3 billion people live in "highly vulnerable" places.

The data in this book shows that we are well on our way to water shortages and droughts, to having one billion people exposed to life-threatening heat waves, and to floods that are already threatening our shorelines and the very existence of some island nations. If climate change is allowed to continue, we can anticipate that it will make parts of the planet unlivable and result in a biblical-scale migration that can destroy civilization. We must simultaneously fight the causes of climate change and adapt to its already occurring impacts. We need immediate action on mitigation (actions taken to curb greenhouse gas emissions) and adaptation (measures that reduce our vulnerability to the effects of climate change).

The disaster has already started. We are already experiencing its first signs. We are also learning that adaptation methods like developing better firefighting helicopters, improving building codes, relocating coastal residents, building coastal dams to protect against flooding, and building new canals to transport water to dry areas or to manage floods will not stop this process. Therefore, while adapting where it is reasonable, we should focus our resources on eliminating the cause. Simultaneously, wealthy nations should help developing ones finance their transition to using green energy and should establish a new fund to help those nations if climate disaster strikes them. This book focused on showing how it can be done.

---

18. United Nations Environment Programme (2022), *Emissions Gap Report 2022.*
19. International Energy Agency (IEA), *World Energy Outlook 2022.*

7.6.2.3 Reasons for Optimism

Even though during the past decades, the fossil lobbies were successful in blocking decisive action by our political leaders; strategies like carbon taxes, cap-and-trade and border taxes turned out to be ineffective; and the conferences and their promises did not turn out to be credible, something did happen! Science made great progress in improving green technology, and the free market discovered the profitability of investing in green energy and came to the rescue of progress. What followed was the immense advance of green technology combined with the unexpected and great drop in green energy prices and the increase in fossil energy costs. This combination made investments in renewable energy technology highly profitable.

Solar electricity costs dropped almost 10-fold during the last decade (Figure 6-2). Solar generating capacity in the United States has increased over 200-fold since 2008, and the US Department of Energy (DOE) envisions that by 2050, 45% of the total electricity in America will come from solar sources. During this same period, the cost of wind power was cut in half, and it was realized that some 90% of the world's population lives in areas where green energy can be cheaper than dirty energy sources.

Also, the conversion of intermittent green energy into a continuously available one has been achieved by developing the required battery and hydrogen technology. The cost of batteries has fallen nearly 90% in the last decade. The issues of hydrogen storage in vessels or caverns, transportation in liquid or pressurized gas form, and distribution by trucks and pipelines have all been solved. Because of its low weight and high energy content, hydrogen is preferred when larger quantities of energy are to be stored and transported. It will also be preferred in the distant future when some or most homes will be self-supporting, using RFCs and hydrogen storage to eliminate the need for a centralized grid and replace it with a fully distributed, local energy supply network.

7.6.2.4 Costs

The estimated dollar value of inaction in fighting climate change or the cost of the temporary mitigation of its effects varies widely based on the assumed peak GW reaches. Still, they all agree that if GW is allowed to rise to about 3°C, human civilization, as we know it, will be destroyed. Naturally, these damage estimates ($50 and $100 trillion) are increased if the cost of temporary mitigation is added to the total damage. Also, these estimates do not include the values of the damage to nature or that of human suffering.

When talking about trillions, we need some reference to appreciate their scale. One reference can be the global GDP, which today is about $100 trillion. The IPCC estimated that the global investment needed to keep GW under 2°C is $40 trillion and to keep it under 2°C is $54 trillion.[20] I have two comments about these numbers:

1. After investing the numbers in the IPCC estimates, our descendants will live in a stable environment, using free, inexhaustible, and emissions-free green energy sources.
2. In my view, the conversion to a solar-hydrogen energy economy will create jobs, growth, and prosperity, exceeding the results of the Marshall Plan at the end of WW2.

My analysis suggests that the yearly investment of 2% or 3% of the global GDP for a couple of decades can probably keep GW under 2°C. Once net zero global emission is reached, we can start removing the gases that accumulated in the air during the industrial age by taking steps such as restoring forests and "whitening" the human footprint to increase heat reflection back into outer space.

7.6.2.5 Next Couple of Years

The main conclusion of this analysis is that the next few years could be the most important ones in the history of humanity. This is because we will soon trigger the tipping points that can speed up or make unstoppable the—up until now—slowly evolving GW process and drastically damage or even destroy our physical environment. Similarly, the consequences of the out-of-control digital age are also approaching the tipping point of turning our cultural environment into a truthless and hate-filled one. Another conclusion is that these do not need to happen; while it will take a gigantic effort, things can still be fixed, and we still have the time and the tools to fix them.

To be successful, it is essential to unify humankind's attitude and understanding of the facts of climate change and to change the often-emotional attitudes into a united and fact-based one. In this effort, calm education is needed, school courses and libraries with books like this one are essential, but even more importantly, a change in the political atmosphere is needed. Both sides must recognize the common good. Caring about the future and the well-being of the coming generation has always been a moral value, and protecting traditions, national cultures, institutions, and spiritual and social values have always been conservative goals. We must reach the point where the fight against truthlessness and climate change denialists is joined by the right and left. Both sides must fight the lack of trust in our scientific and other institutions, and the paralysis caused by polarization must be viewed by all as what it is: the main enemy of all.

---

20. Wu, "The World Was Just Issued 12-Year Ultimatum on Climate Change."

It should also be realized that we all must participate in this effort. Some 30% to 40% of our energy consumption is spent on making our buildings comfortable, and we can reduce that. We all could travel more efficiently, conserve water, reduce waste and make the human footprint more reflective by making our roofs, roads, lawns, and agriculture lighter. We could all vote for leaders who understand what needs to be done. In short, the future is up to us, and we can learn from our Stone Age ancestors who stopped using stone not because they ran out of it but because bronze was better. We must leave fossil fuels where they are and not wait until we run out of them because solar hydrogen is better. Converting to its use will not only protect nature's diversity and create prosperity but it will also bring back our optimism about the future.

Index

Note: "n" refers to the footnote.

Q

R

Printed in the USA/Agawam, MA
September 17, 2026

910661.176